CW01203187

Seismic Amplitude

An Interpreter's Handbook

The oil and gas industries now routinely use seismic amplitudes in exploration and production as they yield key information on lithology and fluid fill, enabling interpretation of reservoir quality and likelihood of hydrocarbon presence. The modern seismic interpreter must be able to deploy a whole range of sophisticated geophysical techniques, such as seismic inversion, AVO (amplitude variation with offset) and rock physics modelling, as well as integrating information from other geophysical techniques and well data.

This accessible yet authoritative book provides a complete framework for seismic amplitude interpretation and analysis, in a practical manner that allows easy application – independent of any commercial software products. Springing from the authors' extensive industry expertise and experience of delivering practical courses on the subject, it guides the interpreter through each step, introducing techniques with practical observations and helping to evaluate interpretation confidence.

Seismic Amplitude is an invaluable day-to-day tool for graduate students and industry professionals in geology, geophysics, petrophysics, reservoir engineering and all subsurface disciplines making regular use of seismic data.

Rob Simm is a Senior Geophysical Adviser for Cairn Energy PLC, and has worked in the oil and gas industry since 1985. He spent the early part of his career working as a seismic interpreter for British independent oil companies including Britoil, Tricentrol and Enterprise Oil. After working in exploration, production and field equity teams, Dr Simm established his own consultancy (Rock Physics Associates Ltd) in 1999, providing project and training services to oil and gas companies. He runs an internationally renowned training course on 'The Essentials of Rock Physics for Seismic Amplitude Interpretation'.

Mike Bacon is a Principal Geoscientist for Ikon Science Ltd, having worked for 30 years in the oil industry with Shell, Petro-Canada and Ikon Science. During that time he has interpreted seismic data from various basins around the world, with particular emphasis on extracting useful information from seismic amplitudes. Dr Bacon has served as Publications Officer of the EAGE (European Association of Geoscientists and Engineers), chairing the editorial board of the journal First Break. He has also co-authored a number of practical texts, including, with Rob Simm and Terry Redshaw, *3-D Seismic Interpretation* (Cambridge University Press, 2003).

'This will be a welcome addition to the library of any geoscientist wanting a firm foundation in state-of-the-art seismic reservoir characterisation. Simm and Bacon cover all of the important topics, in a style that is not overly mathematical, and also illustrate each method with well thought-out and illustrated geological examples from around the world. I highly recommend this book to anyone wishing to understand this important area of geoscience.'

Dr Brian Russell, *Vice President, Hampson-Russell Software: A CGG Company*

'The authors have provided a coherent, comprehensive and, above all, practical guide for interpreters to exploit the information contained in the amplitudes of seismic data. They compare different approaches, highlighting advantages and potential pitfalls, and also explain the terminology. This book will be an invaluable guide for both beginners and experienced professionals and I'd recommend it for all geoscientists working with seismic data.'

Mr Patrick Connolly, *Senior Advisor, Geophysical Analysis, BP*

Seismic Amplitude:
An Interpreter's Handbook

Rob Simm
Cairn Energy PLC

Mike Bacon
Ikon Science Ltd

CAMBRIDGE
UNIVERSITY PRESS

CAMBRIDGE
UNIVERSITY PRESS

University Printing House, Cambridge CB2 8BS, United Kingdom

Cambridge University Press is part of the University of Cambridge.

It furthers the University's mission by disseminating knowledge in the pursuit of education, learning and research at the highest international levels of excellence.

www.cambridge.org
Information on this title: www.cambridge.org/9781107011502

© Rob Simm and Mike Bacon 2014

This publication is in copyright. Subject to statutory exception and to the provisions of relevant collective licensing agreements, no reproduction of any part may take place without the written permission of Cambridge University Press.

First published 2014
Reprinted 2014

Printed in Spain by Grafos SA, Arte sobre papel

A catalogue record for this publication is available from the British Library

Library of Congress Cataloguing in Publication data
Simm, R. (Robert), 1959– author.
Seismic amplitude : an interpreter's handbook / Rob Simm, Cairn Energy PLC, Mike Bacon, Ikon Science Ltd.
　pages cm
ISBN 978-1-107-01150-2 (Hardback)
1. Seismic prospecting. 2. Petroleum–Geology. 3. Amplitude variation with offset analysis. 4. Seismic traveltime inversion. 5. Seismic reflection method. 6. Seismology–Mathematical models. I. Bacon, M. (Michael), 1946– author. II. Title.
TN271.P4S514 2014
622′.1592–dc23   2013040419

ISBN 978-1-107-01150-2 Hardback

Cambridge University Press has no responsibility for the persistence or accuracy of URLs for external or third-party internet websites referred to in this publication, and does not guarantee that any content on such websites is, or will remain, accurate or appropriate.

Contents

Preface ix
Acknowledgements x

1 **Overview** 1
   1.1 Introduction 1
   1.2 Philosophy, definitions and scope 1
   1.3 The practice of seismic rock physics 2

2 **Fundamentals** 3
   2.1 Introduction 3
   2.2 Seismic basics 3
   2.2.1 Seismic geometry 3
   2.2.2 Gathers and stacks 3
   2.3 Modelling for seismic interpretation 6
   2.3.1 The convolutional model, wavelets and polarity 7
   2.3.2 Isotropic and elastic rock properties 10
   2.3.3 Offset reflectivity 14
   2.3.4 Types of seismic models 17
   2.3.5 Relating seismic data to models 19

3 **Seismic wavelets and resolution** 23
   3.1 Introduction 23
   3.2 Seismic data: bandwidth and phase 23
   3.3 Zero phase and minimum phase 24
   3.4 Change of wavelet shape with depth 25
   3.5 Idealised wavelets 28
   3.6 Wavelet phase and processing 29
   3.6.1 Q compensation 29
   3.6.2 Zero phasing 29
   3.6.3 Bandwidth improvement 30
   3.7 Resolution 32
   3.7.1 The problem of interference 32
   3.7.2 Simple models of interference 32
   3.7.3 Estimating vertical resolution from seismic 33
   3.7.4 The effect of wavelet shape on resolution 34
   3.7.5 Lateral resolution 35
   3.8. Detectability 37

4 **Well to seismic ties** 38
   4.1 Introduction 38
   4.2 Log calibration – depth to time 38
   4.2.1 Velocities and scale 38
   4.2.2 Drift analysis and correction 39
   4.3 The role of VSPs 40
   4.4 Well tie approaches using synthetics 43
   4.4.1 Well tie matching technique 43
   4.4.2 Adaptive technique 47
   4.5 A well tie example 47
   4.6 Well tie issues 50
   4.6.1 Seismic character and phase ambiguity 50
   4.6.2 Stretch and squeeze 51
   4.6.3 Sense checking and phase perception 53
   4.6.4 Importance of tie accuracy in horizon mapping 56
   4.6.5 Understanding offset scaling 56
   4.6.6 Use of matching techniques to measure an improving tie 57

5 **Rock properties and AVO** 58
   5.1 Introduction 58
   5.2 AVO response description 58
   5.2.1 Positive or negative AVO and the sign of the AVO gradient 58
   5.2.2 AVO classes and the AVO plot 58
   5.2.3 Introducing the AVO crossplot 59
   5.2.4 Examples of AVO responses 59
   5.3 Rock property controls on AVO 61
   5.3.1 Ranges of parameters for common sedimentary rocks 61
   5.3.2 The role of compaction 62
   5.3.3 The effect of fluid fill 63
   5.3.4 The effects of rock fabric and pore geometry 69
   5.3.5 Bed thickness and layering 72
   5.3.6 The effects of pressure 77
   5.3.7 Anisotropy 82

5.4 The rock model and its applications 88
5.4.1 Examples of rock model applications 88
5.5 Rock properties, AVO reflectivity and impedance 91
5.5.1 AVO projections, coordinate rotations and weighted stacks 92
5.5.2 Angle-dependent impedance 98
5.5.3 Bandlimited impedance 102
5.6 Seismic noise and AVO 107

6 Seismic processing issues 110
6.1 Introduction 110
6.2 General processing issues 111
6.2.1 Initial amplitude corrections 111
6.2.2 Long-wavelength overburden effects 112
6.2.3 Multiple removal 113
6.2.4 Migration 114
6.2.5 Moveout correction 114
6.2.6 Final scaling 115
6.2.7 Angle gathers and angle stacks 116
6.3 Data conditioning for AVO analysis 117
6.3.1 Spectral equalisation 118
6.3.2 Residual moveout removal 118
6.3.3 Amplitude scaling with offset 120
6.3.4 Supergathers 122
6.3.5 Gradient estimation and noise reduction 123

7 Amplitude and AVO interpretation 125
7.1 Introduction 125
7.2 AVO and amplitude scenarios 125
7.2.1 Class II/III hydrocarbon sands and Class I water sands 126
7.2.2 Class III hydrocarbon and water sands 127
7.2.3 Class IV hydrocarbon and water sands 129
7.2.4 Class IIp hydrocarbon sands, Class I water sands 130
7.2.5 Class I hydrocarbon sands, Class I water sands 132
7.2.6 Multi-layered reservoirs 133
7.2.7 Hydrocarbon contacts 135
7.2.8 Carbonates 144
7.2.9 Fractured reservoirs 146

8 Rock physics for seismic modelling 149
8.1 Introduction 149
8.2 Rock physics models and relations 150
8.2.1 Theoretical bounds 150
8.2.2 Empirical models 151
8.2.3 Gassmann's equation 158
8.2.4 Minerals, fluids and porosity 162
8.2.5 Dry rock relations 168
8.2.6 Contact models 173
8.2.7 Inclusion models 175
8.3 Requirements for a rock physics study 177
8.3.1 Data checklist 177
8.3.2 Acoustic logs 178
8.4 Data QC and log edits 181
8.4.1 Bad hole effects 181
8.4.2 V_p and V_s from sonic waveform analysis 182
8.4.3 Log prediction 183
8.4.4 Borehole invasion 187
8.4.5 Sonic correction for anisotropy in deviated wells 190
8.5 Practical issues in fluid substitution 191
8.5.1 Shaley sands 191
8.5.2 Laminated sands 193
8.5.3 Low porosity and permeability sandstones 193
8.6 Rock characterisation and modelling issues 194

9 Seismic trace inversion 197
9.1 Introduction 197
9.2 Deterministic inversion 198
9.2.1 Recursive inversion 198
9.2.2 Sparse spike inversion 198
9.2.3 Model-based inversion 199
9.2.4 Inversion issues 203
9.2.5 Inversion QC checklist 208
9.2.6 Bandlimited vs broadband 208
9.2.7 Inversion and AVO 208
9.2.8 Issues with quantitative interpretation of deterministic inversions 212
9.3 Stochastic inversion 213

10 Seismic amplitude applications 221
10.1 Introduction 221
10.2 Litho/fluid-facies from seismic 221
10.3 Reservoir properties from seismic 223
10.3.1 Reservoir properties from deterministic inversion 223
10.3.2 Simple regression, calibration and uncertainty 225

10.3.3 Reservoir property mapping using geostatistical techniques   228
10.3.4 Net pay estimation from seismic   229
10.4 Time-lapse seismic   235
10.5 Amplitudes in prospect evaluation   246
10.5.1 An interpreter's DHI checklist   247
10.5.2 A Bayesian approach to prospect risking   247
10.5.3 Risking, statistics and other sense checks   249

10.6 Seismic amplitude technology in reserves estimation   251

References   254
Index   270

Preface

The past twenty years have witnessed significant developments in the way that seismic data are used in oil and gas exploration and production. Arguably the most important has been the use of 3D seismic, not only to map structures in detail but also to infer reservoir properties from an analysis of seismic amplitude and other attributes. Improvements in seismic fidelity coupled with advances in the understanding and application of rock physics have made quantitative description of the reservoir and risk evaluation based on seismic amplitude not only a possibility but an expectation in certain geological contexts. It is probably no exaggeration to say that the interpreter has entered a new era in which rock physics is the medium not only for the interpretation of seismic amplitude but also for the integration of geology, geophysics, petrophysics and reservoir engineering. For conventional oil and gas reservoirs, the technology has reached a sufficient state of maturity that it is possible to describe effective generic approaches to working with amplitudes, and documenting this is the purpose of this book.

The inter-disciplinary nature of 'Seismic Rock Physics' presents a challenge for interpreters (both old and new) who need to develop the appropriate knowledge and skills but it is equally challenging for the asset team as a whole, who need to understand how information derived from seismic might be incorporated into project evaluations. This book provides a practical introduction to the subject and a frame of reference upon which to develop a more detailed appreciation. It is written with the seismic interpreter in mind as well as students and other oil and gas professionals. Mathematics is kept to a minimum with the express intention of demonstrating the creative mind-set required for seismic interpretation. To a large extent the book is complementary to other Cambridge University Press publications such as *3-D Seismic Interpretation* by Bacon *et al.* (2003), *Exploration Seismology* by Sheriff and Geldart (1995), *The Rock Physics Handbook* by Mavko *et al.* (1998) and *Quantitative Seismic Interpretation* by Avseth *et al.* (2005).

Acknowledgements

In large part the material and ideas presented in this book are based around the training course 'The Essentials of Rock Physics for Seismic Interpretation', provided to the industry since 1999 by Rock Physics Associates Ltd and Nautilus Ltd (now part of the RPS group) and latterly by Ikon Science Ltd. Many colleagues, past and present, too numerous to mention individually, have indirectly contributed to the work. Special thanks are due to John Chamberlain for discussions, help and guidance and to Roy White for much lively and informative discussion. Rachel Adams at Dreamcell Ltd laboured tirelessly and with great patience over the illustrations. The following software providers are also thanked: Ikon Science Ltd for use of RokDoc® software and Senergy for use of Interactive Petrophysics® software.

We are grateful to the following for permission to reproduce proprietary or copyright material: AAPG for Figs. 3.20, 5.30 and 8.45; Apache Corporation for Fig. 7.15; APPEA for Figs. 7.18, 7.20 and 10.2; Dr P. Avseth for Fig. 5.8; Baird Petrophysical for Figs. 2.6, 2.7, 2.9, 2.21 and 5.69; CGG for Fig. 9.10; Dr P. Connolly for Figs. 5.75, 9.36 and 10.21; CSEG for Figs. 4.6, 6.4, 10.7 and 10.8; CUP for Figs. 2.28, 5.18 and 8.7; EAGE for Figs. 2.5, 3.1, 3.6, 3.9, 3.19, 3.22, 4.20, 4.28, 5.56, 6.1, 6.16, 7.9, 7.13, 7.32, 7.36, 8.28, 8.43, 8.56, 8.58, 8.59, 8.60, 8.61, 9.12, 9.14, 9.15, 9.24, 9.29, 9.30, 9.32, 9.33, 9.34, 9.35, 10.12, 10.14, 10.15, 10.16, 10.20, 10.26, 10.32, 10.34 and 10.35; Elsevier for Fig. 5.43; Dr M. Floricich and Professor C. Macbeth for Fig. 10.30; Dr A. Francis for Figs. 9.11, 10.4 and 10.5; Dr G. Drivenes for Fig. 7.27; the Geological Society of London for Figs. 4.30, 4.31, 7.23, 7.29 and 10.33; Dr S. Helmore for Fig. 3.14; Ikon Science for Figs. 9.25 and 9.26; the Indonesian Petroleum Association for Fig. 6.3; Dr H. Ozdemir for Fig. 9.15; Dr C. Pearse for Fig. 7.31; Dr T. Ricketts for Fig. 10.29a; Dr B. Russell for Fig. 7.4; Rashid Petroleum Company for Figs. 5.26 and 7.1; Rock Physics Associates Ltd for Figs. 2.1, 2.2, 2.3, 2.4, 2.8, 2.11, 2.12, 2.13, 2.14, 2.16, 2.19, 2.20, 2.22, 2.23, 2.25, 2.26, 2.27, 2.28, 2.29, 3.2, 3.3, 3.4, 3.5, 3.7, 3.8, 3.10, 3.12, 3.16, 3.17, 3.18, 4.1, 4.9, 4.10, 4.11, 4.21, 4.22, 4.23, 4.24, 4.25, 4.26, 4.29, 5.1, 5.3, 5.4, 5.5, 5.6, 5.7, 5.9, 5.10, 5.11, 5.12, 5.14, 5.17, 5.29, 5.34, 5.36, 5.38, 5.40, 5.41, 5.45, 5.50, 5.53, 5.54, 5.55, 5.57, 5.58, 5.61, 5.73, 5.74, 6.5, 6.7, 6.10, 6.11, 7.2, 7.3, 7.5, 7.12, 7.21, 8.4, 8.9, 8.10, 8.11, 8.12, 8.15, 8.16, 8.17, 8.18, 8.19, 8.21, 8.22, 8.24, 8.25, 8.26, 8.27, 8.29, 8.30, 8.32, 8.35, 8.36, 8.37, 8.38, 8.44, 8.46, 8.47, 8.48, 8.50, 8.52, 8.53, 9.2, 9.4, 9.6, 9.7, 9.9, 9.13, 9.16, 9.17, 9.18, 9.19, 9.20, 9.21, 9.23, 10.13 and 10.36; Schlumberger for Fig. 8.42; SEG for Figs. 3.11, 3.13, 3.21, 4.7, 4.8, 4.27, 5.2, 5.20, 5.21, 5.25, 5.33, 5.44, 5.46, 5.48, 5.51, 5.52, 5.66, 5.77, 6.2, 6.6, 6.8, 6.9, 6.12, 6.13, 6.15, 6.17, 7.6, 7.7, 7.8, 7.11, 7.14, 7.16, 7.30, 7.33, 7.34, 7.35, 7.37, 7.38, 8.2, 8.3, 8.5, 8.6, 8.20, 8.31, 8.33, 8.34, 8.39, 8.40, 8.51, 8.57, 8.63, 9.5, 9.8, 9.9, 9.23, 10.1, 10.3, 10.10, 10.17, 10.19, 10.23, 10.27, 10.31, 10.37, 10.38, 10.39, 10.40 and 10.41; SIP for Fig. 2.24; SPE for Figs. 5.35, 8.55, 9.31, 10.6, 10.11, 10.24 and 10.29b; Dr J. Smith for Fig. 8.14; Springer for Figs. 5.71 and 9.9; SPWLA for Fig. 8.13; Dr R Staples for Fig. 10.28; Western Geco for Figs. 7.22 and 7.24; Professor R. White for Fig. 4.12; John Wiley for Fig. 8.54; World Oil for Fig. 10.25.

Chapter 1

Overview

1.1 Introduction

This book is about the physical interpretation of seismic amplitude principally for the purpose of finding and exploiting hydrocarbons. In appropriate geological scenarios, interpretations of seismic amplitude can have a significant impact on the 'bottom line'. At all stages in the upstream oil and gas business techniques based on the analysis of seismic amplitude are a fundamental component of technical evaluation and decision making. For example, an understanding of seismic amplitude signatures can be critical to the recognition of direct hydrocarbon indicators (DHIs) in the exploration phase as well as the evaluation of reservoir connectivity or flood front monitoring in the field development phase. Given the importance of seismic amplitude information in prospect evaluation and risking, all technical disciplines and exploration/asset managers need to have a familiarity with the subject.

1.2 Philosophy, definitions and scope

The central philosophy is that the seismic interpreter working in exploration and appraisal needs to make physical models to aid the perception of what to look for and what to expect from seismic amplitude responses in specific geological settings. This usually involves the creation of synthetic seismic models for various rock and fluid scenarios based on available well log data. In rank exploration areas the uncertainties are generally such that only broad concepts, assumptions and analogies can be used. By contrast, in field development settings where data are readily available, physical modelling can lead to a quantification of reservoir properties from seismic (with associated error bars!).

Fundamental to this process of applying models in seismic interpretation is the integration of data from a variety of disciplines including geology, geophysics, petrophysics and reservoir engineering. The core aspect of the data integration is rock physics, which can be defined as the study of 'the relationships between measurements of elastic parameters (made from surface, well, and laboratory equipment), the intrinsic properties of rocks (such as mineralogy, porosity, pore shapes, pore fluids, pore pressures, permeability, viscosity, and stress sensitivity) and overall rock architecture (such as laminations and fractures)' (Sayers and Chopra, 2009). Rock physics effectively provides the rock and fluid parameters for seismic models.

Pennington (1997) describes 'the careful and purposeful use of rock physics data and theory in the interpretation of seismic observations' and calls this approach 'Seismic Petrophysics'. Others commonly refer to it simply as 'Rock Physics' (*sensu lato*), 'Seismic Rock Physics' (Wang, 2001b) or Quantitative Interpretation (QI). The mind-set which drives the approach is not new of course but modern data have provided a new context. More than ever before there is an opportunity, paraphrasing Sheriff (1980), to 'reveal the meaning of the wiggles'. There are numerous workers, past and present, to whom the authors are indebted and whose names occur frequently in the following pages.

The book describes the theory of seismic reflectivity (Chapter 2) and addresses the key issues that underpin a seismic interpretation such as phase, polarity and seismic to well ties (Chapters 3 and 4). On these foundations are built a view of how contrasts in rock properties give rise to variations in seismic reflectivity (Chapter 5). Seismic data quality is an all important issue, controlling to a large extent the confidence in an interpretation, and this is addressed, from an interpreter's perspective, in Chapter 6. Examples of fluid and rock interpretations using Amplitude Versus Offset (AVO) techniques are presented in Chapter 7 for a variety of geological contexts. The key rock physics components that drive seismic models are documented in Chapter 8, whilst the concept of seismic trace inversion is introduced in Chapter 9. Chapter 10 outlines some key applications of seismic amplitudes such as the description of reservoir properties from seismic and the use of amplitude information in prospect evaluation and reserves determination.

1.3 The practice of seismic rock physics

The practice of seismic rock physics depends to a large extent on the application. In some cases, simply fluid substituting the logs in a dry well and generating synthetic gathers for various fluid fill scenarios may be all that is needed to identify seismic responses diagnostic of hydrocarbon presence. On the other hand, generating stochastic inversions for reservoir prediction and uncertainty assessment will require a complete rock physics database in which the elastic properties of various lithofacies and their distributions are defined in an effective pressure context. Either way, the amount of knowledge required to master the art of seismic rock physics is a daunting prospect for the seismic interpreter.

The broad scope of the subject inevitably means that geophysicists need to work closely with petrophysicists, geologists and engineers. Often this is easier said than done. To quote Ross Crain (2013): 'Geophysicists engaged in seismic interpretation seldom use logs to their full advantage. This sad state is caused, of course, by the fact that most geophysicists are not experts in log analysis. They rely heavily on others to edit the logs and do the analysis for them. But, many petrophysicists and log analysts have no idea what geophysicists need from logs, or even how to obtain the desired results'.

Effectively, the use of rock physics in seismic interpretation blurs the distinctions between subsurface disciplines. This book introduces the subject from a practical viewpoint with a description of how it works and how connections are made between the various disciplines. Whilst there is good practice, there is no single workflow to follow. It is hoped that the perspective presented here will be a source of encouragement to those eager to learn the trade as well as providing ideas for creative hydrocarbon exploration and development.

Chapter 2

Fundamentals

2.1 Introduction

Interpreting seismic amplitudes requires an understanding of seismic acquisition and processing as well as modelling for describing and evaluating acoustic behaviour. Separate books have been written about each of these subjects and there is certainly more to say on these issues than can be presented here. The aim of this chapter is to provide a framework of basic information which the interpreter requires in order to start the process of seismic amplitude interpretation.

2.2 Seismic basics

2.2.1 Seismic geometry

Seismic data are acquired with acoustic sources and receivers. There are numerous types of seismic geometry depending on the requirements of the survey and the environment of operation. Whether it is on land or at sea the data needed for seismic amplitude analysis typically require a number of traces for each subsurface point, effectively providing measurements across a range of *angles of incidence*. The marine environment provides an ideal setting for acquiring such data and a typical towed gun and streamer arrangement is illustrated in Fig. 2.1a. Each shot sends a wave of sound energy into the subsurface, and each receiver on the cable records energy that has been reflected from contrasts in acoustic hardness (or impedance) associated with geological interfaces. It is convenient to describe the path of the sound energy by *rays* drawn perpendicular to the seismic *wavefront*; this in turn clarifies the notion of the angle of incidence (θ in Fig. 2.1a). Usually, the reflections recorded on the near receivers have lower angles of incidence than those recorded on the far receivers.

Figure 2.1b illustrates the recorded signal from the blue and red raypaths shown in Fig. 2.1a. The signal recorded at each receiver is plotted against time (i.e. the travel time from source to receiver), and the receiver traces are ordered by increasing source–receiver distance, usually referred to as *offset*. Plotting the traces for all receivers for one particular source position provides a *shot gather* display. In Fig. 2.1 the reflected energy is shown as a wiggle display and the shape of the reflection signal from the isolated boundary describes the shape of the seismic pulse (the *wavelet*) at the boundary. Owing to the difference in travel path, the arrival time of the reflection from the geological boundary increases with offset and, usually, the relation between travel time and offset is approximately hyperbolic. The amplitude of the reflection from the boundary is related to the contrast in acoustic parameters across the boundary, but is also affected by distance travelled, mainly because the energy becomes spread out over a larger area of wavefront. This phenomenon has commonly been referred to as *spherical divergence*, although it is now evident that wavefronts have shapes between spherical and elliptical. An objective of seismic processing is to produce traces where the amplitudes are related only to the contrasts at the reflecting boundary, and all other effects along the propagation path are removed (this is often referred to as *true amplitude* processing). This can be difficult for land data, where there may be large differences from one trace to the next, related to the effectiveness of the coupling of sources and receivers to the surface, as well as rapid lateral variation in the properties of the shallow zone immediately below the surface.

2.2.2 Gathers and stacks

During seismic acquisition, each shot is recorded by many receivers. Figure 2.2 illustrates that each receiver is recording reflections from different subsurface locations for any given shot. The shot gather therefore mixes together energy from different subsurface locations, and is of little direct use for interpretation. If the Earth is made up of relatively flat-lying layers then the various traces relating to source–receiver pairs

Fundamentals

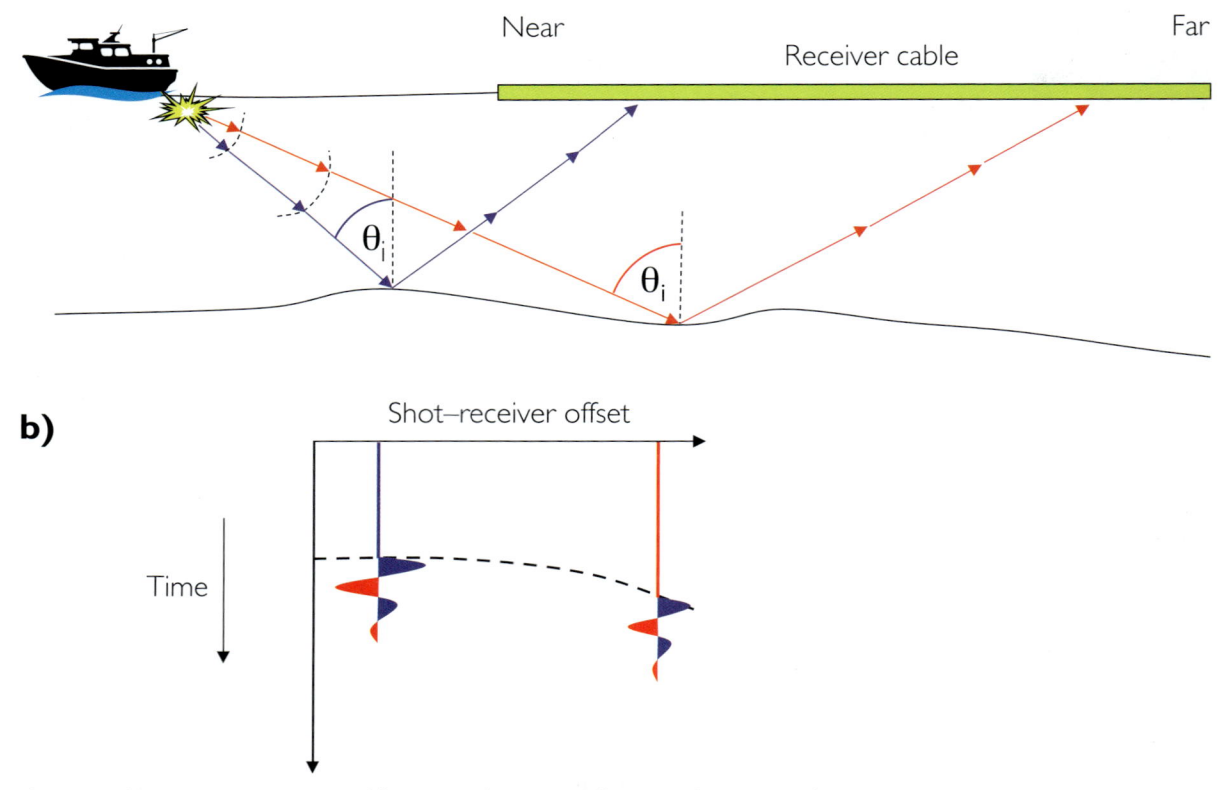

Figure 2.1 Marine seismic geometry 1; (a) source and receiver configuration showing wavefronts, rays (perpendicular to wavefronts) and angle of incidence increasing with offset (b) a shot gather representation of the recorded energy. For a horizontal layered case the reflections form a hyperbola on the gather. Note that the amplitude of a reflection on the gather is related to the rock contrast across the boundary and the decrease in amplitude due to wavefront divergence.

which share a common midpoint (CMP) will also share common subsurface reflection points. These are typically brought together to form a *CMP gather* (Fig. 2.2) and form the basis for further analysis. If the subsurface is not a simple stack of plane layers, it is still possible to create a gather for a common reflection point provided that the subsurface geometry and seismic velocities can be determined reasonably accurately from the data. This is an aspect of seismic *migration*, which attempts to position subsurface reflectors in their true spatial location. There are several different approaches to migration, and a vast literature exists on the subject. Jones (2010) gives a useful overview. For the purpose of this narrative it is assumed that a gather has been produced in which all the traces are related to the same subsurface point at any given reflection time.

In order for the gathers to be interpretable, they need to be processed. Figure 2.3 gives a generalised overview of some of the steps involved. A time-varying gain is applied to remove the effects of wavefront divergence, a mute is applied to remove unwanted signal (typically high-amplitude near-surface direct and refracted arrivals visible on the further offsets at any given time), and pre-stack migration is applied to bring traces into the correct geometrical subsurface location. As shown on the left-hand side of Fig. 2.3, the reflection time of any particular interface on the raw gather becomes later with increasing offset, due to the increased path length. An important step is the application of time-varying time shifts to each trace so as to line up each reflection horizontally across the gather, as shown on the right-hand side of Fig. 2.3. This is needed in conventional processing

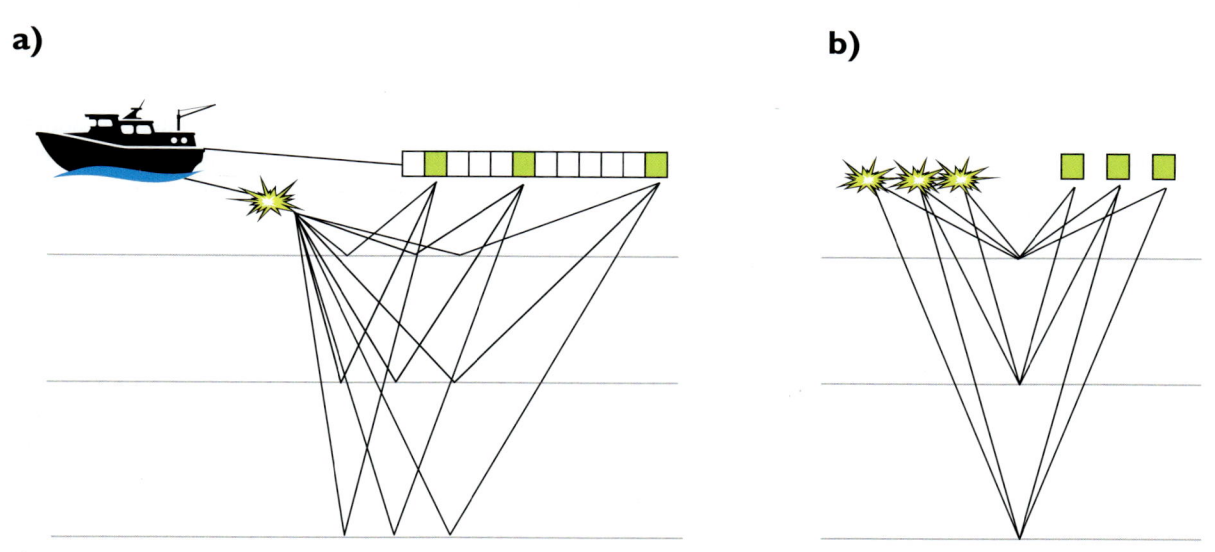

Figure 2.2 Marine seismic geometry 2; (a) acquisition; each shot is recorded at a variety of receivers depending on the depth and the angle of reflection and (b) the common midpoint (CMP); if it is assumed that the Earth is flat the data can be arranged according to reflection location, i.e. different source–receiver pairs sampling the same position in the subsurface. For more complex velocity overburden sophisticated imaging solutions are required.

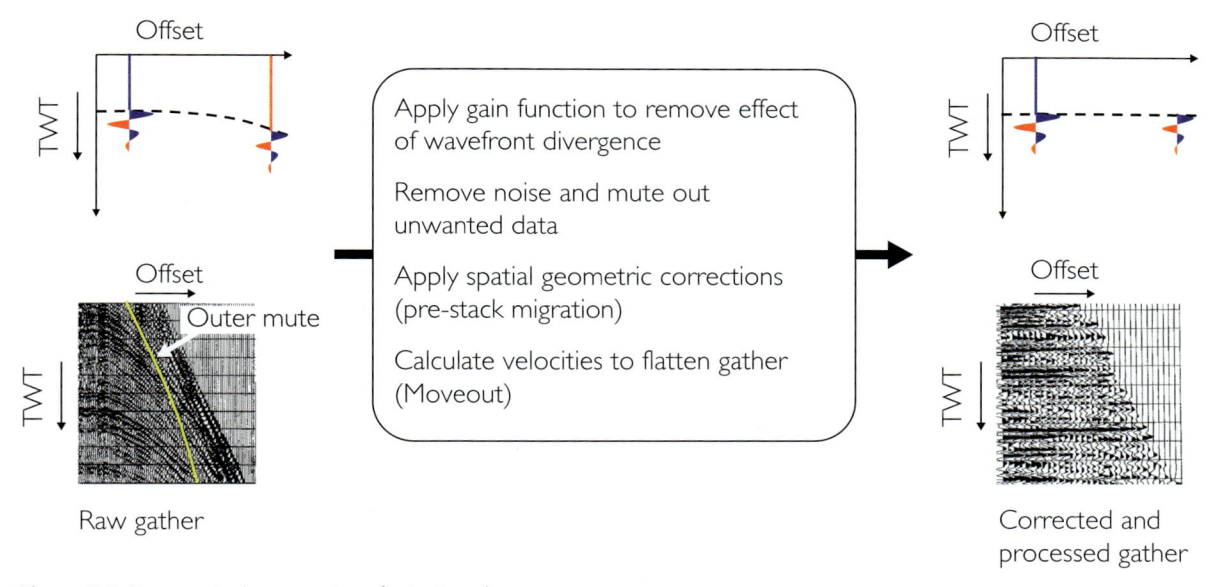

Figure 2.3 Key steps in the processing of seismic gathers.

because the next step will be to *stack* the data by summing the traces of the gather along lines of constant time, i.e. along horizontal lines in the display of Fig. 2.3. This has the important effect of enhancing signal and suppressing noise. Accurate horizontal alignment across the gather is also important for the study of amplitude variation with offset (AVO) described later in this chapter and in Chapter 5. The process of time-shifting to flatten the reflections is called *moveout correction*. A commonly used term is *normal moveout (NMO)*, which refers to the specific case where there is no dip on the reflector.

Figure 2.4 illustrates a stacking methodology that is popular for seismic AVO analysis. Seismic sections have been created by stacking the near-offset data and the far-offset data separately, giving an immediate visual impression of AVO effects but also providing information that can be analysed quantitatively (see Chapters 5 and 7). In this particular case, the difference between near and far trace data is unusually large but it illustrates the point that where there is only a 'full stack' section available (i.e. created by adding together all the traces of the gather) the interpreter may be missing valuable information.

2.3 Modelling for seismic interpretation

Propagation of seismic energy in the Earth is a complex phenomenon. Figure 2.5 shows some of the numerous factors related to geology and acquisition. The goal of course is to relate seismic amplitude to rock property contrasts across reflecting boundaries but there are several other factors besides geology that also have an influence on amplitude. Some of these are associated with the equipment used for the survey; these include variability of source strength and coupling from shot to shot, variability of sensitivity and coupling from one receiver to another, the directivity of the receiver array (i.e. more sensitive at some

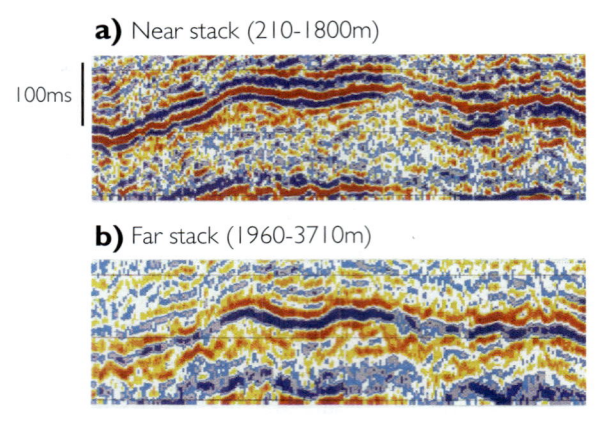

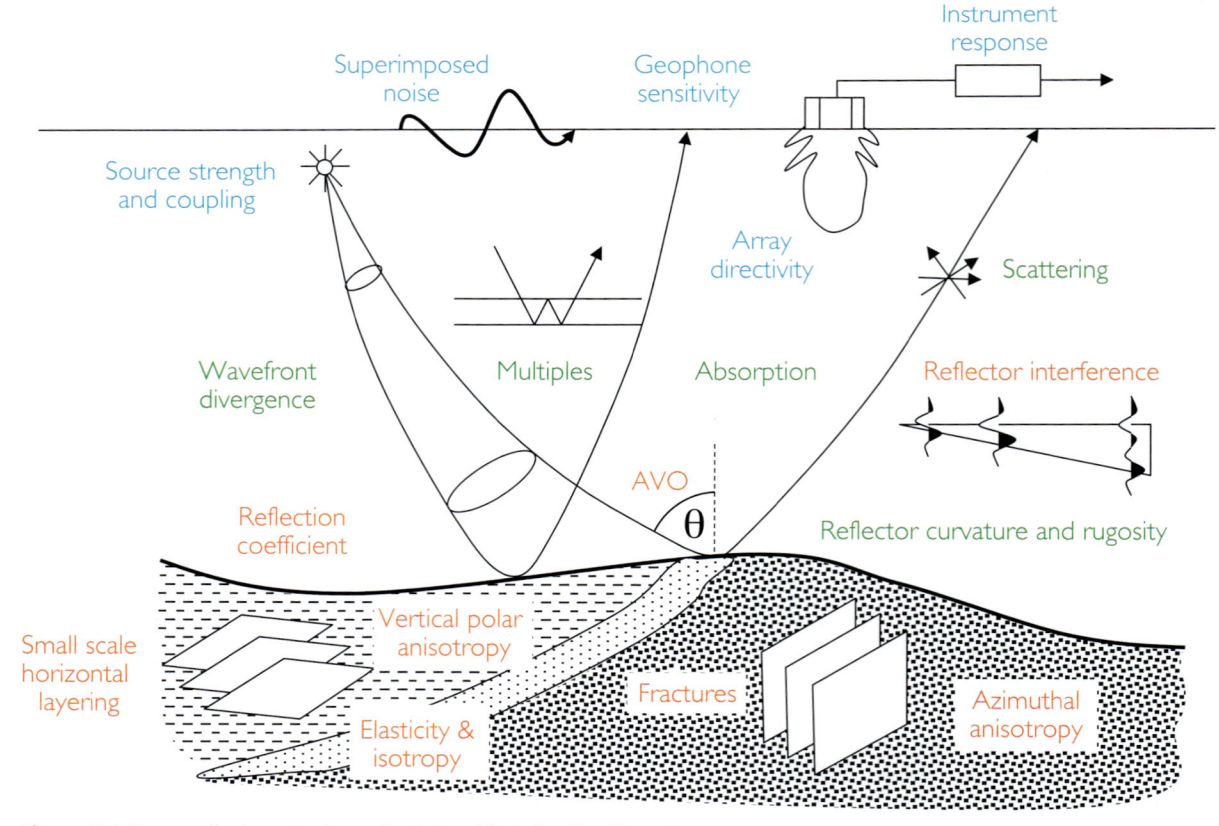

Figure 2.4 Offset stacking example; (a) near stack, 210–1800 m offset, (b) far stack, 1960–3710 m offset. Note that the choice of offsets is dependent on the offset range available and the depth of the target. In this case the offsets have been selected to avoid stacking across a zone of phase reversal.

Figure 2.5 Factors affecting seismic amplitude (modified after Sheriff, 1975).

angles of incidence than others) and the imperfect fidelity of the recording equipment. Marine seismic has the advantage that sources and receivers are very repeatable in their characteristics. This is not true for land data, where the coupling of source and receivers to the ground may be quite variable from one shot to another, depending on surface conditions. However, these effects can be estimated and allowed for by the seismic processor.

Some amplitude effects are features of the subsurface that are of little direct interest and ideally would be removed from the data during processing if possible; these include divergence effects, multiples, scattering, reflection curvature and rugosity, and general superimposed noise. Depending on the individual data set, it may be quite difficult to remove these without damaging the amplitude response of interest. For example, processors often have difficulty in attenuating multiple energy whilst preserving the fidelity of the geological signal. Other effects on seismic amplitude, for example related to absorption and anisotropy, might be a useful signal if their origin were better understood. The processor clearly faces a tough challenge to mitigate the effects of unwanted acquisition and transmission factors and enhance the geological content of the data.

Interpretation of seismic amplitudes requires a model. A first order aspect of the seismic model is that the seismic trace can be regarded as the convolution of a seismic pulse with a reflection coefficient related to contrasting rock properties across rock boundaries. This idea is an essential element in seismic processing as well as seismic modelling. Given that the seismic processor attempts to remove unwanted acquisition and propagation effects and provide a dataset in which the amplitudes have 'correct' relative scaling, the interpreter's approach to modelling tends (at least initially) to focus on primary geological signal in a target zone of interest. Of course, one eye should be on the look out for 'noise' remaining in the section that has not been removed (such as multiple energy and other forms of imaging effects). The presence of such effects might dictate more complex (and more time consuming) modelling solutions and can often negate the usefulness of the seismic amplitude information.

From a physical point of view the geological component of seismic reflectivity can be regarded as having various levels of complexity. For the most part, the geological component can be described in simple physical terms. In the context of the small stresses and strains related to the passage of seismic waves, rocks can be considered perfectly elastic (i.e. they recover their initial size and shape completely when external forces are removed) and obey Hooke's Law (i.e. the strain or deformation is directly proportional to the stress producing it). An additional assumption is that rocks are to first order *isotropic* (i.e. rocks have the same properties irrespective of the direction in which the properties are measured). Experience has shown that in areas with relatively simple layered geology this isotropic/elastic model is very useful, being the basis of well-to-seismic ties (Chapter 4) and seismic inversion (Chapter 9).

There are, however, complexities that should not be ignored. These complexities can broadly be characterised as (a) signal-attenuating processes such as absorption and scattering and (b) anisotropic effects, related to horizontal sedimentary layering (vertical polar isotropy) and vertical fracture effects (azimuthal anisotropy) (see Section 5.3.7). One effect of absorption is to attenuate the seismic signal causing changes in wavelet shape with increasing depth and this is usually taken into account. Attenuation is difficult to measure directly from seismic but at least theoretically this information could have a role in identifying the presence of hydrocarbons (e.g. Chapman *et al.*, 2006; Chapman, 2008). Whilst there is a good deal of theoretical understanding about anisotropy (Thomsen, 1986; Lynn, 2004), there is currently limited knowledge of how to exploit it for practical exploration purposes. One problem is the availability of data with which to parameterise anisotropic models. Practical seismic analysis in which anisotropic phenomena are exploited has so far been restricted to removing horizontal layering effects on seismic velocities and moveout in seismic processing (i.e. flattening gathers particularly at far offsets) (Chapter 6) and defining vertical fracture presence and orientation (Chapter 7).

2.3.1 The convolutional model, wavelets and polarity

The cornerstone of seismic modelling is the convolutional model, which is the idea that the seismic trace can be modelled as the convolution of a seismic pulse with a reflection coefficient series. In its simplest form the reflection coefficient is related to change in acoustic impedance, where the acoustic impedance (AI) is the product of velocity (V) and bulk density (ρ) (Fig. 2.6):

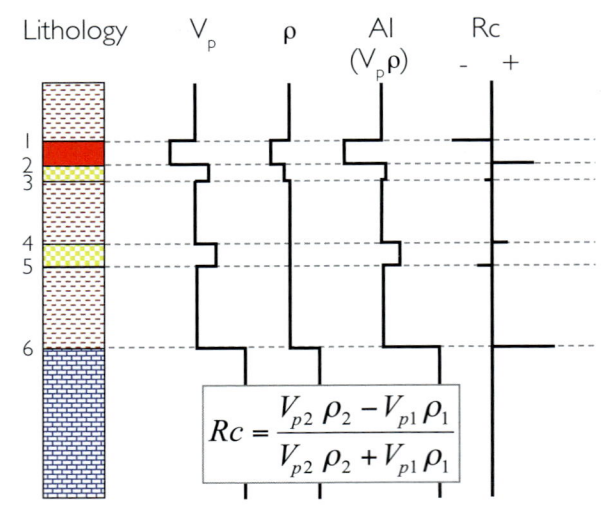

Figure 2.6 The reflection coefficient as defined by the differentiation of the acoustic impedance log (re-drawn and modified after Anstey, 1982).

$$Rc = \frac{V_{p2} \rho_2 - V_{p1} \rho_1}{V_{p2} \rho_2 + V_{p1} \rho_1}$$

$$R = \frac{AI_2 - AI_1}{AI_2 + AI_1}, \quad (2.1)$$

where AI_1 is the acoustic impedance on the incident ray side of the boundary and AI_2 is the acoustic impedance on the side of the transmitted ray. Reflectivity can be either positive or negative. In the model of Fig. 2.6 the top of the limestone (interface 6) is characterised by a *positive reflection* whilst the top of the gas sand (interface 1) is characterised by a *negative reflection*.

It should be noted that the equation above is relevant for a ray which is normally incident on a boundary. The change in reflectivity with incident angle (i.e. offset dependent reflectivity) will be discussed in more detail later in this chapter. A useful approximation that derives from the reflectivity equation and which describes the relationship between reflectivity and impedance is:

$$R \approx 0.5 \Delta \ln AI. \quad (2.2)$$

Effectively, the amount of reflected energy determines how much energy can be transmitted through the section. Following the normal incidence model described above, the transmission coefficient (at normal incidence) is defined by

$$T = 2AI_1/(AI_2 + AI_1). \quad (2.3)$$

Given the boundary conditions of pressure continuity and conservation of energy it can be shown that the amount of reflected energy is proportional to R^2, whereas the transmitted energy is proportional to $(AI_2/AI_1) T^2$. Thus, less energy is transmitted through boundaries with high AI contrasts (e.g. such as a hard sea floor or the top of a basalt layer). In extreme cases, lateral variations in AI contrasts can result in uneven amplitude scaling deeper in the section.

Transmission effects are important in understanding the general nature of recorded seismic energy. O'Doherty and Anstey (1971) noted that seismic amplitudes at depth appear to be higher than can be accounted for with a simple (normal incidence) model of reflection and transmission at individual boundaries. It was concluded that seismic reflection energy is being reinforced by reflections from thin layers for which the top and base reflections have opposite sign (Anstey and O'Doherty, 2002). The cumulative effects of many cyclical layers can be significant and this may provide an explanation for the observation that reflections tend to parallel chronostratigraphic boundaries.

To generate a synthetic seismogram requires knowledge of the shape of the seismic pulse as well as a calculated reflection coefficient series. A recorded seismic pulse typically has three dominant loops, the relative amplitudes of which can vary according to the nature of the source, the geology and the processes applied to the data (Chapter 3). Assuming that no attempt has been made to shape the wavelet or change its timing, a time series representation of the recorded wavelet will start at time zero (i.e. the wavelet is causal). Figure 2.7 shows the reflection coefficient series from Fig. 2.6 convolved with a recorded seismic pulse, illustrating how the synthetic trace is the addition of the individual reflections.

With regard to the polarity representation of the wavelet shown in Fig. 2.7, reference is made to the recommendation of a SEG committee on polarity published by Thigpen et al. (1975). It states that an upward geophone movement or increase in pressure on a hydrophone should be recorded as a negative number and displayed as a trough ('SEG standard polarity') (Fig. 2.8). This definition is almost universally adhered to in seismic recording. The implication is that a reflection from a positive reflection coefficient (a *positive* or 'hard' *reflection*), will start with a trough. Note that a positive reflection is the interpreter's reference for describing polarity. Figure 2.7 conforms, therefore, to the SEG standard polarity convention with positive reflections such as the top of the limestone

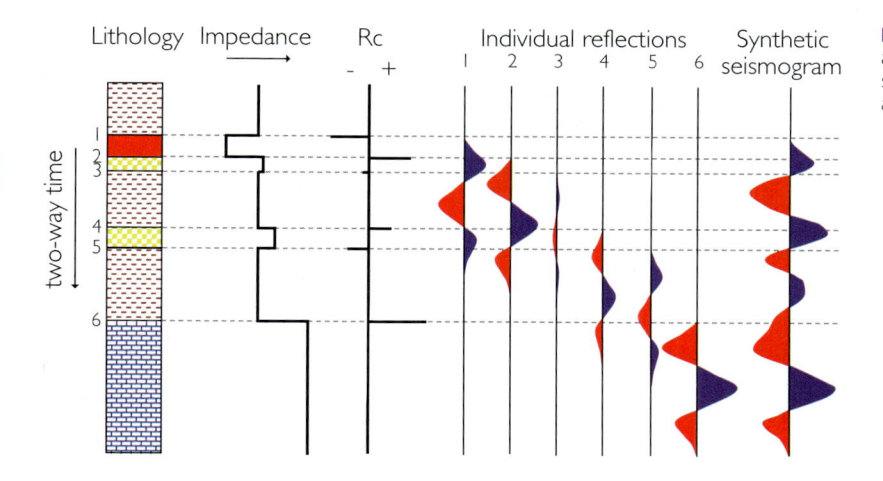

Figure 2.7 Synthetic seismogram using a causal (i.e. recorded) wavelet with SEG standard polarity (re-drawn and modified after Anstey, 1982).

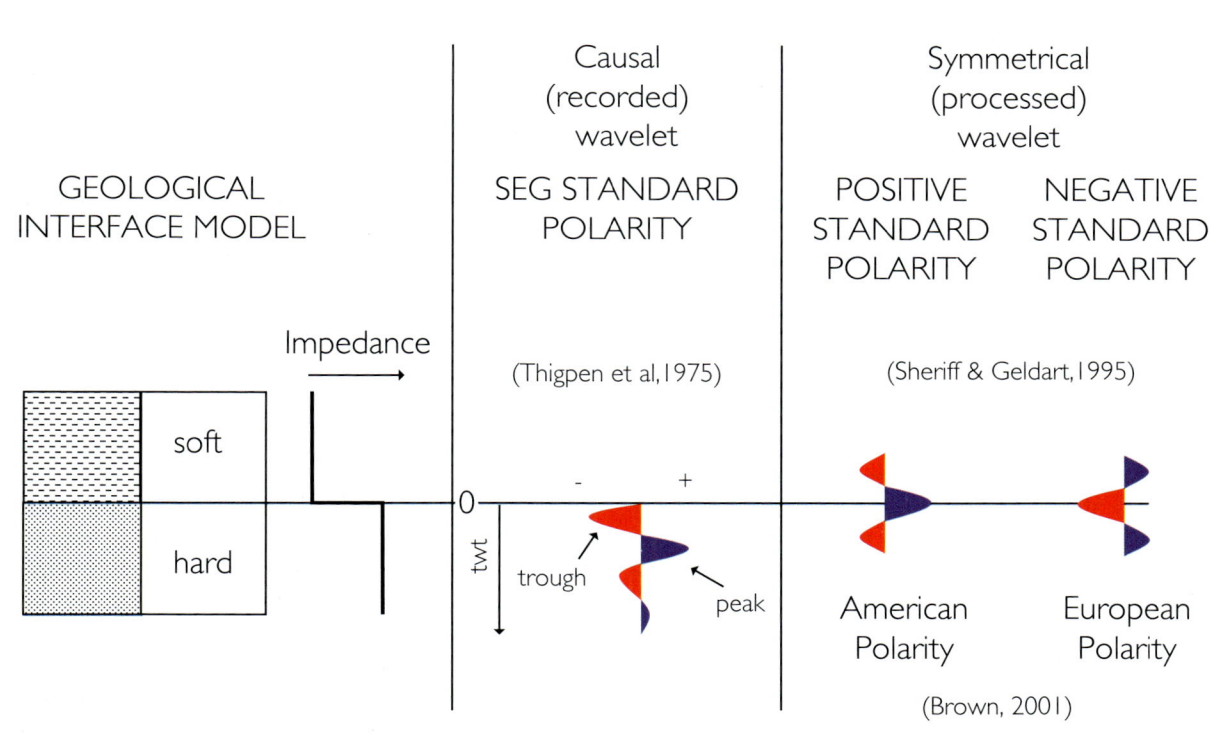

Figure 2.8 Seismic polarity conventions.

starting with a trough and negative reflections (such as the top of the gas sand) starting with a peak.

One problem with causal wavelets (amongst others) is that there is a time lag between the position of the boundary and the energy associated with a reflection from the boundary, making it difficult to correlate the geology with the seismic. Thus, there is a requirement for processing the seismic wavelet to a symmetrical form which concentrates and correctly aligns the energy with the position of geological boundaries. Figure 2.9 shows the same synthetic but now with a symmetrical wavelet. It is now much clearer which loops in the seismic the interpreter needs to pick for the various geological boundaries.

The polarity conventions in common usage that apply to symmetrical wavelets have been defined by Sheriff and Geldart (1995), again with reference to a positive reflection. If a positive reflection is represented

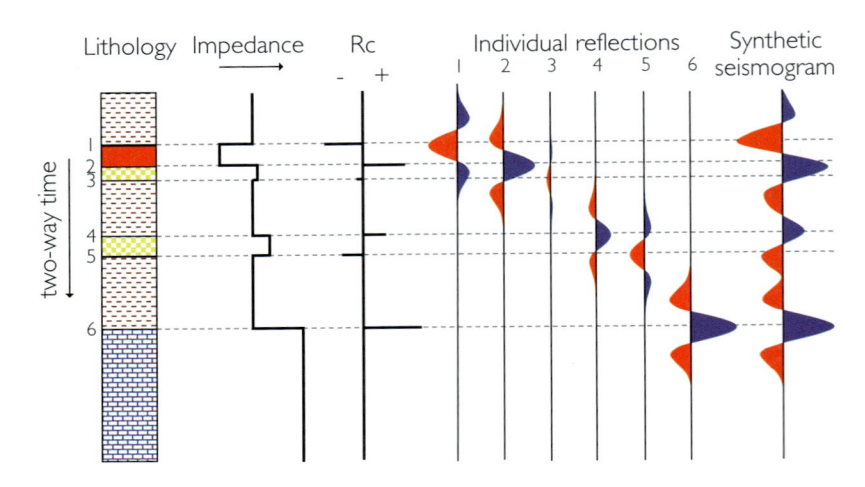

Figure 2.9 Synthetic seismogram using a symmetrical wavelet with positive standard polarity (re-drawn and modified after Anstey, 1982). This illustrates the importance of zero phase to the interpreter; the main layer boundaries are more easily identified with the processed symmetrical wavelet compared to the recorded asymmetrical wavelet shown in Fig 2.7.

as a peak (i.e. positive number) then this is referred to as 'positive standard polarity', whereas if it is represented as a trough (i.e. negative number) it is referred to as 'negative standard polarity' (Fig. 2.8). Historically the usage of these two conventions has been broadly geographical and they have been referred to as 'American' and 'European' (Brown, 2001, 2004). In addition, these conventions are sometimes informally referred to respectively as 'SEG normal' and 'SEG reverse' polarity. It is evident, however, that the use of 'normal' and 'reverse' terms can easily lead to confusion and it is recommended that their use should be avoided.

From the point of view of seismic amplitude and AVO studies it is recommended that positive standard polarity be used as it lessens the potential confusion in the representation of amplitude data. Following this convention will mean that AVO plots will be constructed with positive numbers representing positive reflections and integration type processes (such as coloured inversion (Chapter 5)) will produce the correct sense of change in bandlimited impedance traces (i.e. negative to positive for a boundary with a positive reflection) (see Section 5.5.3).

With respect to colour, it is common for seismic troughs to be coloured red and peaks to be coloured blue or black. There are some notable exceptions, however. For example, in South Africa the tendency has been to adhere to positive standard polarity but colouring the troughs blue and the peaks red. With modern software the interpreter is not restricted to blue and red and can choose from a whole range of colour options. For more discussion of the role of colour in seismic interpretation the reader is referred to Brown (2004) and Froner *et al.* (2013).

For the interpreter who inherits a seismic project it is evident that polarity and colour coding issues (as well as uncertainties concerning the processing of the data) introduce significant potential for misunderstanding and error. It is critical that the interpreter develops a good idea of the shape of the seismic wavelet prior to detailed horizon picking (see Chapters 3 and 4).

2.3.2 Isotropic and elastic rock properties
2.3.2.1 P and S velocities and bulk density

Seismic models for exploration purposes are constructed using velocities and densities, principally from wireline log data (Chapter 8). As will be shown in the following section the calculation of offset reflectivity requires two types of velocity (compressional (P) and shear (S)) as well as the bulk density (ρ). Figure 2.10 illustrates the P and S waves of interest in 3D exploration. The P wave is characterised by particle motion in the direction of wave propagation. The S wave travels in the same direction as the P wave (and at approximately half the speed of the compressional wave) but the particle motion is perpendicular to the direction of wave propagation. Strictly speaking this shear wave is the vertically polarised shear wave. Whilst it is only compressional waves that are recorded in marine seismic, the reason that shear information is important to the interpreter is because changes in amplitude with angle are related to the contrast in the velocities of P waves and the vertically polarised S wave. By contrast, in land exploration a comparison of vertically and horizontally polarised shear waves (recorded with three component sensors)

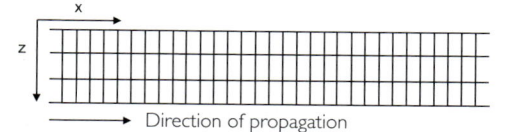

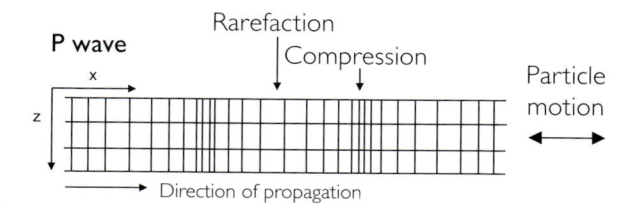

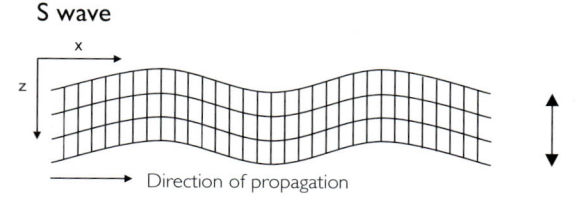

Figure 2.10 Schematic illustration of P and S waves.

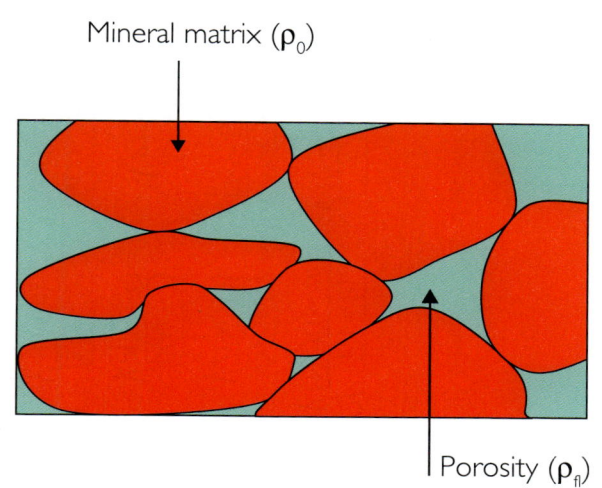

Figure 2.11 Density components of a rock with intergranular porosity.

can indicate the presence and orientation of fractures in the subsurface (Lynn, 2004) (Chapters 5 and 7).

Bulk density ρ_b is a relatively simple parameter, being calculated as the weighted average of the densities of the components (Fig. 2.11):

$$\rho_b = \phi \rho_{fl} + (1 - \phi)\rho_0 \qquad (2.4)$$

where ϕ is the porosity and the ρ_{fl} and ρ_0 parameters are the densities of the fluid in the pore space and of the rock matrix respectively. If more than one fluid or mineral is present the effective densities are calculated simply by weighting the various component densities according to their proportions. Equation (2.4) can be re-written to calculate porosity from bulk density:

$$\phi = (\rho_b - \rho_0)/(\rho_0 - \rho_{fl}). \qquad (2.5)$$

2.3.2.2 Isotropic and elastic moduli

Unfortunately it is insufficient simply to focus on velocity and bulk density. Understanding the isotropic and elastic context of velocities and density is necessary in order to appreciate the rock physics tools at the disposal of the seismic modeller. For example, when calculating the effect of varying fluid fill on sandstone velocities and densities (i.e. fluid substitution) it is necessary to perform the calculation in terms of elastic moduli before the effect on velocities and density can be appreciated (see Chapter 8).

There are a variety of elastic parameters that can be used to describe the isotropic and elastic behaviour of rocks. Table 2.1 provides a useful reference (from Smidt, 2009), illustrating that two independent measurements can be used to calculate any other elastic parameter. Elastic moduli describe rock responses to different types of stress (i.e. force applied over a unit area). The bulk modulus, for example, is the rock response to normal stress applied in all directions on a unit of rock (Fig. 2.12) and relates fractional volume change $\Delta V/V$ to the uniform compressive stress S:

$$K = \frac{S}{\Delta V/V}. \qquad (2.6)$$

As such the bulk modulus is an indicator of the extent to which the rock can be squashed. It is sometimes encountered as its reciprocal, $1/K$, called the *compressibility*. The shear modulus (μ) is the response to a tangential or shearing stress and is defined by

$$\mu = \frac{\text{shear stress}}{\text{shear strain}}, \qquad (2.7)$$

where the shear strain is measured through the shear angle (Fig. 2.12). As such the shear modulus indicates the rigidity of the rock or the resistance of the rock to a shaking motion. Most fluids are not able to resist

Table 2.1. Table of elastic constants for isotropic media (after Smidt, 2009). Note that moduli units = GPa, V=km/s, ρ=g/cc.

Symbol	E	σ	K or κ	M	λ (lambda)	μ (mu)	λ/μ	V_p	V_s	V_p/V_s ratio
Entity	Young's modulus	Poisson's ratio	Bulk modulus	P wave modulus	Lamé parameter	Lamé parameter	Lamé impedance ratio	V-primary	V-secondary	V_p/V_s ratio
(E, σ)			$\dfrac{E}{3(1-2\sigma)}$	$\dfrac{E(1-\sigma)}{(1+\sigma)(1-2\sigma)}$	$\dfrac{E\sigma}{(1+\sigma)(1-2\sigma)}$	$\dfrac{E}{2(1+\sigma)}$	$\dfrac{2\sigma}{1-2\sigma}$	$\sqrt{\dfrac{E(1-\sigma)}{(1+\sigma)(1-2\sigma)\rho}}$	$\sqrt{\dfrac{E}{2(1+\sigma)\rho}}$	$\sqrt{\dfrac{1-\sigma}{1/2-\sigma}}$
(E, κ)		$\dfrac{3\kappa - E}{6\kappa}$		$\dfrac{3\kappa+E}{3}\dfrac{}{9\kappa-E}$	$\dfrac{3\kappa-E}{3}\dfrac{}{9\kappa-E}$	$\dfrac{3\kappa E}{9\kappa-E}$	$\dfrac{3\kappa}{E}-1$	$\sqrt{\dfrac{3\kappa(3\kappa+E)}{\rho(9\kappa-E)}}$	$\sqrt{\dfrac{3\kappa E}{(9\kappa-E)\rho}}$	$\sqrt{\dfrac{3\kappa+E}{E}}$
(E, μ)		$\dfrac{E-2\mu}{2\mu}$	$\dfrac{\mu E}{3(3\mu - E)}$	$\dfrac{4\mu-E}{3\mu-E}$	$\dfrac{E-2\mu}{\mu}\dfrac{}{3\mu-E}$			$\sqrt{\dfrac{\mu(4\mu-E)}{\rho(3\mu-E)}}$	$\sqrt{\dfrac{\mu}{\rho}}$	$\sqrt{\dfrac{4\mu-E}{3\mu-E}}$
(σ, κ)	$3\kappa(1-2\sigma)$			$3\kappa\dfrac{1-\sigma}{1+\sigma}$	$3\kappa\dfrac{\sigma}{1+\sigma}$	$\dfrac{3\kappa}{2}\dfrac{1-2\sigma}{1+\sigma}$	$\dfrac{2\sigma}{1-2\sigma}$	$\sqrt{\dfrac{3\kappa(1-\sigma)}{\rho(1+\sigma)}}$	$\sqrt{\dfrac{3\kappa(1-2\sigma)}{2\rho(1+\sigma)}}$	$\sqrt{\dfrac{1-\sigma}{1/2-\sigma}}$
(σ, μ)	$2\mu(1+\sigma)$		$\dfrac{2\mu(1+\sigma)}{3(1-2\sigma)}$	$2\mu\dfrac{1-\sigma}{1-2\sigma}$	$\mu\dfrac{2\sigma}{1-2\sigma}$			$\sqrt{\dfrac{2\mu(1-\sigma)}{\rho(1-2\sigma)}}$	$\sqrt{\dfrac{\mu}{\rho}}$	$\sqrt{\dfrac{1-\sigma}{1/2-\sigma}}$
(σ, λ)	$\lambda\dfrac{(1+\sigma)(1-2\sigma)}{\sigma}$		$\lambda\dfrac{1+\sigma}{3\sigma}$	$\lambda\dfrac{1-\sigma}{\sigma}$		$\lambda\dfrac{1-2\sigma}{2\sigma}$		$\sqrt{\dfrac{\lambda(1-\sigma)}{\rho\sigma}}$	$\sqrt{\dfrac{\lambda(1-2\sigma)}{\rho 2\sigma}}$	$\sqrt{\dfrac{(1-\sigma)}{1/2-\sigma}}$
(κ, μ)	$\dfrac{9\kappa\mu}{3\kappa+\mu}$	$\dfrac{3\kappa-2\mu}{2(3\kappa+\mu)}$		$\kappa+4\mu/3$	$\kappa-2\mu/3$			$\sqrt{\dfrac{\kappa+4\mu/3}{\rho}}$	$\sqrt{\dfrac{\mu}{\rho}}$	$\sqrt{\dfrac{\kappa+4\mu/3}{\mu}}$
(κ, λ)	$9\kappa\dfrac{\kappa-\lambda}{3\kappa-\lambda}$	$\dfrac{\lambda}{3\kappa-\lambda}$		$3\kappa-2\lambda$		$3(\kappa-\lambda)/2$		$\sqrt{\dfrac{3\kappa-2\lambda}{\rho}}$	$\sqrt{\dfrac{3(\kappa-\lambda)}{2\rho}}$	$\sqrt{2\dfrac{\kappa-2\lambda/3}{\kappa-\lambda}}$
(μ, λ)	$\mu\dfrac{3\lambda+2\mu}{\lambda+\mu}$	$\dfrac{\lambda}{2(\lambda+\mu)}$	$\lambda+2\mu/3$	$\lambda+2\mu$				$\sqrt{\dfrac{\lambda+2\mu}{\rho}}$	$\sqrt{\dfrac{\mu}{\rho}}$	$\sqrt{\dfrac{\lambda+2\mu}{\mu}}$
(V_p, V_s)	$\dfrac{\rho V_s^2(3V_p^2-4V_s^2)}{V_p^2-V_s^2}$	$\dfrac{V_p^2-2V_s^2}{2(V_p^2-V_s^2)}$	$\rho(V_p^2-4V_s^2/3)$	ρV_p^2	$\rho(V_p^2-2V_s^2)$	ρV_s^2	$\left(\dfrac{V_p}{V_s}\right)^2-2$			

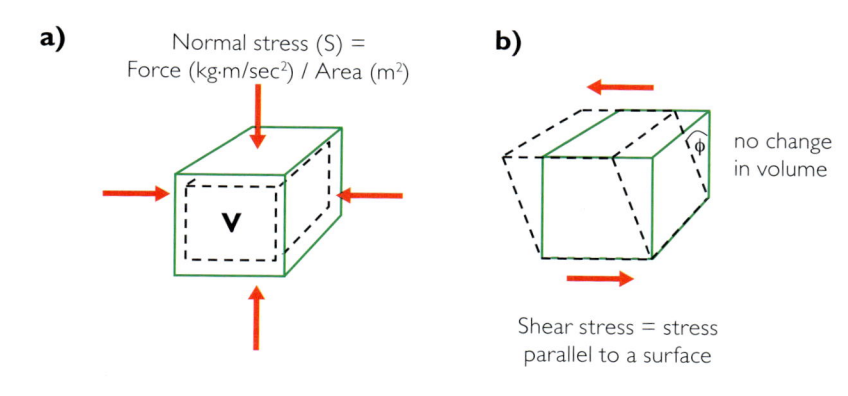

Figure 2.12 Volume and shape changes in rocks under stress; (a) change of volume with stress applied equally from all directions, (b) change in shape associated with shear stress (e.g. stresses applied parallel to bedding boundaries).

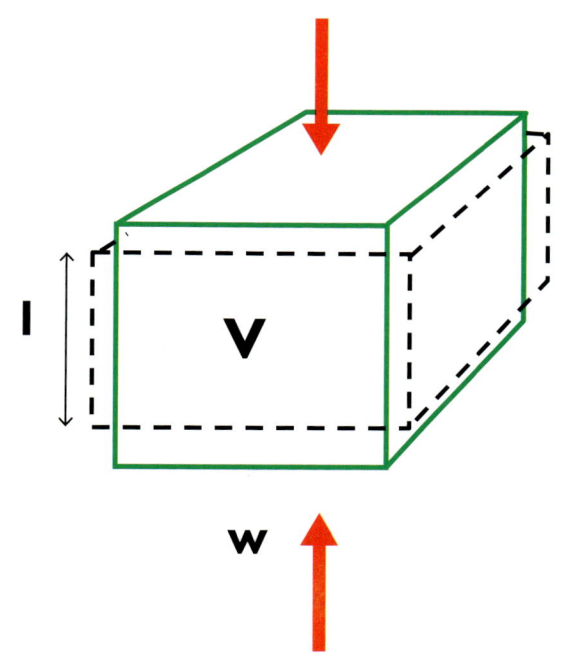

Figure 2.13 Poisson's ratio.

a shear deformation, so it is usually assumed that the shear modulus of fluids is zero.

Some authors (e.g. Goodway et al., 1997) prefer the use of the Lamé parameters (λ and μ) in preference to (K and μ) believing that they offer greater physical insight. The Lamé constant λ is given by

$$\lambda = K - \frac{2\mu}{3}. \quad (2.8)$$

An important elastic parameter in AVO is Poisson's ratio. Poisson's ratio is the ratio of the fractional change in width to the fractional change in length under uni-axial compression (Fig. 2.13). It can be shown that it is given by:

$$\text{Poisson's ratio}(\sigma) = -\frac{\Delta w/w}{\Delta l/l} = -\frac{\text{Transverse strain}}{\text{Longitudinal strain}}$$
$$= \frac{3K - 2\mu}{2(3K+\mu)}. \quad (2.9)$$

The contrast in Poisson's ratio across an interface can have a large control on the rate of change of amplitude with offset.

Elastic moduli are not generally measured directly, for example with downhole logging tools, but they can be calculated from velocity and density measurements. Some key equations relating velocities and densities to elastic properties are as follows:

$$V_p = \sqrt{\frac{K+4\mu/3}{\rho}} \text{ or } V_p = \sqrt{\frac{\lambda+2\mu}{\rho}} \quad (2.10)$$

and

$$V_s = \sqrt{\frac{\mu}{\rho}}, \quad (2.11)$$

where ρ is the density of the material.

It is evident from these equations that the compressional velocity is a more complicated quantity than the shear wave velocity, involving both bulk and shear moduli. Useful equations that illustrate the relationship between P and S velocities and Poisson's ratio (σ) are shown below:

$$\frac{V_p}{V_s} = \sqrt{\frac{2(1-\sigma)}{(1-2\sigma)}} \quad (2.12)$$

$$\sigma = \frac{\gamma-2}{2\gamma-2}, \text{ where } \gamma = \left[\frac{V_p}{V_s}\right]^2 \text{ and } \gamma = \frac{2\sigma-2}{2\sigma-1}. \quad (2.13)$$

The general relationship between V_p/V_s and Poisson's ratio is shown in Fig. 2.14 together with typical ranges

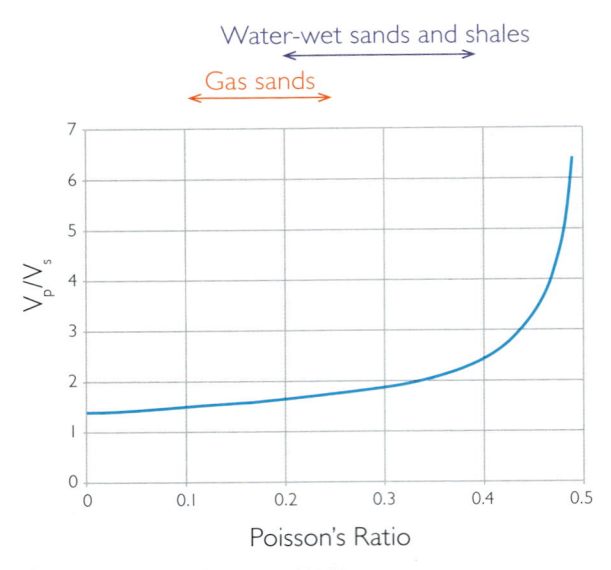

Figure 2.14 Poisson's ratio and V_p/V_s.

for shales and sands with different fluid fill. Sands tend to have a lower Poisson's ratio than shales because quartz has a lower V_p/V_s ratio than most other minerals. Rocks containing compressible fluids (oil and, especially, gas) have lower V_p and slightly higher V_s than their water-wet equivalent. This means that hydrocarbon sands will have a lower Poisson's ratio than water-bearing sands.

2.3.3 Offset reflectivity

The isotropic and elastic behaviour of a P wave incident on a boundary at any angle is described by the Zoeppritz (1919) equations (see Aki and Richards, 1980) (Fig. 2.15). The Zoeppritz equations describe the partitioning of P and S wave energy into reflected and transmitted components. The variation of the P wave reflection coefficient with angle is the key parameter for most seismic interpretation, though S wave reflection coefficients sometimes need to be considered, for example when interpreting marine seismic acquired with cables on the sea floor.

An example of a single boundary calculation using the Zoeppritz equations is shown in Fig. 2.15. The response shown is for shale overlying limestone, not a typical contrast of interest in hydrocarbon exploration but it illustrates an important point. The P wave amplitude of the 'hard' reflection initially decreases with increasing angle, but at a certain point the amplitude increases sharply. This is associated with the point at which the transmitted P wave amplitude is reduced to zero and refractions are generated along the boundary. The angle at which these effects occur is called the *critical angle*. This may be thought of in terms of Snell's Law that defines the relationship between incident and transmission angles (Fig. 2.16). If the upper layer has lower velocity, then the critical angle is given by:

$$\theta_c = \sin^{-1}\left(\frac{V_{p1}}{V_{p2}}\right). \quad (2.14)$$

Beyond the critical angle, the reflected P wave is phase-shifted relative to the incident signal. It is possible that critical angle energy from boundaries with high velocity contrast can be misinterpreted as a hydrocarbon effect so it is important that the interpreter and processor do not include these data in gathers for AVO analysis. A seismic gather example of critical angle energy (at a shale/limestone interface) is shown in Fig. 2.17.

Unfortunately, the Zoeppritz equations are complicated and do not give an intuitive feel for how rock properties impact the change of amplitude with angle. For this reason several authors have derived approximations to the equations for estimating amplitude as a function of angle for pre-critical angles. A popular three-term approximation was developed by Aki and Richards (1980). Various authors have re-formulated the approximation depending on the purpose, but a useful starting point for the interpreter is the formulation generally accredited to Wiggins *et al.* (1983):

$$R(\theta) = A + B\sin^2\theta + C\sin^2\theta \tan^2\theta, \quad (2.15)$$

where

$$A = \frac{1}{2}\left(\frac{\Delta V_p}{V_p} + \frac{\Delta \rho}{\rho}\right)$$

$$B = \frac{\Delta V_p}{2V_p} - 4\left(\frac{V_s}{V_p}\right)^2\left(\frac{\Delta V_s}{V_s}\right) - 2\left(\frac{V_s}{V_p}\right)^2\left(\frac{\Delta \rho}{\rho}\right)$$

and $C = \frac{1}{2}\frac{\Delta V_p}{V_p}$,

where $V_p = \frac{V_{p1}+V_{p2}}{2}$, $V_s = \frac{V_{s1}+V_{s2}}{2}$, $\rho = \frac{\rho_1+\rho_2}{2}$,

$$(V_s/V_p)^2 = \frac{(V_{s1}/V_{p1})^2 + (V_{s2}/V_{p2})^2}{2},$$

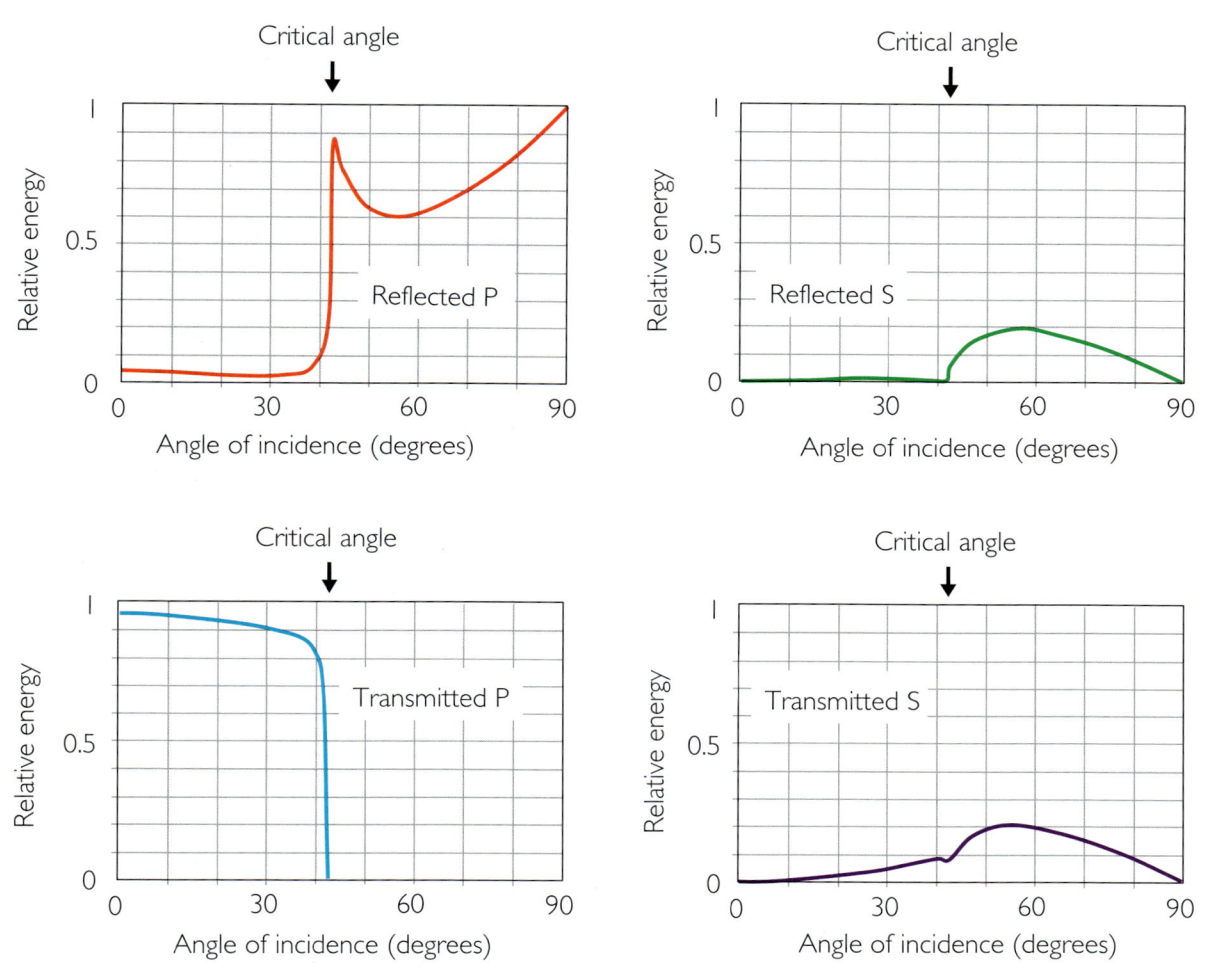

Figure 2.15 Partitioning of P wave energy at a shale/limestone interface as a function of incidence angle. The model uses the Zoeppritz equations for the calculation and the various components sum to one at all angles. Note that the critical angle defines the point at which there is a dramatic rise in reflected P wave energy and a corresponding reduction of transmitted energy to zero. Appreciable S wave energy is generated beyond the critical angle. Not shown in the diagram is energy refracted along the boundary at the critical angle. Elastic values in the model: shale V_p = 2540 m/s, V_s = 1150 m/s, density = 2.35 g/cc; limestone V_p = 3750 m/s, V_s = 1950 m/s, density = 2.4 g/cc. The model was calculated using the Crewes Energy Ratio Explorer software available at www.Crewes.org.

$$\Delta V_p = V_{p2} - V_{p1},\ \Delta V_s = V_{s2} - V_{s1},$$

$$\text{and}\ \Delta \rho = \rho_2 - \rho_1.$$

The first term (*A*) is the zero angle reflection coefficient related to the contrast of acoustic impedance, whilst the second term (*B*) introduces the effect of shear velocity at non-zero angles. A third term (*C*) determines the curvature of the amplitude response near to the critical angle (Fig. 2.18). The interpreter should be aware that there are a variety of symbols for the first and second terms used in the literature. For example, the *A* term is commonly referred to as the *intercept* or by the symbols R_0 or *NI* (for 'normal incidence'), whilst the *B* term is referred to as the *gradient* (*G*) or *slope*. In the past, Shell and Hess corporations have used the symbols *L* and *M* and R_0 and R_1 respectively to refer to intercept and gradient (Fig. 2.18).

The third term can be dropped to give a two term approximation generally accredited to Shuey (1985, although the equation does not actually appear in the paper):

Fundamentals

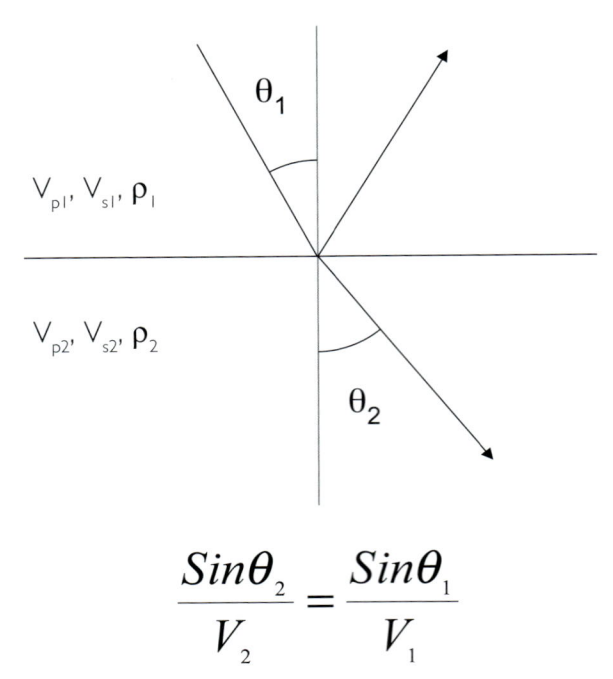

Figure 2.16 Snell's Law.

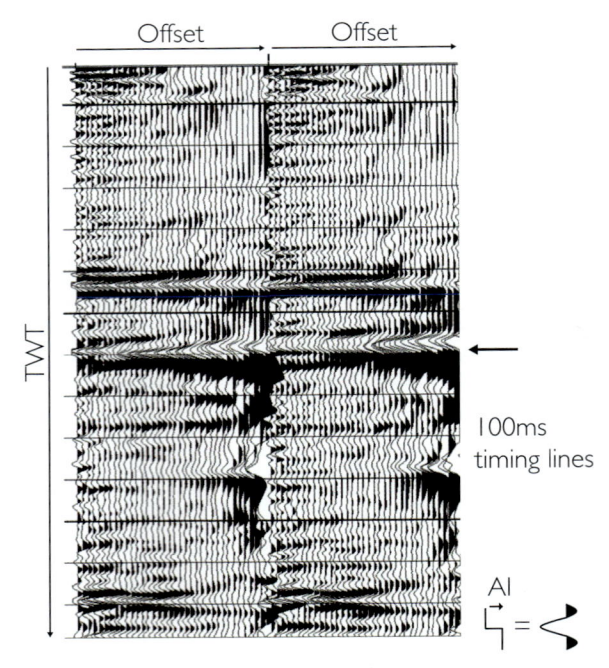

$$\frac{Sin\theta_2}{V_2} = \frac{Sin\theta_1}{V_1}$$

Figure 2.17. An example of critical angle energy on NMO corrected gathers.

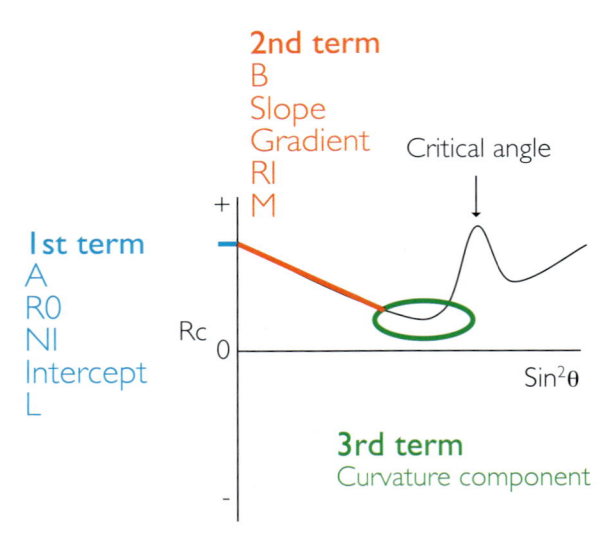

Figure 2.18 The three components of the Aki–Richards (1980) approximation to the Zoeppritz equations.

$$R(\theta) = A + B\sin^2\theta. \qquad (2.16)$$

Shuey's equation is a simple linear regression. For the purpose of describing seismic amplitude variation this approach to linearising AVO is applicable only over a limited range of angles. The angle at which the two-term approximation deviates from the three-term and Zoeppritz solutions depends on the contrasts across the boundary. It is generally safe to assume that the two term approximation holds to an angle of incidence up to 30°; for the case in Fig. 2.19 the second and third order curves start to diverge at around 40°. If intercept and gradient are to be derived from seismic then the interpreter needs to ensure that only traces which show a linear change of amplitude with $\sin^2\theta$ are used (Chapters 5 and 6).

The modern day importance of the Shuey equation is not as a predictor of seismic amplitudes at particular angles but as a tool for analysing AVO data for fluid and lithology effects (described in Chapters 5 and 7). Shuey's equation played a key role in the development of seismic AVO analysis techniques in the 1980s and 1990s. The simplicity of the equation meant that the regression coefficients A and B (intercept and gradient) could be fairly easily derived and a range of AVO attributes defined by various parameter combinations.

Another, rock property oriented, approximation to the Zoeppritz equations has been put forward by Hilterman (2001):

$$R(\theta) = \frac{AI_2 - AI_1}{AI_2 + AI_1}\cos^2\theta + \frac{\sigma_2 - \sigma_1}{(1 - \sigma_{avg})^2}\sin^2\theta. \qquad (2.17)$$

This approximation is effectively the same as the two-term Shuey equation but has been rearranged to

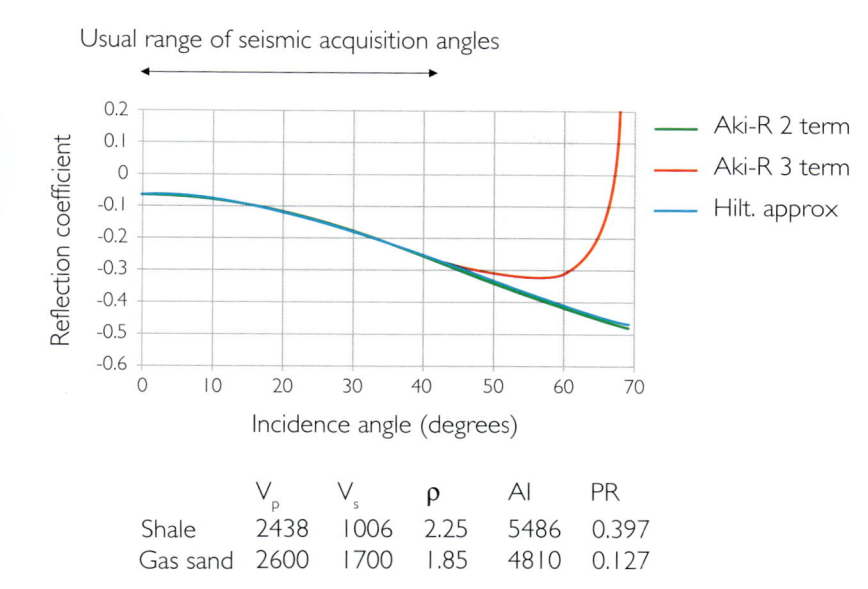

Figure 2.19 Comparison of Aki–Richards two-term and three-term equations and Hilterman equation for an example interface of a shale overlying a gas sand.

	V_p	V_s	ρ	AI	PR
Shale	2438	1006	2.25	5486	0.397
Gas sand	2600	1700	1.85	4810	0.127

highlight the fact that reflectivity is fundamentally related to two rock properties, the acoustic impedance contrast and the Poisson's ratio contrast (see Chapter 5 for rock property controls on acoustic impedance and Poisson's ratio).

2.3.4 Types of seismic models

There are a number of different types of models that can be generated to aid amplitude interpretation (Fig. 2.20). For most applications these utilise relatively simple primaries-only reflectivity. However, there may be occasions when more sophisticated modelling is required. The key problem is that as the complexity or sophistication increases, the time and effort also increases, often without any guarantees that it will be worth the effort. The user needs to select the right degree of complexity for the problem at hand.

2.3.4.1 Single interface model

The simplest model (and sometimes the most important aid to understanding) is the single interface model, where V_p, V_s and ρ for the upper and lower layers are input into an algorithm based on Zoeppritz or its approximations to produce the *AVO plot*, a graph of reflection coefficient versus incidence angle (Fig. 2.20a). This is often the best place to start. If the target comprises thick layers with significant contrasts at top and base, this simple model will give a good idea of what to expect in real seismic data. However, it is less helpful if there is complicated layering or gradational change in properties across a boundary.

2.3.4.2 Wedge model

The wedge model (Fig. 2.20b) is a tool to describe the interaction of reflections from two converging interfaces, and is therefore an important way to understand interference effects. In particular the wedge model is useful for determining vertical resolution (Chapter 3) and can also be used in simple approaches to net pay estimation (Chapter 10). However, like the single interface model the wedge model can be too simplistic for practical purposes, particularly in areas where there are rapid vertical variations in lithology.

2.3.4.3 1D synthetics based on log data

The synthetic seismogram uses wireline log data and a wavelet to calculate either a single normal incidence trace or a range of traces at different angles, simulating the angle variation in a seismic gather. This type of model is important when tying wells to seismic (Chapter 4) and when generating multi-layered models with different fluid fill. One-dimensional (1D) synthetic models are useful in understanding how the seismic response depends on the frequency content of the data (Chapter 3). Following the response of a layer of interest from the fully resolved case with a (perhaps unrealistic) high-bandwidth wavelet through the increasingly complex interference patterns that may arise as the bandwidth decreases is

Fundamentals

a) Single Interface Model

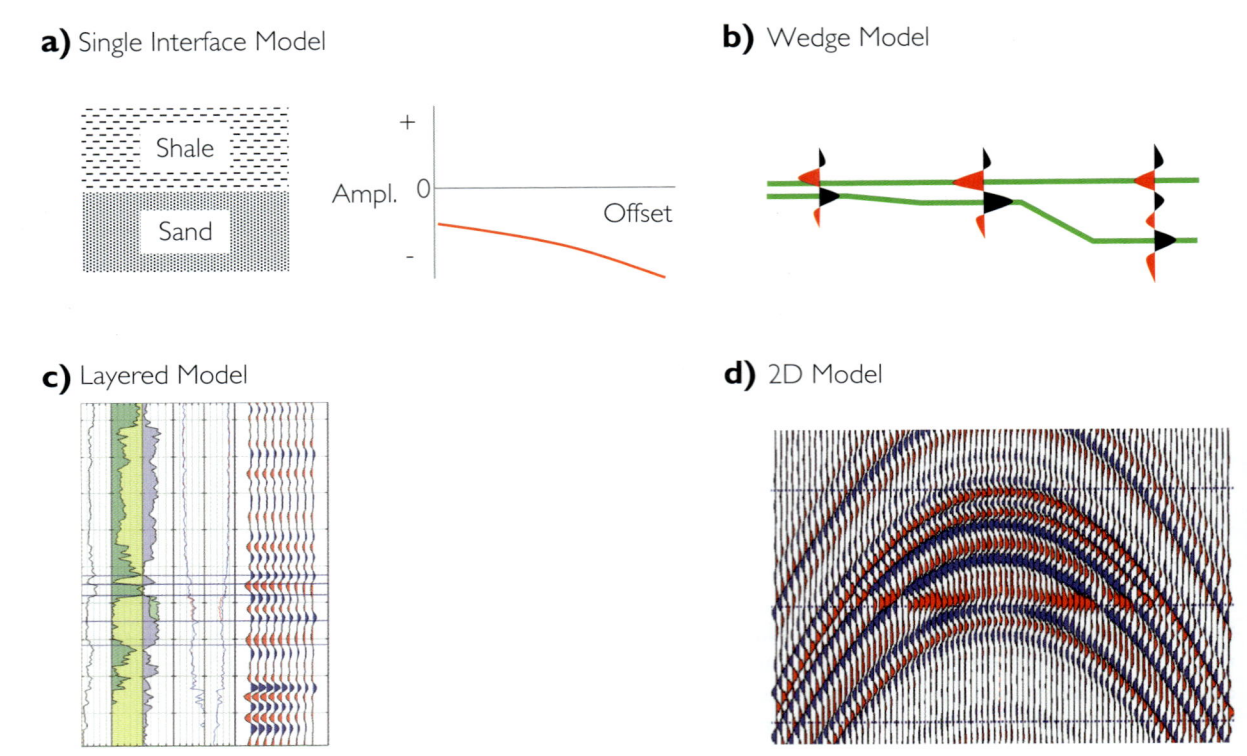

b) Wedge Model

c) Layered Model

d) 2D Model

Figure 2.20 Types of seismic models; (a) single interface model, (b) three-layer wedge model, (c) 1D layered model, (d) 2D model calculated at a given reflectivity angle.

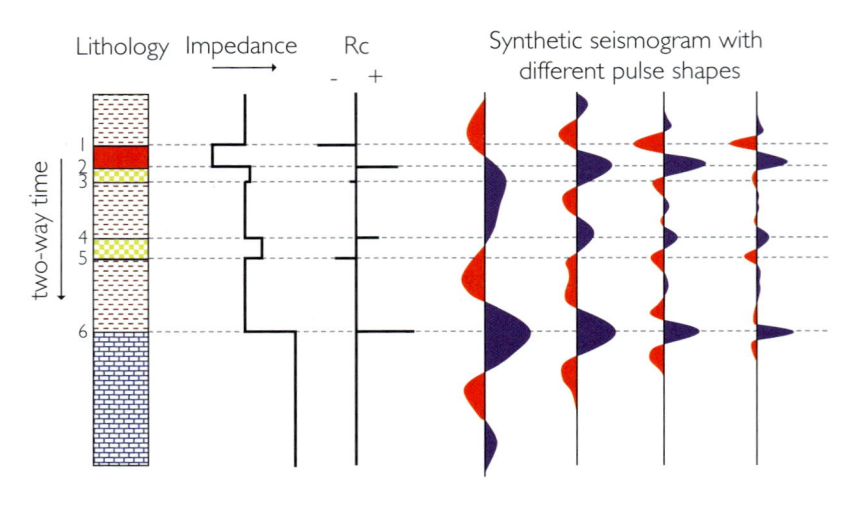

Figure 2.21 Synthetic seismograms with varying symmetrical wavelet shapes from broad to sharp (re-drawn and modified after Anstey, 1982).

often instructive. This can be a useful way to appreciate which geological units can be resolved with the current data and how much improvement in bandwidth (by new acquisition or reprocessing) would be needed to resolve more detail (Fig. 2.21).

Sometimes, multi-layered models can be too complex, combining so many ingredients that it becomes hard to understand how changes in individual layer properties will affect the seismic response without constructing a very large and unwieldy suite of models.

The 1D synthetics can be made more sophisticated by including propagation modes other than primaries-only, for example multiples and mode conversions (Kennett, 1983). In general this results

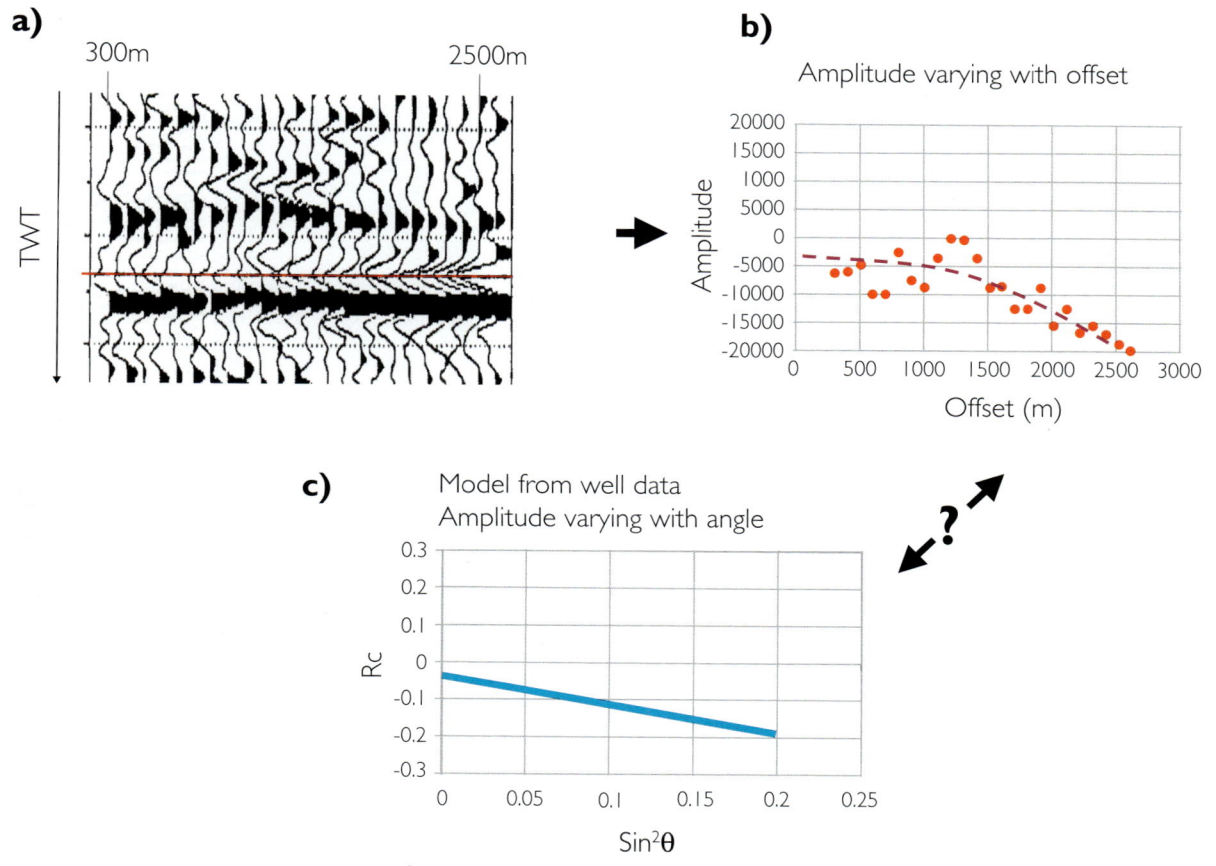

Figure 2.22 The issue of relating offset to $\sin^2\theta$; (a) corrected gather showing a reflection from the top of a thick gas sand, (b) offset versus amplitude crossplot from the reflection shown on the gather, (c) modelled change of reflectivity as a function of $\sin^2\theta$ based on well data.

in longer computer run times and often requires more time to be spent preparing the log data (e.g. in devising a log blocking strategy). Given that in most cases the seismic processor will have tried, with varying success, to remove these events from the data it may be that this type of modelling has limited predictive value.

2.3.4.4 2D models

Two-dimensional (2D) models (Fig. 2.20d) can be extremely useful for understanding the effects of lateral changes in rock properties and/or layer thicknesses on the seismic signature. These types of models are typically created using wireline well log data interpolated along model horizons. Figure 2.20d shows an anticline with gas-bearing sands above a flat gas–water contact. Numerous effects are evident on the section, including changes in the polarity and amplitude of various reflectors at the contact as well as the subtle nature of the interference effects as the fluid contact passes through the geological layering. These types of model can be generated for different angles of incidence to explore which angle gives the clearest fluid response (Chapter 5).

2.3.5 Relating seismic data to models

In order to relate reflectivity models calculated at the wells in terms of angle to seismic data which have been collected as a function of source–receiver offset, a conversion scheme from offset to angle is required (Fig. 2.22). This requires use of a velocity model, usually constructed from velocity information acquired in the course of seismic processing, i.e. in correcting for moveout. If the geology can be approximated as a set of horizontal layers then reflection

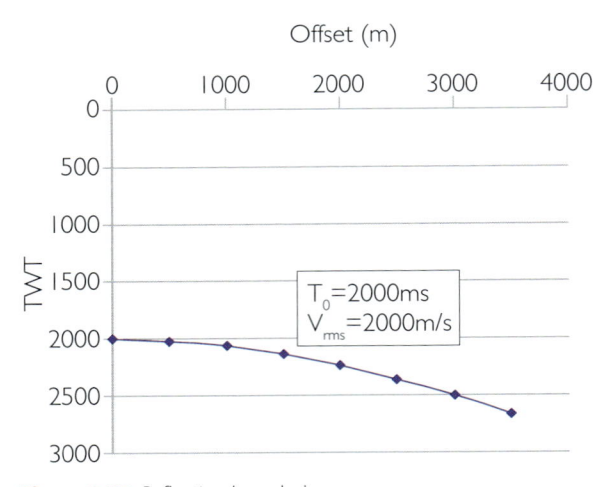

Figure 2.23 Reflection hyperbola.

time increases with offset approximately according to a hyperbolic relation (Fig. 2.23):

$$T_x = \sqrt{T_0^2 + \frac{x^2}{V_{rms}^2}}, \qquad (2.18)$$

where T_x is the time at offset x, T_0 is the time at zero offset, and V_{rms} is the RMS average velocity from the surface to the reflector concerned. Given the assumption that the overburden comprises a stack of layers with velocity V and time thickness t, V_{rms} is given by:

$$V_{rms} = \sqrt{\frac{\sum V_i^2 t_i}{\sum t_i}}. \qquad (2.19)$$

In practice, stacking velocities are picked so that after correcting for moveout the gather is flattened and maximum coherence achieved (Fig. 2.24). These stacking velocities are a first order approximation for RMS velocities (see Al-Chalabi (1974) for a detailed discussion).

The interval velocity between two reflections at times T_1 and T_2 is given by the Dix (1955) equation:

$$V_i = \sqrt{\frac{T_2 V_{rms2}^2 - T_1 V_{rms1}^2}{T_2 - T_1}}. \qquad (2.20)$$

This equation is usually applied to a time interval of at least 200 ms to avoid instability arising from errors in the RMS velocity estimates. Given the RMS velocity down to a reflector and the interval velocity immediately above it, it can be shown that an estimate of the incidence angle at offset x is given by:

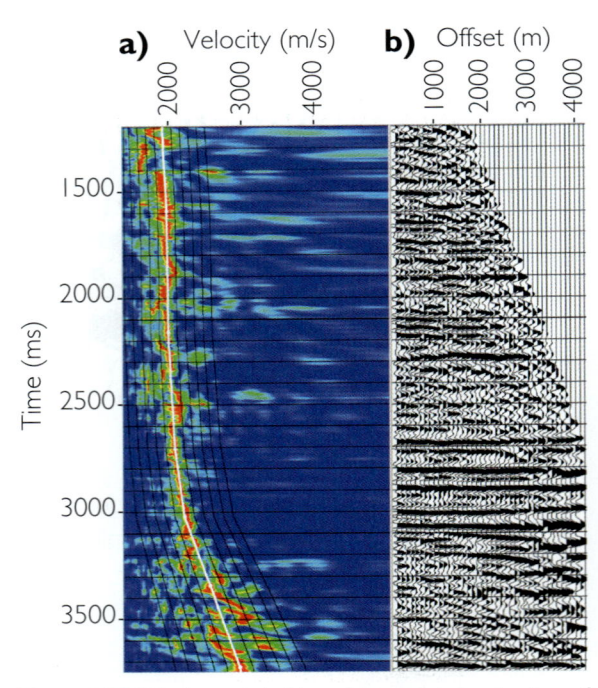

Figure 2.24 Velocity analysis; (a) semblance plot showing lines of equal velocity (black) and stack coherency for each velocity shown in colour (red = high coherency), (b) corrected gather using velocity defined by velocity picks (white trend shown in (a)) (courtesy Seismic Image Processing Ltd).

$$\sin^2 \theta = \frac{x^2 V_i^2}{V_{rms}^2 (V_{rms}^2 T_0^2 + x^2)}, \qquad (2.21)$$

where T_0 is the zero offset travel time. Figure 2.25 shows a moveout corrected gather with angle as a coloured background. An alternative approach that may be useful if velocity information is available (for example sonic and check-shot data from a well) is to calculate offset as a function of incidence angle, using Snell's Law to follow the ray path through the layers to the surface (Fig. 2.26). This requires a model to be constructed for the near-surface where there are likely to be no well data available, but it can give more accurate angle estimates than those obtained from stacking velocities. It is, however, not always clear how to extrapolate well velocities away from a well, as the velocities may show some combination of stratigraphic conformance (if there are big differences in lithology between layers) and increase with depth due to compaction. It is certainly important to take large changes in water depth into account. Ideally the angle conversions obtained from seismic velocities and from well velocities need to be checked against one another and perhaps somehow combined; the

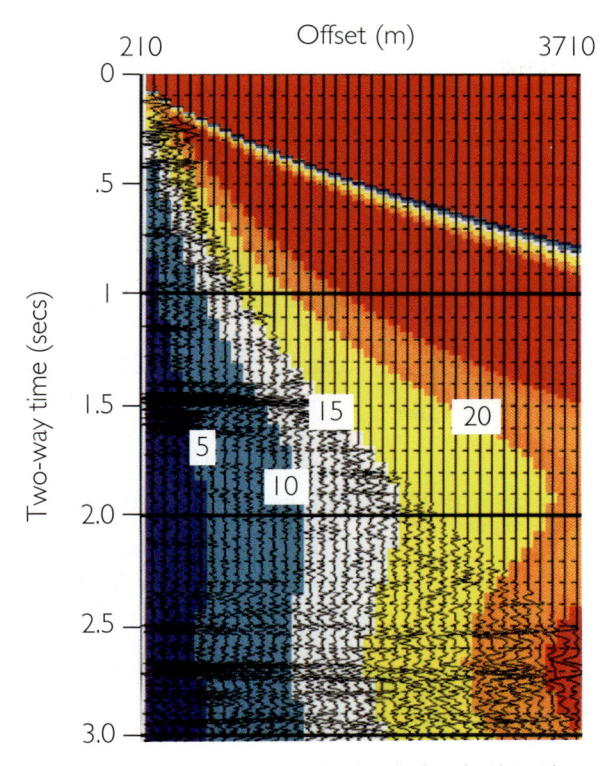

Figure 2.25 A moveout corrected gather displayed with incidence angle as coloured background.

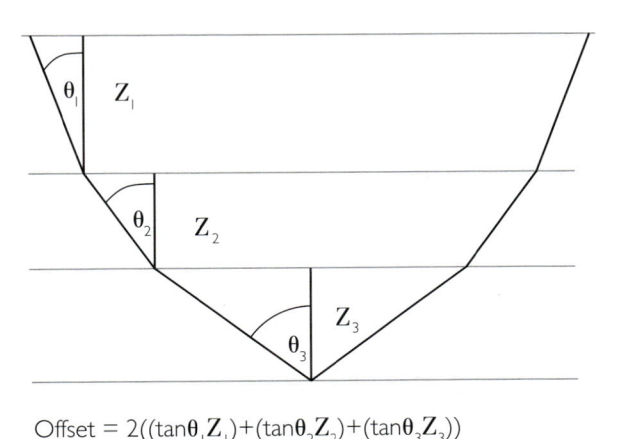

Offset = $2((\tan\theta_1 Z_1)+(\tan\theta_2 Z_2)+(\tan\theta_3 Z_3))$

Figure 2.26 Converting angle to offset using well velocity data; the simplest case is for a horizontal layered model and a straight ray assumption.

accuracy of the offset to angle conversion is often a significant factor for generating partial stacks and for AVO analysis in general.

Once the offsets have been converted to angles then the data in the corrected gathers can be combined in a number of ways. Figure 2.27 illustrates the

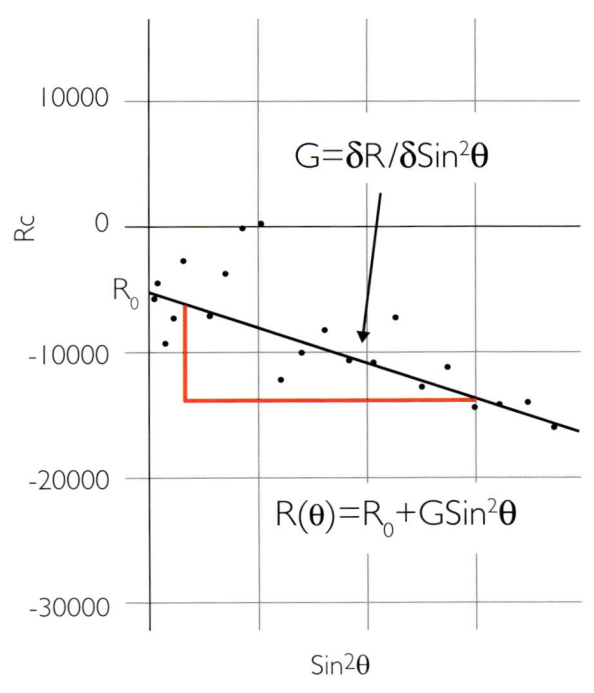

Figure 2.27 Deriving intercept and gradient. Offsets have been converted into $\sin^2\theta$ using a velocity model prior to fitting a robust linear regression through the data.

calculation of intercept and gradient, using the two-term AVO equation. As suggested in the previous section, care must be taken that for the angle range in question a two-term fit is appropriate. Before calculating intercept and gradient, some data conditioning may be needed as it is important that reflections are flat across the moveout corrected gather (Chapter 6).

An alternative to intercept/gradient calculation is to create *angle stacks* (i.e. stack the data according to angle ranges). For the purposes of seismic interpretation the available angle range is usually divided into two or three equal parts and separate stacks created for each. This maintains reasonable signal-to-noise ratio whilst maintaining the key elements of the AVO response. Typically, therefore, the interpreter will have near, mid and far sub-stacks in addition to the conventional full stack. It should be noted that the full stack is seldom the product of adding together the partial stacks. In most instances the full stack is the processor's best data quality stack and frequently has a harsher mute of the outer traces of the gather than the far stack. There may be instances where more than three sub-stacks are generated, for example in simultaneous inversion (Chapter 9).

Fundamentals

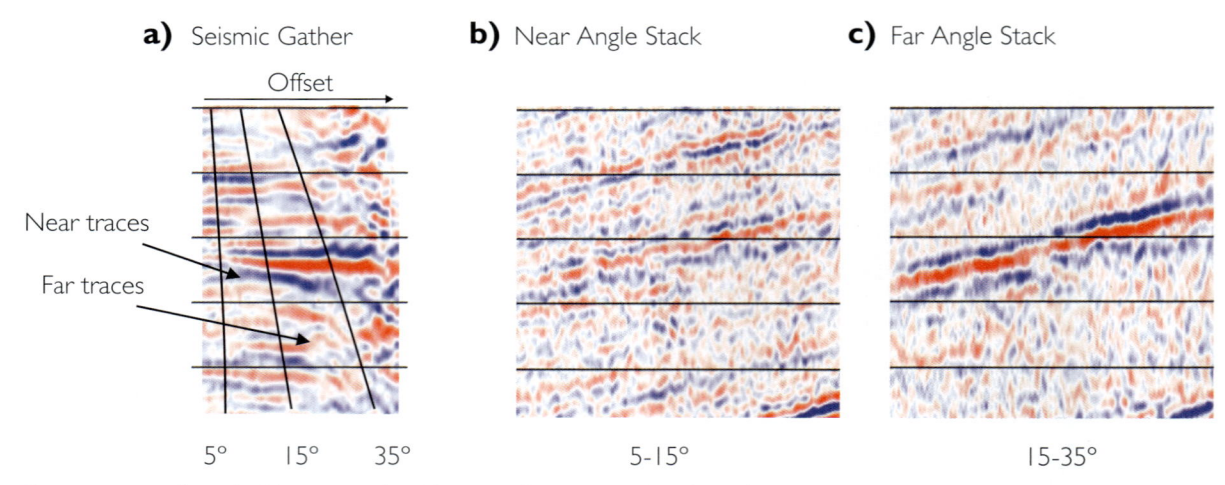

Figure 2.28 Angle stacks; (a) seismic gather, (b) near angle (5–15°) stack, (c) far angle (15–35°) stack. Note that the actual angles stacked will depend on the available angle range and the choice as to number of stacks required (after Bacon et al., 2003).

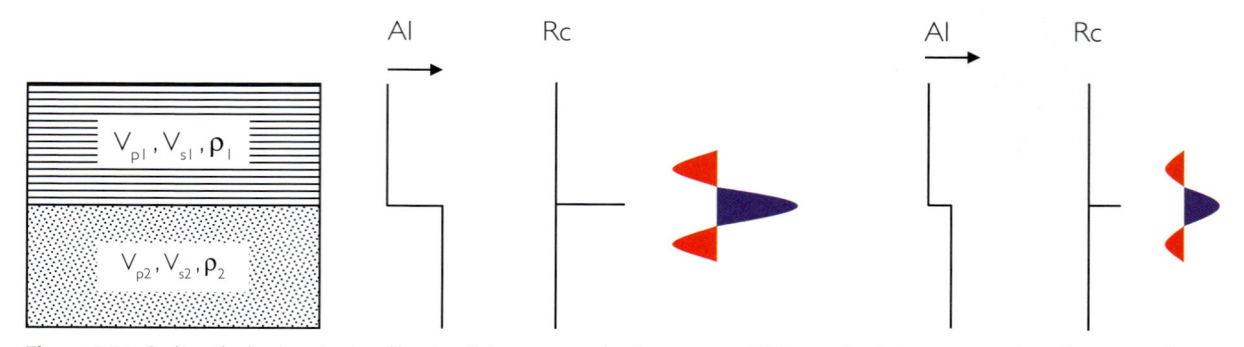

Figure 2.29 Scaling: the key to seismic calibration; if the processor has been successful the amplitude is proportional to reflection coefficient for a unique reflector. This is the key assumption in any inversion (Chapter 9) or quantitative amplitude interpretation.

The sub-stacks can be loaded to a structural interpretation software package in just the same way as the full stack. If the data processing is good the sub-stacks will look similar to the full stack but amplitude changes (for example, an increase from near to far traces) will be immediately apparent when sub-stacks are displayed side by side (Fig. 2.28). This gives the interpreter a way to search through a 3D seismic survey to look for anomalous AVO behaviour, which can then be followed up by looking at CMP gathers. As will be discussed, AVO differences are often made more apparent by combining the data from partial stacks using a weighting procedure (Chapter 5).

For the purpose of calibration of the angle stack to a well-based AVO plot or indeed for seismic inversion of individual stacks (Chapter 9), it can be considered that each angle stack represents reflectivity at a particular (effective) angle. Theoretically, the effective angle should be derived from the average of $\sin^2\theta$ across the sub-stack, but in practice (especially since the offset to angle transformation contains a degree of uncertainty) it can usually be assumed to be the midpoint of the range of angles included.

The goal of quantitative interpretation is to relate changes in seismic amplitude to changes in rock properties. A key aspect of this is the scaling of the seismic amplitude to the reflection coefficient (Fig. 2.29). This is done principally through well ties (Chapter 4). Correct scaling is particularly critical for seismic inversion (Chapter 9), but problems with scaling can seriously compromise all quantitative amplitude interpretation.

Chapter 3

Seismic wavelets and resolution

3.1 Introduction

A fundamental aspect of any seismic interpretation in which amplitudes are used to map reservoirs is the shape of the wavelet. This chapter presents introductory material relating to the nature of seismic wavelets; how they are defined, described and manipulated to improve interpretability. Seismic resolution, in terms of recognising the top and base of a rock layer, is controlled by wavelet properties. However, owing to the high spatial sampling of modern 3D seismic, resolution in the broadest sense also includes the detection of geological patterns and lineaments on amplitude and other attribute maps.

3.2 Seismic data: bandwidth and phase

The seismic trace is composed of energy that has a range of frequencies. Mathematical methods of Fourier analysis (e.g. Sheriff and Geldart, 1995) allow the decomposition of a signal into component sinusoidal waves, which in general have amplitude and phase that vary with the frequency of the component. An example is the seismic wavelet of Fig. 3.1, which can be formed by adding together an infinite set of sine waves with the correct relative amplitude and phase, of which a few representative examples are shown in the figure. The *amplitude spectrum* shows how the amplitude of the constituent sine waves varies with frequency. In Fig. 3.1 there is a smooth amplitude variation with a broad and fairly flat central peak. This is often the case as the acquisition and processing have been designed to achieve just such a spectrum. The *bandwidth* of the wavelet is usually described as the range of frequencies above a given amplitude threshold. With amplitudes that have been normalised, such as those shown in Fig. 3.1, a common threshold for describing bandwidth is half the maximum amplitude. In terms of the logarithmic decibel scale commonly used to present amplitude data this equates to -6 dB (i.e. $20 \log_{10} 0.5$).

In practice, the amplitude spectrum is calculated from the seismic trace using a Fourier transform over a given seismic window, usually several hundred milliseconds long. It is assumed to first order that the Earth reflectivity is random and that the wavelet is invariant throughout the window. The amplitude spectrum of the wavelet is then assumed to be a scaled version of the amplitude spectrum of the seismic trace. Amplitude spectra are also commonly estimated over short seismic segments using a range of different approaches (e.g. Chakraborty and Okaya, 1995). Such analysis is an essential component of spectral decomposition techniques for use in detailed stratigraphic interpretation (e.g. Burnett *et al.*, 2003).

In addition to the amplitude spectrum, the other piece of information that is needed to define the shape of the wavelet uniquely is the relative shift of the sine wave at each frequency (i.e. the phase). The wavelet in Fig 3.1 shows frequency components that have a peak aligned at time zero. Such a wavelet is termed *zero phase*. As described in Chapter 2 a zero phase wavelet is ideal for the interpreter because it has a strong dominant central trough or peak at zero time. If this is the wavelet in a processed seismic dataset, then an isolated subsurface interface between layers of different impedance will be marked by a correctly registered trough or peak (depending on the sign of the impedance contrast and the polarity convention used). This makes it fairly easy to relate the seismic trace to the subsurface layering, even in 'thin-layered' situations with overlapping reflections. The interpretation is made more difficult if the seismic wavelet is not symmetrical but, for example, has several loops with roughly the same amplitude; then the interference between adjacent closely spaced reflectors will be difficult to understand intuitively. The importance of zero phase wavelets to the interpreter becomes clear when considering well to seismic ties (Chapter 4). Zero phase is a condition that requires processing of the seismic data (see Section 3.6.2).

Seismic wavelets and resolution

a) Zero phase wavelet and selected frequency components

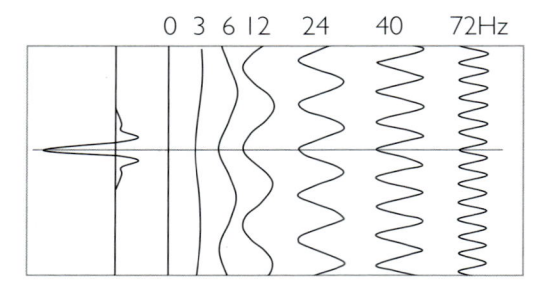

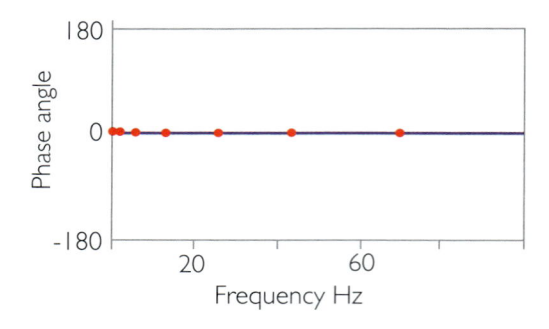

b) Phase spectrum

c) Amplitude spectrum

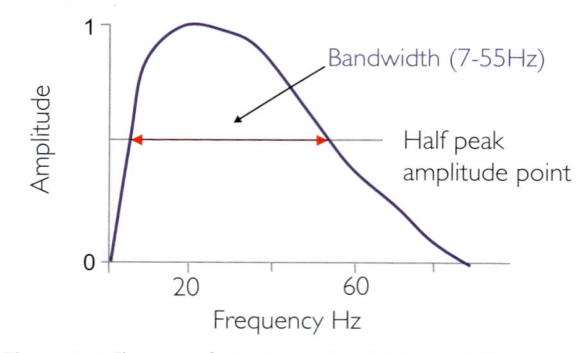

Figure 3.1 Elements of seismic wavelets; (a) sinusoidal frequency components, (b) phase spectrum, (c) amplitude spectrum (after Simm and White, 2002).

Phase can be thought of as a relative measure of the position of a sinusoidal wave relative to a reference point, and is measured in terms of the phase angle (Fig. 3.2). The sine waves in Fig 3.2 all have the same frequency but are out of phase. At the time zero reference point, they vary from peak (red wave) to zero-crossing (blue wave) to trough (green wave) to zero-crossing (brown wave). The turning wheel idea is useful for describing the angular differences in phase.

Seismic wavelets can sometimes be approximated as constant phase across the dominant bandwidth. In such a case (Fig. 3.3), the phase of each frequency component is the same. Figure 3.3 shows a wavelet in which the phase is 90°. Thus, a zero phase wavelet is a special case of a constant phase wavelet. Another case often seen is linear phase, where phase is related linearly to frequency (Fig. 3.4). In effect, a wavelet with linear phase is equivalent to a time-shifted constant phase wavelet.

As will be discussed in detail below, the bandwidth and phase of the seismic signal emitted by the source is modified in its passage through the subsurface by the 'earth filter' (Fig. 3.5). Shallow targets will generally have higher bandwidth than deeper targets. Higher bandwidth essentially means greater resolving power.

3.3 Zero phase and minimum phase

From the interpreter's point of view the ideal wavelet is zero phase, i.e. symmetrical and with the amplitude centred on time zero. Real sound sources such as explosives and airguns typically have *minimum phase* signatures. Generally, a minimum phase wavelet is one that has no energy before time zero and has a rapid build-up of energy. Each amplitude spectrum can be characterised by a unique minimum phase wavelet (i.e. the wavelet that has most rapid build-up of energy for that given spectrum). It is possible to find the minimum phase wavelet for a given spectrum using methods based on the z transform (see e.g. Sheriff and Geldart, 1995) but there is no intuitive way to determine whether a given wavelet is minimum phase or not. Conversely, to say that a dataset is minimum phase does not guarantee a particular shape. Figure 3.6 shows an example of two minimum phase wavelets with similar bandwidth but slightly different rates of change of frequency in the low-frequency range. Thus, if an interpreter is told that the seismic data are minimum phase then it is not clear exactly what response should be expected from a given interface. What is important is that the interpreter is able to assess and describe wavelet shape in seismic data. Determining wavelet shape from seismic is described in Chapter 4.

A useful semi-quantitative description of wavelet shape is in terms of a constant phase rotation of an

Change of wavelet shape with depth

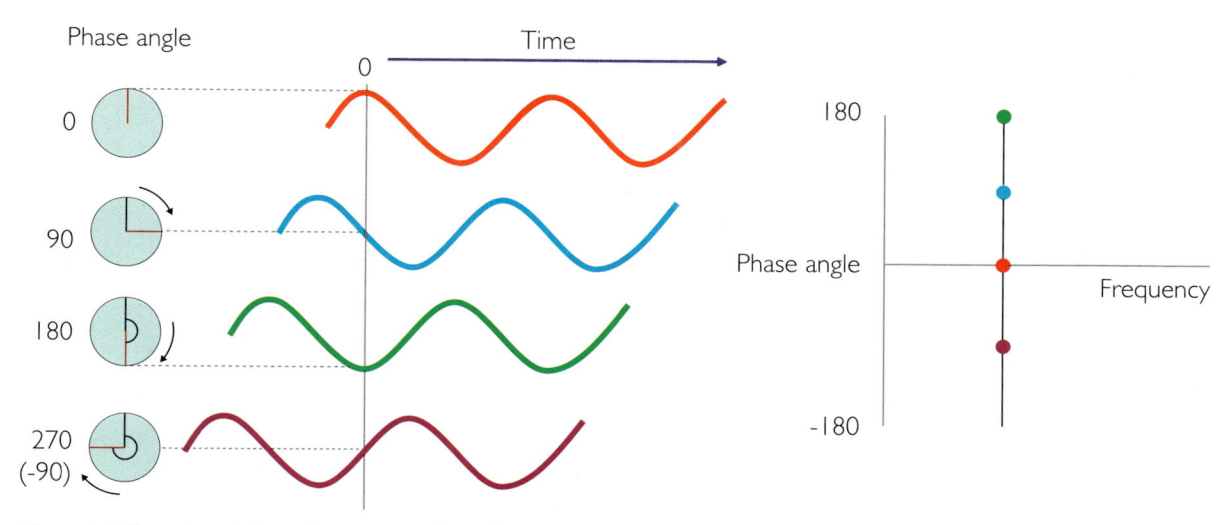

Figure 3.2 Illustration of phase; showing several waveforms with the same frequency but different phase (i.e. different timing of the waveforms relative to the zero reference point). The turning wheel model describes the angular relations intrinsic to the advancement of the waveform.

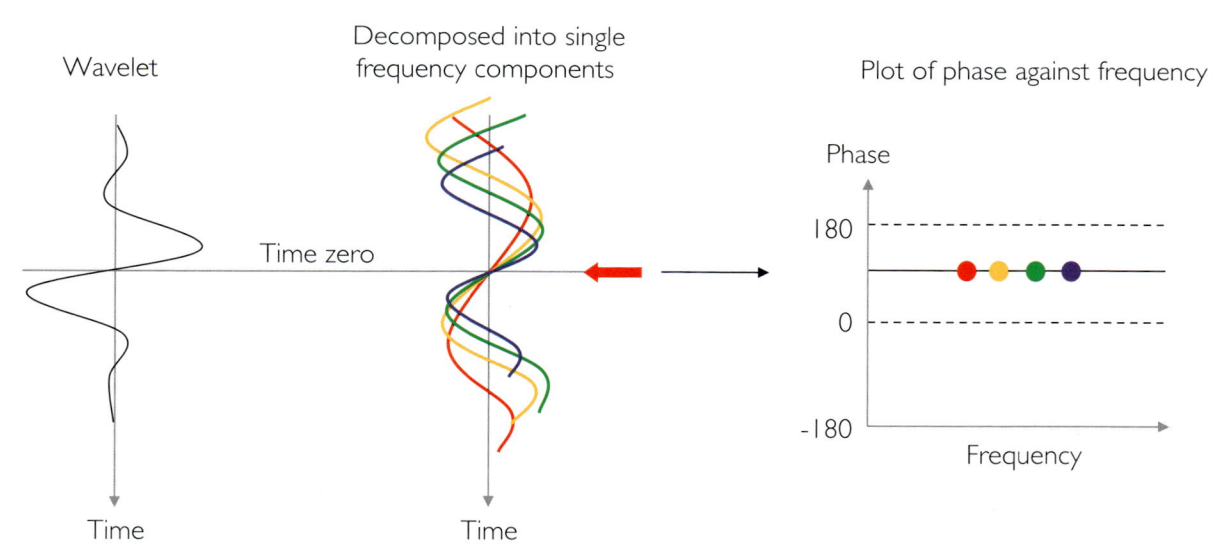

Figure 3.3 Constant phase wavelet (+90°) (J. Chamberlain, personal communication).

initially zero phase wavelet (Fig. 3.7). As the phase rotates, the relative amplitudes of the peak and trough loops of the wavelet change. At a phase angle of 90° the two loops have the same amplitude. Note that the description of phase is dependent on the reference polarity. It is most common to reference phase descriptions to positive standard polarity (Chapter 2).

3.4 Change of wavelet shape with depth

The Earth filter can have significant effects on the amplitude and phase spectrum of the wavelet. Figure 3.8 shows an example where the seabed response corresponds to a −60° phase rotated zero phase wavelet, whereas at the target the optimum wavelet for the well

Seismic wavelets and resolution

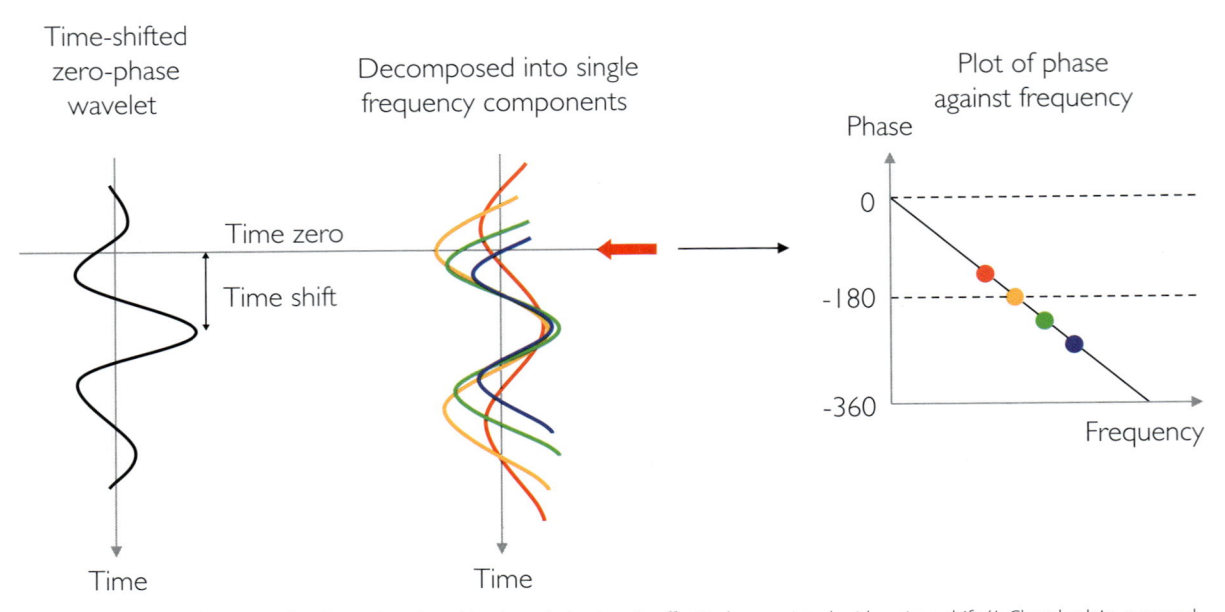

Figure 3.4 Linear phase wavelet; illustrating that this phase behaviour is effectively associated with a time shift (J. Chamberlain, personal communication).

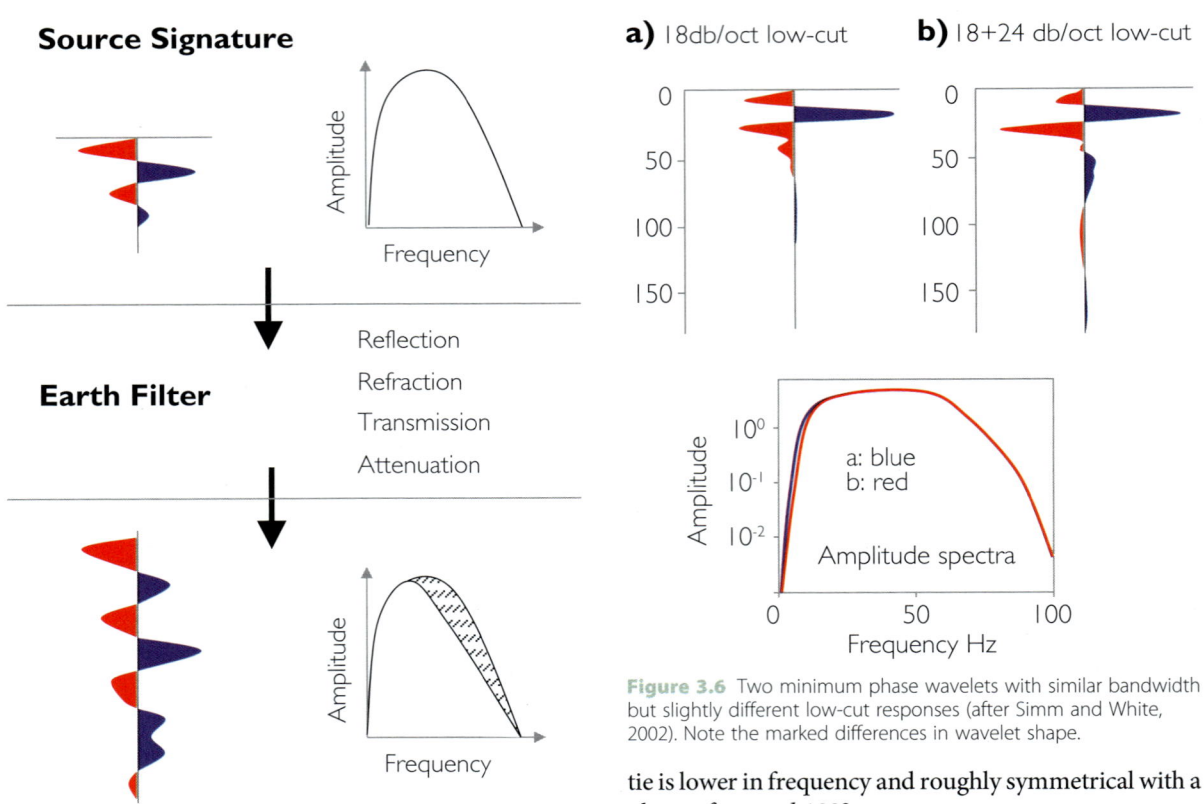

Figure 3.5 Seismic bandwidth and the Earth filter.

Figure 3.6 Two minimum phase wavelets with similar bandwidth but slightly different low-cut responses (after Simm and White, 2002). Note the marked differences in wavelet shape.

tie is lower in frequency and roughly symmetrical with a phase of around 180°.

These effects are controlled by attenuation, which varies with the lithology and state of consolidation.

Change of wavelet shape with depth

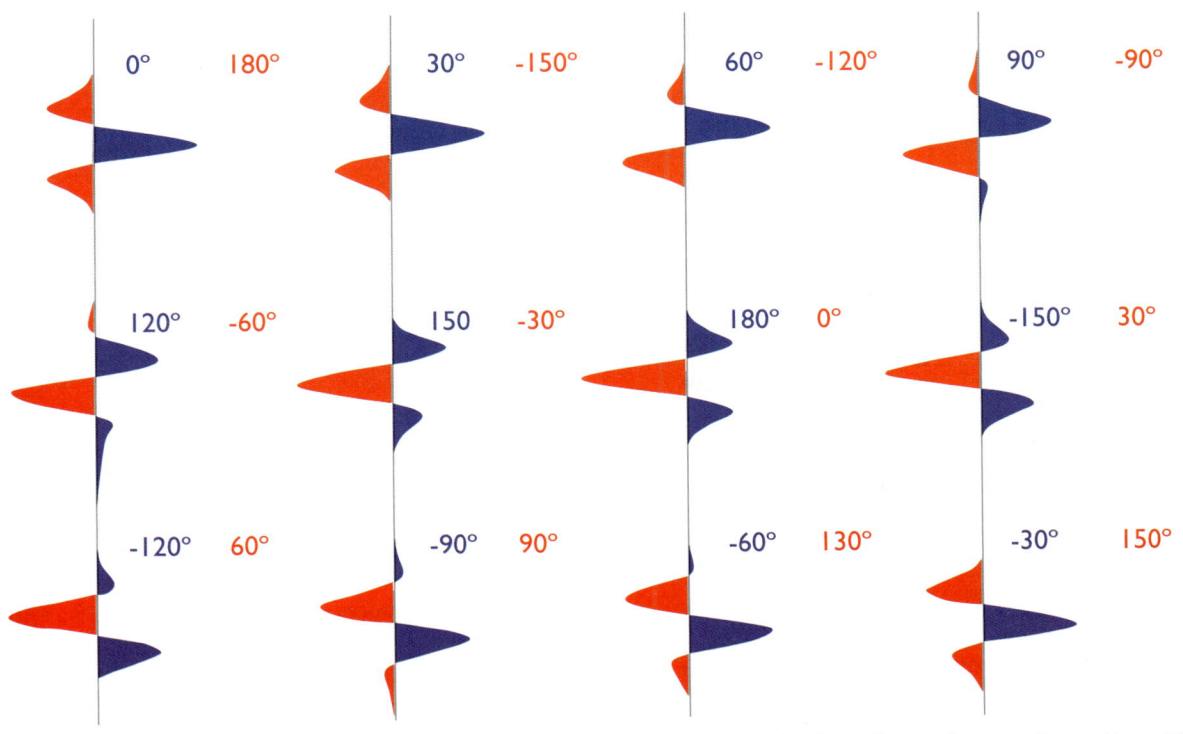

Figure 3.7 Constant phase rotation of a zero phase wavelet – a useful description for wavelet shape. Blue numbers are referenced to positive standard polarity and red numbers are referenced to negative standard polarity. Note it is usual to use the positive standard convention when describing wavelet phase.

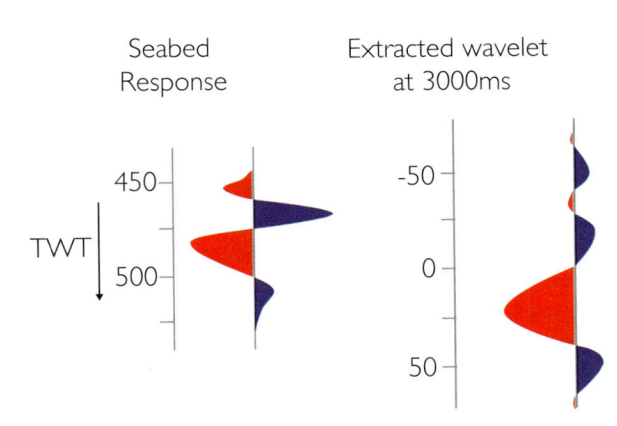

Figure 3.8 Example of change in wavelet shape with depth.

Attenuation is parameterised by the quantity Q defined as

$$Q = 2\pi/\text{fraction of energy lost per cycle.} \quad (3.1)$$

The effect of attenuation is to reduce amplitude at high frequencies more than at low frequencies, because in any given subsurface path there will be more cycles (i.e. wavelengths) at a higher frequency than at a lower one. Attenuation also causes seismic wave propagation to be dispersive (i.e. the seismic velocity varies with frequency), and thus changes to the phase spectrum depend on distance travelled. For a Q value greater than about 10, velocity should vary with frequency according to

$$(C_2 - C_1)/C_1 = \frac{\ln(f_2/f_1)}{\pi Q}, \quad (3.2)$$

where C_1 is the velocity at frequency f_1 and C_2 is the velocity at frequency f_2 (O'Brien and Lucas, 1971). An example of a modelled effect of Q on wavelet phase is shown in Fig. 3.9. Essentially low values of Q will give greater phase rotation for a given sediment thickness than higher values of Q.

Q can be measured on core samples in the laboratory (e.g. Winkler, 1986), but it is not clear whether the results obtained are representative of in-situ values. At the seismic scale the easiest way to measure Q is from a VSP (e.g. Stainsby and Worthington, 1985; see also Chapter 4). The method compares direct (down-going) arrivals at two different depths in a vertical well using vertical incidence geometry

Seismic wavelets and resolution

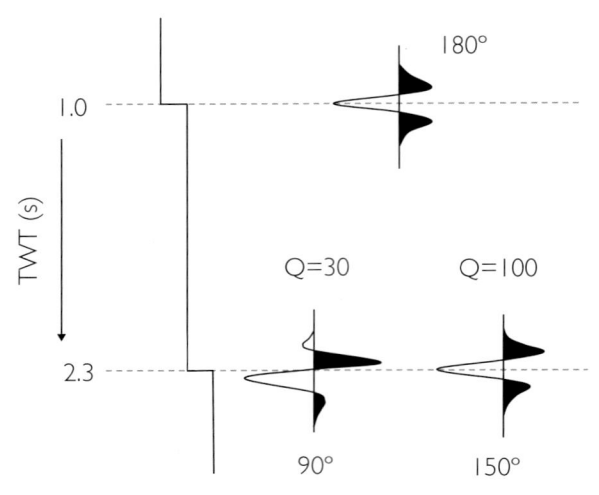

Figure 3.9 Modelled effect of absorption on wavelet shape over a 1.3 s zone (modified after Angeleri and Loinger, 1984). In a real example the bandwidth would also change. Note that $Q = 30$ is characteristic for example of shales whereas $Q = 100$ would be characteristic of limestones and sandstone.

Table 3.1. Some typical values of Q (after Sheriff and Geldart, 1995)

Lithology	Q
Sedimentary rocks	20–200
Sandstone	70–130
Shale	20–70
Limestone	50–200
Chalk	135
Dolomite	190
Rocks with gas in pore space	5–50
Metamorphic rocks	200–400
Igneous rocks	75–300

(i.e. near-zero horizontal offset between source and receiver). This source–receiver geometry is highly favourable, because raypaths for the top and base of an interval are essentially coincident. Thus spectral differences between the arrivals can be reliably assumed to be due only to the interval properties between the two depths.

Determining Q from VSPs is of course only possible where there is a borehole, so it would be desirable to extend lateral coverage by measuring Q from surface seismic. This is not easy to do and it is not routine practice. Q cannot be derived accurately from stacked traces owing to mixing of traces with different path length, the spectral distortion due to moveout and the incorporation of AVO effects. Dasgupta and Clark (1998) have devised a method for using prestack data in which amplitude spectra are computed for a particular reflection for each trace of a CMP gather individually. They are then corrected for moveout stretch (Chapter 6) and compared to the source signature to obtain an estimate of Q.

As can be seen from Table 3.1, the presence of gas in a sandstone can give rise to anomalously low Q (high absorption), particularly at intermediate saturations. This has driven interest in the use of Q in direct hydrocarbon detection. Chapman *et al.* (2006) have developed a theoretical framework to explain abnormally high attenuation in hydrocarbon reservoirs. Dvorkin and Mavko (2006) suggest that the gas effect (i.e. high absorption in the presence of gas) may be due to cross-flow of liquids between fully liquid-saturated pores and surrounding rock with partial gas saturation. These developments are interesting, but there is as yet no single widely accepted view of the significance of absorption effects as a direct hydrocarbon indicator. Sometimes gas sands show an obvious shift to low-frequency signal at and for some distance below them but sometimes they do not.

3.5 Idealised wavelets

There are a number of types of idealised wavelets that are in common use, for example to make well synthetics when the exact wavelet is not known. Some examples are shown in Fig. 3.10. In each type the wavelets can be zero phase, constant phase, or minimum phase, and the frequency content is specified by the user. The Butterworth wavelet is defined by the lower and upper bandpass frequencies and the slopes of the response, usually in dB per octave. The Ormsby wavelet is defined by four frequencies: low-cut, low-pass, high-pass and high-cut. Both these wavelets develop oscillatory side lobes if the cut-off slopes are set too steep. The Ricker wavelet (Ricker, 1940) is defined by a single central frequency and has only two side lobes. Hosken (1988) advised strongly against the use of Ricker wavelets, largely because real seismic generally has a flat topped as opposed to a peaked amplitude spectrum (Fig. 3.10). For detailed work it is probably best to use a custom wavelet created by the methods to be described in Chapter 4,

Wavelet phase and processing

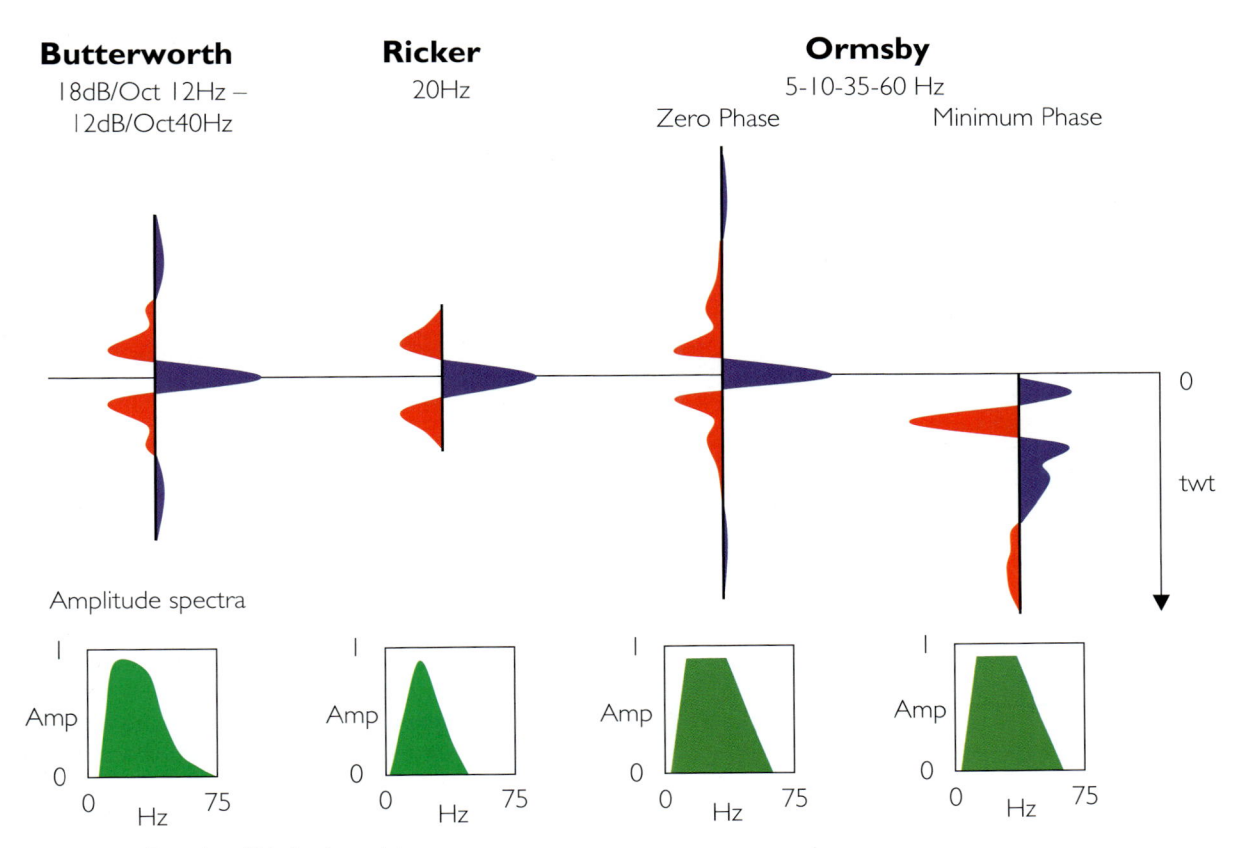

Figure 3.10 Examples of idealised wavelets.

Butterworth — 18dB/Oct 12Hz – 12dB/Oct 40Hz
Ricker — 20Hz
Ormsby — 5-10-35-60 Hz (Zero Phase and Minimum Phase)

but the Ricker wavelet is often a useful tool for first-pass synthetics, given its simplicity.

3.6 Wavelet phase and processing

In order to enhance the interpretability of seismic a number of processes can be applied to the data. Typically these relate to the zero phasing of the seismic, compensation for absorption effects and bandwidth improvement. These are not mutually exclusive and the processor/interpreter needs to evaluate appropriate approaches for the data at hand.

3.6.1 Q compensation

Inverse Q filtering (e.g. Wang, 2006) is a way of correcting for the exponential loss of high-frequency energy and increasing wavelet distortion with increasing two-way time. Figure 3.11 shows an example where this type of amplitude and phase compensation has been applied. In this case the result is fairly dramatic but the usefulness of the correction needs to be evaluated, for example with well ties (Chapter 4). It is essential that this process does not boost noise levels in the data. In the worst case scenario the apparent improvements suggested by the high-frequency look to the data may not be real. When correctly applied the amplitude corrections give rise to data that have true relative amplitude and the phase correction enhances vertical resolution, particularly with increasing depth (e.g. Gherasim et al., 2010). Amplitude and phase corrections can be applied individually or in combination.

3.6.2 Zero phasing

The process of zero phasing seismic is conceptually straightforward. If the wavelet is known (for example from wavelet extractions (Chapter 4)) then a convolution operator can be applied to transform it to zero phase (Fig. 3.12). In practice, this process is not always carried out in this way owing to uncertainties in the definition of the wavelet. One problem is that different wavelet extraction methods can give different

Seismic wavelets and resolution

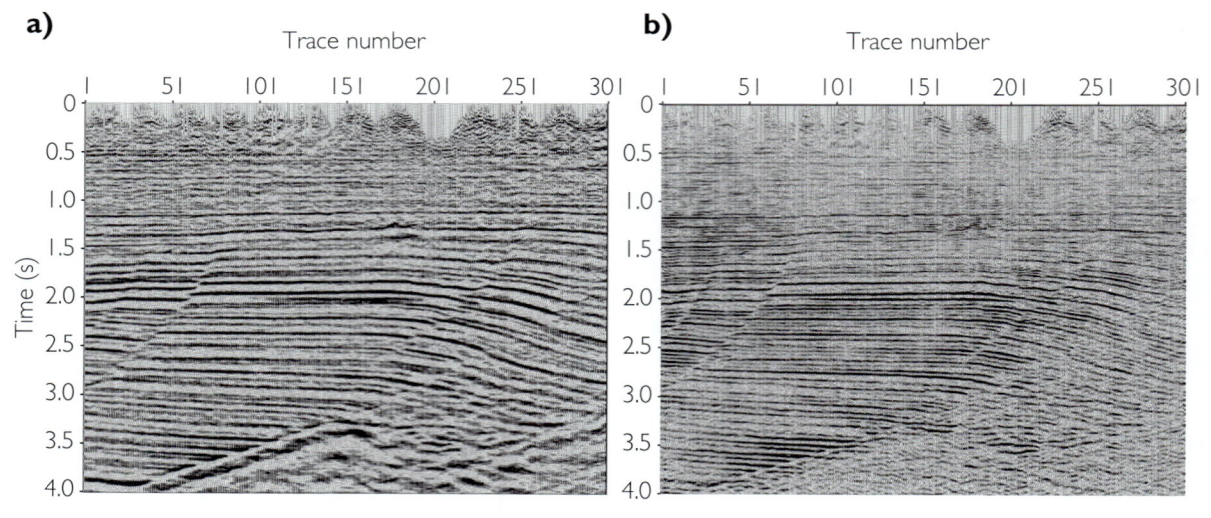

Figure 3.11 An example of inverse Q filtering; (a) before and (b) after inverse Q filtering (after Wang, 2006).

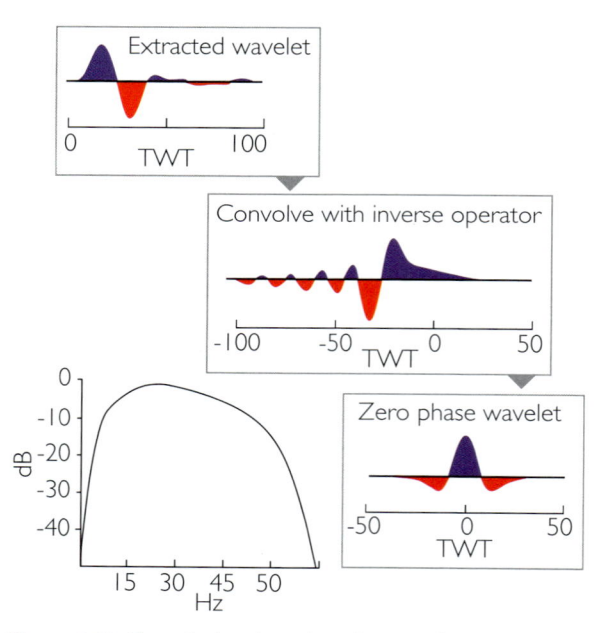

Figure 3.12 Zero phasing through application of an inverse operator.

results and the interpreter needs to determine which method is best for the dataset at hand. Often the various wavelet extractions are used to 'de-phase' the seismic data (i.e. apply a 'phase-only' correction) followed by an evaluation of the results. In situations where Q is poorly understood or where inverse Q has not been applied to the seismic, it may only be possible to create a zero phase result at a specific target rather than over the whole section.

If the interpreter does not have access to software for detailed wavelet shaping, an approximation to zero phase can often be achieved by a simple phase rotation. The amount of rotation is determined by rotating the extracted wavelet until it is approximately symmetrical with the desired polarity. This rotation is then applied to the seismic dataset. Software for such simple phase rotation is widely available.

In the absence of well data to establish phase it may not be possible to accurately zero phase the seismic data. Sometimes it is possible to check wavelet phase from seismic reflectors whose characteristics are well known, such as the top of a thick soft shale (such as the Kimmeridge shale in parts of the North Sea) or the top of a thick hard limestone. The problem of phase determination without wells is discussed in Chapter 4.

3.6.3 Bandwidth improvement

Convolution operators can also be used to broaden the bandwidth of seismic data. In principle, increasing the bandwidth gives better vertical resolution and also an improved result in trace inversion. Historically the aim was to achieve a flat amplitude spectrum, usually called a *white* spectrum by analogy with visible light, where low frequencies are red and high frequencies are blue. Effectively the amplitudes of all frequencies

Wavelet phase and processing

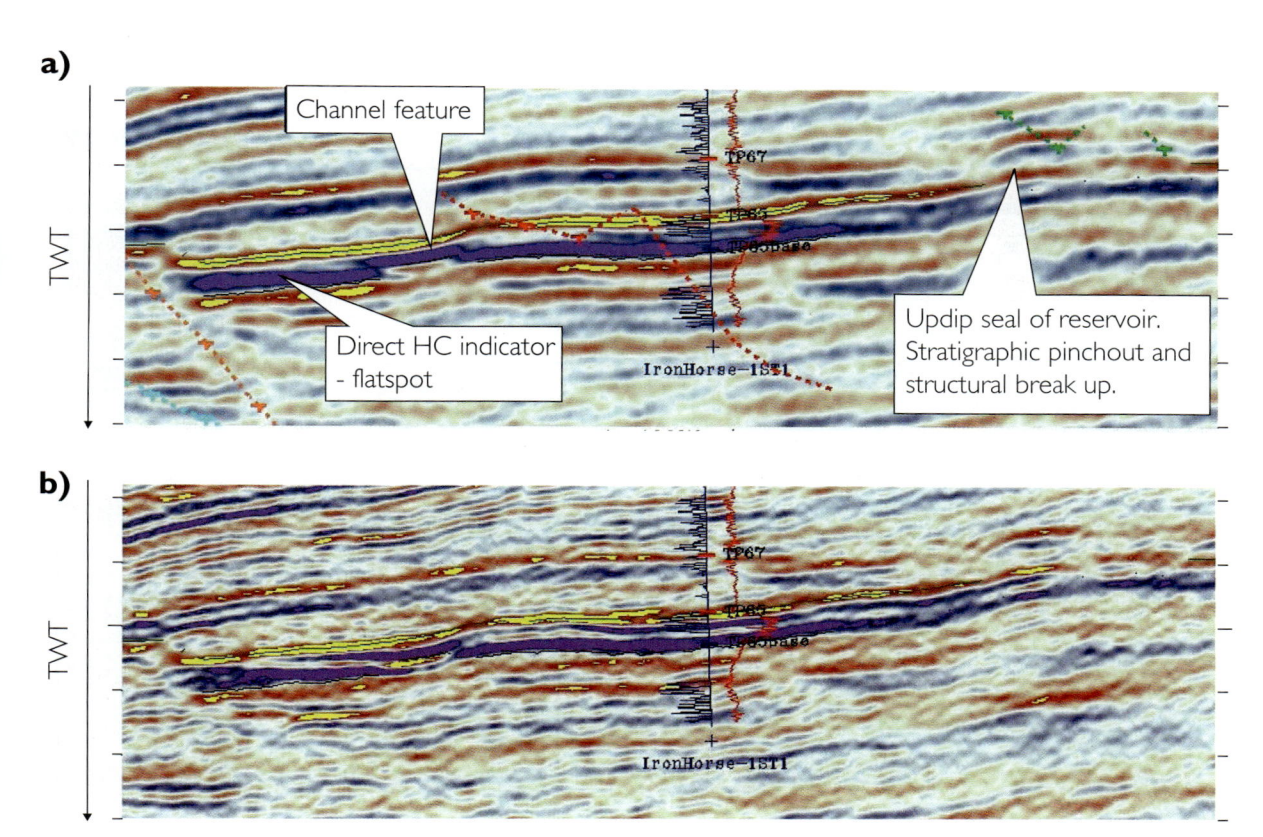

Figure 3.13 Example of spectral blueing; (a) before spectral blueing, (b) after spectral blueing (after Blache-Fraser and Neep, 2004). In this example the blueing has been targeted at enhancing the interpretability of the channel feature.

are boosted to match the maximum amplitude in the original spectrum. A difficulty with this is the effect of noise in real data. At frequencies where there is very little signal, boosting the amplitude will simply increase the noise content. In practice, therefore, tests need to be run to find out what frequency range can be whitened without making the trace data excessively noisy.

A more recent and more geophysically justifiable approach is to design an operator to shape the reflectivity spectrum of the seismic to match the spectrum of reflectivity derived from well data. As shown by Walden and Hosken (1985), well-based reflectivity tends to be blue with amplitude increasing with frequency across the frequency range usually found in seismic data. The slope of the amplitude–frequency relation varies from one depositional environment to another and Lancaster and Connolly (2007) have shown that the spectral decay of impedance as a function of frequency is related to the fractal distribution of layer thickness in the Earth. Matching the seismic spectrum to the well spectrum is thus sometimes called *spectral blueing*. An example is shown in Fig. 3.13, where blueing has enhanced the resolution and facilitated the identification of a subtle channel feature. As is sometimes the case, improved definition of a target level has led to an increase in noise elsewhere in the section. The scope for blueing is thus still limited by the need to avoid undue boosting of the noise content.

It may be useful to improve the signal to noise ratio before bandwidth improvement, for example by the application of particular types of filtering techniques. Helmore et al. (2007) have proposed that the data should be divided into frequency bands and filtering adapted to local dip (referred to as structure-oriented filtering) is applied to each band separately, followed by their recombination. In this way the filter can be adapted to the noise level in each frequency band. As can be seen from the example in Fig. 3.14, after this noise reduction it is possible to apply quite aggressive bandwidth improvement (in this case spectral whitening) without introducing excessive noise.

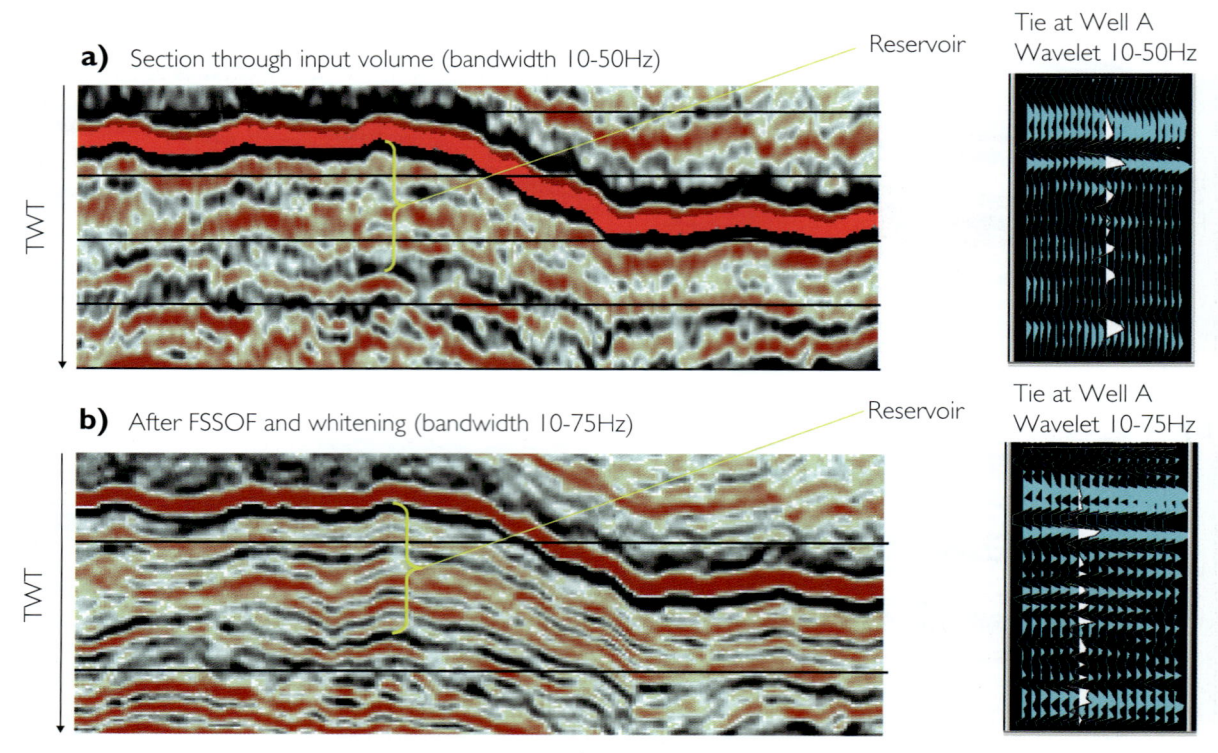

Figure 3.14. Seismic section (a) before and (b) after processing with frequency split structurally oriented filtering and trace whitening (after Helmore et al., 2007).

3.7 Resolution

A key issue in the interpretation of bandlimited seismic data is the fact that there is a lower limit (imposed by bandwidth and wavelet shape) to the bed thickness that can be uniquely resolved. This limit defines the boundary between 'thick' beds (above the resolution limit) and 'thin' beds (below the resolution limit). The uncertainty in the interpretation of 'thin' beds that results from resolution limits is a major issue for quantitative interpretation and this discussion will be developed in Chapters 5, 9 and 10. Although vertical resolution is a key concept for the interpreter to consider, in the context of 3D amplitude and other seismic attribute maps the idea of resolution also tends to incorporate the notion of detectability.

3.7.1 The problem of interference

The modelled seismic trace in Fig. 3.15 illustrates the fact that a seismic section is a complex interaction of the wavelet and reflecting surfaces. Figure 3.15 shows a high-impedance sand encased in low-impedance shales. With a positive standard polarity zero phase wavelet, a peak would be expected at the top of the sand and a trough at the base and in a general way this is what is seen. In detail, however, the acoustic impedance curve shows variability within the sand. The effect of this fine layering is to modify the seismic response. Owing to the fact that the seismic wavelet is longer than the spacing between impedance contrasts, the character of the top and base sand reflections is dependent not only on the contrast at the reflecting interface but also on the layering close to the boundary. The amount of interference is controlled by the length of the seismic pulse and the spacing in two way time of the impedance contrasts (a function of the interval velocity).

3.7.2 Simple models of interference

A simple way of looking at wavelet interference is via the 'wedge' model (Widess, 1973). This is usually thought of as a wedge of sand encased in shale (the sand may be higher or lower impedance than the

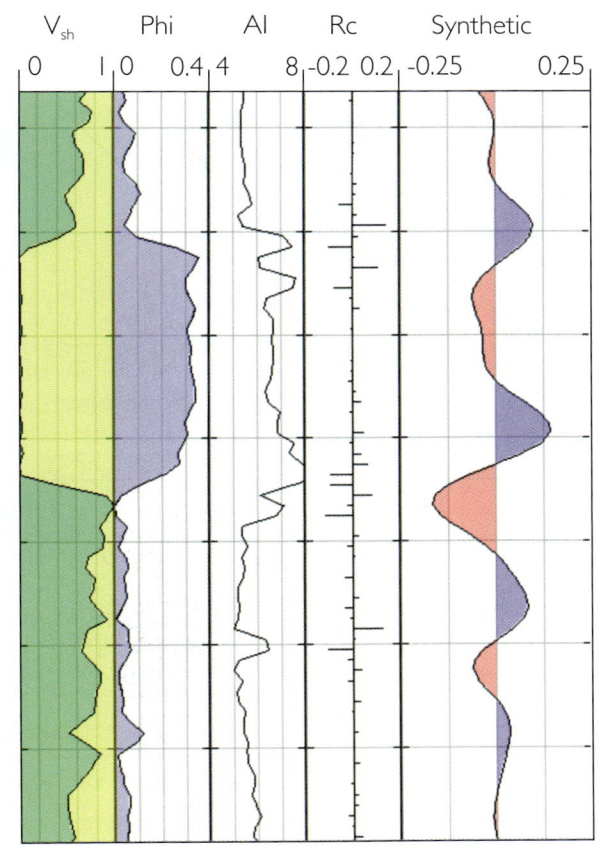

Figure 3.15 Synthetic seismogram illustrating how geology relates to seismic.

shale) such that the reflection coefficients at top and base of the wedge are equal but of opposite polarity. It is particularly applicable for example to isolated gas sands in unconsolidated basin fill sections. Figure 3.16 (upper part) shows an example of this type of model constructed using a positive standard polarity zero phase wavelet. The figure shows the modelled seismic response together with crossplots of thickness and amplitude, and illustrates the key problem for vertical resolution (in the sense of uniquely identifying the top and base of the wedge). Owing to the finite seismic bandwidth there is a thickness below which the seismic loops effectively have a constant separation. Thus, when the sand is very thin, estimating its thickness from the separation of trough and peak seismic loops will result in a significant overestimate (Fig 3.16c).

The tuning curves in Figs. 3.16b,c show how the trough amplitude changes with both actual sand thickness and with apparent sand thickness (defined as the trough-to-peak time separation). It is evident that the character of the tuning curve is related to the shape of the wavelet (Fig. 3.16b), with the actual thickness vs amplitude relationship having a shape similar to a wavelet half cycle. The amplitude tuning curve is a maximum at the sand thickness below which the peak and trough separation remains constant. This is commonly referred to as the *tuning thickness*. Below the tuning thickness, the amplitude decreases in response to the thinning sand. A comparison of Figs. 3.16b,c indicates that, for a limited range of thicknesses above tuning thickness, estimating sand thickness using trough-to-peak separation would result in a slight underestimate. Note that although tuning imposes limits on the interpretation of thicknesses the changes in seismic amplitude and apparent thickness can be used in net pay interpretation (Chapter 10).

Interference related to two reflections of the same polarity is shown in Figs. 3.16e,f. In this model the two reflections converge to a single loop at the tuning thickness. The tuning thickness is characterised by maximum destructive interference (i.e. low amplitude) and as the wedge thins below this point there is an increase in amplitude. This type of interference pattern might be characteristic of angular unconformities.

3.7.3 Estimating vertical resolution from seismic

For practical purposes the tuning thickness can be considered as an indication of the vertical resolution. Fundamentally, the tuning thickness is determined by the compressional velocity of the unit and the wavelength of the seismic pulse (λ). A first order approximation for making quick calculations:

$$\text{tuning thickness} = \lambda/4 \text{(Widess, 1973)} \quad (3.3)$$

$$\lambda = V_p(\text{m/s})/F_d(\text{Hz}) \quad (3.4)$$

$$F_d = 1/T, \quad (3.5)$$

where

λ = wavelength (m), F_d = dominant frequency and T = 'period' (measured in seconds from trough to trough or peak to peak) on a seismic section (e.g. Fig. 3.17).

A worked example based on Fig. 3.17:

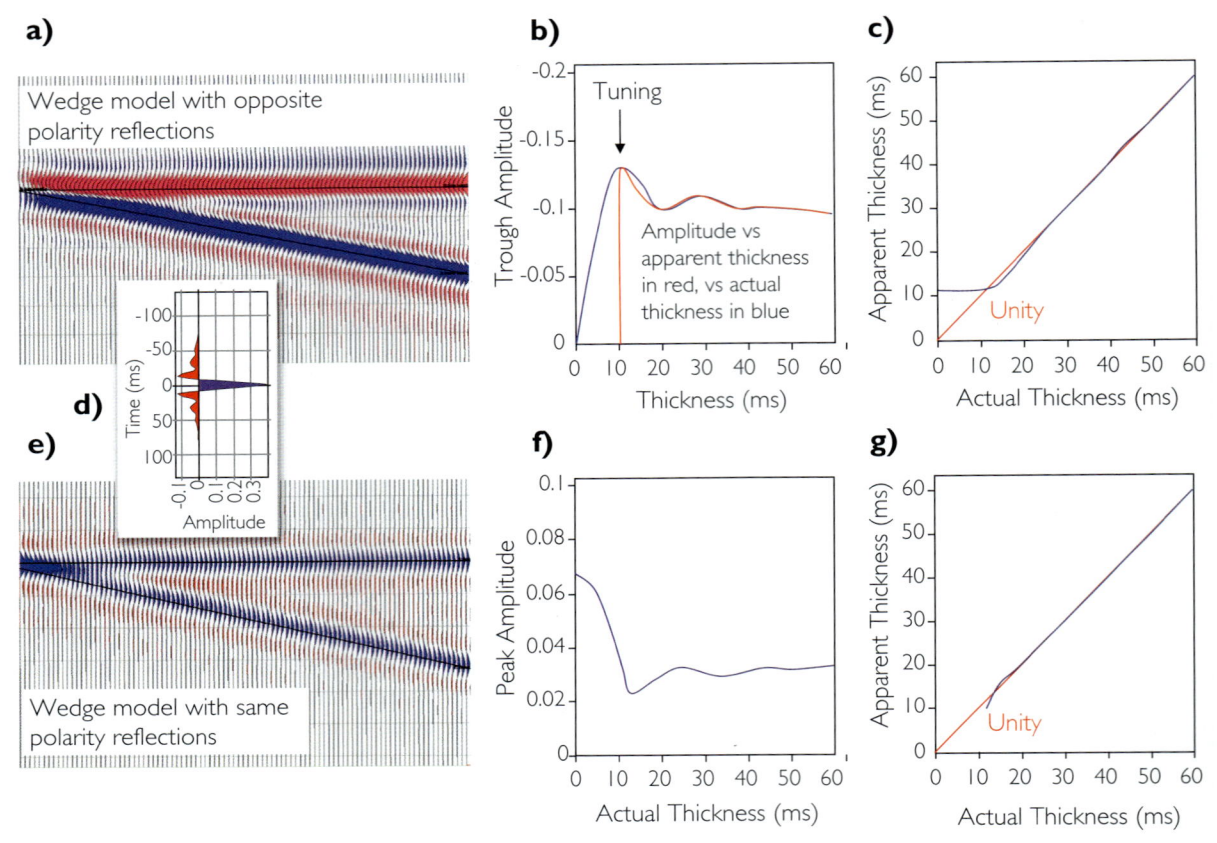

Figure 3.16 Simple interference models: (a–c) wedge model with opposite polarity reflections and thickness vs amplitude characteristics, (d) seismic wavelet, (e)–(g) wedge model with same polarity reflections.

$T = 0.027$ s
$F_d = 37$ Hz
$\lambda = 60$ m (if $V_p = 2220$ m/s)
tuning thickness (m) = 15 m
tuning thickness (ms) = 0.0135 s.

Theoretically, the temporal (or vertical) resolution (i.e. the point at which the two reflections become a composite) is slightly less than tuning thickness and for a simple pulse is defined as $T_r = 1/(2.31 F_d)$ s. Of course in reality resolution is dependent on the wavelet in the data. A useful reference is Kallweit and Wood (1982) where the resolving properties of Ricker wavelets are discussed. It is worth noting that the 'peak' frequency (F_p) (i.e. the frequency descriptor for the Ricker wavelet, and the frequency having the largest amplitude) is related to the dominant frequency by $F_p = F_d/1.3$.

A common first order effect evident on seismic data is the decrease in frequency content with depth (Fig. 3.18). This is due to the various effects of the earth filter. In the shallow part of the section the seismic loops are close together whilst in the deep section they are further apart. This change in frequency combined with the general increase in velocities with depth clearly has an effect on vertical resolution.

3.7.4 The effect of wavelet shape on resolution

The work of Koefoed (1981) was instructive in illustrating the fact that wavelets with the same bandwidth can give different seismic traces and differences in perceived resolution. Figure 3.19 shows two zero phase wavelets with nominally the same bandwidth

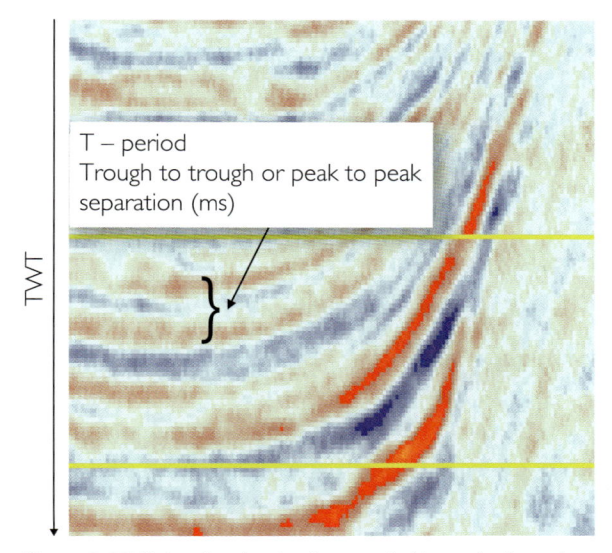

Figure 3.17 Estimating the dominant period in a seismic section.

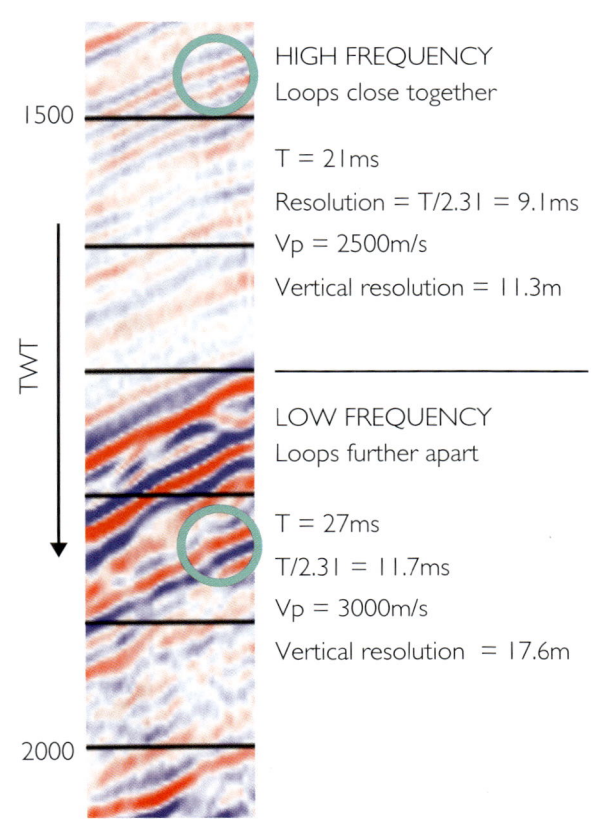

Figure 3.18 Frequency, velocity and vertical resolution.

but which have different slopes on the amplitude spectra, together with synthetics generated from a reflection series with different types of thin beds. The wavelet with steep slopes has a narrow main lobe but significant side lobe energy. This wavelet seems to show some detectability of thin beds with same polarity reflections but is poor at identifying thin beds with opposite polarity reflections, owing to the presence of side lobe energy. By contrast, the opposite appears to be the case with a wavelet with shallower slopes, a broad main lobe and minimal side lobe energy. In practice, seismic processors make judgements on the trade-off between the desire for a narrow main lobe and the effects of reverberations due to high filter slopes.

3.7.5 Lateral resolution

For most practical purposes lateral seismic resolution is not as important for amplitude interpretation as vertical resolution, but it is still worth considering briefly as it may be a factor in evaluating the significance of lateral shifts in quantitative well ties (Chapter 4). The starting point for considering lateral resolution is the fact that reflections are the result of constructive interference over an area of the wavefront called the *Fresnel zone* (Sheriff, 1977) (Fig. 3.20). The reader is referred to a useful discussion of the Fresnel zone by Lindsey (1989). For an un-migrated seismic image the Fresnel zone diameter is given by (Sheriff, 1977):

$$F_d = 2\sqrt{\left(z+\frac{\lambda}{4}\right)^2 - z^2}, \quad (3.6)$$

where z = depth and λ = wavelength.

Un-migrated seismic objects that are smaller than the Fresnel zone are, therefore, not uniquely identified. In the days when prospect maps were generated with un-migrated seismic it was important to calculate the Fresnel zone. It was used, for example, to determine an adequate stand-off distance from faults and to ensure that the well was drilled on the up-thrown side. Migration, however, plays a significant role in enhancing lateral seismic resolution. Figure 3.21 shows how the Fresnel zone is compressed in the inline direction for a 2D migration. Note that 3D migration collapses the Fresnel zone to a small circle (ideally around one half of the wavelength in

Seismic wavelets and resolution

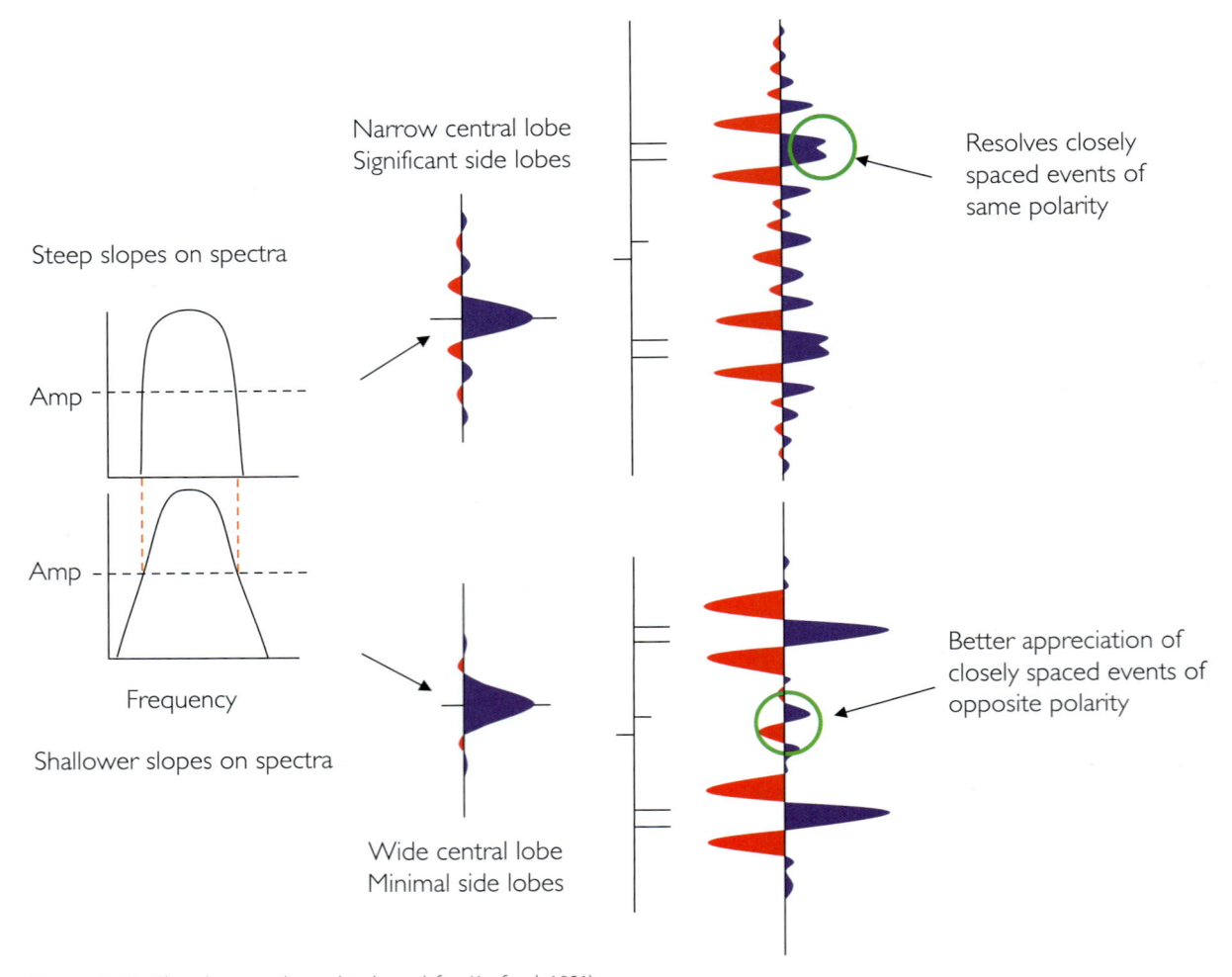

Steep slopes on spectra

Narrow central lobe
Significant side lobes

Resolves closely spaced events of same polarity

Shallower slopes on spectra

Wide central lobe
Minimal side lobes

Better appreciation of closely spaced events of opposite polarity

Figure 3.19 Filter slopes and wavelet shape (after Koefoed, 1981).

$\lambda/4$ FOR LOW FREQUENCY

$\lambda/4$ FOR HIGH FREQUENCY

High Frequency Zone

Low Frequency Zone

Post migration Fresnel zone

Pre migration Fresnel zone

CMP

Line direction

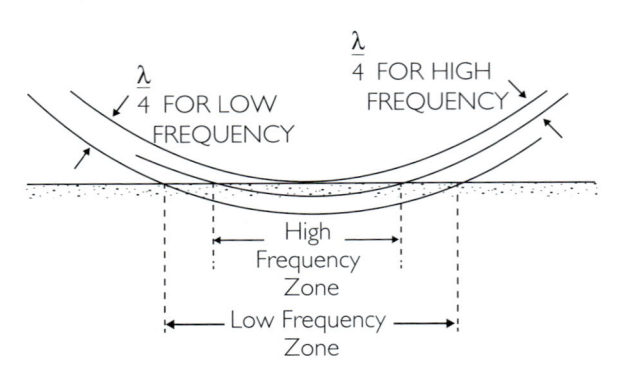

Figure 3.20 The 'Fresnel' zone; the area at the acoustic boundary from which the dominant part of the reflection originates (after Sheriff, 1977; AAPG © 1977, reprinted by permission of the AAPG whose permission is required for further use). The Fresnel zone varies with frequency.

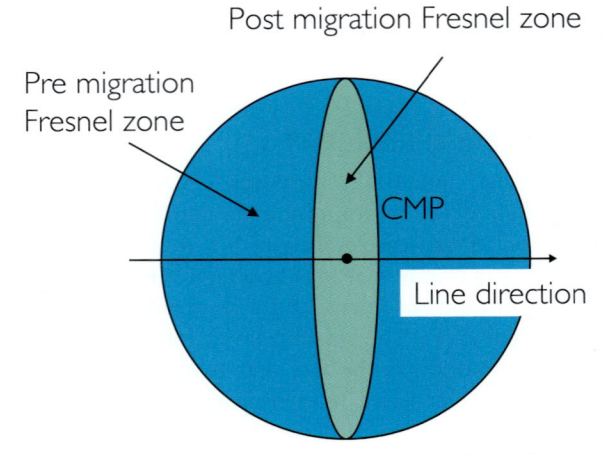

Figure 3.21 The effect of migration on the Fresnel zone for a 2D line (after Lindsey, 1989).

diameter). Key factors in lateral resolution for 3D seismic surveys are bandwidth, accuracy of the migration velocity model, and adequate sampling of steep-dip information. Owing to the dense sampling of modern 3D data, lateral resolution (in terms of the Fresnel zone) is not usually an issue.

3.8 Detectability

Although a gas sand may have a thickness below tuning thickness, it will still potentially have amplitudes higher than background. Thus the sand is detectable without being fully resolved. The limit of detectability (i.e. the thickness at which the amplitude merges with the background) depends essentially on the acoustic impedance contrast and the signal-to-noise ratio of the seismic. As a general rule of thumb, for low impedance hydrocarbon sands with reasonable data quality the limit of detectability can be around $\lambda/20$ to $\lambda/30$ (Sheriff and Geldart, 1995; Sheriff, 2006). Whenever net pay is determined from seismic (Chapter 10) there is always the uncertainty that there may be pay in this 'below detectability' zone. Whether this hydrocarbon is producible of course depends on the connectivity of the reservoir and is a question for the reservoir engineer.

Much of the power of 3D seismic is in detecting subtle features below seismic resolution. The notion of detectability is illustrated in Fig. 3.22, which contrasts the interpretability of subtle effects on amplitude maps and vertical sections. A 3D model is shown of a formation defined by two bounding surfaces with a hydrocarbon contact (Fig. 3.22a). The 3D synthetic models were generated to simulate the seismic before and after production and a difference dataset was generated (Fig. 3.22b). It is evident that the change in the contact can be seen effectively as a composite of two interfering reflections (see also Section 10.4 on time lapse seismic). For realism, noise was added and the corresponding section is shown in Fig. 3.22c. In the presence of noise the time lapse signature is not easily interpretable on the vertical section. Time slices were also generated at the level of the original contact (Figs. 3.22d,e) and it is clear that even in the noisy data the outline of the contact change is clearly visible. The effective dynamic range of amplitude maps (and other seismic attribute maps) is much greater than vertical sections.

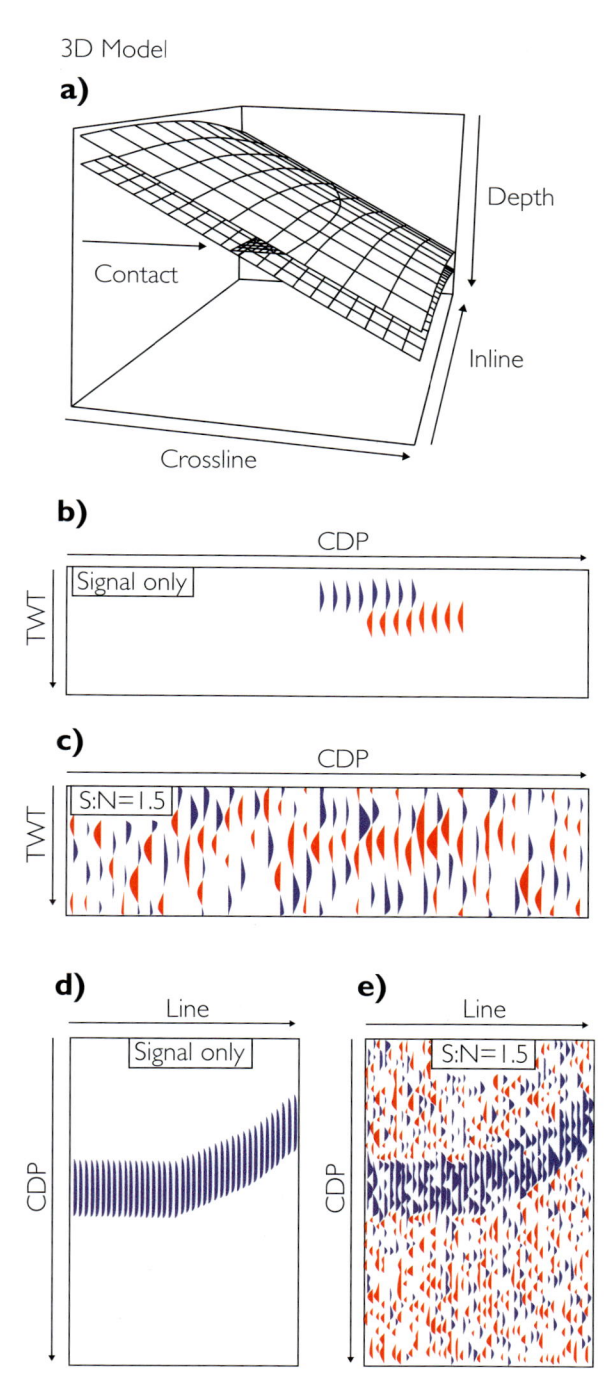

Figure 3.22 Detectability in seismic sections and maps; (a) 3D model of top and base Oseberg reservoir and a hydrocarbon contact, (b) vertical difference section after pre-production and monitor models are subtracted, (c) vertical difference section with realistic levels of noise added, (d) time slice through model at level of contact, (e) time slice through model with added noise. Note how the time lapse signature is evident in the noise prone map (e) but quite indistinct on the noise prone section (c) (re-drawn after Archer et al., 1993).

Chapter 4
Well to seismic ties

4.1 Introduction

There are several possible objectives in performing well ties:

- zero phasing: checking whether data are zero phase, and helping to adjust the phase if required,
- horizon identification: relating stratigraphic markers in the well to loops on the seismic section,
- wavelet extraction for seismic inversion or modelling,
- offset scaling: checking whether the seismic data have been 'true amplitude' processed to have the correct AVO behaviour, and adjusting amplitudes if necessary.

Achieving these objectives requires integration of regional interpretation experience with well tie analysis, linking surface seismic to synthetic seismograms and vertical seismic profiles (VSPs), as shown in Fig. 4.1. One way of looking at the well tie is that it is the interpreter's chance to conduct an experiment to test the connection between the geology and the seismic data. In practice, there are numerous issues to consider and an analytical approach is useful. There are some situations where the tie and the phase of the seismic data are not in doubt; in other cases there is significant uncertainty. Estimates of well tie and wavelet accuracy frame the context for assessing the quality of calibration in an interpretation and amongst other benefits they provide insight into the feasibility of a good quality trace inversion (Chapter 9).

4.2 Log calibration – depth to time

An important element in any well tie technique is the conversion of depth to time. Typically this will involve the use of checkshot or VSP depth and time data. These data are derived from the direct arrivals of shots from seismic sources suspended over the side of the drilling rig. In the past it was not uncommon for interpreters to re-interpret these data from paper copies. In today's digital world the individual shot data are seldom included in well data packages, so the interpreter is usually reliant on the final contractor compilation of depth–time data presented in the final well report or well seismic report. Understanding the datum of the data is critical for land data. Marine data are invariably referenced to mean sea level.

Determining a continuous function of time and depth requires the integration of checkshot and log data. Given that velocity is dispersive (i.e. it is dependent on the frequency), thought should be applied to how these datasets are integrated. Effectively this means that log based velocities should be upscaled to the seismic scale prior to performing the well tie.

4.2.1 Velocities and scale

Different velocity averaging methods are appropriate for different wavelength and layer spacing situations (Marion *et al.*, 1994). When the ratio of wavelength (λ) to layer spacing (d) is ≤1, as it tends to be at logging frequencies, then an appropriate averaging method is to sum the travel times through each layer. This is usually referred to as the *time average* and for regularly spaced data such as sonic logs this equates to a simple arithmetic average of the slowness over a sliding window. When the ratio of wavelength (λ) to layer spacing (d) is ≥1 (e.g. at seismic wavelengths and with typical rock layering) the situation becomes complicated by the fact that the layering gives rise to anisotropy. In this case the average or 'effective medium' velocity is dependent on the nature of the layering and the path that the acoustic energy takes through the rock (e.g. Backus, 1962; Mavko *et al.*, 1998).

The *Backus average* (Backus, 1962) is an appropriate way to describe average P and S wave velocities through finely layered media. Essentially this method averages moduli rather than velocity. For the purposes of routine well ties, where vertical acoustic propagation and isotropy are usually assumed, an appropriate

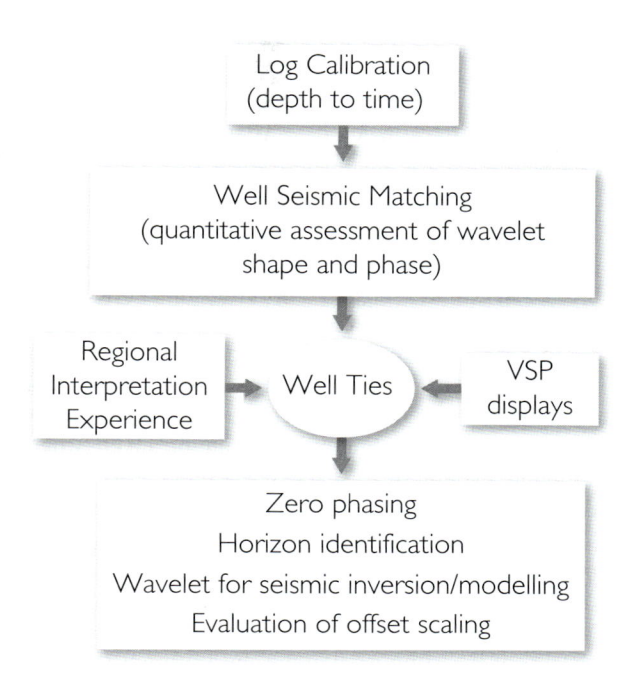

Figure 4.1 The well tie process.

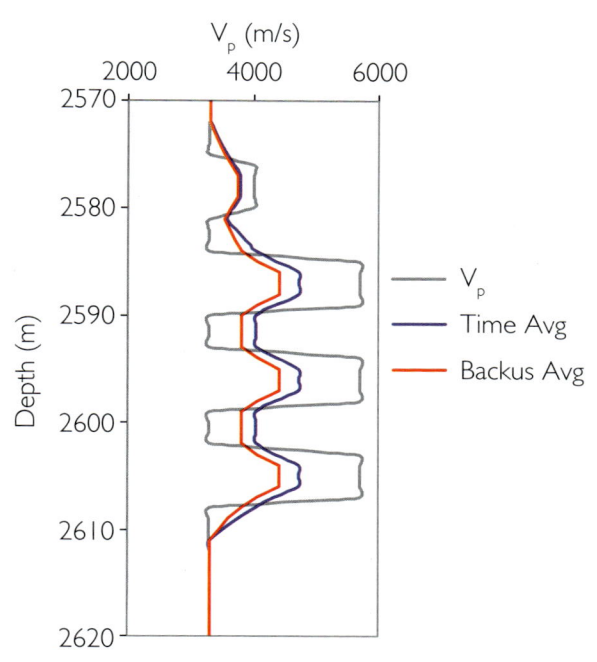

Figure 4.2 Model showing effect of averaging V_p logs; grey curve = V_p model, blue curve = time average over 7 m window, red = Backus average over 7 m window.

average V_p and V_s can be determined from the harmonic averages of the P wave modulus (M) and the shear modulus (μ) as follows.

(1) Determine the M modulus and shear modulus from P wave velocity, S wave velocity and density (V_p, V_s and ρ) using $\mu = V_s^2 \rho$ and $M = V_p^2 \rho$.
(2) Over an averaging length to be defined below, calculate the arithmetic average of ρ and the harmonic average of M and μ, e.g. $M_{bavg}^{-1} = n^{-1} \sum M^{-1}$.
(3) Use these averaged parameters to calculate the Backus averaged velocities using: $V_{p\_bavg} = \sqrt{\frac{M_{bavg}}{\rho_{avg}}}$ and $V_{s\_bavg} = \sqrt{\frac{\mu_{bavg}}{\rho_{avg}}}$.

Various suggestions have been made for the averaging length required in the Backus average. Liner and Fei (2007), who provide a useful general background to the theory of the method, propose a maximum value of $V_{s\_min}/3f$, where V_{s_min} is the minimum shear velocity after Backus averaging and f is the dominant frequency in the wavelet (Chapter 3). In general, Backus upscaling is important where there are strong contrasts in velocity, such as in a sequence of interbedded shales and limestones, but has a much less marked effect where the velocity contrasts are small, as is often the case for a sand/shale sequence (Fig. 4.2).

4.2.2 Drift analysis and correction

Log calibration seeks to analyse and resolve the differences in times derived from the sonic log and from checkshots. The general workflow is described below and with reference to Fig. 4.3.

(1) Integrate velocity log from the uppermost or lowermost checkshot that ties the log.
(2) Calculate the *drift* (i.e. seismic time minus time from integrated log).
(3) Evaluate quality of checkshots using drift points and checkshot velocities.
(4) Repeat (2) if checkshots are de-selected
(5) Fit a curve to the drift points such that the differences between the final time–depth relation and the integrated calibrated velocity log are less than about 2 ms.
(6) Apply the drift correction to the time–depth curve from the integrated velocity log. The corrections may also be applied to the velocity log to generate the *calibrated velocity log*.
(7) Evaluate the effect of corrections on the velocity log (large differences are not expected).

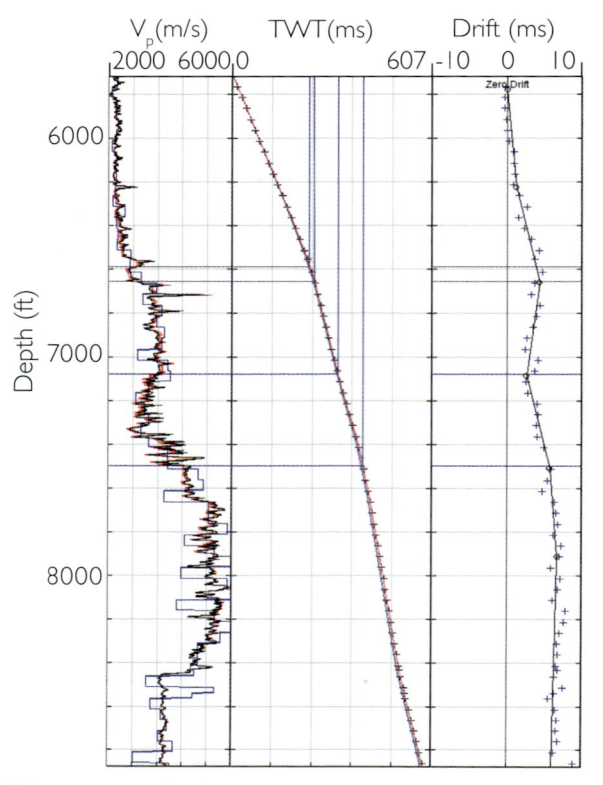

Figure 4.3 Log (depth–time) calibration: column 1, black – V_p log, red – calibrated velocity log, blue – velocities from checkshots; column 2, blue – integrated depth–time curve from V_p, red – calibrated depth-time curve (i.e. with drift applied); column 3, blue crosses – drift points, black – drift curve fitted to the data using linear segments with knee points.

Drift, in general, is a real effect of the subsurface, principally related to velocity dispersion between log and seismic frequencies (e.g. Stewart et al., 1984). Each formation usually has its own drift characteristics. Usually, sonic log velocities are higher than seismic velocities, giving rise to *positive drift* (i.e. shorter integrated log times compared to the seismic) (Fig. 4.3). In cases where the sonic velocity is less than the seismic (i.e. *negative drift*), the log data should be checked for hole problems such as washouts. It has been common practice for positive drift corrections to be applied linearly to the data whereas negative drift corrections are preferentially applied to the lower velocities. This is based on the assumption that the negative drift is likely to be due to erroneously low sonic velocities in zones where hole conditions are poor. In deviated wells, drift may also be in part caused by velocity anisotropy (for example shale velocities in deviated wells are often higher than in vertical wells). It should be noted that in deviated wells rig source checkshots are not particularly useful, as it can be difficult to accurately correct the slant ray times to vertical. Walk-above VSP shots are the best dataset for this purpose.

Commonly used drift functions are linear trends with knee points (e.g. Fig. 4.3), polynomial fits and spline fits. The knee points for linear trends should be located at major changes in lithology or at unconformities where there are changes in velocity in order to avoid generating spurious reflections if the calibrated velocity log is used for synthetic generation. Spline fits are reasonable if there are a large number of points such as in a VSP, but care should be taken when applying a spline fit to sparse checkshots. Polynomial fits are useful in basin fill sequences where there is a simple compaction trend and linear fits with knee points are appropriate in basins with a number of prominent unconformities and variations in lithological style. If linear fits are made between each checkshot (i.e. simply using the checkshots as the time vs depth function) it is important to check the effect on the velocities.

When using the time–depth relations in a well tie the interpreter has a choice of whether to use the calibrated velocity log or the original log in the well tie match. Given that the differences should not be large the choice is not usually important. Many workers favour using the calibrated time–depth relationship with the original (upscaled) velocity log, although benchmark tests appear to show slightly better ties when using the calibrated velocity log (Roy White, personal communication).

4.3 The role of VSPs

Vertical seismic profiles (VSPs) are useful in the well tie process because they provide a link between wells and seismic at the correct scale. The essence of the VSP method is to record a surface seismic source using down-hole geophones. The simplest geometry is for a vertical well (Fig. 4.4). A string of geophones is deployed in the well and, by shifting them up between shots, it is possible to record signals at a large number of levels. For example, records might be obtained at 50 ft spacing over an interval of 5000 ft in the well, to give 100 levels in all. At each level, the geophone will record down-going waves (such as the direct arrival, the leftmost raypath in Fig. 4.4) and up-going waves (such as the two reflections in Fig. 4.4). Multiples are

The role of VSPs

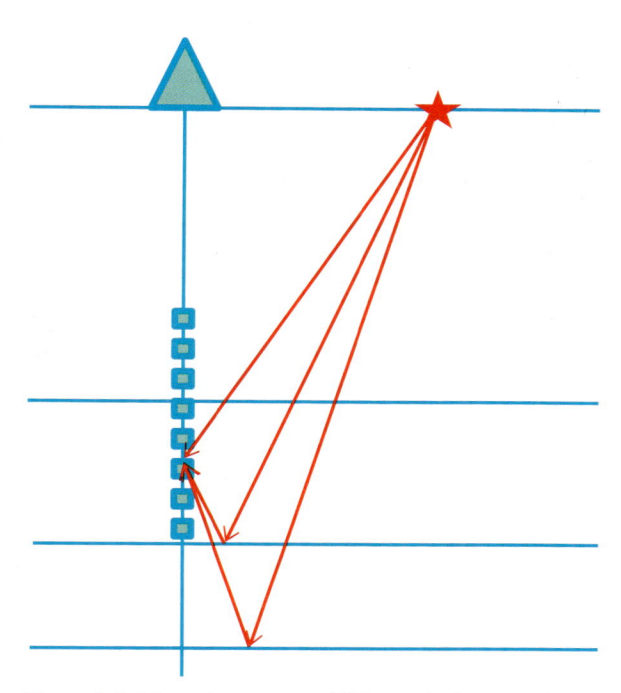

Figure 4.4 Schematic geometry of VSP raypaths from a surface source to borehole geophones. The separation of the source from the borehole would in practice be very small for the case of a zero offset VSP, but is here exaggerated for clarity.

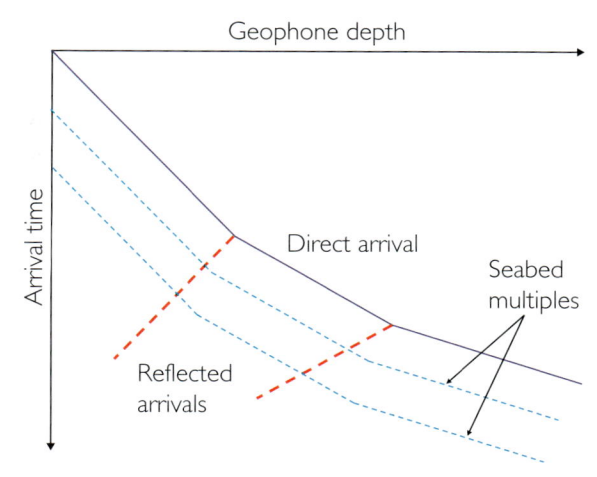

Figure 4.5 Schematic graph of VSP arrival time against geophone depth for VSP data.

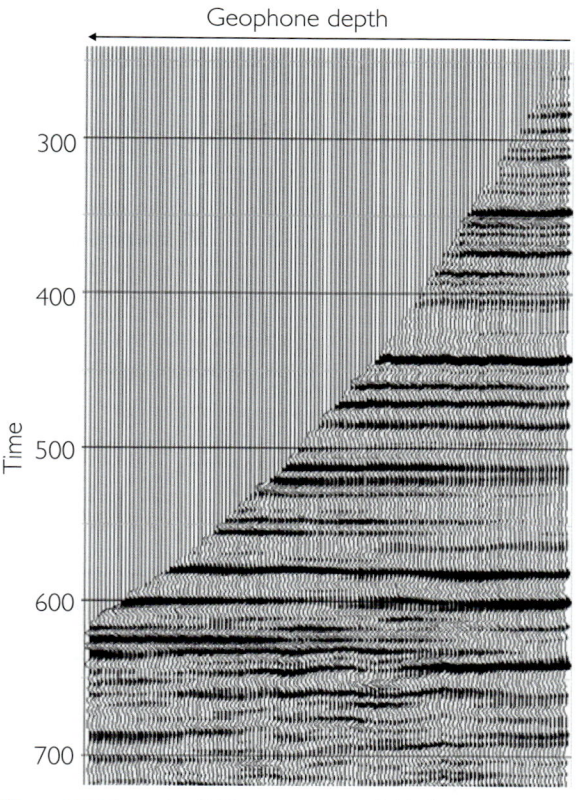

Figure 4.6 Example of VSP up-going wavefield after signal enhancement (after Chopra et al., 2004; reprinted with permission of the author and the CSEG Recorder).

also present but these arrive after the direct wave and single bounce up-going reflections. When the direct arrival can be clearly defined it is possible to accurately convert the direct arrival and the reflections that immediately follow it to zero phase.

A schematic travel-time display for VSP data is shown in Fig. 4.5. Key steps in the processing of VSP data include the following.

(1) Measurement of the arrival time for the direct wave arrival (energy onset or max/min of first loop).
(2) Zero phasing operator design on the direct arrival.
(3) Determination of the down-going wavefield by horizontal alignment of the data at the direct arrival time followed by filtering to enhance laterally continuous events.
(4) Estimation of up-going wavefield by subtraction of downgoing wavefield from the data.
(5) Re-alignment of upgoing wavefield to position reflections at their two-way times from surface and enhancement by various processing methods (Fig. 4.6).
(6) Generation of the *corridor stack* from the processed upgoing wavefield. This involves

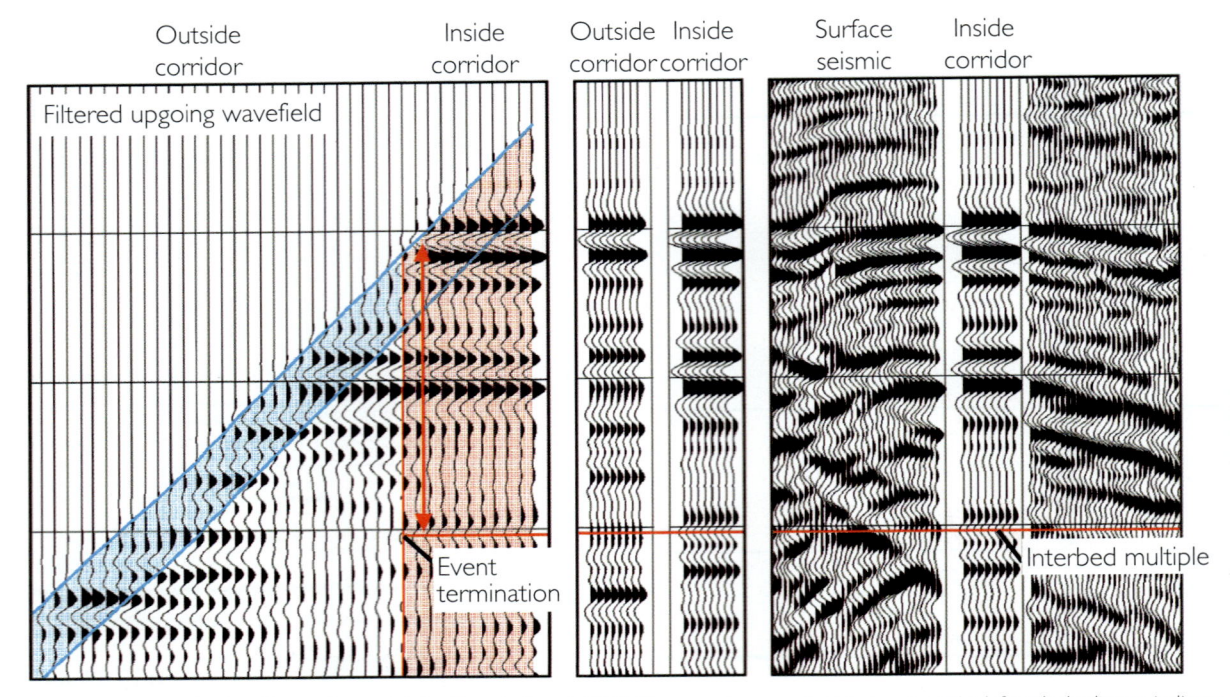

Figure 4.7 Example of a VSP corridor stack (after Campbell *et al.*, 2005). The up-going wavefield is shown on the left with shading to indicate the zones stacked to make the outside corridor stack (blue) and inside corridor stack (red).

stacking the data in a time window immediately following the first arrival. This should be free from multiples and is therefore the ideal reference trace to compare with both the well synthetic and the actual surface seismic.

Figure 4.7 compares a multiple-free 'outside' corridor stack with an 'inside' corridor stack which is contaminated by interbed multiples. At the horizontal red line there is a strong event on the inside corridor stack that is absent from the outside stack, and is thus inferred to be an intra-bed multiple. The event is absent from the surface seismic, implying that the processor has been successful in removing the multiple.

The VSP can be particularly useful in highly deviated wells. It is common for well synthetics to tie surface seismic poorly for such wells, perhaps because of anisotropic effects (see Section 8.4.5). In a walk-above acquisition, the surface source is positioned vertically above the geophone at a series of levels in the deviated borehole. This makes acquisition more time-consuming and costly, as the source has to be moved to an accurately determined location for each level. However, the benefit is that the raypaths are constrained to be nearly vertical. An image is produced of the subsurface directly below the borehole, which can be compared with the surface seismic along the well trajectory. An example is shown in Fig. 4.8.

The benefits of the VSP are numerous (Campbell *et al.*, 2005).

- High-density sampling gives good control on the time–depth relationship.
- Control on phase: trace by trace deconvolution of the upwave using the down-going wavefield results in a zero phase upwave corridor stack.
- The upwave corridor stack is largely multiple free; comparison with seismic can highlight potential multiple problems.
- In favourable situations the absorption parameter Q can be estimated from wavelet shape changes in the downgoing wavefield (e.g. Tonn, 1991; Harris *et al.*, 1997).
- It may be possible to apply inverse Q filtering to optimise resolution as well as using the VSP to determine zero phasing operators for surface seismic.

However, there are several reasons why VSPs may not tie exactly to seismic.

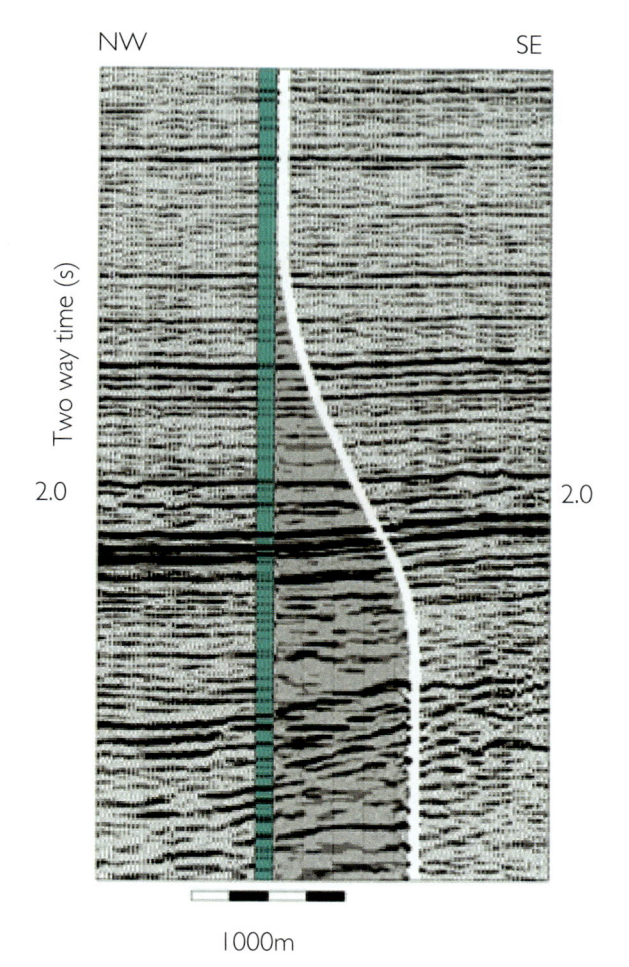

Figure 4.8 A walk-above VSP image from a deviated well inserted into surface seismic data. VSP data are the greyed zone between the blue and white lines. Also shown in blue is a corridor stack for a vertical borehole drilled from the same surface location (after Kaderali et al., 2007).

- The volume of rock sampled is different.
- Difference in frequency content (the VSP will generally be higher frequency and will need to be filtered back to match the seismic bandwidth).
- Differences in wave propagation effects, e.g. anisotropy and attenuation, due to differences in raypath.
- Migration effects.

There are some practical issues that affect the way that the seismic interpreter makes use of VSPs.

- In general the key step is to compare the corridor stack with the surface seismic.
- If this comparison is documented in the VSP processing report, it is important to check that the seismic that was used is still the current version; if not, new displays need to be created.
- If a VSP processing report compares the VSP with a well synthetic, the wavelet used to create the synthetic needs to be checked. It is likely to have been a simple idealised wavelet.
- Routine digital manipulation of VSPs by the interpreter is less easy than creation of synthetics from log data, because more specialised software is needed.
- Greater integration of VSPs into standard interpretation workflows would be beneficial, including VSP inversion to impedance.

4.4 Well tie approaches using synthetics

When tying wells to seismic it is quite indefensible simply to use a time–depth relation from checkshots, post the well tops on the seismic and start picking. Whilst the timing of the seismic and the checkshots is likely to be close they are probably not the same and making assumptions about wavelet phase, as will be shown below, is prone to error. The well tie is a basic tool to analyse the connection of geology and seismic and two approaches will be discussed that can be implemented with most seismic interpretation software. The first, referred to as the 'well matching technique', requires good control on depth and time and seeks to extract a wavelet from the seismic without making any assumptions about phase and timing. In the absence of good time–depth control the second approach, referred to here as the 'adaptive technique', involves more trial and error.

4.4.1 Well tie matching technique

In order to estimate the correct wavelet for the purposes of picking seismic or designing zero phasing or inversion operators, a pragmatic approach is to estimate the wavelet directly from the seismic data. The technique described here is that of White (1980) and White and Simm (2003). The wavelet is extracted from the data through a least squares technique (Fig. 4.9), treating the well tie as a noisy input–noisy output problem (Walden and White, 1998). Vertical well ties are the most straightforward but most software now offers the possibility of extracting the

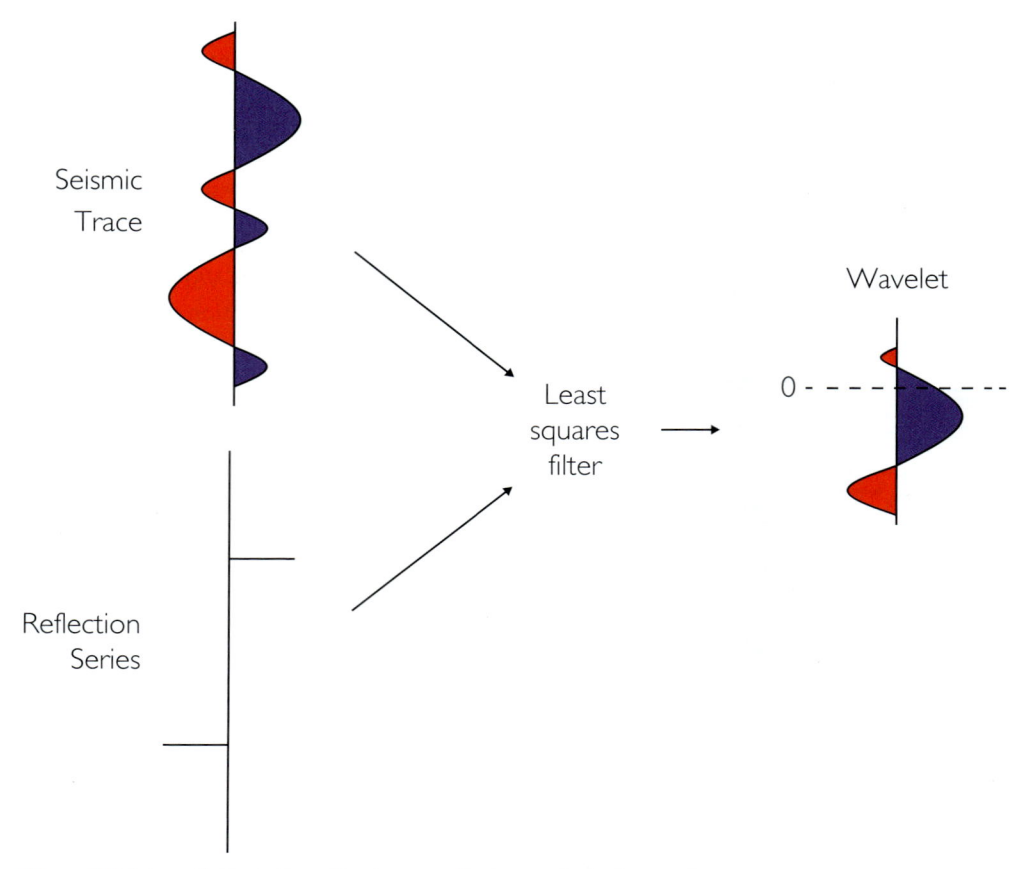

Figure 4.9 Schematic illustration of least squares filtering to obtain the wavelet.

wavelet along the borehole path. To achieve a reliable extraction the quality of the log data has to be good and the time–depth conversion has to be accurate. Matching involves extracting a wavelet operator (of length L) from a time window of seismic data (of length T). In signal processing terms L is the 'lag window length'. Confusingly, L is often referred to as the 'wavelet length' but it is approximately twice the length of a three-loop wavelet. The time segment length should be around 500 ms. If it is any longer the stationarity assumption concerning phase may be invalid and phase rotation with depth could adversely affect the results. If the segment is shorter, the chances of a statistically valid tie are reduced. If possible it is best to choose a time segment for which the start and end have muted reflectivity; wavelet extractions are likely to be distorted if the time segment truncates close to a large reflection.

In order for matching measures such as the cross-correlation to be statistically meaningful L and T (Fig. 4.10) need to satisfy the following relations (White, 1980; Walden and White, 1984; White, 1997):

$bT = (3.408\ T) / L$,

where b = analysis bandwidth and is related to L by: b = constant / L (for correlations tapered with a Papoulis window, constant = 3.408);

bT is referred to as the spectral smoothing factor. This parameter needs to be >6;

b/B should between 0.25 and 0.5 (where B = statistical bandwidth estimated from the data).

Effectively these relationships mean that the time segment to (three-loop) wavelet length ratio should be around 3 for low-bandwidth data ($B \sim 25$ Hz) and around 6 for high-bandwidth data ($B \sim 50$ Hz).

4.4.1.1 Well tie measures: goodness of fit and accuracy

In the well matching technique goodness of fit is measured as the proportion of trace energy predicted (*PEP*) by the synthetic seismogram:

$PEP = 1 - $ (energy in the residuals/trace energy),

where the *energy* of a trace is the sum of the squares of the time segment and the residuals are the difference between a seismic trace and its matched synthetic.

PEP can be related to the cross-correlation coefficient *R*, a more traditional goodness of fit measure, by $PEP \approx R^2$. Thus, a cross-correlation of 0.7 effectively means that just less than 50% of the energy is being matched. Goodness of fit alone is not a definitive measure of accuracy. The cross-correlation increases with increasing wavelet length, but as the wavelet gets longer there is more chance that noise is being matched. A crude measure of accuracy is the normalised mean square error (*NMSE*) which can be approximately related to the phase error:

$$NMSE = \frac{1}{bT} \frac{1 - PEP}{PEP},$$

and the approximate phase standard error in radians will be $\sqrt{\frac{NMSE}{2}}$. Good well ties have $PEP > 0.7$ and $NMSE < 0.1$.

4.4.1.2 Analysis of wavelet phase and timing

An appreciation of phase uncertainty can be derived from an analysis of the phase variation with frequency and the estimated standard errors (Fig. 4.11). The phase and timing of the wavelet is calculated by making a linear fit to the phase spectrum over the seismic bandwidth between its standard errors. The intercept gives the average phase and the slope of the line gives the timing of the wavelet, where timing in ms = slope

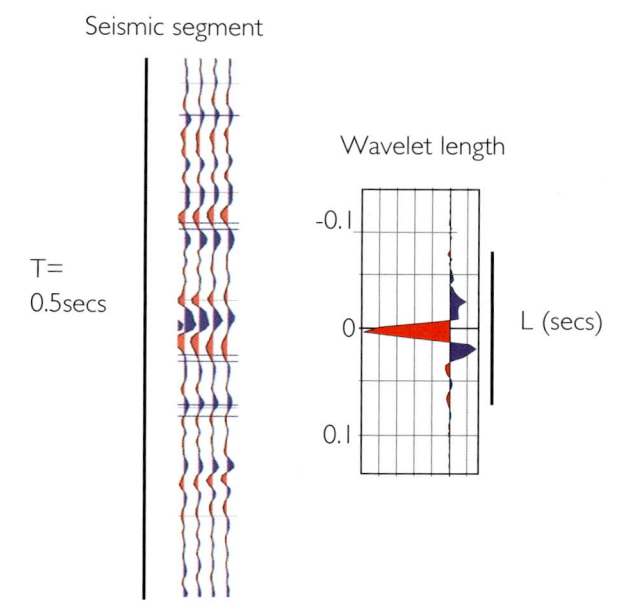

Figure 4.10 Seismic segment (*T*) and wavelet length (*L*).

a) High bandwidth data

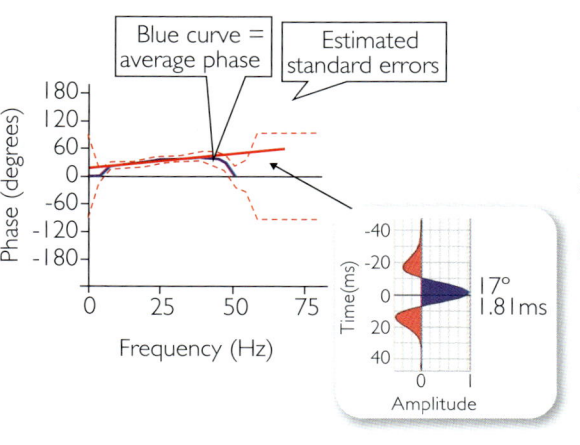

b) Low bandwidth data

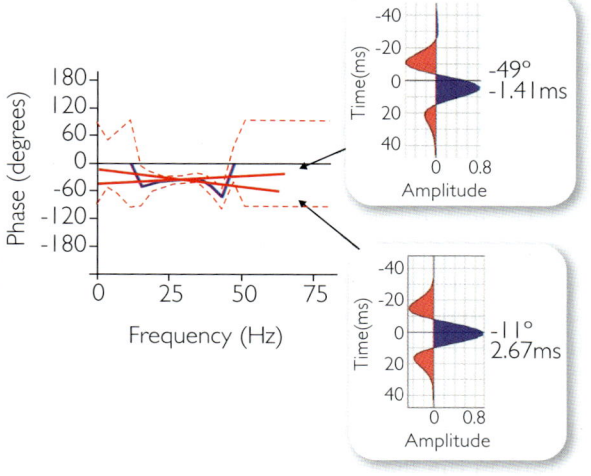

Figure 4.11 Evaluating wavelet phase ambiguity through line fitting between standard errors; (a) high-bandwidth data with accurate estimation of phase, (b) low-bandwidth data with appreciable phase ambiguity (modified after White and Simm, 2003).

L (ms)	Estimated wavelet	PEP	NMSE (%)	(b/B)	(bT)	Comment
60		0.5	3.5	0.85	28.7	Oversmoothed
100		0.6	3.8	0.63	17.4	Oversmoothed
140	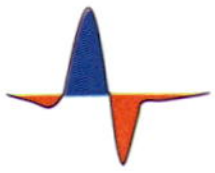	0.65	4.3	0.50	12.5	About right
200	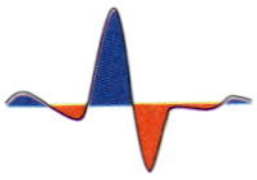	0.69	5.1	0.38	8.8	Undersmoothed?
248	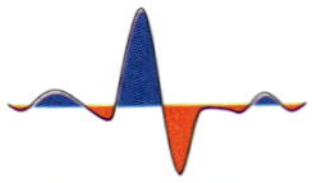	0.71	5.7	0.32	7.2	Undersmoothed?
300	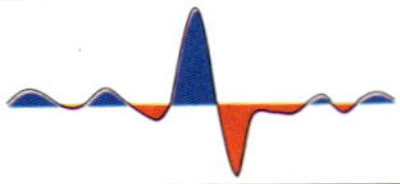	0.73	6.3	0.27	6.0	Too noisy

Figure 4.12 Extracted wavelets from marine seismic with varying wavelet length (L) from a 500 ms time segment. A wavelet of about 140 ms is probably the best trade-off in this case between noise and oversmoothing (R. E. White, personal communication).

(in deg/Hz) × (1000/360), with negative values corresponding to a time delay and positive values to a time advance. In the high-bandwidth example in Fig. 4.11a there is little scope to vary the fit and the wavelet phase is constrained to within about 10°, whereas in the low-bandwidth example in Fig. 4.11b a wider range of fits are possible and there is around 40° of phase ambiguity.

4.4.1.3 How long is a wavelet?

The length of a seismic wavelet is a compromise between goodness of fit (as expressed by the *PEP* or the cross-correlation) and the estimated phase accuracy. An example of wavelets extracted from a 500 ms time segment of seismic data is shown in Fig. 4.12. It is evident that the long wavelets have been over-fitted to the data with the incorporation of noise, whereas the short wavelets have been oversmoothed. The choice of a wavelet length of 140 ms in this case is based on a number of factors including visual appreciation and a compromise between the cross-correlation and phase error.

4.4.1.4 Best match location

It is usual to run the matching technique on a cube of seismic around the well location rather than just on the single trace at the location. The reason for this is that the migration applied to the seismic data may not have positioned the data correctly, particularly in the presence of steep dips and significant lateral velocity variation. There are two separate possible causes of error. One of them is the use of time migration rather than depth migration. Time migration assumes that the overburden velocity is laterally invariant on a length scale similar to the migration aperture. This effectively means that in the presence of dip the best match location is usually up-dip of the well with time-migrated data. The other problem is that the velocity model used to perform the migration is never perfectly determined. The positioning error is likely to be greater as target depth and reflector dips increase; shifts of 100 m are quite possible. It is also possible that the positioning is wrong, for example in old wells where the deviation surveys are inaccurate, or with seismic that has been loaded with the wrong coordinate system.

An example of a best match location being different from the well location is shown in Fig. 4.13. The *PEP* map (Fig 4.13a) shows that the best match (red area) is up-dip of the well, and at this location the time delay map (Fig. 4.13c) shows zero delay. In deciding whether to accept such a shift, it is important

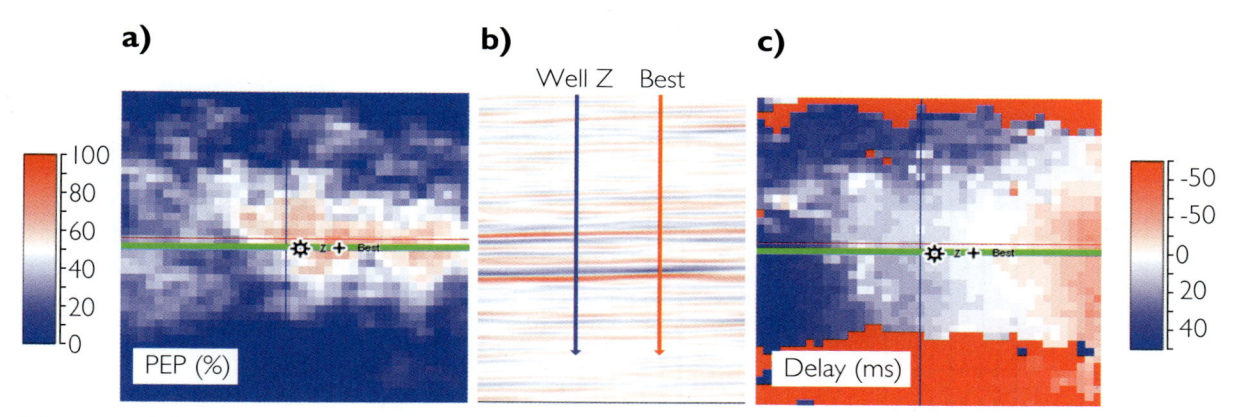

Figure 4.13 Well tie mapping; (a) *PEP* map (%), (b) seismic crossline (green line on maps) through well location and best match location, (c) delay map (ms). Note that the distance between the well and the best match location is 110 m.

to examine the *PEP* and delay maps carefully. The amount of lateral shift has to be reasonable, it should be consistent when using different time windows for the match, and the *PEP* map should show a distinct area of maximum values.

4.4.2 Adaptive technique

The well matching technique described above is applicable in areas such as the North Sea where data acquisition tends to be maximised and checkshot data are regularly acquired. In cases where good checkshots are not readily available, for example in areas such as onshore Canada or in parts of the Gulf of Mexico, the well tie process involves a little more trial and error. Without checkshots the time–depth relation is based on the integrated sonic log sonic tied to the two way time of a prominent seismic reflector. The wavelet used in the synthetic is derived from the amplitude spectrum of the data and an initial (zero) phase assumption. There is usually a downward stretch of the well synthetic required to tie the seismic. In most cases a single stretch is all that is required. Most software has an option to squeeze the synthetic, although often there is no justification for doing this. The effects of stretch and squeeze on the velocity log should be carefully reviewed before proceeding. The workflow, which is described here as an 'adaptive approach' is usually applied at the well location and comprises:

(1) depth to time conversion (e.g. sonic integration),
(2) estimate the amplitude spectrum and derive the zero phase equivalent wavelet,
(3) generate the synthetic,
(4) cross-correlate the synthetic trace with the seismic trace and evaluate the shape of the cross-correlation function,
(5) phase rotate the wavelet so that the cross correlation function is symmetrical,
(6) apply a bulk shift or single stretch to correlate the synthetic to the seismic,
(7) extract wavelet as a final QC step.

With this method it would be possible to scan for the best match location but it is likely that the best match would depend to some extent on the stretch applied.

4.5 A well tie example

Both the well matching and adaptive techniques have been applied to a North Sea dataset and the results are shown in Figs. 4.14–4.18. Figure 4.14a illustrates the well tie using the extracted wavelet (from a 500 ms time segment). A time lag of +16 ms was calculated in the wavelet estimation and this has been applied to the wavelet to make the tie. It is a good tie with a *PEP* of 71%. The wavelet (Figs. 4.14b, e) has an unambiguous phase of about $-70°$ (using SEG positive standard polarity as the reference). It has been extracted using the borehole deviation and ties to a location 88m away from well location (Fig. 4.14d). The shape of the cross-correlation confirms that the wavelet estimation is good. Although this is a good tie, at this point the interpreter should be wary of associating the horizons on the logs with the seismic trace (Fig. 4.14a) owing to the lag and phase rotation of the wavelet.

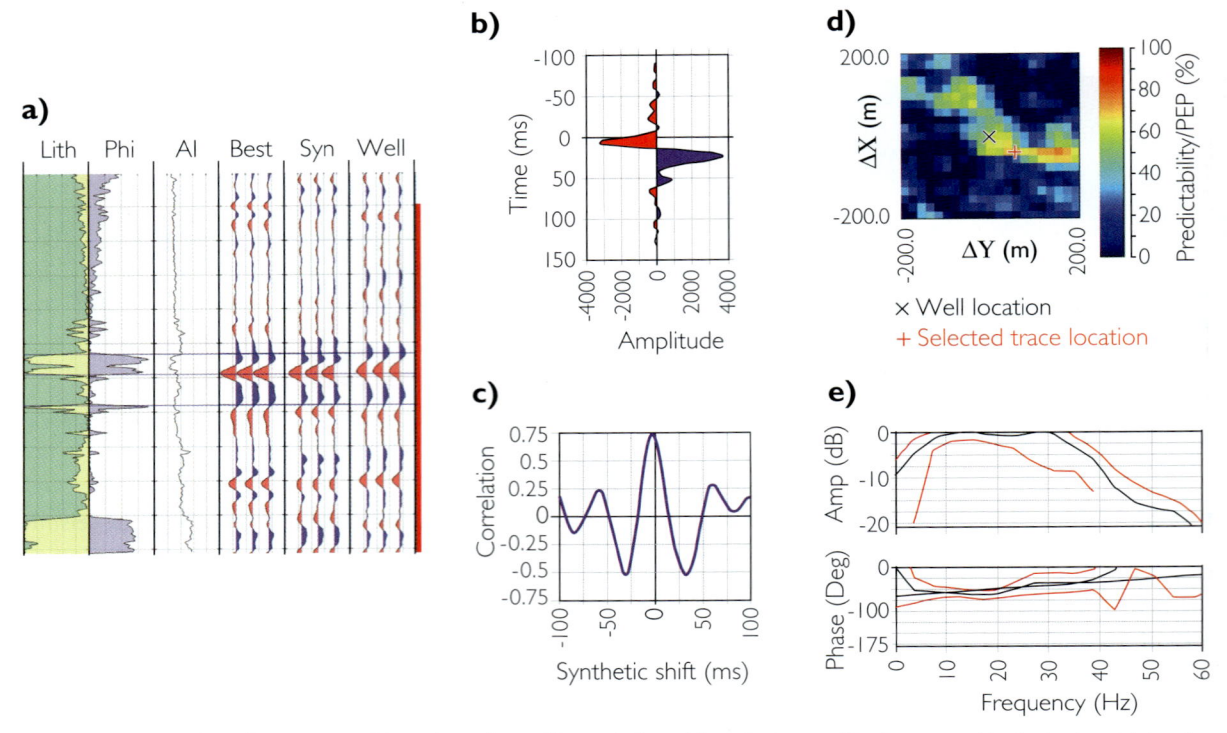

Figure 4.14 Example well tie generated from the well matching technique; (a) synthetic trace ('Syn') compared to the best match location trace ('Best'); red line defines the time segment for the analysis, (b) extracted wavelet, (c) cross-correlation over 500 ms time segment, (d) *PEP* map showing well location and best match location, (e) wavelet spectra.

Based on the well tie in Fig. 4.14 a phase shift of +70° and +16 ms time shift was applied to the seismic. A subsequent well tie now shows the extracted wavelet being generally symmetrical (Fig. 4.15) and the top and base of the low impedance sand are identifiable as a trough and peak respectively. The asymmetry in the wavelet may be due to the incorporation of noise into the wavelet estimation and it is questionable as to whether this should be included for example in a zero phase operator design. The wavelet has been edited to give the zero phase equivalent of the amplitude spectrum and the resulting tie is shown in Fig. 4.16. There is clearly no degradation in the tie and the modelled AVO gather is unambiguous in the responses expected from the principal lithological boundaries.

Figures 4.17 and 4.18 show how the adaptive technique might work with this dataset. The same time–depth relationship (based on checkshots) was used as in the previous figures. It is evident that the top of the main sand appears to be represented by a peak on the original dataset (Fig. 4.17a). On the assumption that this might indicate that the data has negative standard polarity an initial tie can be made with a 180° wavelet (with constant phase) derived from the estimated amplitude spectrum (Fig. 4.14e). The cross-correlation is 0.65 and at this point the interpreter might be happy enough with the result. However, the cross-correlation plot shows significant asymmetry (Fig. 4.17c), indicating that the wavelet is unlikely to be zero phase. Rotating the phase to −50° and time shifting the synthetic relative to the seismic gives the tie in Fig. 4.18, with an improved cross-correlation of 0.71. As would be expected, it is a similar tie to that shown in Fig. 4.14 based on the well matching technique.

In most situations similar results are obtained using the well matching technique and the adaptive technique. It might be argued, however, that when good time–depth data are available the well matching technique offers greater flexibility in evaluating the data as well as describing the improvements in a tie following log editing (Chapter 8) or seismic conditioning (Chapter 6).

A well tie example

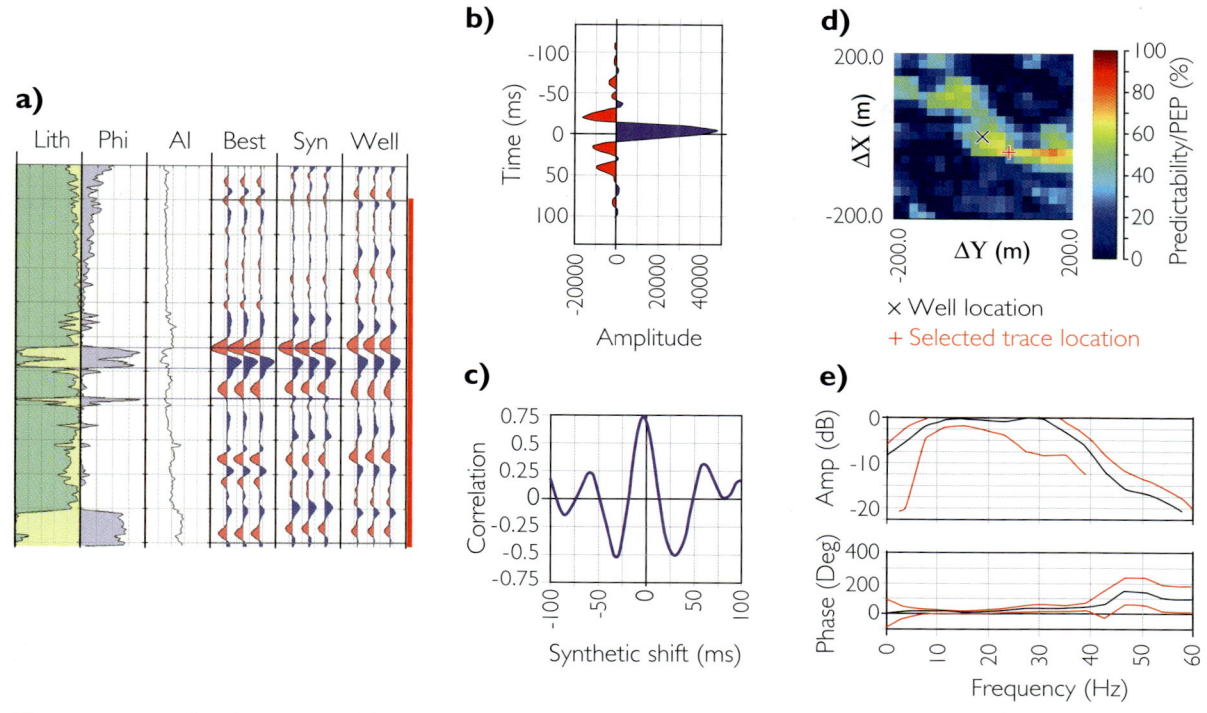

Figure 4.15 Example well tie generated from the well matching technique after 70° phase rotation and +16 ms time shift; (a) synthetic trace ('Syn') compared to the best match location trace ('Best'), (b) extracted wavelet, (c) cross-correlation over time segment, (d) PEP map showing well location and best match location, (e) wavelet spectra.

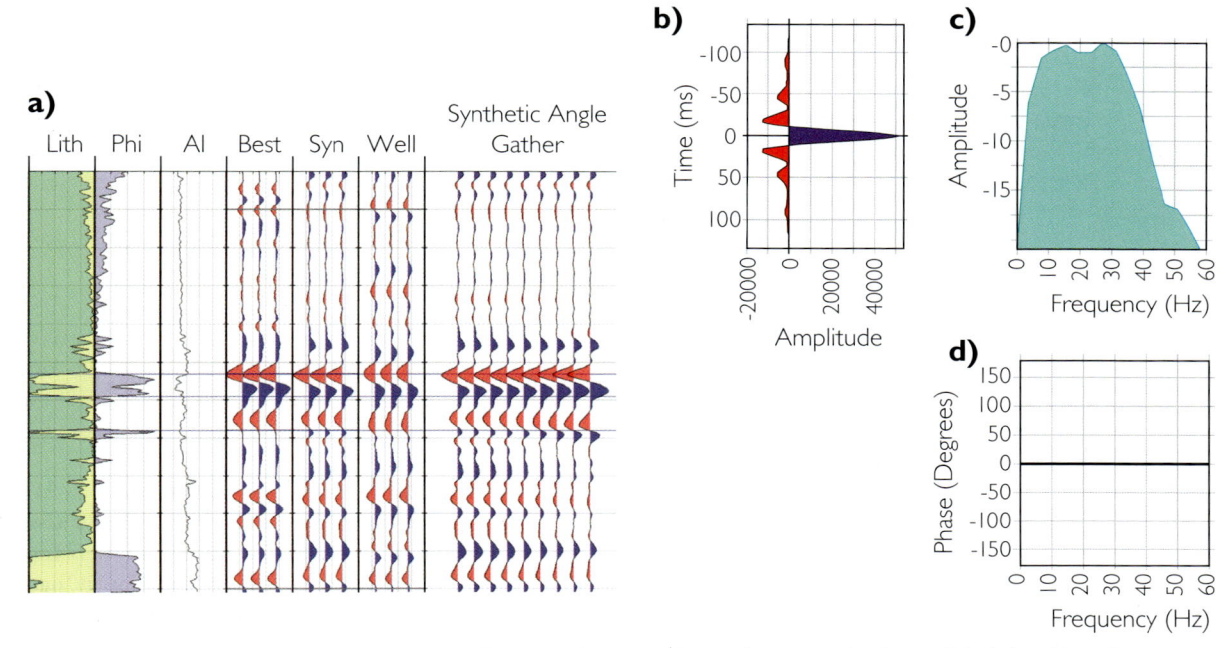

Figure 4.16 Example well tie generated using modified (zero phase) wavelet on phase rotated and time shifted data; (a) synthetic trace ('Syn') compared to the best match location trace ('Best') and synthetic gather (0–40°), (b) zero phase wavelet, (c) amplitude spectra, (d) phase spectra.

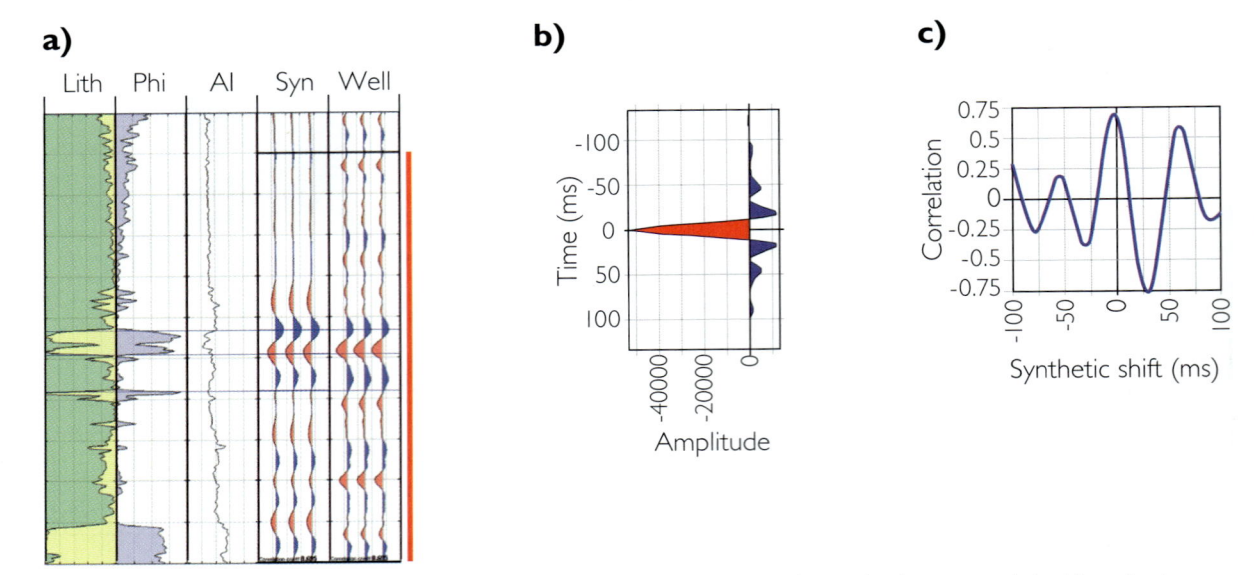

Figure 4.17 Example well tie generated using 180° constant phase wavelet based on the amplitude spectrum derived from the data; (a) synthetic trace ('Syn') compared to the well location trace ('Well') and, (b) 180° wavelet, (c) cross-correlation over 500 ms window.

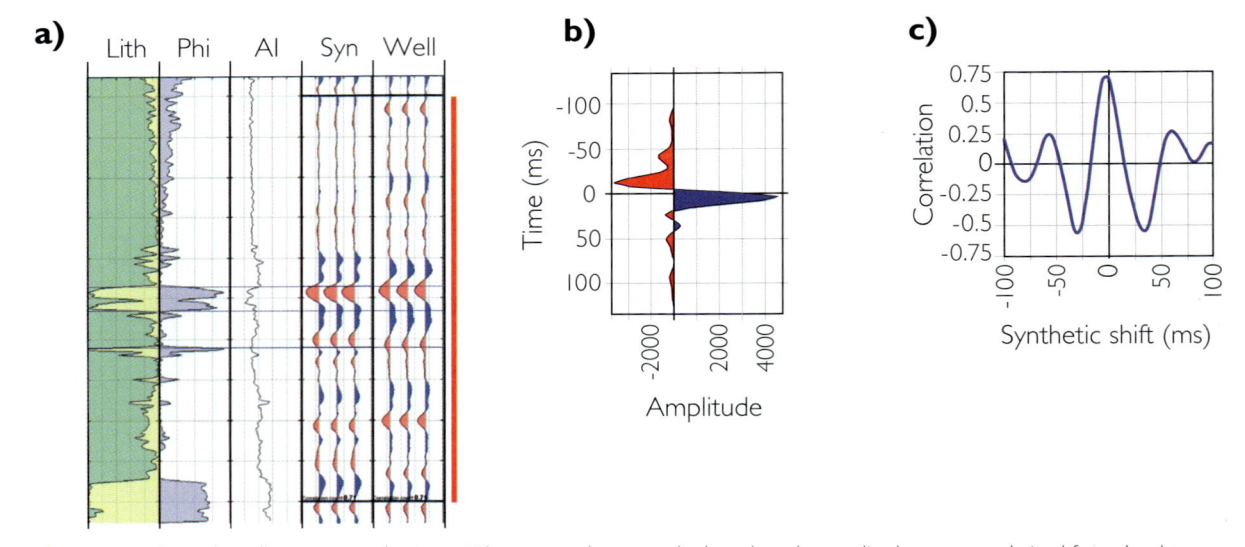

Figure 4.18 Example well tie generated using −50° constant phase wavelet based on the amplitude spectrum derived from the data; (a) synthetic trace ('Syn') compared to the well location trace ('Well') and, (b) −50° wavelet, (c) cross-correlation over 500 ms window.

4.6 Well tie issues

4.6.1 Seismic character and phase ambiguity

In situations where the seismic data show a distinctive character, there is often little ambiguity about the phase of the data. Figure 4.19 shows an example of a very good tie ($PEP = 83\%$, cross-correlation $= 0.91$) where the seismic can be tied only with a zero phase wavelet. Indeed, it probably wouldn't matter too much which wavelet was used; so long as it had an appropriate bandwidth and was zero phase a very good tie would result in this case. In general, this type of clarity depends on the presence of a number of strong reflections from layers with varying

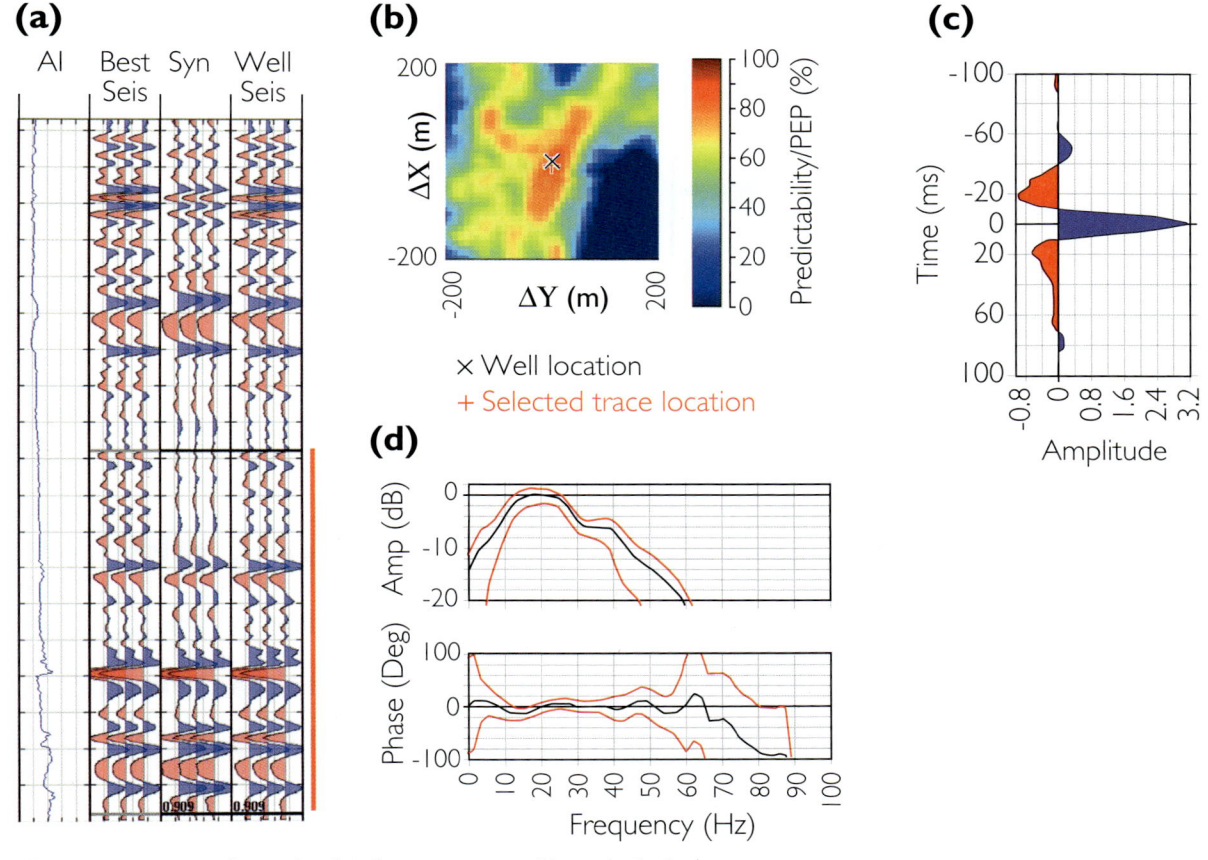

Figure 4.19 Exceptionally good well tie from a sequence with good seismic character.

properties and thicknesses. High data bandwidth and good signal/noise are very important.

In contrast, there are other situations where there is significant phase ambiguity, due to low bandwidth and monotonous seismic character. Phase ambiguity means that data can be tied equally well with wavelets that have different phase and timing. Figure 4.20 shows an example in which equally good ties can be made with wavelets that have almost 90° phase difference. The main problem here is cyclical geology. Ties can be made above and below this zone with much less phase ambiguity.

4.6.2 Stretch and squeeze

A common reason for stretching and squeezing synthetics is to generate background models for seismic inversions so that the inversion results fit the well. The interpreter needs a degree of scepticism about modifying the depth–time relationship and should investigate the possible reasons for the mistie (such as structural imaging), as well as the implied velocity changes, before deciding to make adjustments. Figure 4.21 shows an example of a mistie that appears to be resolvable with a simple stretch. The up-dip tie (Fig. 4.22) suggests that the underlying problem lies in the migration of the seismic data. In most instances reworking the migration is not an option so an ad-hoc solution is to change the well location or apply a stretch. Either way, events other than the two strong reflections tie quite poorly.

Of course, the quality of the checkshot data that were used to construct the synthetic is a major factor. If there are none, or if they are of dubious quality and best ignored, then it would certainly be justifiable to apply a stretch to the synthetic to allow for dispersion effects on velocity between sonic and surface seismic frequencies. Application of a squeeze would require

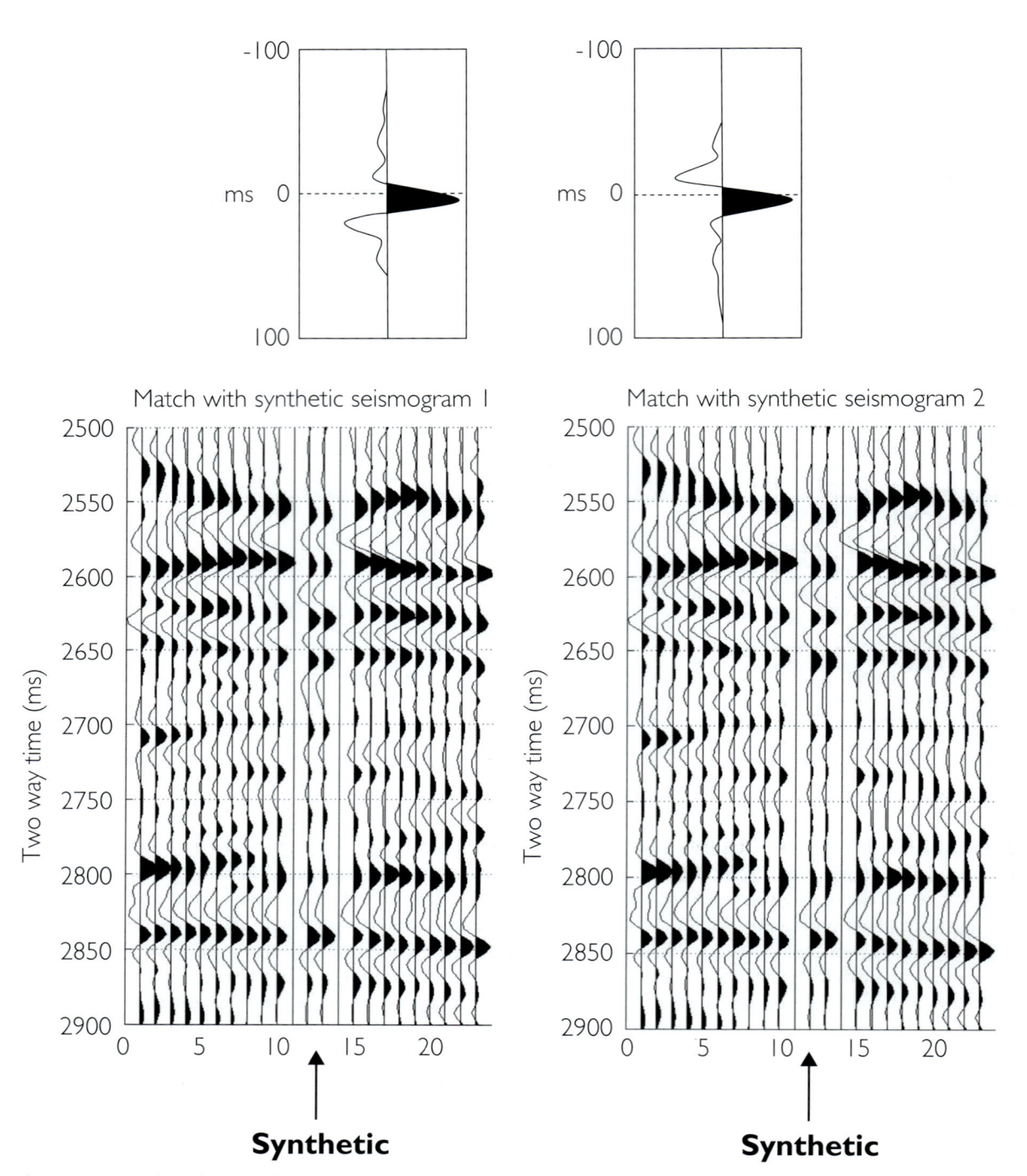

Figure 4.20 Wavelet ambiguity with poor-bandwidth data. The phase of the wavelet on the left is ~30° whereas the phase of the wavelet on the right is ~−60° (after White and Simm, 2003).

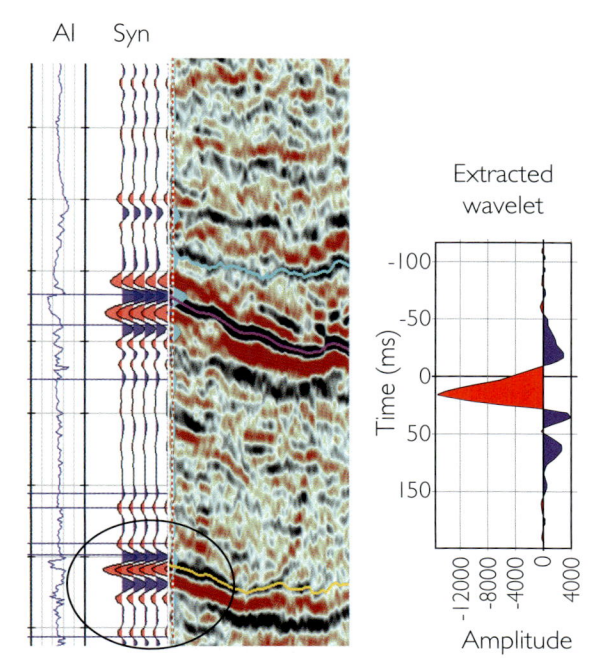

Figure 4.21 A well tie where the synthetic apparently needs to be stretched (seismic timing lines have 100 ms separation).

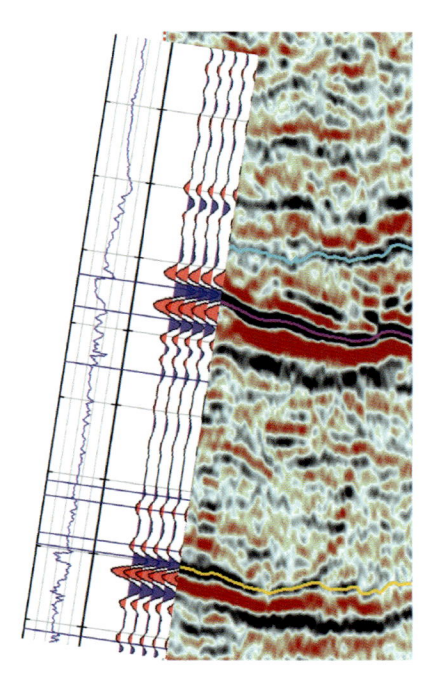

Figure 4.22 Stretch or residual migration effect? This is Fig. 4.21 with the synthetic rotated to achieve the tie (seismic timing lines have 100 ms separation).

more justification, perhaps to account for poor log quality. Where there are high-quality checkshots, perhaps from a vertical-incidence VSP, it is still possible for there to be residual timing differences between synthetic and seismic related to the way that arrival times are picked. If the onset of the wavelet (the *first break*) is picked, then the effect of absorption is to make the travel time less than it would be if picking is based on the first peak of the waveform (Fig. 4.23). This effect is usually not large, amounting to only a few milliseconds.

What certainly has to be avoided is a loop-by-loop stretch and squeeze, which can be made to produce an excellent but spurious tie. At this level of detail, there are many reasons why the tie will not be perfect, including issues with both well log and seismic data. If available, a VSP may help to identify the underlying problems, but in many cases the exact cause of a mistie will remain unknown; its presence is a warning against trying to over-interpret the data.

4.6.3 Sense checking and phase perception

The interpreter should not necessarily believe all the results of well matching or indeed any other well tie approach. Geological sense gained from the interpretation should be brought to bear on the choices made in well ties. Figure 4.24 shows an example where the wavelet extracted is peak-dominated and the visual tie is shown. The character match is quite good, though relative amplitudes of events are not well matched. However, other wells in the area are characterised by trough-dominated wavelets. The synthetic tie with the trough-dominated wavelet (extracted from other wells in the area and consistent with polarity of major sequence boundaries) generally looks better in terms of matching peak and trough amplitudes (Fig. 4.25), except for the highlighted high-amplitude (red) trough at a major unconformity. It is probable that this has biased the wavelet estimation in Fig. 4.24. The reasons for this are not clear but it may be that a short-period interbed multiple has modified the appearance of the reflection at the unconformity.

In some circumstances distinctive geological features might be used to infer the phase of the data ('zero phaseness' according to Brown, 2004). For example, a gas sand wedge model, based on a well log and added noise, shows a variety of signatures as a result of constant phase rotations (Fig. 4.26). At the thin end of the wedge the zero phase and 30° phase

Well to seismic ties

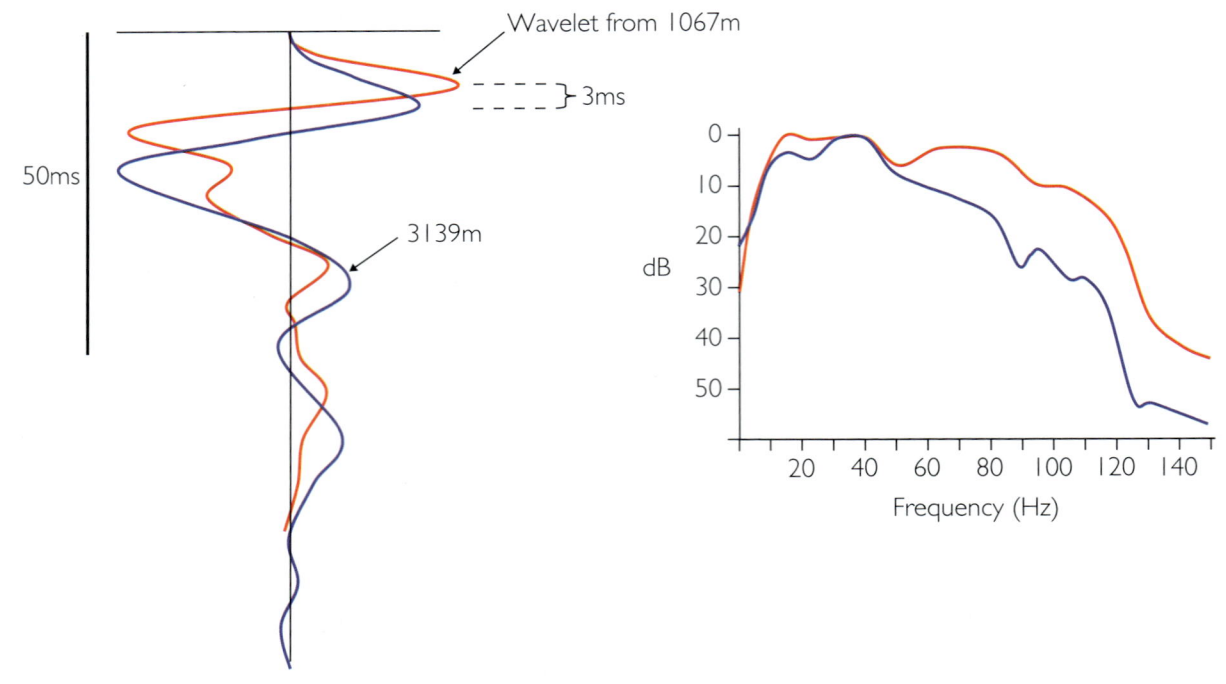

Figure 4.23 Two wavelets from different depths in a VSP, aligned on the first break.

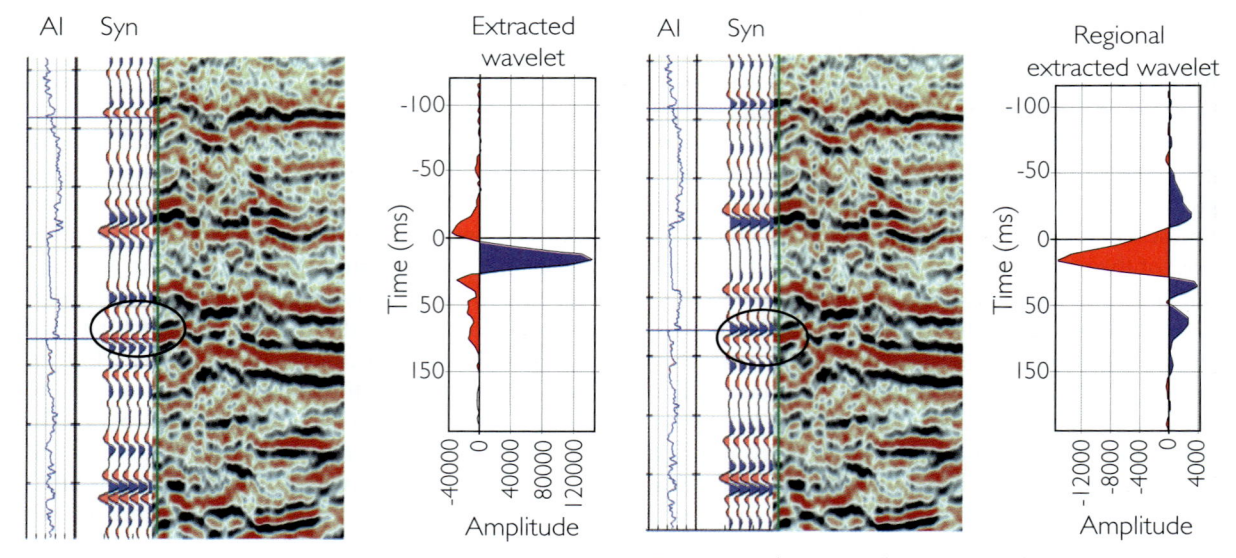

Figure 4.24 Visual well tie made with peak-dominated zero phase wavelet (seismic timing lines have 100 ms separation).

Figure 4.25 The same well tie as Fig. 4.24 but with a trough-dominated wavelet (seismic timing lines have 100 ms separation).

pictures show the expected trough/peak (red/blue) signature of a low-impedance thin bed. This contrasts with the 60° and 90° pictures that have a red/blue/red signature as well as a wedge-shaped geometry owing to the interference of the blue peaks. The difference between the 0° and 60°cases is clear, but the difference between the 0° and 30°cases is subtle. A general rule of thumb is that a phase rotation of 30° is about the least

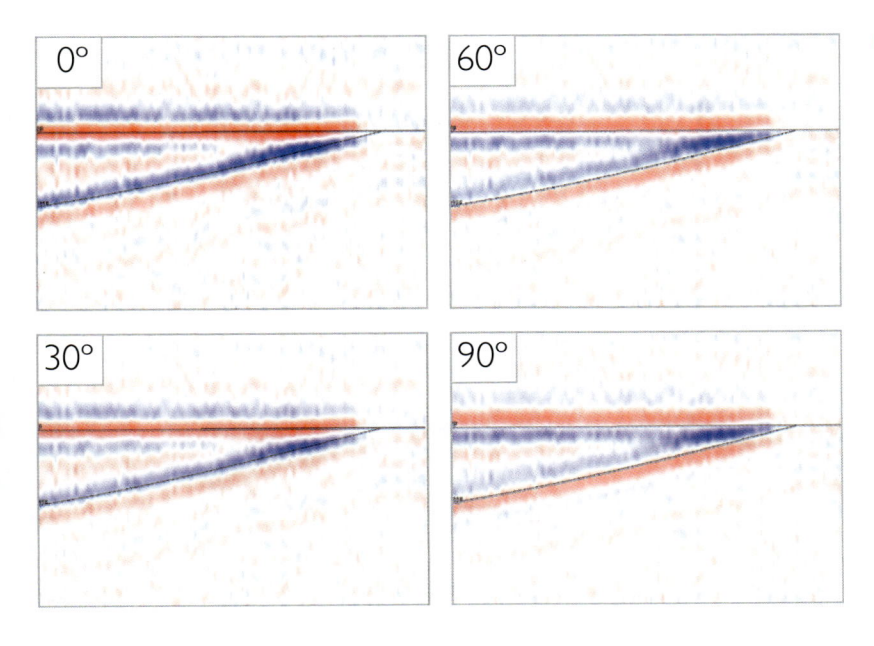

Figure 4.26 Wedge models of a gas sand with various constant phase rotations.

that is required to make changes to a seismic display that are apparent to the naked eye.

When there is no well control the problem of determining wavelet phase becomes much more difficult. In principle the answer is to look at the response of reflectors of known impedance contrast, such as the (hard) seabed or the tops of thick (hard) igneous intrusions or thick (hard) carbonates overlain by shales. A problem in using the seabed is that it is usually well above any target of interest, so the phase of the wavelet may change significantly between the seabed and the target due to absorption effects (Chapter 3). Lithological contrasts are also problematic, as the reflection from the top can easily be modified by tuning effects in thin layers. Assuming that the top reflection is truly the signature of a single interface, then a method of phase rotation and measurement of amplitude may give a clue to phase (Fig. 4.27). The idea is that the phase rotation that gives the highest amplitude produces seismic data that are close to zero phase. Of course, once phase has been addressed in this way there is still the additional problem of inferring polarity using available geological knowledge.

Maximum reflection stand-out is also the basis for wavelet extraction methods that use higher order statistics. Given that the Earth's reflectivity is non-Gaussian (Walden and Hosken, 1986) one approach is to rotate the phase of the seismic, with a constant

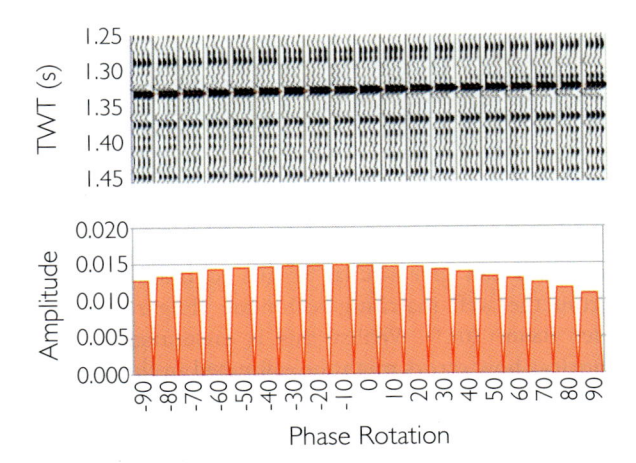

Figure 4.27 Method of determining phase through phase rotation and amplitude measurement (after Roden and Sepulveda, 1999). Amplitudes are measured on the peak at ~1.33 s (considered to be a unique reflector) at phase rotations from −90° to 90° relative to the initial phase of the seismic.

phase assumption, and measure the kurtosis (or deviation from gaussianity) (e.g. Levy and Oldenburg, 1987). The wavelet phase is the opposite (i.e. 180° difference) of the phase rotation that maximises the kurtosis. Early application of the maximum kurtosis technique suffered from drawbacks such as bias toward large reflections and instablility with poor bandwidth data (e.g. van der Baan and Pham, 2008).

Well to seismic ties

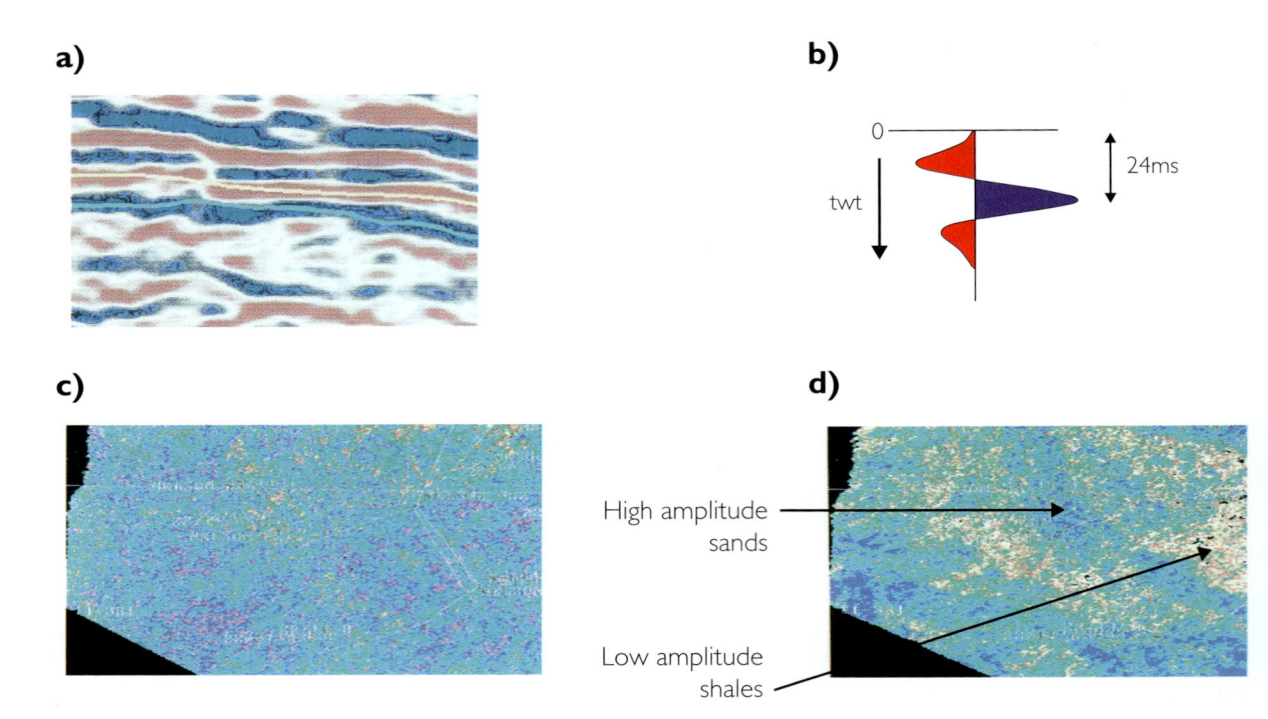

Figure 4.28 Which loop to pick? It can matter (after White and Simm, 2003): (a) seismic section showing trough and peak picks, (b) the seismic wavelet, (c) amplitude map of red trough (minimum value in short window around yellow horizon), (d) amplitude map of blue peak (maximum value in short window around light blue horizon).

However, with modern hi-fidelity data and advances in algorithm design it has been shown that statistical methods can give similar results to those obtained from the well matching technique (Edgar and van der Baan, 2011). Whilst this is encouraging, it is unclear how errors might be estimated in the absence of wells (i.e. to determine the level of confidence in the results). As yet, statistical methods are not routinely available in commercial software packages but it is evident that they could prove useful for quality control purposes.

4.6.4 Importance of tie accuracy in horizon mapping

Figure 4.28 shows an example where understanding the polarity of the data is vital to identifying the loop that carries the amplitude information about the top reservoir. Lithological variations at the top of the reservoir are evident on the map generated from the blue peak pick (Fig. 4.28d) but not on the map of the red trough above (Fig. 4.28c). The red trough is not simply a side lobe of the peak; it interferes with the side lobe of the top of the sealing formation. Accurate understanding of the wavelet shape is necessary to make the correct interpretation.

4.6.5 Understanding offset scaling

A useful aspect of well matching techniques is that not only is the likely wavelet shape identified but also the scaling of the wavelet is calculated. If the amplitudes of near and far stacks have been balanced correctly (i.e. to be consistent with the amplitude variation in the well-based angle synthetic), then wavelet extractions of each stack should give not only similar wavelet shapes but also the same scaling. Figure 4.29 shows an example where a difference in scaling between near and far stacks was identified through wavelet extraction. Prior to AVO analysis a scalar would be applied to correct the balancing. Many inversion software packages apply such corrections semi-automatically, but it is still a good idea to check the scaling manually as gross differences between actual seismic and well-based expectations may be a pointer to problems in the seismic processing sequence.

Well tie issues

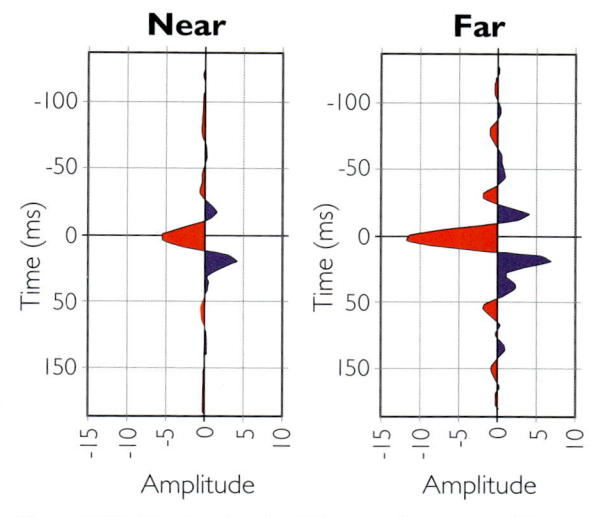

Figure 4.29 Wavelets showing different scaling extracted from near and far seismic cubes.

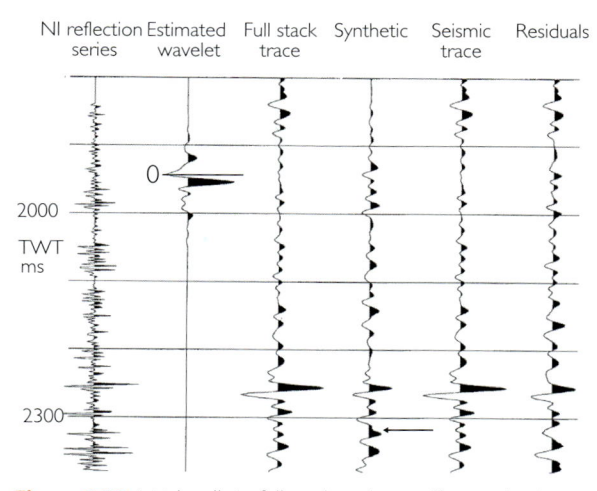

Figure 4.30 Initial well tie: full stack and zero offset synthetic (after Simm et al., 1999).

4.6.6 Use of matching techniques to measure an improving tie

Historically, it was common practice to tie migrated full stack seismic sections to zero offset synthetics, usually because only a full stack dataset was available and a shear velocity log had not been acquired. It is generally accepted now that the reflectivity series used in well matching should be calculated at the effective angle of the stack (Chapter 2). Figure 4.30 shows an example (from Simm et al., 1999) where an initial tie was made to full stack data using a zero offset reflectivity series, although the top reservoir is characterised by Class IIp reflectivity. The tie is not a good one: *PEP* = 43% and estimated phase error is 16°. The top reservoir is not clearly identified on the full-stack seismic. Subsequent re-processing of the seismic to create an intercept (i.e. zero offset) stack and log conditioning to remove the effects of fluid invasion resulted in the tie shown in Fig. 4.31. The tie is improved (*PEP* = 72% and estimated phase error is 9°), and the top reservoir is now clearly tied from synthetic to seismic.

It is evident from this discussion that well ties are a critical part of seismic interpretation and that

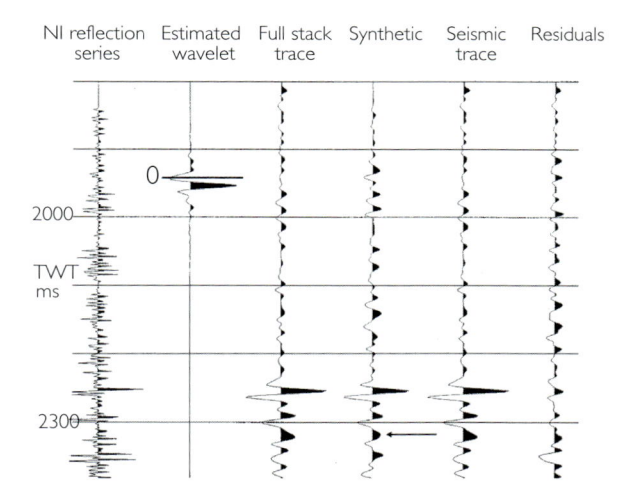

Figure 4.31 Improved well tie: intercept stack and zero offset synthetic calculated from invasion-corrected logs (after Simm et al., 1999). Note the arrow marks the top reservoir on the synthetic.

analytical approaches give valuable information to the interpreter. They also provide a useful tool in guiding seismic processing (Barley, 1985).

Chapter 5
Rock properties and AVO

5.1 Introduction

This chapter is an overview of how rock properties relate to seismic amplitude and AVO. Much of the information presented is the basis for further discussion of rock physics (Chapter 8), seismic inversion (Chapter 9) and approaches to reservoir characterisation (Chapter 10). The principal idea is to show the reader how different types of information fit together.

5.2 AVO response description

There are a number of terms that are generally used to describe AVO responses. Unfortunately, they can easily create confusion for the non-expert.

5.2.1 Positive or negative AVO and the sign of the AVO gradient

First of all, there is an important distinction to be made between 'positive (or negative) AVO' and positive (or negative) AVO gradient. As discussed in Chapter 2, the *AVO gradient* is calculated as the slope of the change in amplitude with $\sin^2\theta$. On the AVO plot, negative AVO gradients are inclined from upper left to lower right, whereas positive gradients are inclined from upper right to lower left (Fig. 5.1). On the other hand *positive AVO* (also sometimes called *rising AVO*) is a term used to describe the case where there is an increase in the absolute value of the amplitude with increasing offset (or angle). Thus, *positive AVO* is applied to either positive or negative amplitudes which are increasing in magnitude with angle. Accordingly, *negative AVO* describes the case where the absolute value of the amplitude decreases with increasing offset (or angle).

5.2.2 AVO classes and the AVO plot

At the next level of detail, AVO response is conventionally described in terms of a number of classes. Rutherford and Williams (1989) first classified shale/gas sand interface responses into three types (I, II and III). Class I responses are characterised by a positive impedance contrast (i.e. the sand impedance is larger than the shale impedance), together with a negative AVO gradient, so that the reflection coefficient is positive and decreases with angle. Class II responses have small normal incidence reflection coefficients (which may be positive or negative) and a negative gradient so that the AVO effect leads to large negative reflection coefficients at far offsets. Ross and Kinman (1995) suggested that the small positive normal incidence coefficient Class II responses should be termed Class IIp, owing to the phase reversal that is inherent in the response, and that the term Class II should be reserved for the small negative normal incidence coefficient case. Class III responses have large negative impedance contrasts and a negative gradient, leading to increasing amplitude with angle (Fig. 5.2). There appears to be no precise definition of where the boundary lies between Class II and Class III responses. It depends on what is meant by 'small' normal incidence reflection coefficient in the definition of Class II. Typically, Class II might be applied to the case where the amplitudes on the nearest traces of the moveout corrected gather have virtually no amplitude. A further class of AVO response, Class IV, was introduced by Castagna and Swan (1997). This has a large negative normal

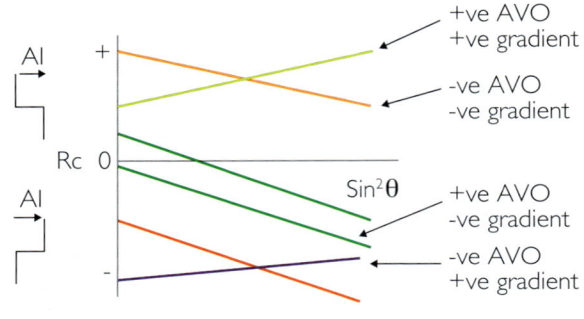

Figure 5.1 Some AVO terminology.

AVO response description

Figure 5.2 The AVO classes (modified after Rutherford and Williams, 1989; Ross and Kinman, 1995; Castagna and Swan, 1997).

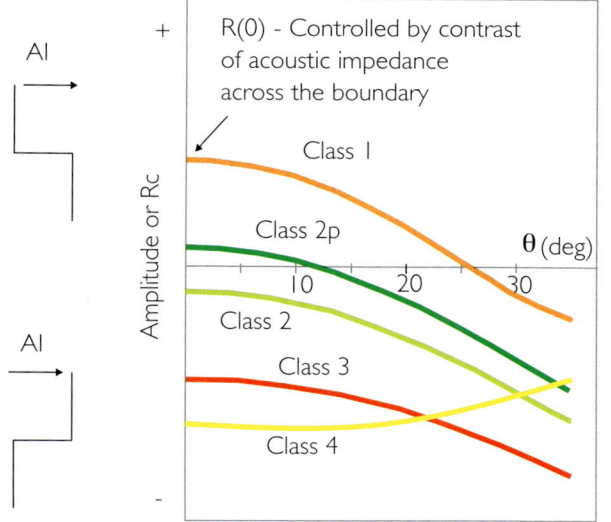

Class	Gradient	Absolute Amplitudes	Notes
I	−	Far < R(0)	+ve R(0), may have phase reversal
IIp	−	Far > R(0)	+ve R(0), phase reversal
II	−	Far > R(0)	Low amplitude at R(0)
III	−	Far > R(0)	Relatively high amplitude at R(0)
IV	+	Far < R(0)	Very high amplitude at R(0)

incidence reflection coefficient and overall decreasing amplitude with offset.

As previously discussed in Chapter 2, the first order control on the intercept is the acoustic impedance contrast. On the other hand, the sign and magnitude of the AVO gradient is determined principally by the contrast of shear velocity across the boundary (Castagna et al., 1998; Castagna and Smith, 1994). Negative gradients are associated with a positive shear velocity contrast (i.e. lower shear velocity in the upper layer) whilst positive gradients are associated with negative shear velocity contrast (i.e. higher velocity in the upper layer).

Given the origins of the AVO classes it could be argued that they should only be applied to hydrocarbon sands. Clearly this is restrictive and most workers tend to use the class definitions simply as descriptors of AVO behaviour. The AVO classes are a useful general description but in practice it is often necessary to look at AVO plots from gathers to understand the response in more detail.

5.2.3 Introducing the AVO crossplot

Whilst the AVO plot shown in Fig. 5.2 is a useful way of visualising AVO responses it is generally of little use in analysing the multitude of responses from seismic data. These limitations are overcome by plotting the intercept and gradient of each response as a single point on an *AVO crossplot* (e.g. Foster et al., 1993; Castagna and Swan, 1997; Sams, 1998; Smith 2003). Figure 5.3 illustrates how the various AVO classes described above occupy different areas of the plot. The AVO crossplot is an important tool for understanding lithology and fluid discrimination and this is discussed in more detail in Section 5.5 and Chapter 7.

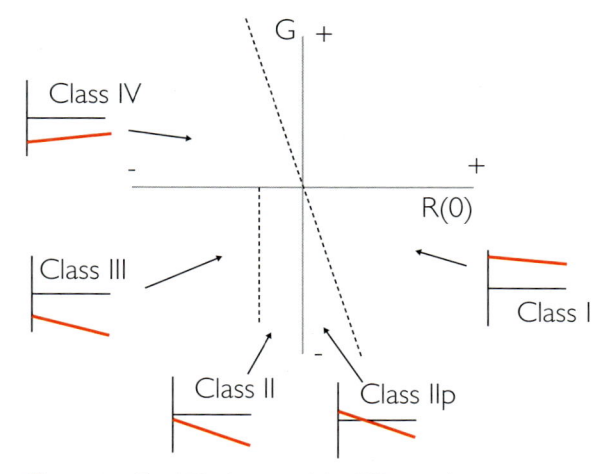

Figure 5.3 The AVO classes and the AVO crossplot.

5.2.4 Examples of AVO responses

Figure 5.4 shows some real examples of the different classes of AVO response. It should be noted that there is nothing about these responses that allows us to

Rock properties and AVO

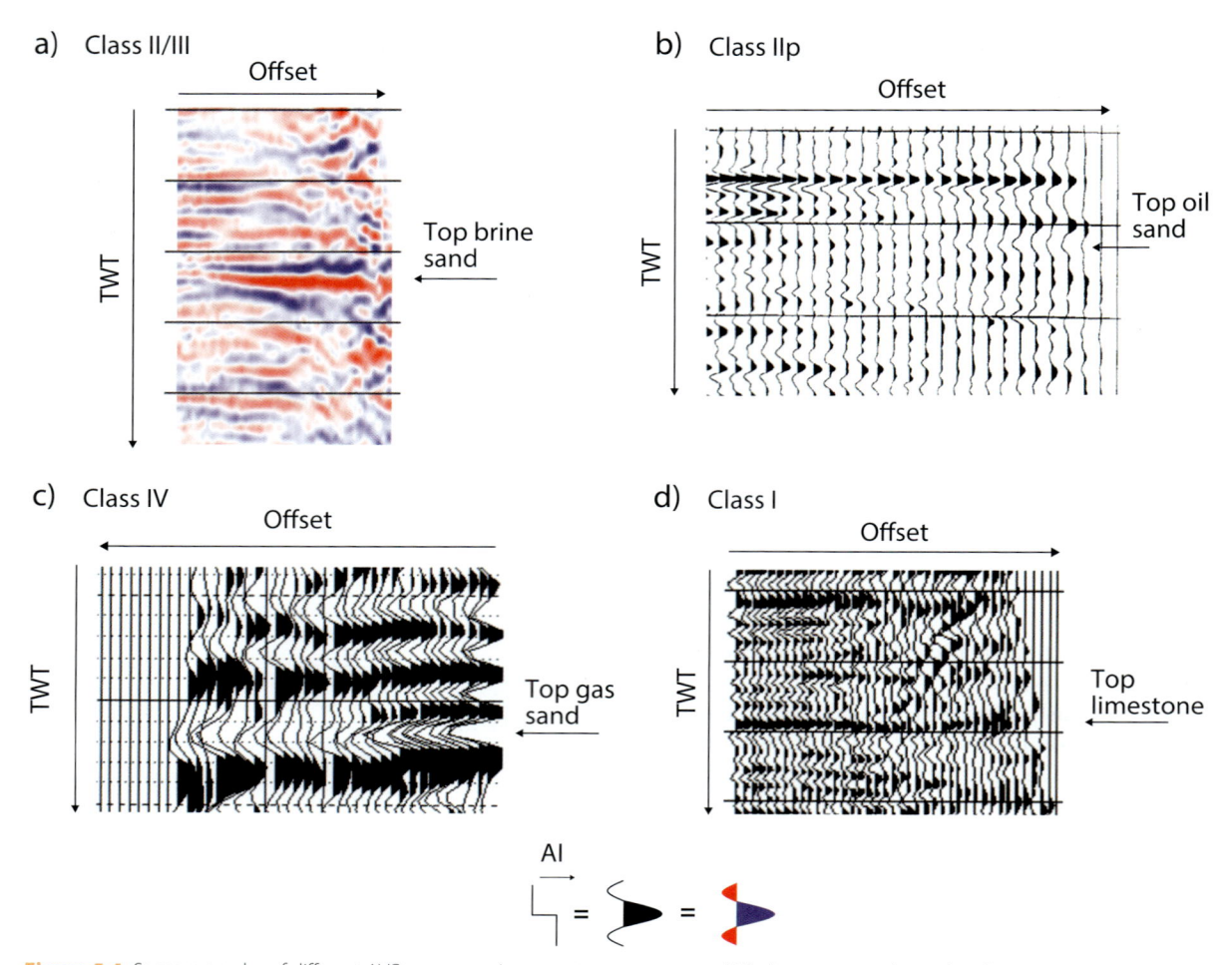

Figure 5.4 Some examples of different AVO responses. In current common usage AVO classes are simply used as descriptors – only in specific circumstances may they have a diagnostic significance. Note that all examples have the same polarity; (a) Class II/III, (b) Class IIp, (c) Class IV, (d) Class I.

generalise about their significance. It is often believed that 'rising AVO means pay', possibly because of the descriptions in Ostrander's (1984) original paper on AVO, but this is clearly refuted by the brine sand response in Fig. 5.4a. Clearly the geological context is paramount in AVO analysis. To assess the significance of different AVO responses the interpreter needs to make models for different scenarios (for example by varying fluid fill and/or porosity) and compare the results with seismic observations. It is usually not the class of response that is significant, but the relative change in AVO response, for example from an up-dip to a down-dip location or a change in a target reflector in comparison to a background reflection. Examples of different AVO scenarios are given in Chapter 7.

The AVO classes as published describe AVO responses only for the top sand interface. In terms of a sand/shale system, reflections associated with the base of sands and hydrocarbon contacts typically have AVO responses with positive gradients. Base of brine sand responses can have various styles of response including Class IV, or a phase reversal from negative to positive amplitude with increasing offset or indeed a positive reflection with increasing amplitude with offset (but in this case the gradient is usually quite low) (Fig. 5.5).

Positive AVO with a positive intercept is generally characteristic of modelled hydrocarbon contacts (Fig. 5.5). This can be a good diagnostic, as lithological changes with positive intercept usually have a negative gradient (i.e. Class I). In real data, however, contacts may not have strong AVO characteristics owing to interference of the fluid boundary reflections with bedding reflections or the presence of

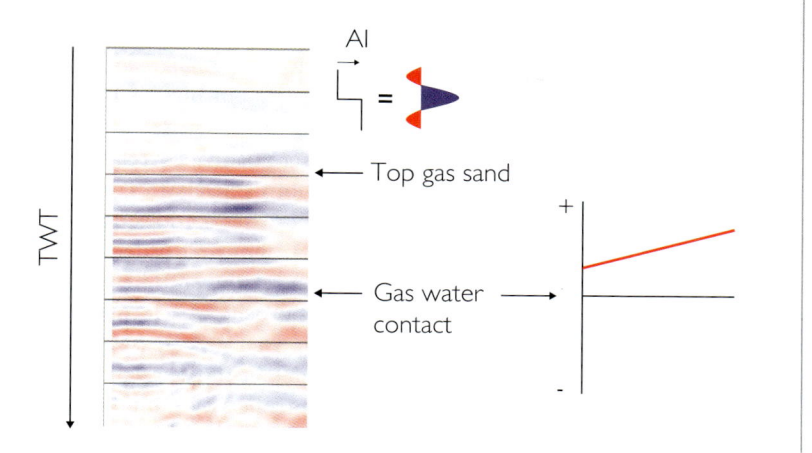

Rock property controls on AVO

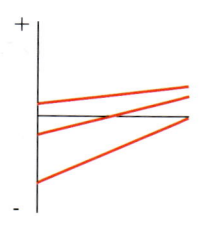

Figure 5.5 Examples of (a) hydrocarbon contact and (b) base sand AVO responses.

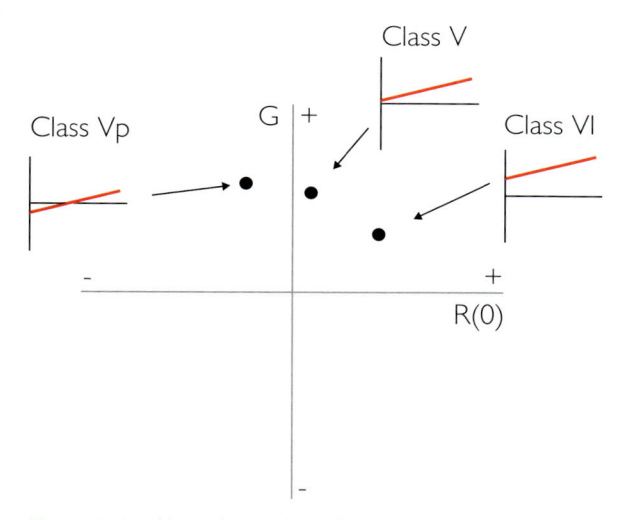

Figure 5.6 Additional AVO classes for responses with positive gradients (such as base of sands and hydrocarbon contacts).

amplitude decay with offset unrelated to the target geology (see Chapter 7). It is possible using the AVO crossplot to extend the descriptive scheme of AVO classes to cover AVO responses with positive gradients (i.e. Classes Vp, V and VI) (Fig. 5.6).

5.3 Rock property controls on AVO

Each sedimentary basin has a particular style of deposition, compaction, diagenesis, hydrocarbon generation and fluid flow that gives rise to seismic amplitude signatures. It is not the intention here to give a detailed description of how basin processes affect the elastic parameters of rocks and detailed styles of AVO behaviour, but rather to develop a general discussion to draw the interpreter into thinking about how geology affects the seismic response. For each basin the seismic interpreter needs to work closely with the petrophysicist, geologist and engineer to formulate knowledge of how the petroleum system is characterised and how this knowledge can be exploited in seismic analysis.

5.3.1 Ranges of parameters for common sedimentary rocks

The sedimentary rocks of most interest in exploration are siliciclastics (i.e. sands and shales) and various forms of carbonate deposits. Further definition of these terms will be given where relevant but it is worth stating that the term 'shale' is used to describe a non-reservoir rock with low permeability characterised predominantly by clay minerals (usually greater than 60% by volume) but also containing silt and electrostatically bound water (e.g. Katahara, 2008).

Chapter 2 highlighted the importance of compressional velocity, shear velocity, bulk density and the composite terms of acoustic impedance and Poisson's ratio to the modelling and interpretation of AVO responses. Figure 5.7 shows some generalised crossplots (largely based on data from Castagna et al., 1993) which illustrate how these parameters vary in different sedimentary rocks.

Rock properties and AVO

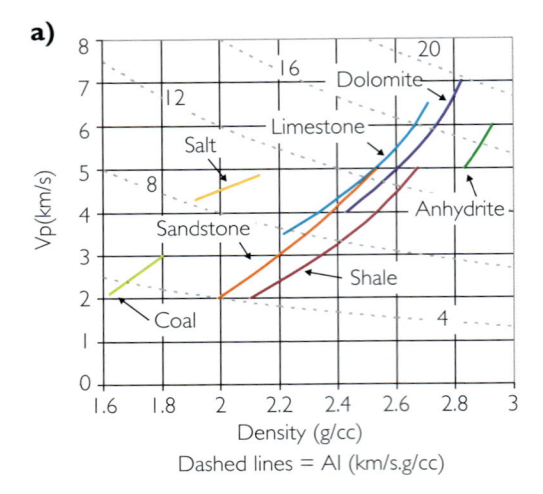

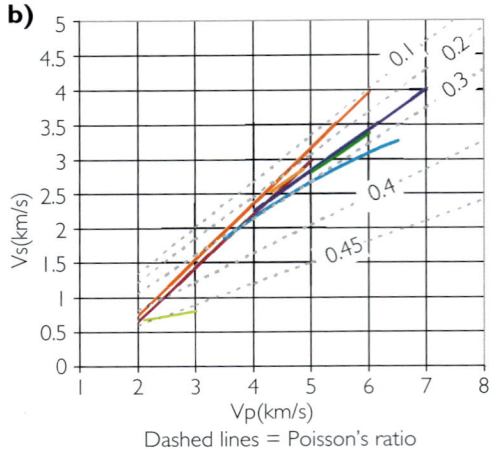

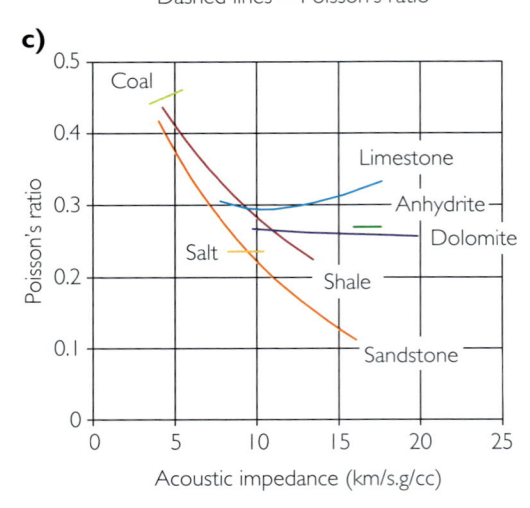

Figure 5.7 Ranges of parameters for common sedimentary rocks (brine-bearing), based on data in Castagna et al. (1993): (a) density vs P wave velocity (dashed lines = AI (km/s.g/cc), (b) V_p vs V_s (dashed lines = Poisson's ratio), (c) acoustic impedance vs Poisson's ratio.

Several comments can be made on these plots.

- The various rock types have a large degree of overlap in velocity and density (Fig. 5.7). In general, carbonates tend to have higher density and velocity than siliciclastics but high-porosity chalks overlap with sands.
- In general there is a strong positive correlation of velocity and density for both siliciclastics and carbonates. As velocity increases so does density. This gives rise to the important first order observation that there is a relationship between acoustic impedance and porosity. As porosity increases the acoustic impedance decreases. As will be seen, the specific details of the relationship for each lithology are controlled by mineralogy and pore geometry.
- Despite the overlapping nature of lithologies in terms of acoustic impedance there are significant differences in Poisson's ratio characteristics (Fig. 5.7) and these can be a basis for determining lithology from seismic.
- A few rock types are very distinctive, such as coal which has very low density and high Poisson's ratio. Coals will cause high-amplitude soft reflections that could potentially be confused with gas sands. Coals tend to generate Class IV AVO responses and this may prove a good discrimination in cases where gas sands have Class III AVO.
- The Poisson's ratio trends of siliciclastic and carbonate lithologies are quite different (Fig. 5.7). Sands and shales exhibit steep sub-parallel trends such that the Poisson's ratio of both sands and shales decreases with decreasing porosity (higher AI). Limestones and dolomites tend to have fairly flat (constant Poisson's ratio) trends on the AI–PR plot. Thus, the interpreter must be careful in applying rock physics rules of thumb developed on siliciclastic rocks to carbonate settings.

5.3.2 The role of compaction

The most important control on the elastic parameters of sedimentary rocks is porosity. During burial, rocks undergo both mechanical and chemical changes which give rise to a general decrease in porosity (and increase in density) with depth. Burial is also an important factor in the porosity and acoustic impedance of Chalks, but it could be argued that it is less important for other types of carbonates, for example where pore characteristics are intimately associated with depth independent diagenetic processes. The following discussion will focus on siliciclastic (sand/shale) lithologies.

Sands and shales are originally deposited as suspensions but become load bearing rocks upon burial. The depositional or *'critical' porosity* (i.e. the point at which there is no cohesion between grains) varies significantly between 36%–40% for sands to 60%–80% porosity in shales. Shales tend to de-water rapidly and, providing that it is possible for expelled water to escape to surface, generally follow an exponential decrease in porosity with increasing depth. Athy (1930) defined the relationship:

$$\phi = \phi_0 e^{-cz},$$

where ϕ_0 is the critical porosity, c is a constant and z = depth.

Similar functions have been derived from different basins (e.g. Ramm and Bjørlykke, 1994; Dutta *et al.*, 2009). Clean sands generally have a more linear relationship of porosity decrease with increasing depth. Mechanical compaction, characterised by grain rearrangement and crushing in sands and collapse of platy mineral structures in shales, is the dominant mechanism in the upper part of the section. These processes reduce porosity and stiffen the rock, increasing acoustic impedance and decreasing Poisson's ratio. Mechanical compaction gives way to chemical compaction largely through the influence of temperature. Generally at around 70–80°C smectite transforms to illite (which is harder than smectite) with the production of both water and silica. This change in the nature of shale is often observed to be associated with the onset of cementation of sands (Avseth *et al.*, 2008). Figure 5.8 shows an example of relative changes in compressional velocity between sands and shales related to both mechanical and chemical compaction.

Figure 5.9 shows schematic illustrations of several effects of compaction on clean sands and shales. Typically in the shallow parts of basins shales have higher *AI* than the sands, whereas in the deeper (more compacted) section the sands have higher *AI* than shales. Thus, at a certain depth the sand and shale impedance trends cross-over, with very low *AI* contrast between them. Above this depth, shales are harder than sands (Fig. 5.9a) and generally have a slightly higher Poisson's ratio than sands. Below the cross-over, the sands are harder than the shales and the high *AI* (low porosity) sands start to have markedly low *PR* values.

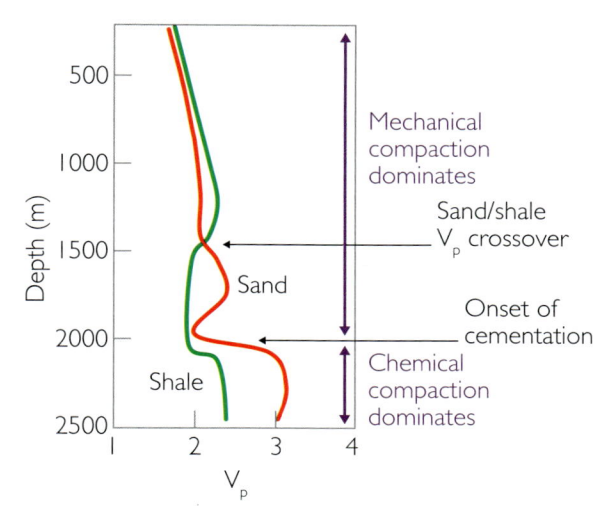

Figure 5.8 Generalised velocity–depth trends for sands and shales in the Palaeogene offshore Norway (modified after Avseth, 2000).

Thus for the shallow zone, a shale–sand interface is marked by a negative *AI* change and a modest negative *PR* change, giving rise to a negative zero offset reflection coefficient and a modestly negative AVO gradient (i.e. a Class III AVO response). For the deep zone, the shale–sand interface has a positive *AI* change and a more marked negative *PR* change than the shallow section, giving a positive zero offset reflection coefficient and a markedly negative AVO gradient (i.e. a Class I response). Class II and IIp responses will be found in the vicinity of the cross-over.

The schematic trends shown in Fig. 5.9 are seldom quite so clear in real data. Impedance crossovers may occur within a fairly restricted depth range or they may be an extensive zone hundreds of metres thick. Indeed there may be more than one crossover. Sedimentology and pressure variations are key factors in modifying the trends. For example, the presence of silts and carbonates (and other lithologies such as volcanic tuffs) can dramatically modify the impedance contrasts. Clearly, it is particularly important when analysing numerous wells for rock physics trends that detailed rock characterisation is done before averaging the data (e.g. Avseth *et al.*, 2008).

5.3.3 The effect of fluid fill

In general, the effect of replacing brine with hydrocarbon is to reduce the P wave velocity and the density of the rock, whilst having little effect on the shear velocity (see Chapter 8). The bulk modulus

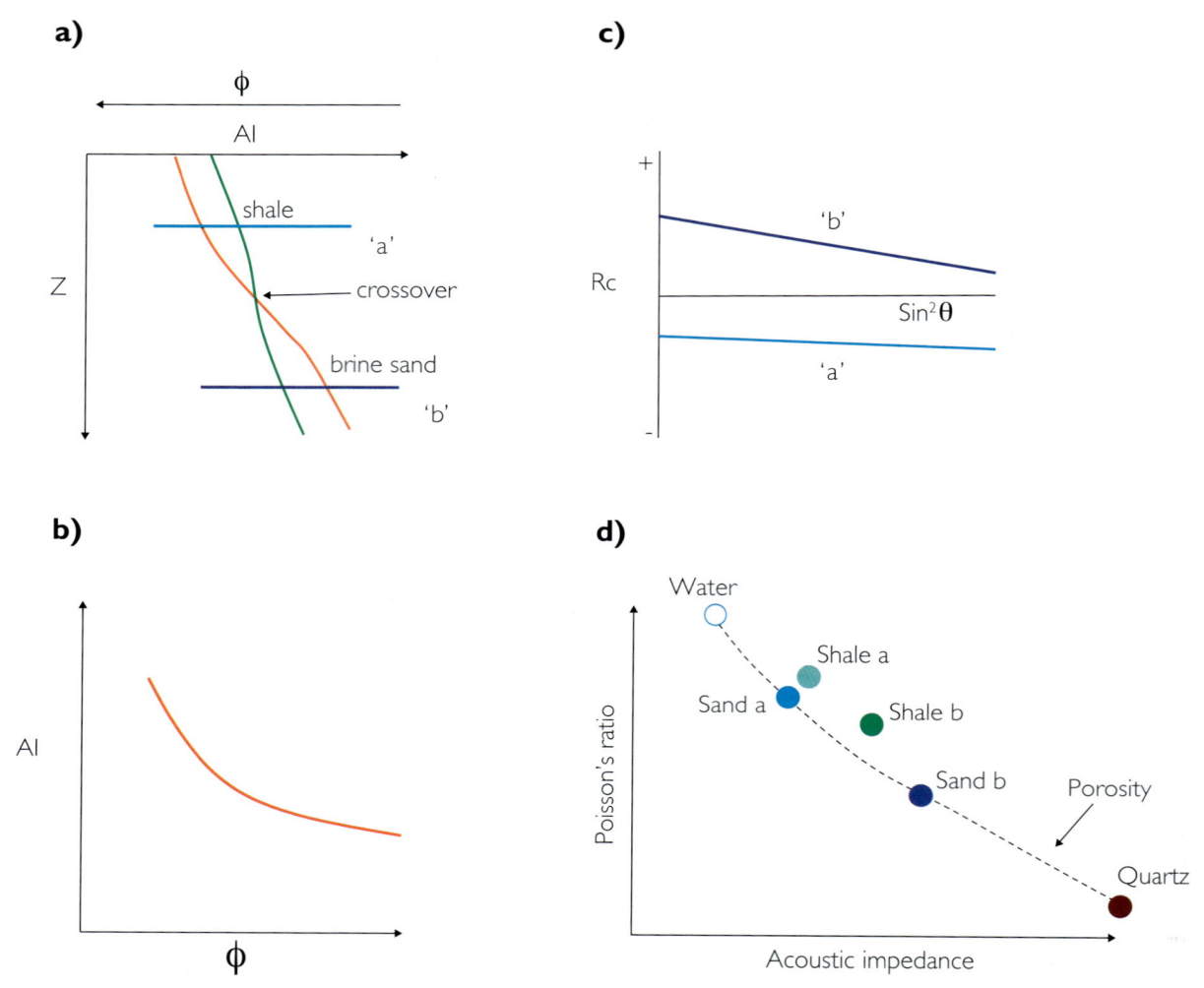

Figure 5.9 The effect of compaction on brine-filled sands and shales, (a) depth vs AI, (b) sand porosity vs AI, (c) AVO plot showing shale/sand AVO responses, (d) acoustic impedance vs Poisson's ratio.

(i.e. K_{sat}), and hence the P wave velocity, can be significantly affected by the replacement of relatively incompressible brine with highly compressible gas or condensate. The shear modulus of the rock is unaffected by fluid replacement because liquids have no rigidity. Since the density of the rock is decreased when hydrocarbons replace brine (hydrocarbons are invariably less dense than brine), the effect (following Eq. (2.11) in Chapter 2) is to increase the shear velocity slightly. Overall the effect of replacing brine with hydrocarbon in a rock is to reduce both the acoustic impedance and the Poisson's ratio. It was originally thought, based on the properties of stock tank oils, that oil would have limited or no effect on seismic amplitude (e.g. Gregory, 1977). However, the work of Clark (1992) and Batzle and Wang (1992) showed that live oil (i.e. oil with dissolved gas at reservoir temperature and pressure) can have significantly lower moduli and density than dead oil at surface temperature and pressure. Gas generally has a bigger effect on seismic amplitude than oil, though in deep basins light oils and condensates may have an effect similar to that of gas. Owing to the higher moduli of calcite and dolomite compared to quartz, the magnitude of fluid substitution effects tends to be significantly less for carbonates compared to siliciclastics.

Figure 5.10 shows a schematic illustration of the effect of replacing brine with gas for the sand in Fig. 5.9. It is evident that the reduction in AI and PR becomes less with decreasing porosity.

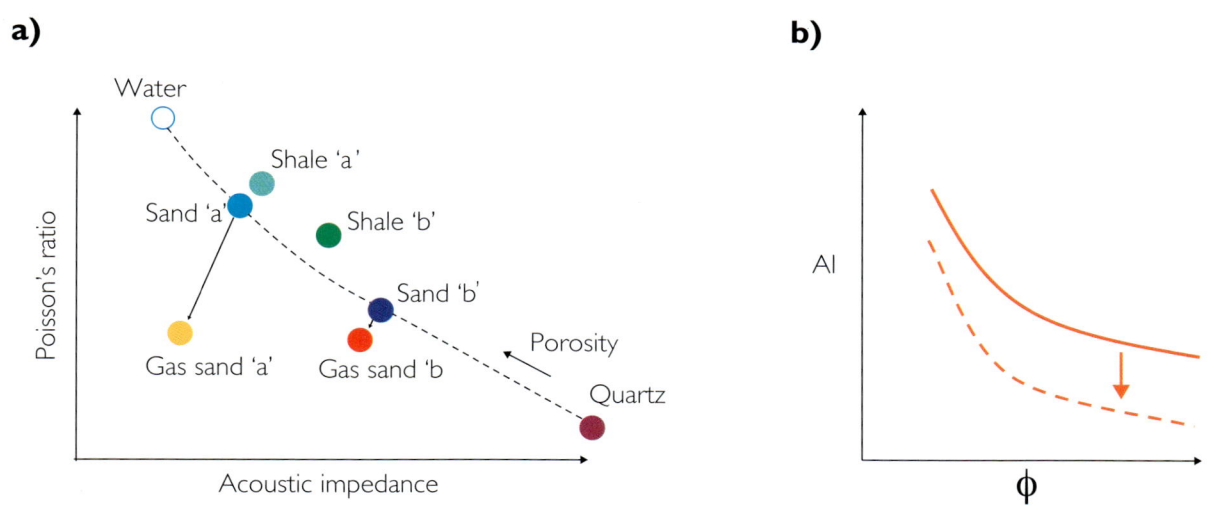

Figure 5.10 Effect of fluid fill at different porosities in sandstones, (a) acoustic impedance vs Poisson's ratio, (b) sand porosity vs *AI*.

The underlying concept of how AVO might be used to distinguish between a gas sand and a high-porosity brine sand that have similar intercept amplitudes is shown in Fig. 5.11. As a starting point, consider a sand (the blue point in Fig. 5.11a) which has a negative acoustic impedance contrast with an overlying shale (the green point) (this will be referred to as the reference sand). The AVO response of the top of this sand is Class III and is shown in Fig. 5.11b. If the porosity of the sand is increased relative to the reference sand then the acoustic impedance contrast becomes more negative and the intercept moves down the AVO plot. At the same time, the AVO gradient is less negative than the shale/reference sand response because the shear velocity of the high-porosity sand is less than the reference sand (but still greater than the shale above). The effect of replacing brine with hydrocarbon (Fig. 5.11c) has a similar effect on the intercept (i.e. it is made more negative) but the AVO gradient is also made more negative owing to the increased shear velocity of the gas sand relative to the brine filled reference sand.

Thus, the high-porosity sand can be differentiated from the gas sand on the basis of the relative change in AVO (Figs. 5.11c and 5.11f). The amplitude ratio of the high-porosity sand response to the reference sand response decreases with increasing angle whilst the gas sand/reference ratio increases (Chiburis, 1984). This idea of AVO analysis determining a relative effect is an important one and in areas where a laterally persistent reference horizon can be identified with confidence, the 'relative AVO' method can work well (e.g. Chiburis, 1993) (see also Chapter 7).

Another way of visualising the difference in AVO effect between the gas sand and high-porosity brine sand is to use the AVO crossplot. The data from the AVO plot in Fig 5.11d are shown on the AVO crossplot in Fig 5.12. Note that the brine filled points define a trend which is inclined from left to right whereas the gas sand plots away from this trend and down to the left. This 'down-to-the-leftness' is typical of top hydrocarbon sand responses irrespective of the AVO class of the brine filled sand. This is the primary effect that is exploited in AVO analysis.

Unfortunately, it is possible for a small amount of gas to give rise to large seismic amplitudes and AVO anomalies. Certainly the effect is large enough to give rise to uncertainty and ambiguity in interpretation, particularly when considering the uncertainties in other parameters such as porosity. Figure 5.13 shows the effect of variable gas and oil saturation on the AVO response at the top of a high-porosity sand. A small amount of gas has a dramatic effect in lowering the saturated bulk modulus, although it has little effect on the bulk density. The bulk modulus change has a significant effect in lowering the velocity (Fig. 5.13a). With increasing gas saturation the effect on the modulus is negligible but the density decreases. As the density term is in the denominator of the velocity equation (Eq. (2.10)), the velocity increases with increasing gas saturation. The acoustic impedance and Poisson's ratio plots (Figs. 5.13b, c) show

Rock properties and AVO

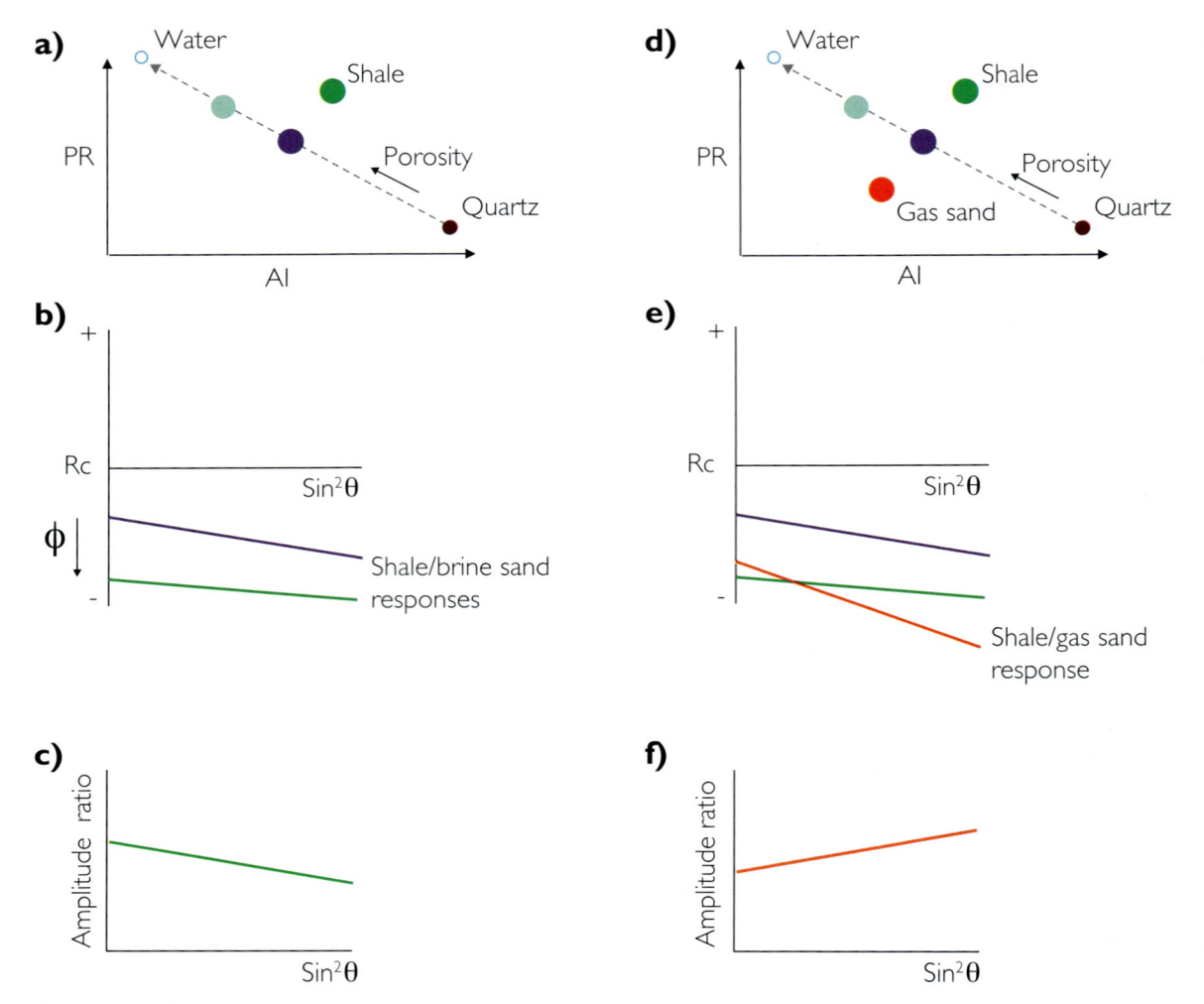

Figure 5.11 Effect of porosity and fluid fill in sandstones; (a) AI vs PR showing reference sand (blue), high-porosity sand (green) and shale, (b) AVO plot showing top sand responses of water bearing sands, (c) variation of amplitude ratio (i.e. green/blue in (b)) with respect to $\sin^2\theta$, (d) AI vs PR plot showing the reference sand (blue) with gas (red), (e) AVO plot including the top reference sand with gas, (f) variation of amplitude ratio (i.e. red/blue in (e)) with respect to $\sin^2\theta$. Note how with increasing angle the amplitude difference between the reference sand and the high-porosity brine sand decreases whereas with gas the difference increases.

that most of the gas effect occurs with the first 5%–10% of added gas. In terms of the AVO response (Fig. 5.13d) the low gas saturation case has a large magnitude Class III signature. Tuning of this response would give similar amplitudes to the 30% water saturation response. In contrast to the gas responses, the oil responses are more linear.

It should be noted that this 'fizz gas' effect is only significant at relatively low pressure and temperature (Han and Batzle, 2002). The distinctive convex downward shape in the saturation vs velocity plot is effectively related to the way water and gas are mixed and this in turn depends on the pressure and temperature (Chapter 8).

Porosity is an important parameter in the magnitude of expected AVO anomalies. Figure 5.14 shows an AVO crossplot with data from two sands of different porosity with the same overlying shale. Both brine filled and gas filled points are shown for each case. It is evident that the magnitude of the separation between brine and gas points is much less in the case of the deep (compacted) 18%

Rock property controls on AVO

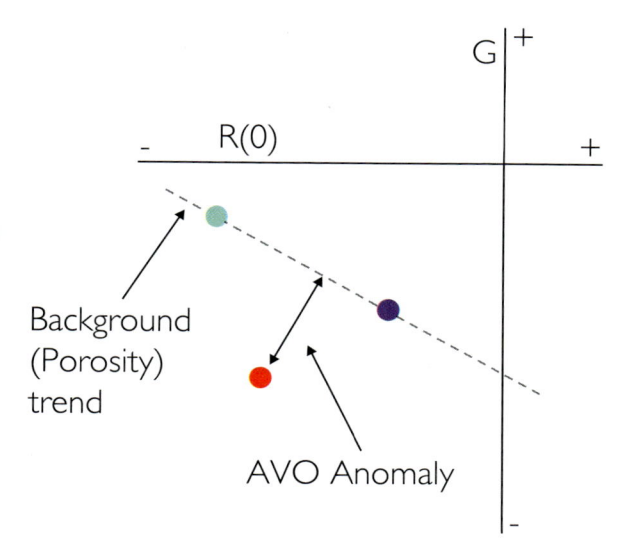

Figure 5.12 An AVO anomaly defined on the AVO crossplot.

porosity sand. Accordingly, it is potentially much more difficult to use AVO analysis techniques with low-porosity rocks. Inevitably there will be an *AVO floor*, i.e. a point below which in practice it is not possible to confidently use AVO techniques. It is often the case in siliciclastic environments that AVO analysis becomes increasingly more difficult below a porosity of around 15%.

It is generally held that AVO techniques are less applicable in carbonate reservoirs. In carbonates mineralogy often has a more dominant role in controlling elastic properties. This has the effect of limiting the relative contribution of the fluid modulus to the overall rock modulus (e.g. Wang, 1997b). The magnitude of porosity change can also be substantially greater than in siliciclastic settings so that deciphering porosity changes from fluid fill effects can be very difficult. However, many carbonate AVO models display clear differential AVO effects related to varying fluid fill

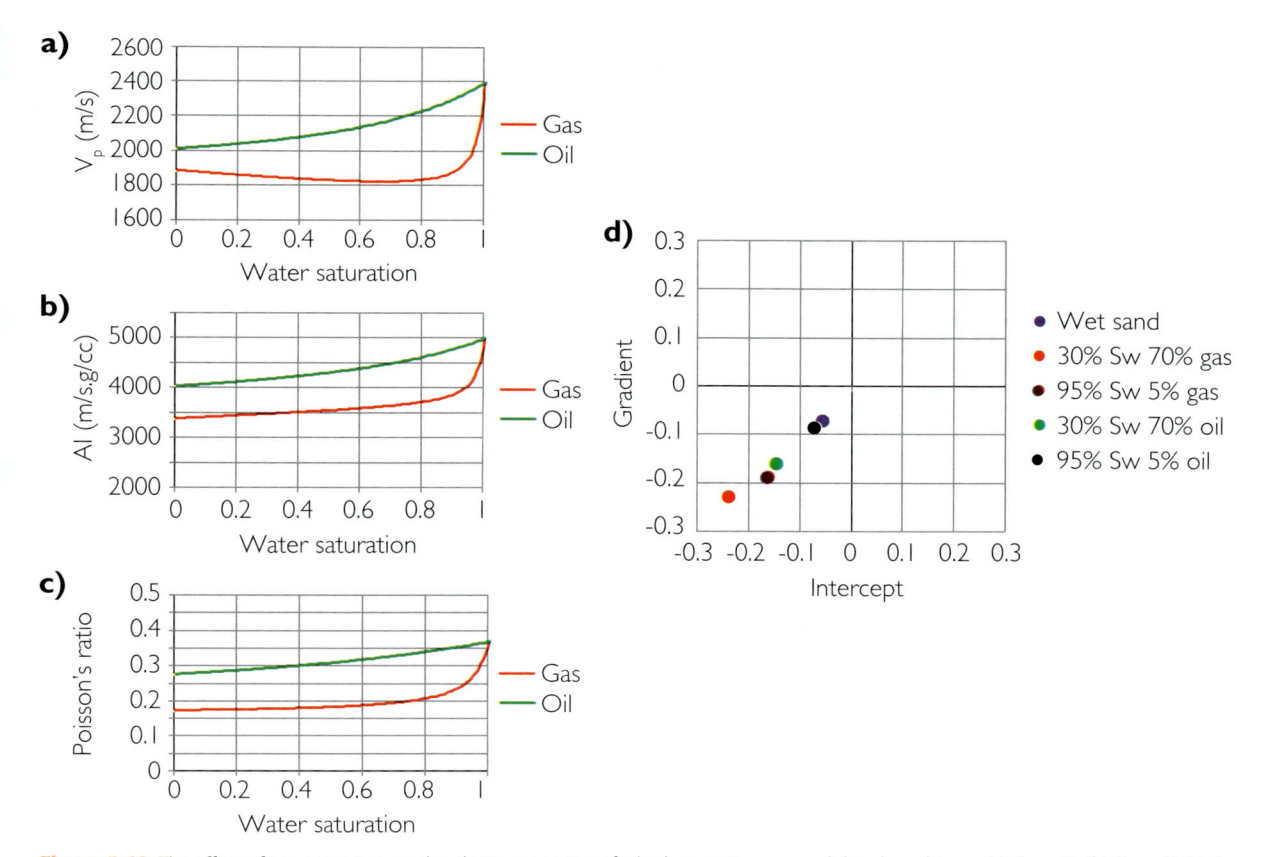

Figure 5.13 The effect of gas saturation on the elastic properties of a high-porosity unconsolidated sandstone, (a) S_w vs V_p, (b) S_w vs AI, (c) S_w vs PR, (d) intercept vs. gradient crossplot.

Rock properties and AVO

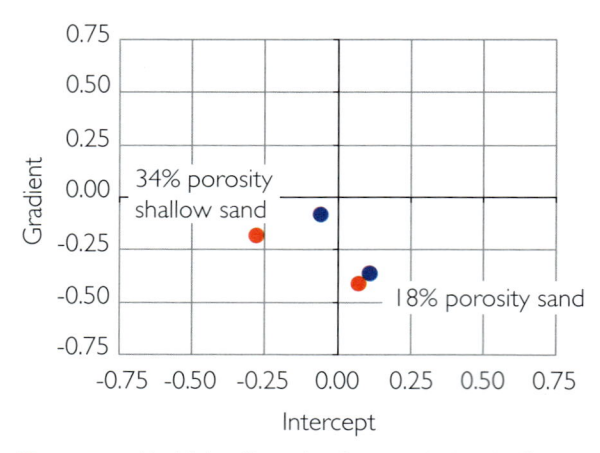

Figure 5.14 Model data illustrating that porosity is a significant factor in the magnitude of an AVO anomaly. Note: blue = shale/brine sand, red = shale/gas sand.

implying the possibility that AVO could be a useful technique.

Figure 5.15 shows Chalk data with the same porosities as the sands in Fig. 5.14. The AVO plots (Figs. 5.15a and (b)) show clear differential AVO effects between brine and gas responses over angles in the range 0–30°. However, it is evident from the AVO crossplot (Fig 5.15c) that the magnitude of the AVO difference at 34% porosity is much less than for the sandstone with the same porosity shown in Fig. 5.14. Interestingly, in this example, the fluid AVO difference for the 18% porosity chalk is greater than for the 18% porosity sandstone. Figure 5.16 shows a model for the same shale but now overlying a 20% porosity dolomite. The fluid related differential AVO is now much less

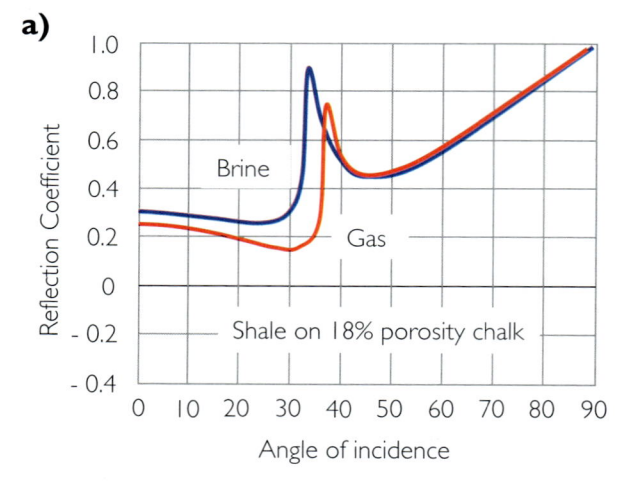

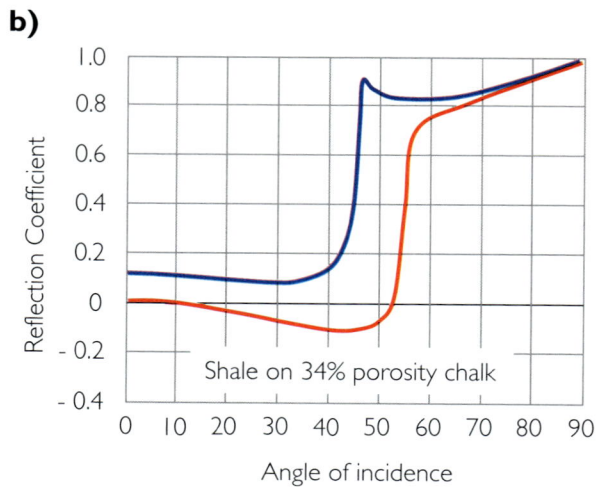

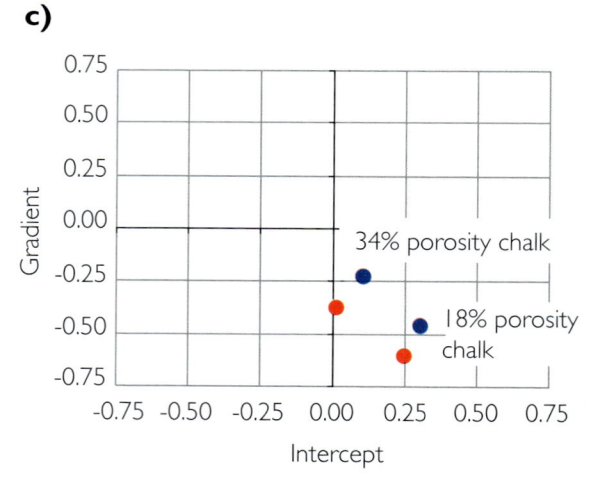

Figure 5.15 AVO models of North Sea Chalks with different porosities: (a) shale overlying 18% porosity chalk with brine and gas, (b) shale overlying 34% porosity chalk with brine and gas, (c) AVO crossplot showing data from pre-critical reflections in (a) and (b).

Rock property controls on AVO

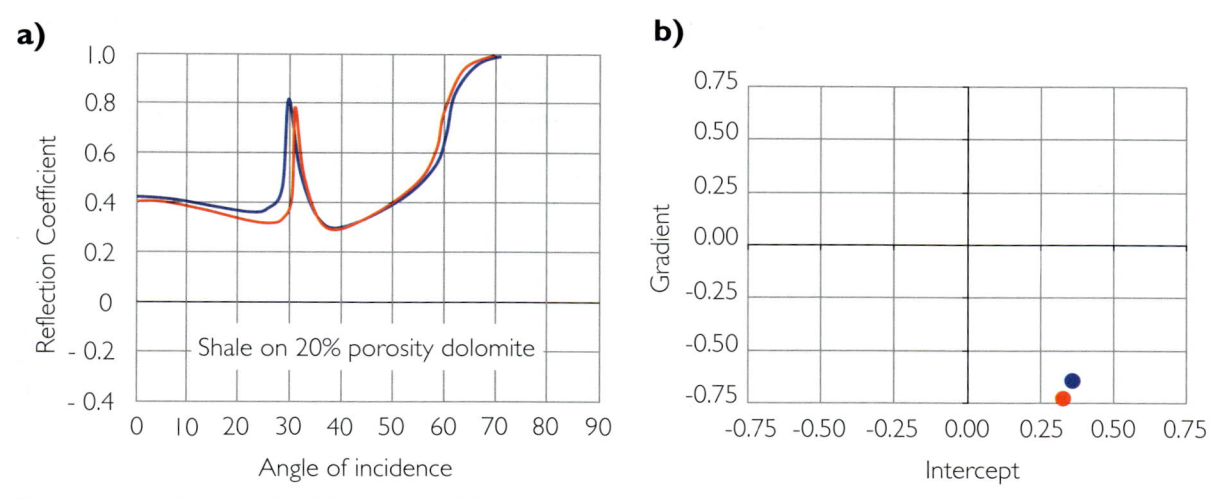

Figure 5.16 Simple AVO model of shale overlying dolomite with 20% porosity: (a) AVO plot, (b) AVO crossplot showing the AVO response for pre-critical angles.

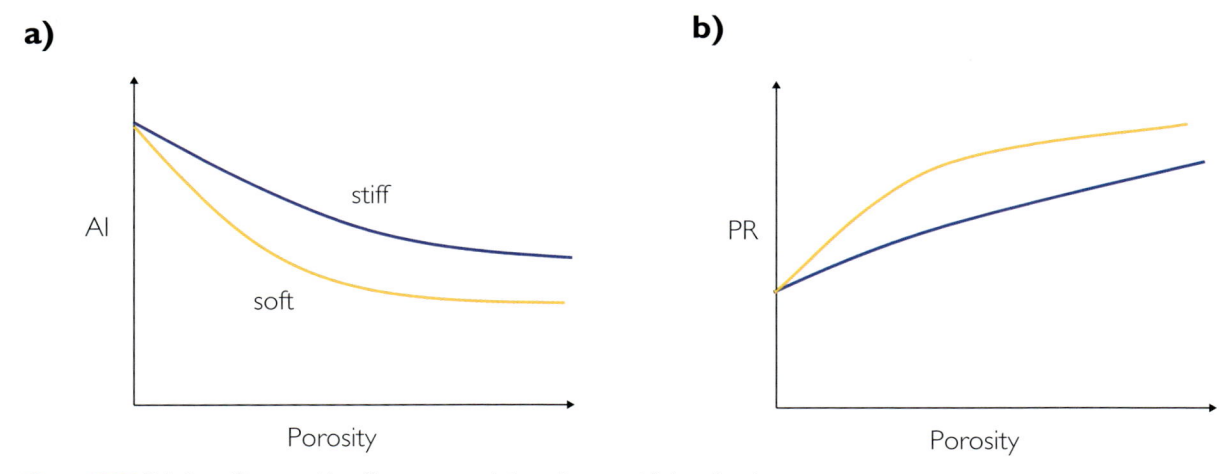

Figure 5.17 Relative stiffness and its effect on acoustic impedance and Poisson's ratio.

than for the 18% porosity chalk case. Importantly, these observations indicate that the fluid effect is not simply related to porosity, it is also related to a combination of the mineral composition and the stiffness of the rock frame. Another observation from these carbonate models is how the critical angle is often within the acquired angle range of seismic data and that the critical angle changes with fluid fill and porosity. Mapping the onset of critical angle energy in carbonate environments may provide a useful seismic attribute for lithology or fluid interpretation.

5.3.4 The effects of rock fabric and pore geometry

Rock fabric and pore geometries are important factors in the stiffness (and velocity) of all rocks. Stiffness has a key influence on the rock frame, critical in determining the velocity and magnitude of fluid effects on compressional velocity (Chapter 8). The concept is straightforward, with 'soft' rocks tending to give low acoustic impedance and high Poisson's ratio whilst 'stiff' rocks give high acoustic impedance and low Poisson's ratio (Fig. 5.17).

Rock properties and AVO

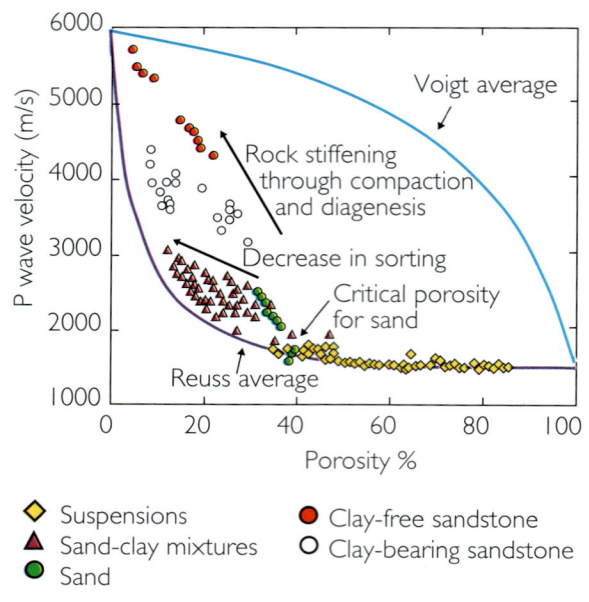

Figure 5.18 Velocity–porosity characteristics of brine-bearing siliciclastic sediments (data from Han, 1986; Yin, 1992; Hamilton, 1956; modified after Marion et al., 1992 Avseth et al., 2005).

Variations in stiffness of sandstones can be related to a number of factors such as the number of grain contacts, the amount and type of cement, clay/shale content and the distribution of pore shapes throughout the rock.

The broader context for the effects of pore geometry and rock fabric on elastic properties is illustrated in Fig. 5.18. On this plot of compressional velocity vs porosity, sand/shale rocks fall within an envelope defined by two bounds, the Reuss (1929) and Voigt (1910) bounds. These bounds are a useful type of rock physics model, essentially representing different ways of mixing rock and fluid; the softest possible mix is the Reuss bound (harmonic average) and the stiffest is the Voigt bound (arithmetic average) (see Chapter 8 for more discussion on Voigt and Reuss bounds). It is evident from Fig. 5.18 that the Reuss bound (also referred to as Wood's (1955) relation) effectively describes the behaviour of suspensions. Note that the yellow points refer to marine ooze data acquired at or close to the seabed.

It is also clear from Fig. 5.18 that the Voigt bound in this context is not especially useful to describe trends of compressional velocity change with changing porosity. It is possible to draw a more effective upper bound in a number of ways. Nur et al. (1998) derived a *modified Voigt* bound, with a curve drawn from the mineral point at zero porosity to the point where the critical porosity intersects the Reuss bound (the so-called *critical porosity model*). Other possibilities include the Hashin–Shtrikman (HS) upper bound modified to include the critical porosity (Avseth et al., 2005) and the contact cement model (Dvorkin and Nur, 1996) (Chapter 8). Within the envelope defined by the elastic bounds, sandstones tend to fall along trends either sub-parallel to the Reuss bound (with porosity increase related to improving sorting) or steeper near linear (modified Voigt or critical porosity) trends related to diagenesis (i.e. cementation).

A good example of these trends is shown in Fig. 5.19. The data are from an 80 m section in which unconsolidated sands and sandy shales (sometimes referred to as *heterolithics*; Avseth et al., 2003) overlie cemented sands. The low angle sorting trend is clear in the cluster associated with the uncemented sands as is the high angle trend associated with the change from uncemented to cemented sands (these sands have roughly 2% cement). Presumably, with additional chemical diagenesis the clean sands would follow the modified HS bound to the mineral point at about 6000 m/s. Figure 5.19b shows that the cemented sands have higher values of AI and lower values of Poisson's ratio. The rock fabric differences shown here have a significant impact on the AVO response. Taking average values for the shales above the sand, two single interface AVO crossplots have been generated for each of the sand types. The uncemented sands show Class II behaviour in which oil sands would be expected to give a brightening of the negative amplitude response at top sand. In contrast, the cemented sands show Class I behaviour and the presence of oil would be evident as a dimming of the positive amplitude.

The inference from Fig. 5.19a is that the initial compaction phase is characterised by a steep slope on the velocity vs porosity crossplot. Figure 5.20 illustrates some data from 12 wells offshore West Africa and Gulf of Mexico for arenites (sands and sandstones with 2%–12% clay content). The figure clearly shows the change in slope on the velocity vs porosity plot marking the transition from dominantly mechanical compaction and initial pressure solution in unconsolidated sands (high rate of velocity increase with decreasing porosity) to the zone of lower porosities in consolidated sands within

Rock property controls on AVO

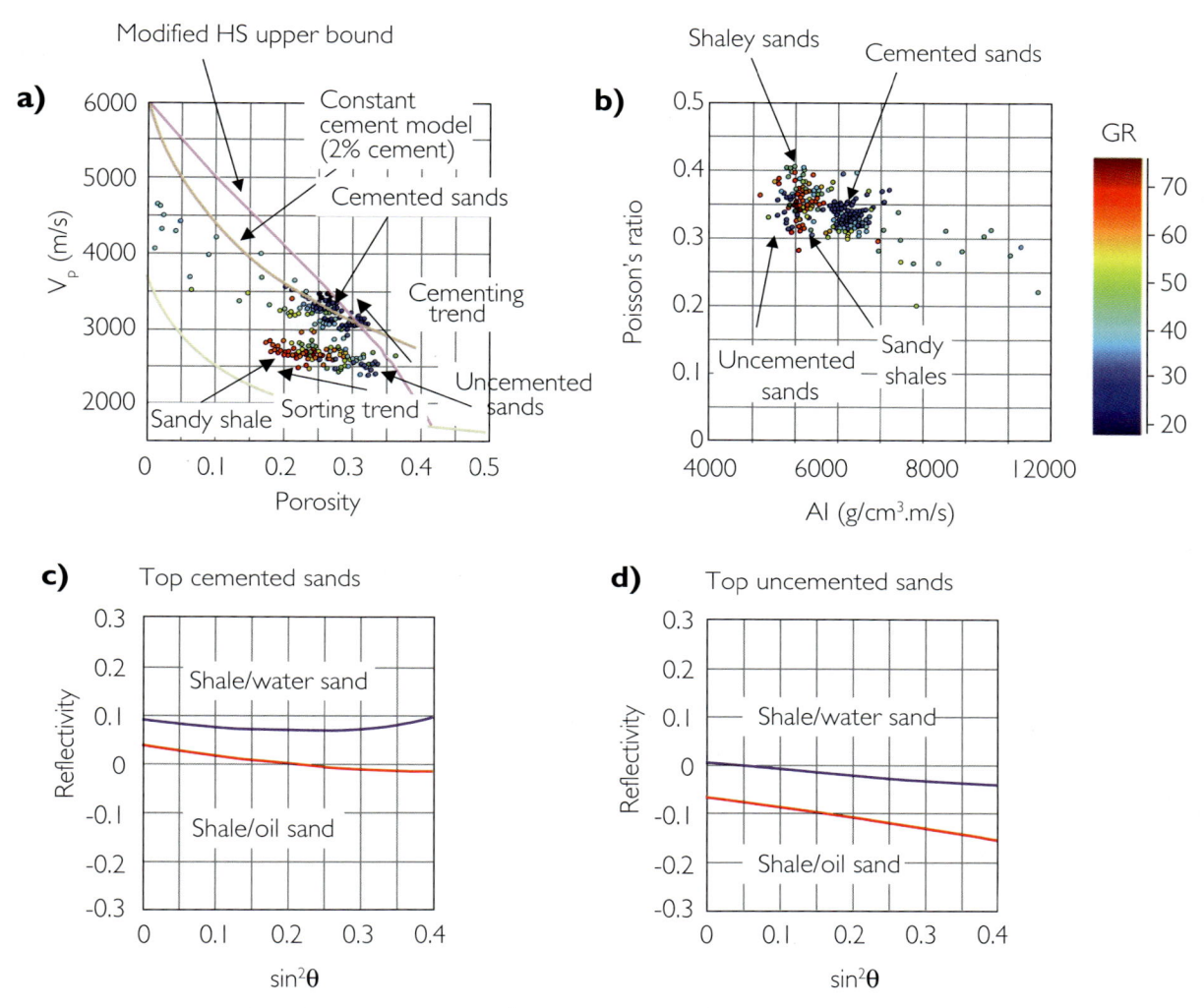

Figure 5.19 Brine saturated data from a logged section of high-porosity sands – offshore Norway; (a) porosity vs V_p, (b) AI vs PR, (c) AVO plot showing top sand responses for brine filled (blue) and gas filled (red) in cemented sands, (d) AVO plot showing top sand responses for brine (blue) and gas filled (red) uncemented sand.

which porosity reduction is through advanced cementation (lower rate of velocity increase with decreasing porosity). The porosity at which the change occurs has been referred to as the *consolidation porosity* (Vernik, 1998). The consolidation porosity is commonly between 22%–30% and principally depends on grain sorting (e.g. clay content) and stress state (Vernik and Kachanov, 2010). It effectively defines the point at which sands change from behaving as a granular material to sandstones behaving as a solid with pores and cracks.

Shale content is an important element to consider in the acoustic properties of sandstones. In general two sand/shale situations are common: (a) rocks in which sand and shale are mixed, with a gradation from matrix supported shales and sandy shales to grain supported sands and shaley sands and (b) rocks composed of alternating laminae of sand and shale (i.e. laminated sands). In shaley sands the clay may be pore filling (often referred to as 'dispersed' clay) or structural (i.e. part of the load bearing framework) or indeed both, dependent on the relative sizes of sand and clay/shale 'grains'. This simple picture is made complicated by the fact that clays in sandstones may also be authigenic in origin (i.e. derived in-situ from the breakdown of other minerals such as feldspars). While some generalised discussion of the effects of clays in sandstones is given below each reservoir

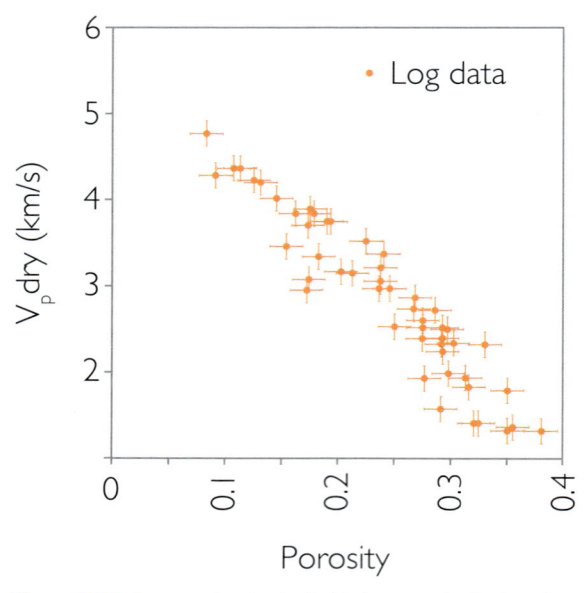

Figure 5.20 Compressional velocity (dry) vs porosity for logs from 12 wells in deep offshore West Africa and Gulf of Mexico (re-drawn after Vernik and Kachanov, 2010).

needs to be specifically analysed. Ultimately, the importance of recognising the effects of clay type and location is that these effects have an intimate relationship to reservoir porosity and permeability.

The experiments of Marion *et al*. (1992) are a useful illustration (Fig. 5.21) of the effects of mixing sand and clay. Coarse grained Ottawa sand was mixed with fine kaolinite powder and Fig. 5.21b shows the effect of gradually increasing the kaolinite content on the porosity and acoustic impedance. The effect of a small amount of clay is primarily to occlude the pore space and increase the acoustic impedance (note that this effect is evident in the log data in Fig. 5.19a). AI continues to increase until the pore space is fully occupied by shale. Further increments of clay then add to the rock frame and soften it, thus reducing the acoustic impedance (Fig. 5.21b). The rotated 'V' signature in Fig. 5.21b is often observed on sonic vs porosity crossplots (e.g. Katahara, 2008).

In terms of Poisson's ratio, Fig. 5.21c shows that the transition from sands to shaley sands causes a slight decrease in Poisson's ratio, largely in response to the lowering of porosity. The transition from shaley sands to shales is characterised by an increase in Poisson's ratio, responding to the higher Poisson's ratio of clay minerals compared to quartz (Fig. 5.21c). In terms of the intercept vs gradient crossplot these transitions also give rise to a rotated 'V' signature. The AVO model assumes that the overlying shale is similar to the shale in the sand/shale composite, so the shale point is coincident with the origin of the crossplot (Fig. 5.21d).

A log data example of a 'dispersed' shale scenario is shown in Fig. 5.22. Note the rotated 'V' shape signatures. Laminated sands behave in a more linear fashion on crossplots (Figs. 5.23 and 5.24). In a structural clay scenario where individual 'grains' of shale form part of the load-bearing frame, the clays will have a softening effect on the rock frame and the elastic effects are likely to be similar to those shown on the laminated sand/shale crossplots.

Pore geometry effects on elastic properties are important, particularly in consolidated rocks (e.g. Xu and White, 1995; Vernik, 1998). Flat pores, characteristic of clay mineral fabrics and microcracks for example, are highly compressible whereas cuspate, triangular or rectangular shaped pores characteristic of sandstones (e.g. Vernik and Kachanov, 2010) are much stiffer. This explains why in Fig. 5.25 for a given porosity increased clay content has the effect of lowering compressional velocity. The presence of flat pores and microcracks would also have an effect of reducing shear rigidity and increasing the Poisson's ratio relative to clean sandstones with similar porosity and no crack like pores. Additional factors that may be important in the understanding of the elastic properties of sand/shale rocks are the anisotropic properties of clays (e.g. Sams and Andrea, 2001) and anisotropy induced by stress (e.g. Xu, 2002).

5.3.5 Bed thickness and layering

In most of the AVO models presented so far the assumption has been that reflectivity is associated with 'thick' reservoirs (i.e. greater than tuning thickness) and that each reflection is associated with a sharp boundary contrast. Tuning of course alters the amplitude response (Chapter 3) as does the presence of transitional boundaries, for example associated with upward fining or upward coarsening sand reservoirs.

5.3.5.1 Reflector interference

A common issue for the interpreter to deal with is the variation of amplitude with thickness. Figure 5.26 shows an example of a gas sand seismic response with

Rock property controls on AVO

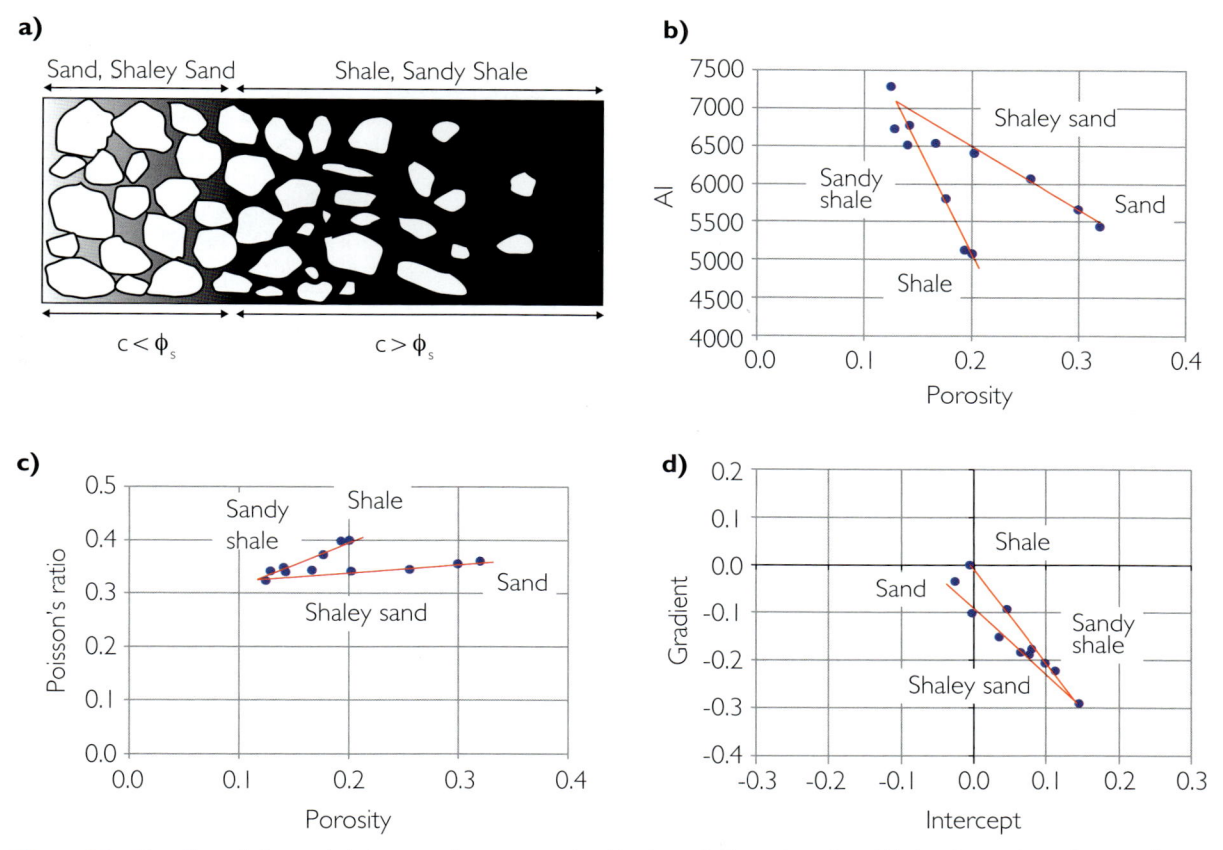

Figure 5.21 The effect of dispersed shale on sandstone properties; (a) schematic illustration of sand/shale mixtures (note that c = clay content and ϕ_s = sand porosity), (b) porosity vs AI plot, (c) porosity vs Poisson's ratio plot, (d) AVO crossplot. Note that these figures relate to experimental results with coarse quartz grains and fine kaolinite powder at an effective pressure of 40 MPa (modified after Marion et al., 1992).

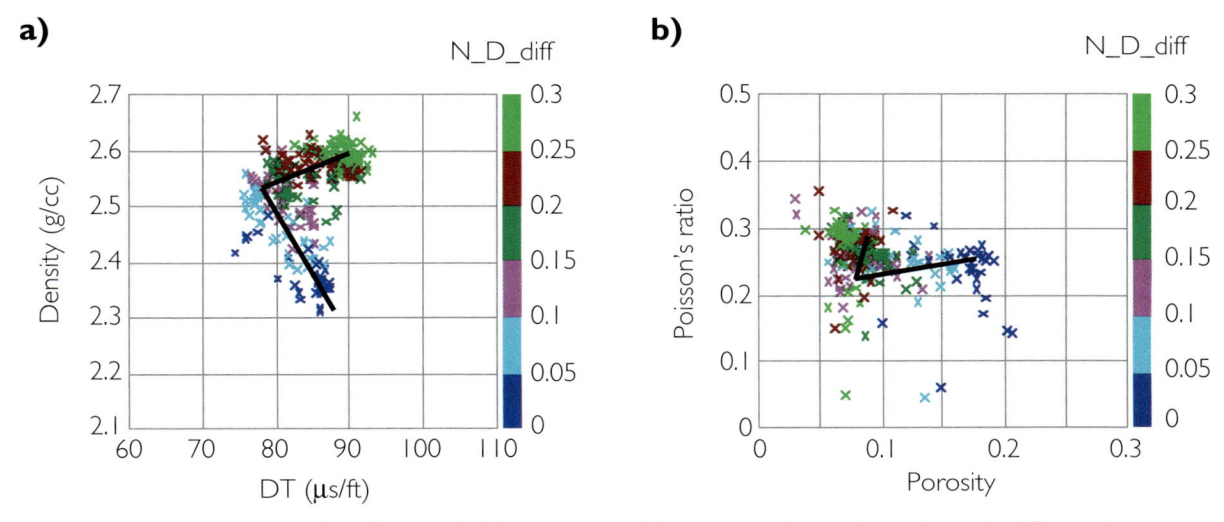

Figure 5.22 An example of the effect of dispersed shales in a Jurassic deltaic sequence. Note colour coding is the difference between neutron porosity and density porosity, a good indicator of clay content in sandstones (Katahara, 2008); (a) sonic slowness vs density plot, (b) density porosity vs Poisson's ratio plot.

Rock properties and AVO

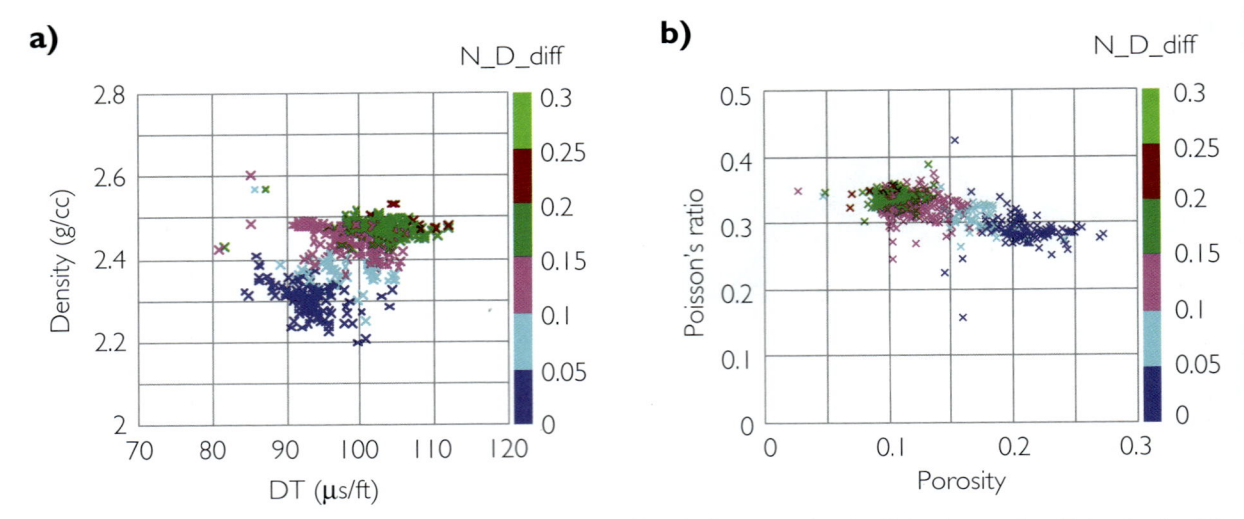

Figure 5.23 A log data example of a laminated sand/shale sequence; (a) sonic slowness vs density, (b) density porosity vs Poisson's ratio plot. Colour coding as in Fig. 5.22.

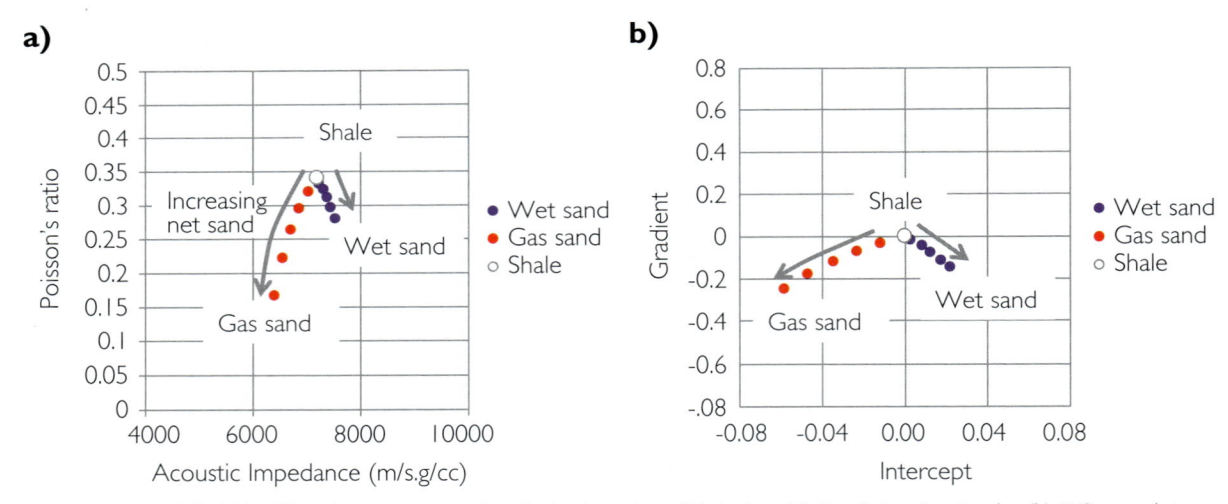

Figure 5.24 Modelled AI vs PR and intercept vs gradient for laminated sand/shale data; (a) AI vs Poisson's ratio plot, (b) AVO crossplot. Katahara's (2004) model was used to calculate upscaled velocities (see Section 8.5.2).

tuning effects present (Cowan *et al.*, 1998). The tuning is evident at the edges of the bright reflections in Fig. 5.26a where the black and red loops parallel one another. This is manifested on the top sand amplitude map (Fig 5.26b) as a distinct high amplitude lineament parallel to the edge of the accumulation. Clearly the bright amplitudes were responsible for the discovery but the thickness effects on amplitude suggest that the amplitudes are not an indication of pay thickness and that the amplitude map should not be used as the only dataset on which to plan appraisal wells. Using seismic amplitude and apparent thicknesses from seismic in order to predict net pay is discussed in Chapter 10.

In terms of AVO, tuned responses will be more anomalous than isolated reflections or responses from beds less than tuning thickness. Figure 5.27 shows the idealised AVO signature from the top reflection of a low impedance wedge, modelled using a simple one-dimensional reflectivity model. Clearly, a possible pitfall is that a tuned water bearing sand with a Class III AVO response may be mistaken for a hydrocarbon bearing sand. In real seismic data the problem is further complicated by the fact that interference effects are not the same at all offsets.

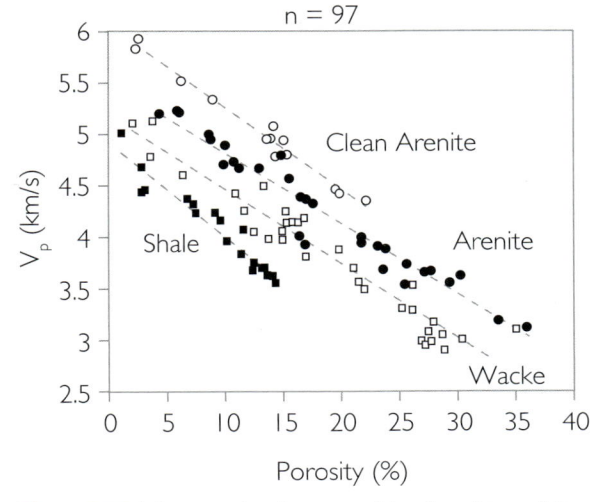

Figure 5.25 Laboratory data from consolidated sandstones (after Vernik, 1994). Note sand–shale categories refer to clay volume (as percentage of mineral component), clean arenite <2%, arenite 2%–15%, wacke 15%–35%, shales >35%.

Reflection hyperbolae, for example from the top and base of a low impedance sand, are not parallel and this gives rise to variable interference effects across a seismic gather as bed thickness changes. This can give rise to a spiraling effect on the intercept vs gradient crossplot. Figure 5.28 shows a modelled example (incorporating timing stretch) of a prominent Class III response. When the gradients are low it is possible to change a Class III response into a Class IV response simply through offset dependent interference.

5.3.5.2 Thin beds, and the non-uniqueness of composite responses

One consequence of the bandlimited nature of seismic is that below tuning different combinations of impedance and thickness can give rise to the same seismic signature (Fig. 5.29). Thus, there is uncertainty in thin bed interpretation. Fortunately, in many sand/shale environments the amplitude of the composite response corresponds in general to the net sand component (Meckel and Nath, 1977). Figure 5.30 illustrates Meckel and Nath's model result with a near linear relationship between net sand and composite

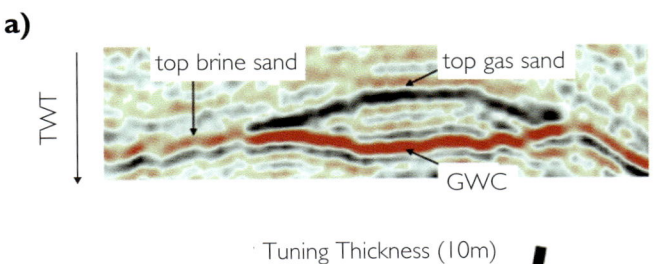

Figure 5.26 A gas sand seismic response with amplitude signature varying with thickness; (a) migrated stack section (black = soft response), (b) amplitude map at top reservoir with a high amplitude (orange) lineament parallel to the edge of the gas accumulation and related to tuning between the top sand and gas water contact. Courtesy of Rashid Petroleum Company.

Rock properties and AVO

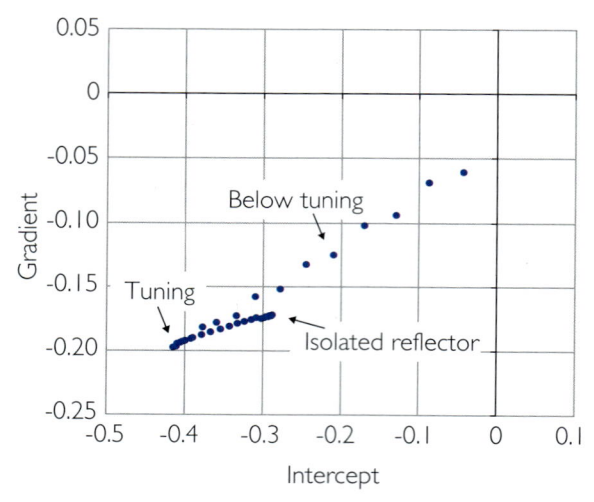

Figure 5.27 Idealised AVO signature from the top reflection of a low impedance wedge, modelled using a simple one-dimensional reflectivity model.

amplitude. Whilst this observation gives some justification for interpreting net pay from seismic amplitude at and below the tuning thickness, a critical observation from Fig. 5.30b is that there is scatter around the trend (i.e. there is uncertainty).

Uncertainty becomes important in field development work where it can have a significant impact on field development planning. Figure 5.31 shows another example, putting the notion of calibration into the context of uncertainty. It shows a schematic illustration of the reservoir (two sands with a shale unit in between) represented by a composite amplitude signature. Five wells have been drilled giving a near linear relation between amplitude and net pay and a high correlation coefficient. It gives the impression that the net pay error prediction is about $\pm/-9$ ft.

It should be remembered that wells tend to be drilled in sweet spots and possibly are not statistically

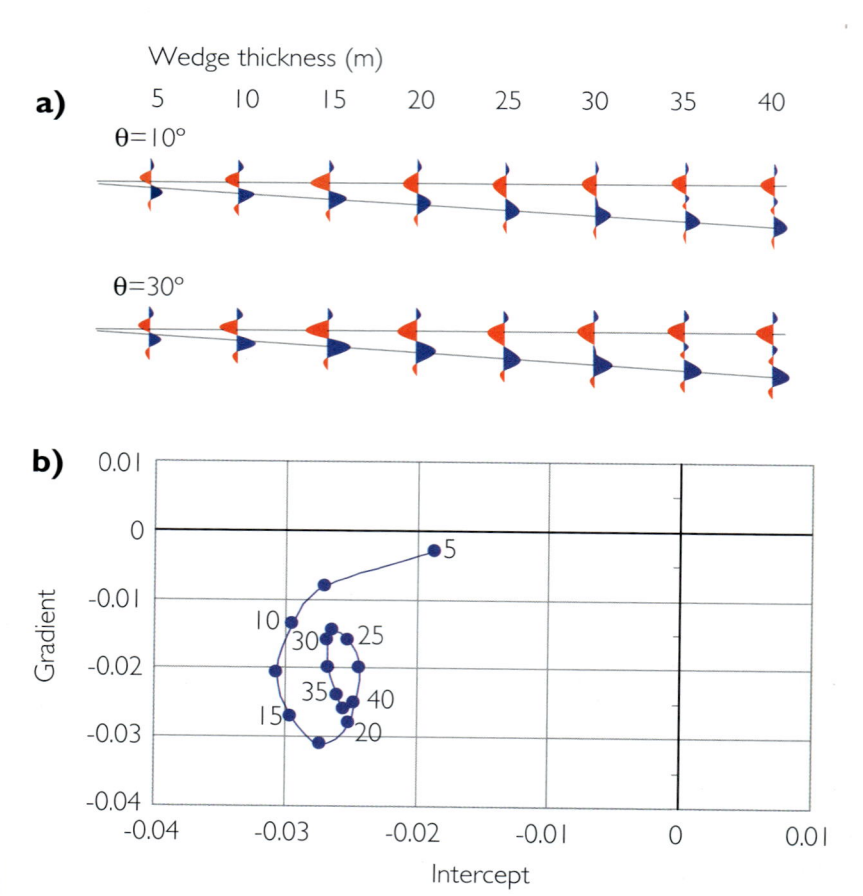

Figure 5.28 Model data showing effect of tuning on the AVO response; (a) two angle stacks from the model, $\theta = 10°$ and $\theta = 30°$ and (b) intercept/gradient crossplot. Note numbers are thicknesses in metres (data from Roy White, personal communication).

Rock property controls on AVO

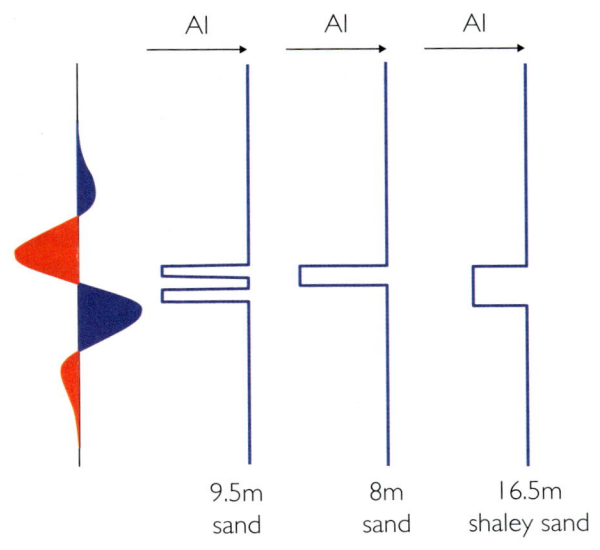

Figure 5.29 The non-uniqueness of thickness and impedance below tuning.

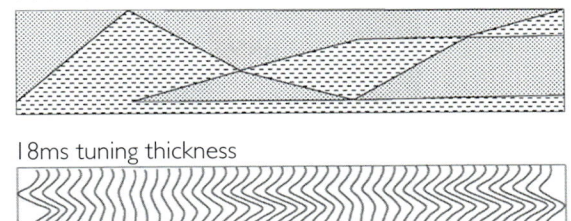

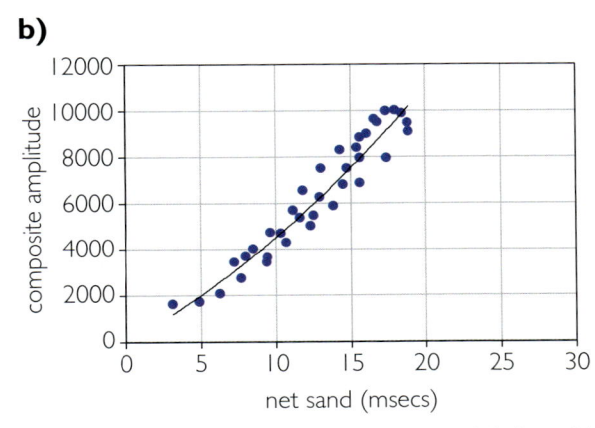

Figure 5.30 Sand/shale model with constant (tuning) thickness; (a) form of the model with synthetic responses, after Meckel and Nath (1977; AAPG © 1977, reprinted by permission of the AAPG whose permission is required for further use); (b) relationship between composite amplitude (i.e. sum of absolute amplitudes of trough and peak) and net sand.

representative. Various models were created by changing the thicknesses of the sands and shale unit and the results are shown in Fig. 5.31. Assuming that the modelled variability is geologically reasonable it is evident that 'calibration' based on the wells (Fig. 5.31a) is not an adequate description of the uncertainty. An appropriate approach for deriving estimates of uncertainty in this instance would be to use Monte Carlo simulation with probability density functions describing the distribution of elastic properties and thicknesses of sand and shale as well as incorporating the effect of seismic noise (e.g. Mukerji and Mavko, 2008; Saussus and Sams, 2012). This topic is discussed in more detail in Chapters 9 and 10.

5.3.5.3 Transitional boundaries

Amplitudes respond differently to gradual versus sharp boundary changes (e.g. Anstey, 1980). Figures 5.32a–c illustrate how the thickness of a transition zone can have a direct impact on the amplitude and signature. Increasing the thickness of the transition zone results in a decrease of amplitude and imparts a lower-frequency look to the response. When the transitional zone is associated with a thin bed, for example with coarsening or fining upward sands the presence of the transitional zone will impart asymmetry to the composite response (Figs. 5.32d–f).

5.3.6 The effects of pressure

Pressure is an important factor in the acoustic properties of rocks, and seismic reflectivity and AVO responses may vary from those expected simply through the effects of pressure. There are two aspects of interest: (1) the effect of pressure variations with depth and, in particular, the differences between seismic signatures in *normally pressured* sections compared to those from *over-pressured* regions, and (2) the effect of pressure variations due to reservoir production.

5.3.6.1 Rock velocity and effective pressure

Laboratory studies give valuable information on the response of sandstones and other rocks to pressure variations. Experiments are performed on core plugs

Rock properties and AVO

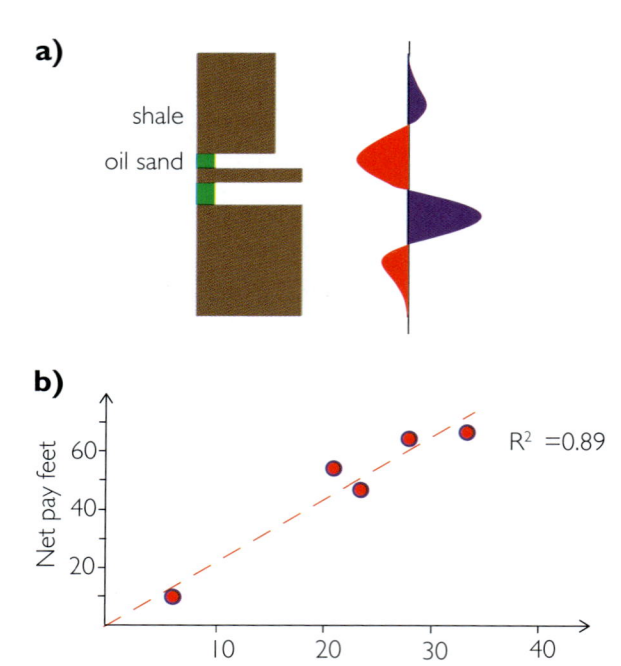

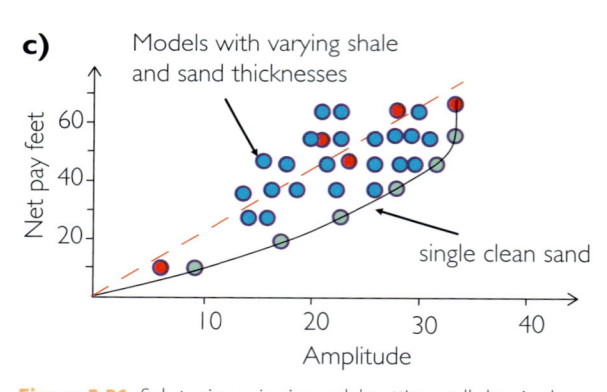

Figure 5.31 Sub-tuning seismic model putting well data in the context of uncertainty; (a) schematic model of 'thin-bed' reservoir, (b) well data showing a high correlation coefficient between amplitude and net pay thickness at wells, (c) model data generated by varying thickness of sands and intervening shale.

with various combinations of internal pore pressure and external confining pressure. A key observation is that the compressional velocity of a rock varies according to the difference between the two pressures (Fig. 5.33) (i.e. the higher the *differential* or *effective pressure*, the higher the velocity). Figure 5.33 also shows a high degree of variability in the rate of change of velocity with pressure between different sandstones. This is discussed in Chapter 8.

In sedimentary basins the differential or effective pressure is the difference between the overburden pressure (related to the weight of rock above the reservoir) and the pore pressure (Fig. 5.34). Provided that a reservoir is in porous communication with the surface (i.e. water expelled as the rock compacts is free to flow to surface) the pore pressure is hydrostatic (i.e. the pressure is equivalent to the weight of a column of water above the reservoir). Typically, hydrostatic gradients are around 0.45 psi/ft but may vary depending on salinity of formation waters. The overburden pressure gradient is typically about 1 psi/ft but can vary depending on average rock density. If all rocks are hydrostatically pressured (a situation commonly referred to as *normally pressured*), the effective pressure (and rock velocity) will increase with increasing depth.

Pore pressure can depart from hydrostatic for a number of reasons (e.g. Mouchet and Mitchell, 1989; Swarbrick and Osborne, 1998; Mukerji *et al.*, 2002), the two most important being related to (1) the entrapment of pore waters as the rocks are buried (i.e. the rate of burial exceeds the rate at which pore water can be expelled), which is typically referred to as *disequilibrium compaction*, and (2) the generation and movement of fluids related to hydrocarbon maturation at depth. In each case the pore pressure will be abnormally high (i.e. significantly greater than hydrostatic). In the first case, compaction is retarded and rocks will have porosity greater than expected for a given depth whilst in the second case the rocks have already undergone some degree of compaction before being overpressured.

Given the effect of pressure on rock velocity, the sonic log is commonly a good indicator of pressure. Figure 5.35 shows an example from the Gulf of Mexico. The sonic has been plotted on a log scale, revealing the 'normally compacted' shales lying on a straight line trend. With the onset of overpressure the sonic values depart from the trend (i.e. increase in slowness). The magnitude of the departure of the sonic log from the 'normal' trend can be calibrated to drilling mud weight and the degree of overpressure (e.g. Shaker, 2003).

5.3.6.2 Overpressure and amplitude signatures

Overpressuring will lower the effective pressure, soften the rock and reduce the velocity compared to a hydrostatically pressured rock. In some cases the top of overpressure (also referred to as *geopressure*) is

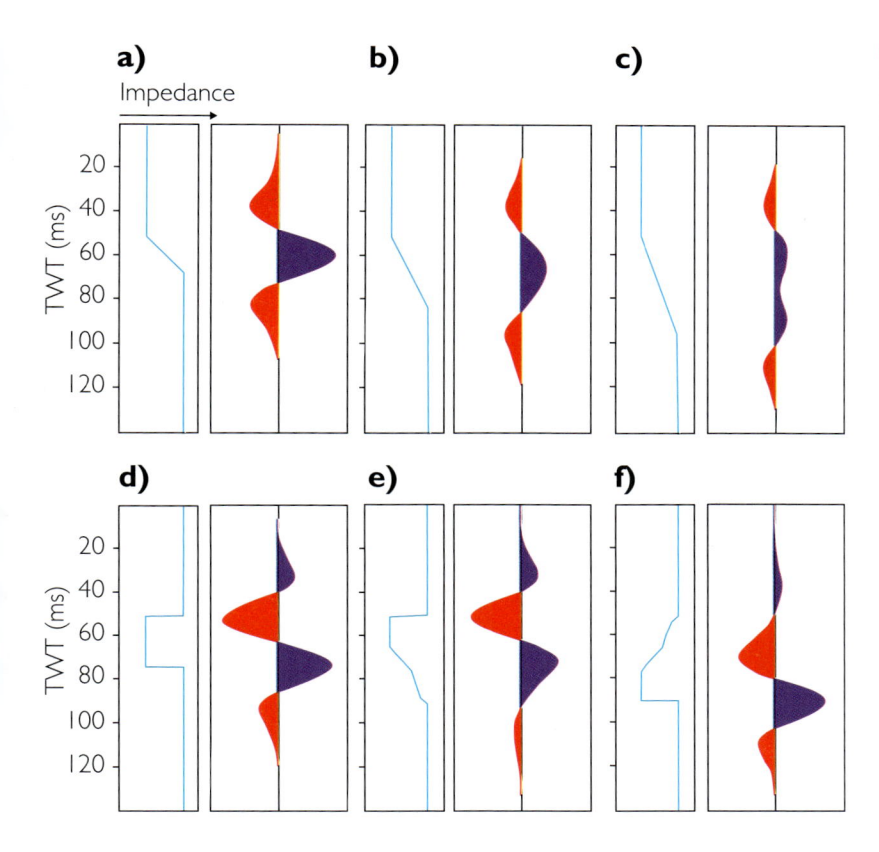

Figure 5.32 Transitional boundary effects on seismic signatures; (a–c) amplitude decreasing with increasing thickness of transitional zone, (d) thin bed with sharp boundaries, (e)–(f) asymmetric responses related to coarsening upward and fining upward layering.

defined by a strong negative amplitude response. Top geopressure boundaries defined by normally pressured shale overlying overpressured shale typically have a Class IV AVO response. However, in many cases the top of geopressure is transitional and it may not be visible as a distinct reflection. The difference between reflection responses from normally pressured and overpressured sands depends on many factors. It is possible that the overall style may be similar between the two regions (e.g. strong Class III gas sand responses), particularly if the increase in overpressure with depth is gradual, such as appears to be the case in many parts of the Gulf of Mexico (Hilterman, 2001). In other cases there can be significant differences in AVO response.

Figure 5.36 shows a modelled AVO crossplot example of three sands (based on a real case study) in which the middle sand is overpressured. The upper and lower sands are normally pressured and show Class I to Class II behaviour whereas the middle sand has a Class IV signature (Fig. 5.36a). The top sand full stack response changes depending on the stratigraphic interval and the fluid fill, so that sand 1 shows hard responses when brine filled but soft responses when hydrocarbon filled; sand 2 shows a negative reflection when brine filled and bright spots with hydrocarbon, and sand 3 shows dimming of a hard event when hydrocarbon filled. It is interesting that even though the reflection character changes, oil and gas responses for each case show increasing absolute amplitude difference with offset compared to the brine cases. In practice, however, these types of relative AVO comparisons are often hampered by uncertainty in identifying the appropriate water sand signature on seismic. Dvorkin et al. (1999) have noted that the Poisson's ratio of gas bearing rocks can dramatically reduce at low effective pressures whilst water filled rocks do not show this effect. A dramatic drop in Poisson's ratio may be a useful indicator of hydrocarbons in overpressured sediments.

Pressure changes can have a significant effect on reflectivity during production. To illustrate this, Fig. 5.37 shows the modelled top reservoir reflectivity from an initially overpressured reservoir. Note that it is assumed in the model that there is no change in porosity and that the overburden properties do not

Rock properties and AVO

Two sandstone cores
8864 A = 17.6% porosity
8293 B = 29% porosity
ΔP = external pressure minus internal pressure

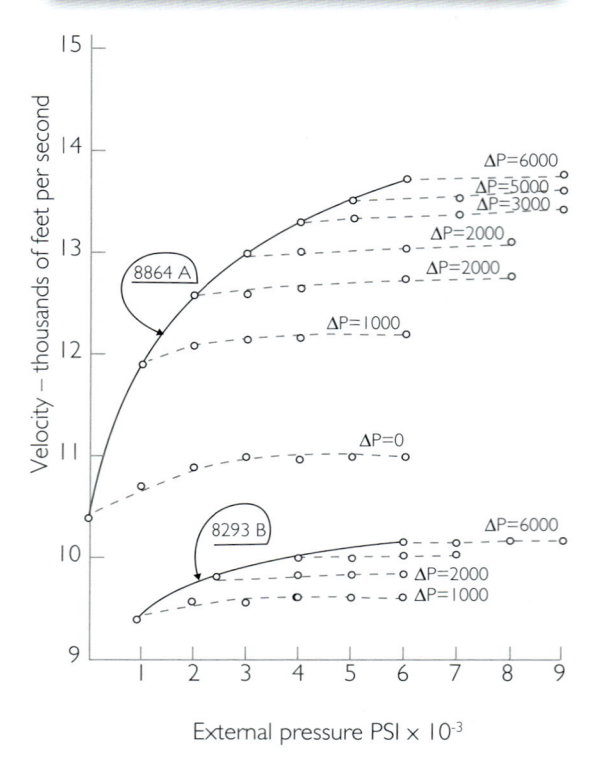

Figure 5.33 Laboratory measurements from two sandstone cores showing velocity as a function of differential pressure (after Hicks and Berry, 1956). Note how the sands have very different pressure sensitivities.

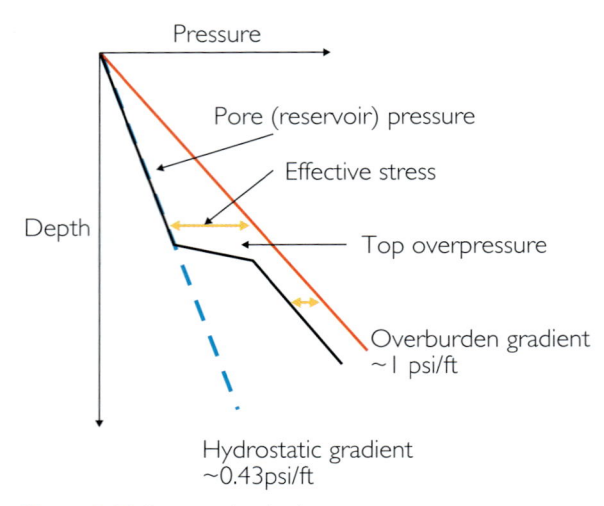

Figure 5.34 Pressure–depth relations.

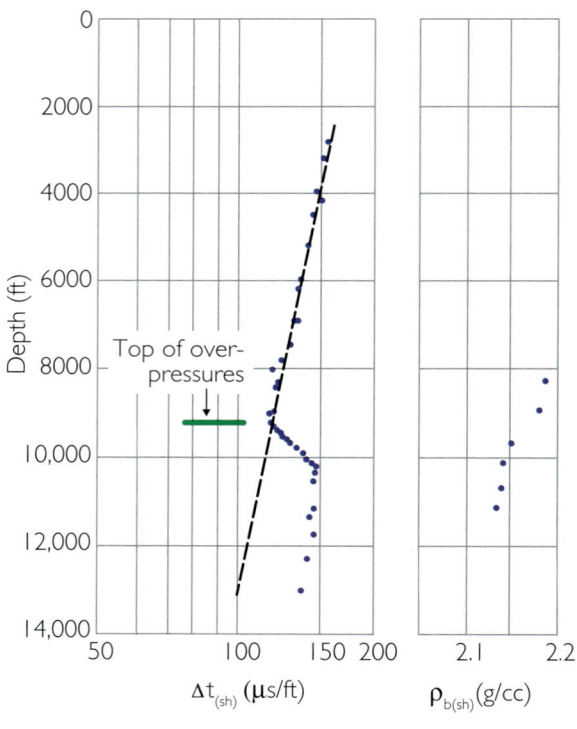

Figure 5.35 Sonic log example plotted on log scale to highlight the 'normal' shale compaction trend and onset of overpressure, from Hottmann and Johnson (1965). Copyright 1965 SPE; reproduced with permission of SPE, further reproduction prohibited without permission.

change. In practice the shale cap rock may undergo stress effects that alter physical properties (e.g. Gray et al., 2012). In the model (Fig. 5.37) the pre-production AVO style is Class III but the AVO style changes with production. Changes in water saturation and pressure drawdown (i.e. increase in effective pressure) stiffen the reservoir rock. Thus, the intercept amplitude becomes more positive as the reservoir is produced. The largest effect occurs where there is both pressure and fluid change. If pressure is maintained during production the gradient decreases, but if the effective pressure increases the gradient becomes more negative. These types of dynamic effects are most evident in unconsolidated and overpressured reservoirs whilst the reflectivity associated with 'normally pressured' reservoirs is usually less susceptible to pressure change. There is often an asymmetry to pressure effects, for example increasing pressure in normally

Rock property controls on AVO

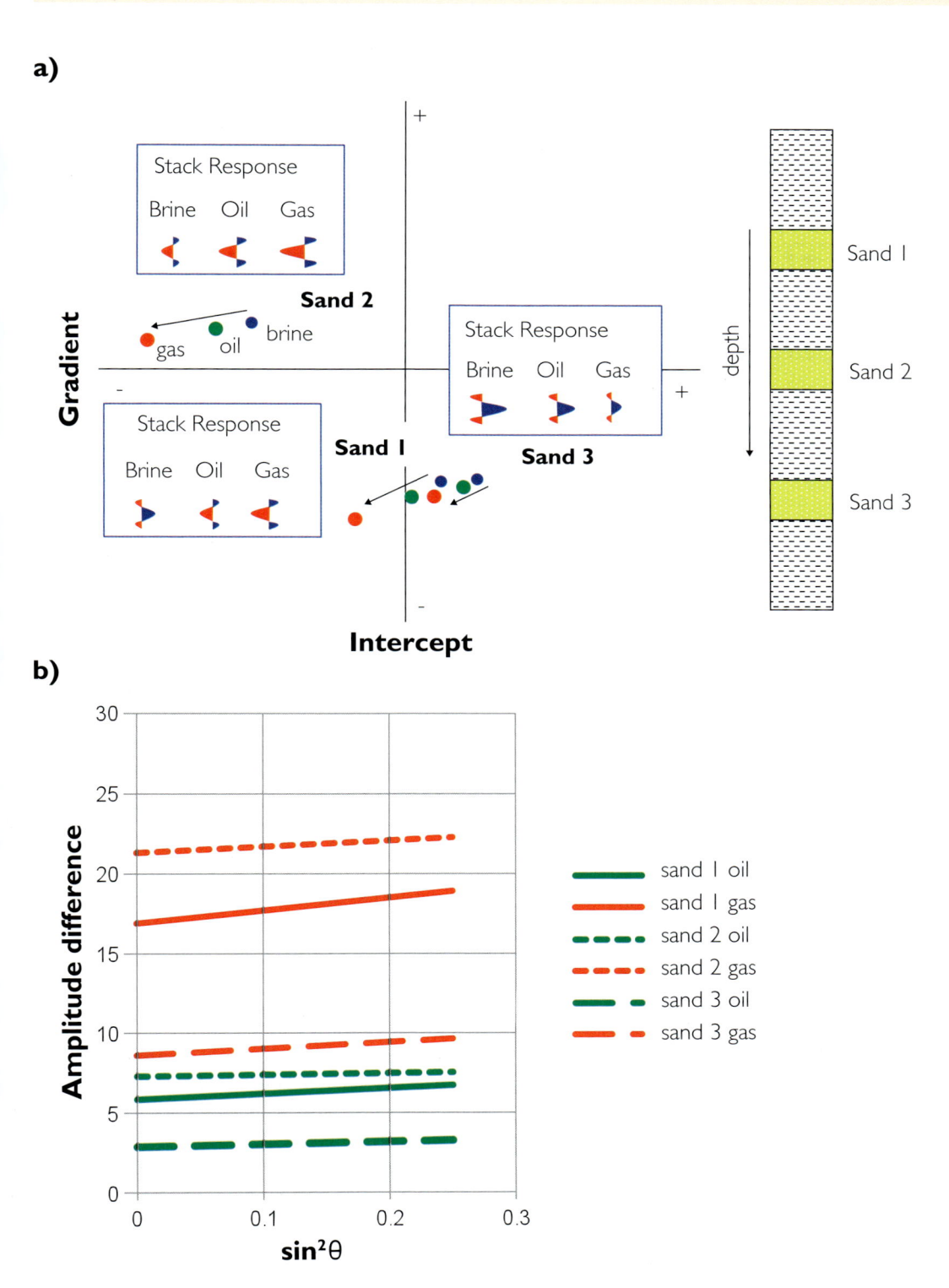

Figure 5.36 The effect of pressure on AVO. Sands 1 and 3 are normally pressured whilst sand 2 is overpressured by around 500 psi: (a) intercept/gradient crossplot illustrating the variation in seismic signatures (including Class 1, 2p, 2 and 4) at the top sand, (b) absolute amplitude difference plot showing how all hydrocarbon sands show increasing amplitude difference with angle.

81

Rock properties and AVO

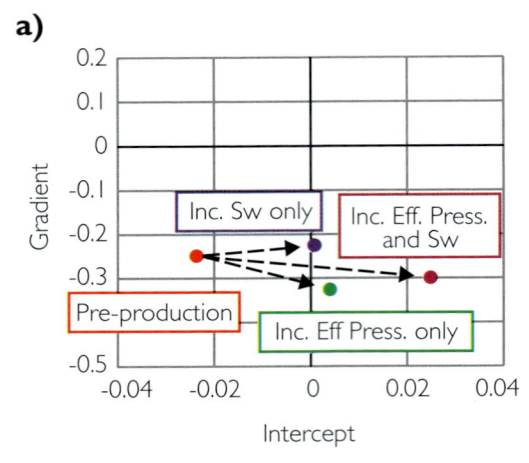

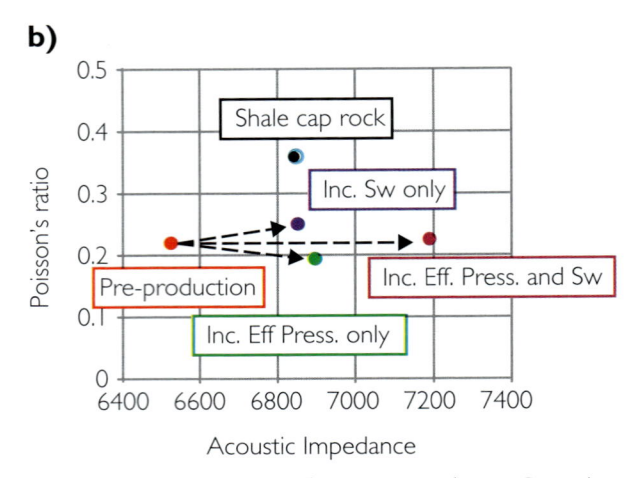

Figure 5.37 Modelled production effects on an initially overpressured reservoir; (a) intercept vs gradient, (b) acoustic impedance vs Poisson's ratio.

pressured reservoirs tends to have a greater effect than decreasing reservoir pressure.

5.3.7 Anisotropy

Whilst the amplitude interpreter's model tends to be focussed on an isotropic view of the world, an understanding of anisotropy is important in seismic processing and AVO modelling. The aim of the following discussion is designed to introduce concepts and present useful first order observations and is not meant as a detailed treatment of the subject. Readers are referred to the works of Thomsen (2002), Lynn (2004), Tsvankin et al. (2010), Grechka (2009) and Lynn and Michelena (2011) for a more detailed treatment.

By definition the characteristic of anisotropic rocks is that velocities measured in different directions (e.g. vertical vs horizontal) are not the same. In essence this is due to the effect of preferred alignments of geological components. For example, at the pore scale, shale-prone rocks are anisotropic owing to the alignment of clay minerals. Anisotropic effects can also occur at the seismic scale owing to sedimentary layering and vertical fracturing.

There are essentially two types of anisotropy that are relevant for the seismic interpreter (Fig. 5.38). These are as follows.

(1) Anisotropy related to layering. Examples of this type include shale units with aligned particles as well as horizontally organised beds or layers on a scale much smaller than the seismic wavelength. For this type of anisotropy rock properties vary with angle of incidence (for example in a flat layer case horizontal velocities are faster than vertical velocities) but within a plane perpendicular to the axis of symmetry do not show any variation with direction. As this layering conforms to polar symmetry (i.e. having a single axis of rotational symmetry) the anisotropy that results from this type of configuration is referred to as *polar anisotropy* (Thomsen, 2002). If the layering is flat then it might be referred to as *vertical polar anisotropy*. It is also commonly called *vertical tranverse isotropy* or VTI, given that perpendicular to the axis of symmetry (i.e. the transverse direction) the rock is essentially isotropic. Another common acronym is TTI (tilted transverse isotropy), which refers to the case where the polar axis is not vertical.

(2) Azimuthal anisotropy – in which velocity varies with horizontal direction. Examples of this type can include zones of dipping thin beds, but more usually the application is to zones of near-vertical fractures. The simplest form of azimuthal anisotropy is polar anisotropy with a horizontal axis of symmetry or transverse isotropy with horizontal symmetry (HTI).

5.3.7.1 (Vertical) polar anisotropy

Whilst isotropic materials require only two moduli to describe them, e.g. K (bulk modulus) and μ (shear

Rock property controls on AVO

Polar Anisotropy (with vertical symmetry) or Transverse isotropy with vertical symmetry (VTI)

E.g. shale units with aligned particles or horizontally organized beds or layers (scale $<\lambda$).

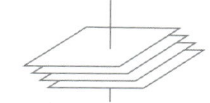

Vertical axis of symmetry. Amplitude varies with angle but no azimuthal variation in properties.

Azimuthal anisotropy

E.g. dipping zones of thin beds or (more usually) fractured rock.

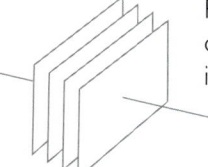

Rock properties vary depending on angle of incidence <u>and</u> direction.

The simplest form of azimuthal anisotropy is Polar anisotropy with horizontal symmetry or HTI (transverse isotropy with horizontal symmetry).

Figure 5.38 Two types of anisotropy important to the seismic interpreter.

modulus), a polar anisotropic (VTI) material requires five parameters for its description (Thomsen, 1986). These are usually taken to be two velocities in the direction of the symmetry axis (i.e. velocities at normal incidence V_{p0} and V_{s0}), ε (epsilon), γ (gamma) and δ (delta). Epsilon (ε) delta (δ) and gamma (γ) are defined according to combinations of elasticity tensors and these are shown with respect to the 3D coordinate system in Fig. 5.39. It should be noted that this model can be applied to seismic given the assumption of weak anisotropy (i.e. ε, γ and δ are each much less than unity). The reader is referred to Mavko et al. (1998) for a detailed explanation of the elastic stiffness tensor system.

With respect to Fig. 5.39, the Thomsen parameters are defined by the following tensor combinations:

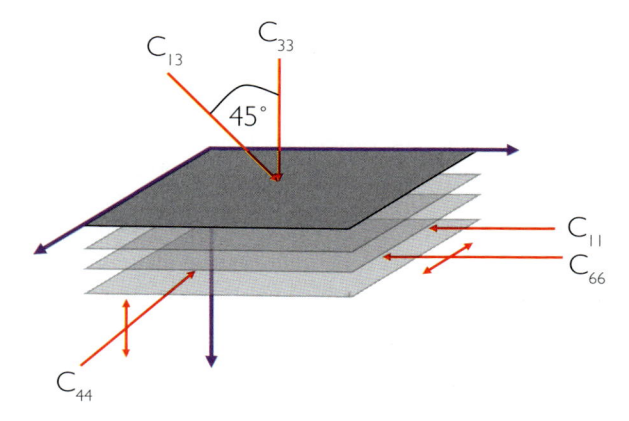

$$\varepsilon = (C_{11} - C_{33})/2C_{33}$$
$$\gamma = (C_{66} - C_{44})/2C_{44}$$
$$\delta = [(C_{13}+C_{44})^2 - (C_{33} - C_{44})^2]/[2C_{33}(C_{33} - C_{44})].$$
(5.1)

Figure 5.39 Polar anisotropy with vertical axis of symmetry (or vertical transverse isotropy (VTI)). Tensor parameters (Voigt notation) shown with reference to a vertical symmetry axis. Note that for horizontal layers: C33 is the M modulus which determines the vertical compressional velocity ($V_p(0°)$), C11 determines the horizontal compressional velocity $V_p(90°)$, C44 is the μ modulus which controls the vertically travelling shear wave perpendicular to the layering ($V_s(0°)$) and also the horizontally travelling shear wave with displacement perpendicular to the layering ($V_{sh}(0°)$), C66 controls the horizontally travelling shear wave with displacement parallel to the layering ($V_{sh}(90°)$), and C13 controls propagation at oblique directions for both P and S waves.

In the horizontal layering case the tensors C_{44} and C_{66} define the vertically polarised shear wave and the horizontally polarised shear waves, respectively. These have commonly been referred to as S_v and S_H in the literature. However, given that the thin layers may have dip a more general notation is $S_\perp$ for the shear wave polarised perpendicular to the symmetry plane (i.e. the plane perpendicular to the axis of symmetry) and $S_\parallel$ for the shear wave polarised parallel to the symmetry plane. Some useful relations for describing velocities with respect to incidence angle can be derived (Thomsen, 2002):

$$V_p(\theta) \approx V_{p0}[1+\delta \sin^2\theta \cos^2\theta + \varepsilon \sin^4\theta]$$
$$V_p(90°) \approx V_{p0}[1+\varepsilon]$$
$$V_{s\perp}(\theta) \approx V_{s0}\left[1+\left(\frac{V_{p0}}{V_{s0}}\right)^2 (\varepsilon - \delta) \sin^2\theta \cos^2\theta\right]$$
$$V_{s\parallel}(\theta) \approx V_{s0}[1+\gamma \sin^2\theta]$$
$$V_{s\parallel}(90°) \approx V_{s0}[1+\gamma].$$
(5.2)

Additionally, common descriptions of P wave and S wave anisotropies are:

$$\varepsilon \approx \frac{(V_{p90} - V_{p0})}{V_{p0}} \quad \text{P wave anisotropy,} \quad (5.3)$$

$$\gamma = \frac{(V_{s\parallel 90} - V_{s0})}{V_{s0}} \quad \text{S wave anisotropy.} \quad (5.4)$$

Polar anisotropy can have a large effect on seismic velocities rendering simple assumptions in seismic processing inadequate. Ignoring the problem causes mis-positioning and degraded focusing of reflections (e.g. Leaney, 2008). In particular, accounting for polar anisotropy can be a key issue in the flattening of seismic gathers (see Chapter 6). Incomplete correction for anisotropic effects has the effect of stretching and squeezing the relative values of the AVO gradient with respect to the intercept (Thomsen, 2002). This may not necessarily have a detrimental impact on qualitative definition of AVO anomalies but it is likely to cause problems for AVO calibration at far angles.

In terms of modelling the effect of polar anisotropy on seismic amplitude an approximation published by Rüger (1997) gives useful insight. This is an anisotropic equivalent of the Aki–Richards approximation:

$$R_p(\theta) = \frac{1}{2}\left[\frac{\Delta Z_0}{Z_0}\right] + \frac{1}{2}\left[\frac{\Delta V_{p0}}{V_{p0}} - \left(\frac{2V_{s0}}{V_{p0}}\right)^2 \frac{\Delta G_0}{G_0} + \Delta\delta\right] \sin^2\theta + \frac{1}{2}\left[\frac{\Delta V_{p0}}{V_{p0}} + \Delta\varepsilon\right]\sin^2\theta \tan^2\theta,$$
(5.5)

where
θ = angle of incidence,
V_{p0} = average P wave (vertical) velocity,
V_{s0} = average S wave (vertical) velocity,
$Z_0 = \rho V_{p0}$ (vertical P impedance),
$G_0 = \rho V_{s0}^2$ (vertical shear modulus),
δ and ε are the Thomsen parameters, equal to zero for the isotropic case.

It is evident from Eq. (5.5) that in order to model the effect of polar anisotropy on P wave AVO it is not necessary to know γ (gamma). This is due to the fact that horizontally polarised shear waves are not excited by P waves or vertically polarised V_s waves. The main problem with using the Rüger approximation for practical purposes is obtaining relevant measurements for the input parameters. Available data are sparse and often ambiguous. There are no real rules of thumb for the interpreter to draw on or indeed convincing and accessible case studies that illustrate the issue of anisotropy as a problem for amplitude interpretation. Unless there is a specific need for such information, the relevant data are not acquired in most exploration wells. Published measurements of polar anisotropy or VTI have been derived in the laboratory and from the analysis of walkaway and multi-component VSPs (e.g. Thomsen, 1986; Vernik and Liu, 1997; Ryan-Grigor, 1997; Macbeth, 2002). Figures 5.40–5.43 show a compilation of measurements.

Some general observations can be made:

- δ may not be zero (as is often assumed) in sands;
- δ values are generally positive for sands but can be negative as well as positive for shales;
- ε is usually positive for sands and shales:
  ε values in shales are generally higher than in sands, and may be higher than δ by a factor of 2 or more.

In an attempt to derive a useful rule of thumb in the absence of accurate data, Ryan-Grigor (1997) suggested that the relationship between δ and V_p/V_s can

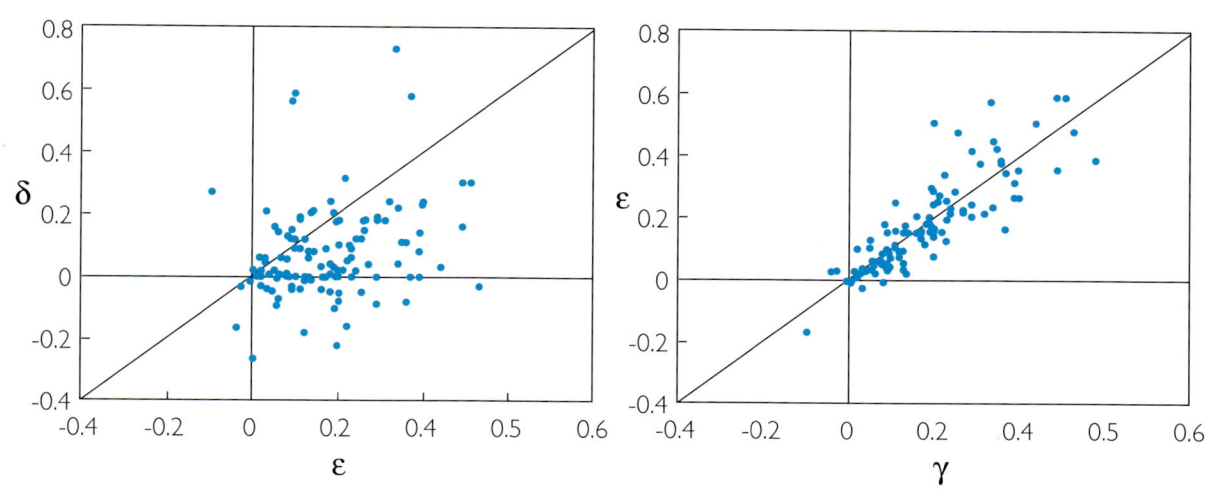

Figure 5.40 Crossplots of vertical polar anisotropy parameters listed by Thomsen (1986), Vernik and Liu (1997) and Ryan-Grigor (1997).

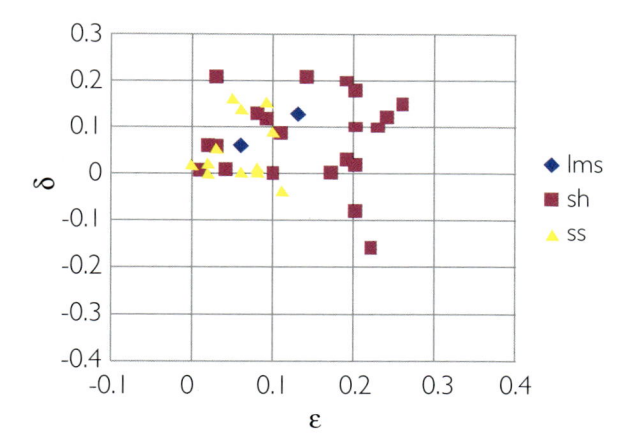

Figure 5.41 V_p/V_s vs δ crossplot based on data selected by Ryan-Grigor (1997).

be simplified by assuming a relationship between C_{13}/C_{44} and V_v/V_s (Fig. 5.44). Using published laboratory data the following relationship can be derived:

$$\frac{C_{13}}{C_{44}} = 3.61 \frac{V_p}{V_s} - 5.06. \quad (5.6)$$

This allows the calculation of δ from

$$\delta = \frac{\left[1 + \left(\frac{C_{13}}{C_{44}}\right)\right]^2 - \left[\left(\frac{V_p}{V_s}\right)^2 - 1\right]^2}{2\left(\frac{V_p}{V_s}\right)^2 \left[\left(\frac{V_p}{V_s}\right)^2 - 1\right]}. \quad (5.7)$$

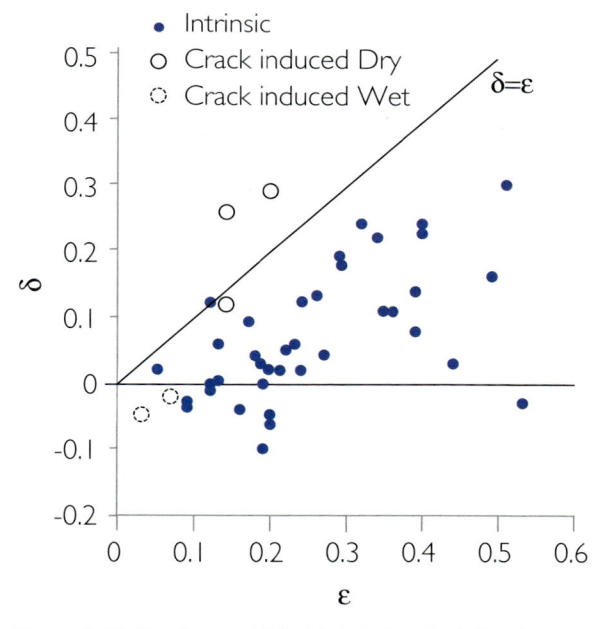

Figure 5.42 Data for ε and δ for black shales – intrinsic anisotropy is inferred from data at high confining pressure of 70 MPa and crack induced anisotropy is inferred from data at low confining pressure (after Vernik and Liu, 1997).

Effectively, at small to moderate angles of incidence it is the difference in δ across the boundary that is the controlling factor in the effect of (vertical) polar anisotropy on the AVO response (Banik, 1987). $\Delta\delta$ is defined as $\delta_2 - \delta_1$, where δ_1 and δ_2 are the delta values on the upper and lower sides of

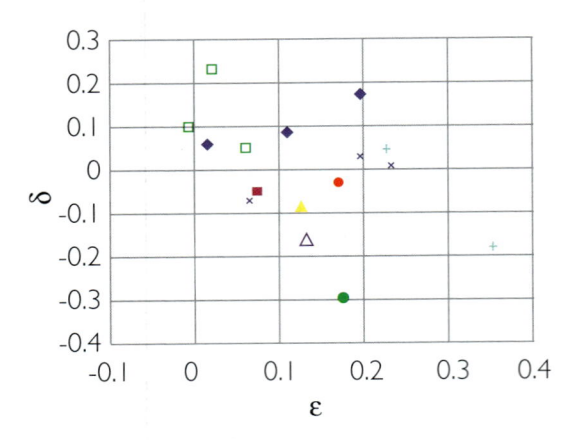

Rock properties and AVO

Figure 5.43 Vertical polar anisotropy data derived from multi-offset VSPs (after Macbeth, 2002) ©Elsevier and reproduced with permission.

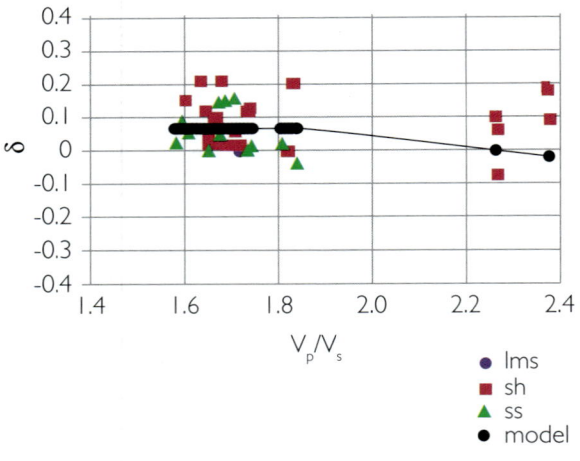

- ◆ Pierre shale
- ■ East Texas sand & chalk
- ▲ East Texas shale
- ● East Texas marl
- △ S. China sea shale
- ● North sea shale
- □ North sea chalk
- × Java sea shale
- + W. Africa unattrib

Figure 5.44 Relationship of δ with V_p/V_s for rock with weak polar anisotropy. Data from Ryan-Grigor (1997).

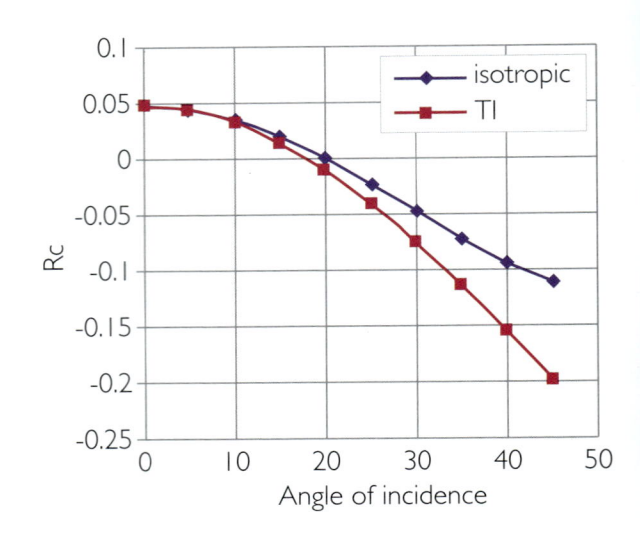

	V_p	V_s	ρ	δ	ε
layer 1	2.438	1.006	2.250	0.150	0.020
layer 2	2.953	1.774	2.036	0.000	0.000

Figure 5.45 Shale on oil sand model – the effect on the top reservoir response is to steepen the gradient if the difference in δ (i.e. $\delta_2 - \delta_1$) is negative.

the interface respectively. When $\Delta\delta$ is negative the gradient is made more negative by the anisotropic contribution. Conversely, when $\Delta\delta$ is positive the gradient is made more positive. The observation that δ is most likely to be positive in shales and close to zero in sands might suggest that the 'normal' effect of vertical polar anisotropy is to enhance negative gradients and stretch the gradient relative to the intercept. Figure 5.45 shows an example comparing a Class IIp top oil sand response with and without including polar anisotropy in the overlying shale. The anisotropy becomes important in terms of the amplitude and the AVO gradient beyond an angle of incidence of around 30°. In Fig. 5.45 the amplitude of the anisotropic case at 40° is 50% higher in absolute terms than the isotropic model.

It is possible therefore that this type of anisotropy could generate false positive hydrocarbon indicators, for example in the case of Class II wet sands, although, as previously stated, there are no definitive published examples. Equally, it is possible that layered sections of shales may have variations of δ values, for example owing to changes in organic content or pressure, that could lead to AVO anomalies dominated by the anisotropic component. This is perhaps one explanation of some AVO failures in which no reservoir was found (e.g. Allen and Peddy, 1993).

5.3.7.2 Azimuthal anisotropy

Most rock formations have sub-vertical fracture systems caused by stresses within the crust. Fractures

Rock property controls on AVO

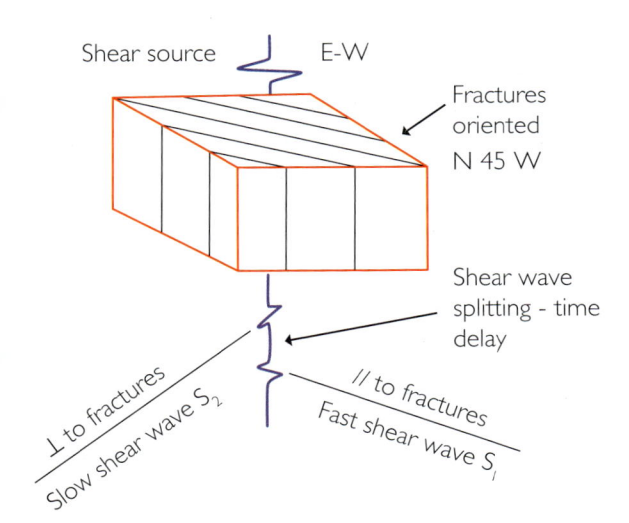

Figure 5.46 Shear wave splitting in azimuthally anisotropic media (after Martin and Davis, 1987; Crampin, 1990).

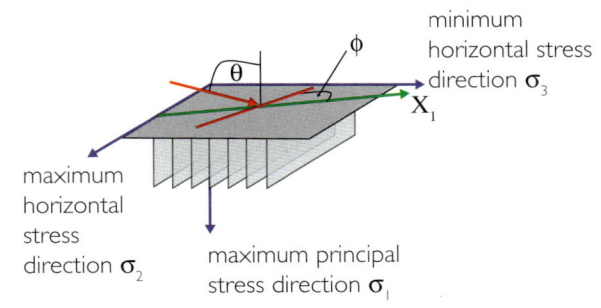

Figure 5.47 Modelling fractures as polar anisotropy with horizontal axis of symmetry or horizontal transverse anisotropy (HTI). Reflectivity depends on the angle of incidence (θ) and the azimuth (ϕ). Fracture orientation is interpreted with respect to changes in the AVO gradient.

that relate to the extensional parts of the current stress regime are likely to be open and fluid filled. It is these fractures that effectively contribute to azimuthal anisotropy. Closed fractures do not contribute significantly because of a lack of impedance contrast across the fracture. Some oil and gas fields rely almost entirely on fractures for their production. It follows then that exploiting azimuthal anisotropy may provide direct information on fracture presence and possibly on permeability. There are generally two approaches to exploiting azimuthal anisotropy for detection of fractures, namely the analysis of 'shear wave splitting' on land and the analysis of P wave AVO differences with azimuth (commonly referred to with an array of different acronyms such as AVOA, AzAVO, AVD, AVOZ).

Shear wave techniques commonly employed for fracture detection on land are based on the fact that a shear wave entering an azimuthally anisotropic rock unit at an angle oblique to the fractures splits into two waves with different polarisations (Fig. 5.46). A fast shear wave (S1) is polarised parallel to the fractures whereas the slower shear wave (S2) is polarised perpendicular to the fractures. The anisotropy is commonly described by the time delay (i.e. $(T_{s1}-T_{s2})/T_{s2}$) and an estimate of relative crack density can be made on the basis of the delay (e.g. Crampin et al., 1986). An example of shear wave splitting is shown in Chapter 7 and the reader is referred to Lynn (2004) for a thorough introduction.

In areas with significant azimuthal anisotropy the P wave AVO response is dependent on the azimuth of the seismic. A simple model that is commonly used to describe this situation is polar anisotropy with horizontal axis of symmetry. Figure 5.47 illustrates that the AVO at an interface between an isotropic layer and one with vertical cracks will vary depending on the azimuth.

Rüger (1998) derived a reflectivity equation, again similar to the Shuey equation, to characterise this situation. Following Jenner (2002), for angles up to about 35° it can be written:

$$R(\theta,\phi) = I + \left[G_1 + G_2 \cos^2(\phi-\beta)\right]\sin^2\theta, \quad (5.8)$$

where

$R(\theta,\phi)$ = reflection coefficient as a function of incidence angle (θ) and ϕ (source–receiver azimuth with respect to a pre-defined direction, such as true north),

β = angle between chosen zero azimuth and either the isotropy or symmetry axis plane

I = P wave impedance contrast divided by 2

G_1 = isotropic AVO gradient

G_2 = anisotropic gradient defined by:

$$G_2 = \frac{1}{2}\left[\Delta\delta^{(v)} + 2\left(\frac{2}{g}\right)^2 \Delta\gamma\right], \quad (5.9)$$

where $g = V_p/V_s$ (average P wave velocity / average S wave velocity),

$\Delta\delta^{(v)}$ = the difference in the Thomsen style parameter δ for HTI,

$\Delta\gamma$ = change in the shear wave splitting parameter γ (note that γ is directly related to the crack density).

Rock properties and AVO

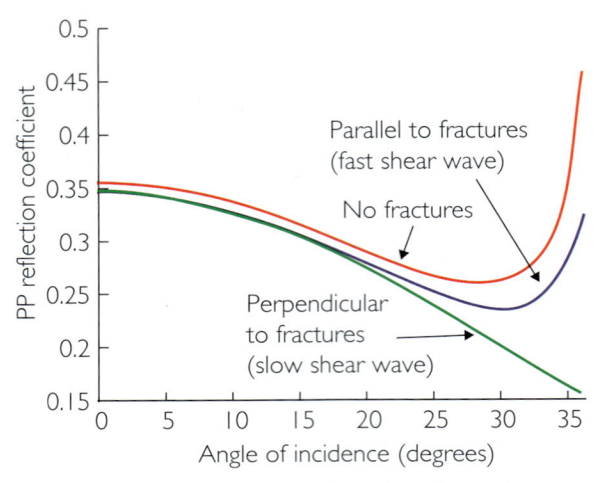

Figure 5.48 AVO model of VTI shale overlying fractured granite (after Leaney et al., 1995).

Figure 5.48 illustrates a modelled solution for anisotropic shale overlying a fractured granite using the HTI assumption. The AVO responses perpendicular and parallel to the fracture direction show significant amplitude differences on far angles. In this particular case the dimming of far offset amplitudes (i.e. steepest negative AVO gradient) occurs perpendicular to fractures and is consistent with Thomsen's (1995) suggestion that the lesser AVAZ gradient is parallel to fracture strike (Goodway et al., 2010). There are numerous case studies in which fractures have been interpreted from multi-azimuth seismic (e.g. Hall and Kendall, 2003; Neves et al., 2003; Gray et al., 2002) but there does not appear to be a rule of thumb with regard to the interpretation of fracture orientation. Far offset dimming may occur perpendicular or parallel to fractures. Clearly, whilst the AVO provides the all-important spatial context, integration with other data such as borehole image logs, mud loss measurements and estimates of regional and local stresses (e.g. borehole breakout studies) is required. Modelling of course is also a key component in interpretation but it is clear that care needs to be taken in how the models are applied. Goodway et al. (2010) have documented potential ambiguities in the application of Rüger's two-term equation, suggesting that a three-term solution is required.

There are clearly some challenges in extracting fracture information from seismic even assuming that seismic signal-to-noise, azimuth distribution and fold are optimised (Johns et al., 2008).

- If the fracture density is low the anisotropy might be too weak to detect.
- Fractures in multiple azimuths can diminish the magnitude of azimuthal anisotropy.
- Steeply dipping beds can induce azimuthal anisotropy as well as vertical fluid filled fractures.
- Anisotropy at the reservoir level might be influenced by anisotropy from above.

5.4 The rock model and its applications

The previous discussion illustrates that in order to effectively interpret seismic amplitude data the interpreter not only needs to understand the geology of the basin (e.g. sedimentology, stratigraphy and structural history) but also the rock physics that explains the nature of the reflectivity. It is important therefore to generate a database of rock properties from an analysis of wireline log and other data. This work involves detailed integration of petrophysical and rock physics analyses, including log quality control and conditioning, fluid substitution and application of various rock physics models (Chapter 8). As such, the job of putting together a rock physics database is commonly done by specialists rather than seismic interpreters. From the rock physics database a rock model (Fig. 5.49) can be established which defines:

(1) seismic lithofacies (based on crossplots and log plots (upscaled data)) and statistical variation (e.g. mean, standard deviation);
(2) thicknesses and stratigraphic ordering of lithofacies;
(3) first order depth trends of
  (a) elastic parameters of seismic lithofacies (i.e. V_p, V_s, density and porosity)
  (b) effective pressure and temperature
  (c) fluid properties.

The rock model can be used in numerous applications including the generation of seismic rock models for reflectivity interpretation (Chapter 7), seismic trace inversion (Chapter 9) and net pay analysis (Chapter 10).

5.4.1 Examples of rock model applications

The time and effort put into generating a rock physics database and rock model to some extent depends on

The rock model and its applications

a) Seismic Lithofacies

b) Elastic parameter pdfs

c) Parameter correlations

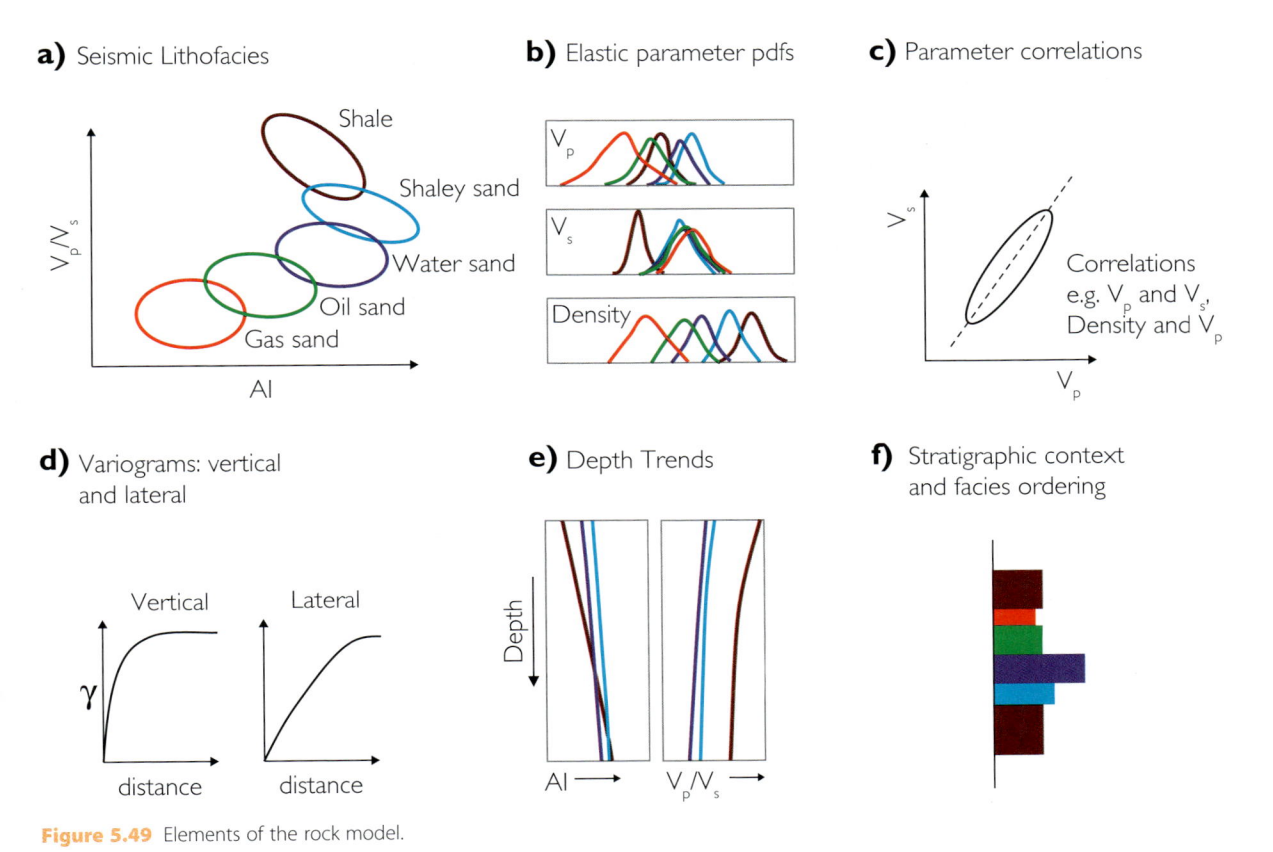

Correlations e.g. V_p and V_s, Density and V_p

d) Variograms: vertical and lateral

e) Depth Trends

f) Stratigraphic context and facies ordering

Figure 5.49 Elements of the rock model.

the application. In partially explored areas it is often the case that a relevant seismic model can be established fairly quickly by an analysis of wells close to the prospect. In such cases it can reasonably be assumed that the stratigraphy and rock types have been adequately sampled in the offset wells. In other cases, such as regional evaluations or detailed statistical applications, a considerable amount of time is spent in generating the rock model and understanding variance.

Figure 5.50 illustrates an example of an application of a simple seismic rock model in interpreting a potential oil prospect. A migrated full stack seismic section from an area in which the reservoir sandstones are relatively widespread and thick is shown in Fig. 5.50a. A top sand interpretation has been made with a consistent black peak (hard event), tying the water wet well down-dip to the right of the section (Fig 5.50a). A basic workflow was rapidly implemented using the data from the down-dip well, comprising fluid substitution of water wet sands to hydrocarbon (nearby fields provide information on likely fluid parameters), extracting average properties from the logs and generation of a single interface AVO plot. The AVO plot (Fig 5.50b) suggests that at the effective angle of the seismic the presence of hydrocarbons would result in a dramatic dimming of the top reservoir reflection. So, either the initial interpretation is correct and the prospect is water filled or there is another interpretation.

The AVO plot suggests that an oil water contact would have a similar polarity and level of amplitude to the top of the water filled sand (Fig. 5.50b). So there is the possibility that the initial interpretation has deviated from the top of the water filled sand outside closure and onto the contact reflection over the culmination. Several observations would appear to justify this interpretation. The peak is fairly flat over the central portion of the culmination consistent with a contact. In addition, the isochron between the picked peak and the prominent peak at the top of

Rock properties and AVO

the section shows a thickening over the culmination. This is unusual as there is no evidence of structural inversion in this part of the basin. Ghosting-in the top reservoir based on isochronous stratigraphic layering (Fig. 5.50c) would make sense of the flat portion of the picked peak as the oil water contact. Further evidence for the interpretation of a contact is also given by the presence of the apparent reflector terminations below the peak (Fig. 5.50c). These features are characteristic of modelled contact effects in thick sands in this area. The feature was drilled and found to be oil bearing.

An example of a more detailed application of a rock model has been described by Avseth et al. (2003) in which the statistics of the rock model have been used to predict the probability of seismic responses representing particular lithofacies scenarios. The workflow applied is as follows.

(1) Determine lithofacies and elastic parameter statistics at the depth of interest.
(2) Establish layer configurations (boundary types and vertical facies associations) based on an understanding of the geology – in this case the target is in a deep sea submarine fan setting.
(3) For each boundary type use Monte Carlo simulation to generate intercept vs gradient scatter plots (a statistical analysis of these results provides the probability of a given value of intercept and gradient representing a particular boundary type).
(4) Generate intercept and gradient from the seismic trace data and crossplot.
(5) Perform AVO calibration (i.e. scale the intercept and gradient to the model values). As the gradient is not a directly scalable quantity this step in essence is performed by scaling reflectivities and recalculating the gradient (see Section 6.3.3). The scalers are determined for the 'background' responses.
(6) Blind test at wells (determining acceptable misclassification error).
(7) Use the probability model to predict boundaries from seismic (the number of interface categories can be reduced to reflect pay (oil and gas) vs non-pay outcomes).

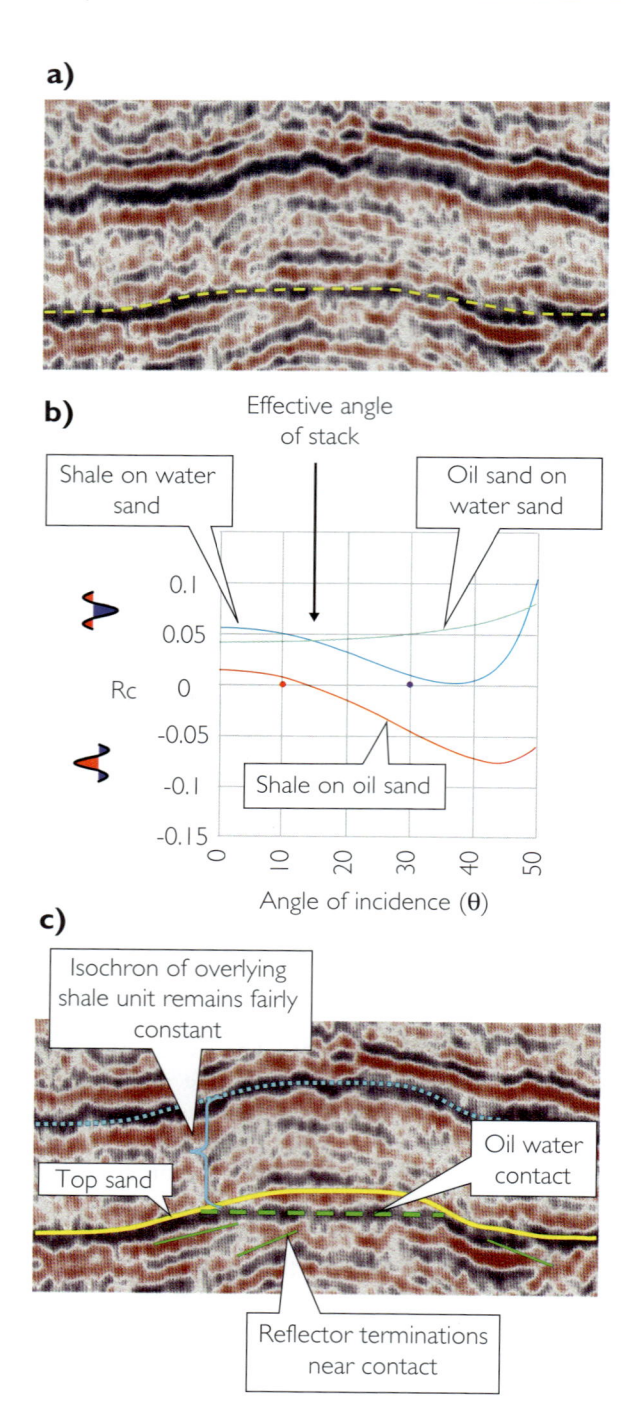

Figure 5.50 Migrated stack seismic interpretation aided by an AVO plot constructed from data in a nearby well; (a) original interpretation of top reservoir, (b) AVO plot showing responses for water sand, oil sand and oil water contact, (c) revised interpretation taking into account the AVO model and other seismic observations.

Figure 5.51 illustrates some key components of statistical AVO, including Monte Carlo simulation of modelled AVO responses and the need to 'calibrate' the background trend. Whilst accounting for rock physics variability is a good thing there are

Rock properties, AVO reflectivity and impedance

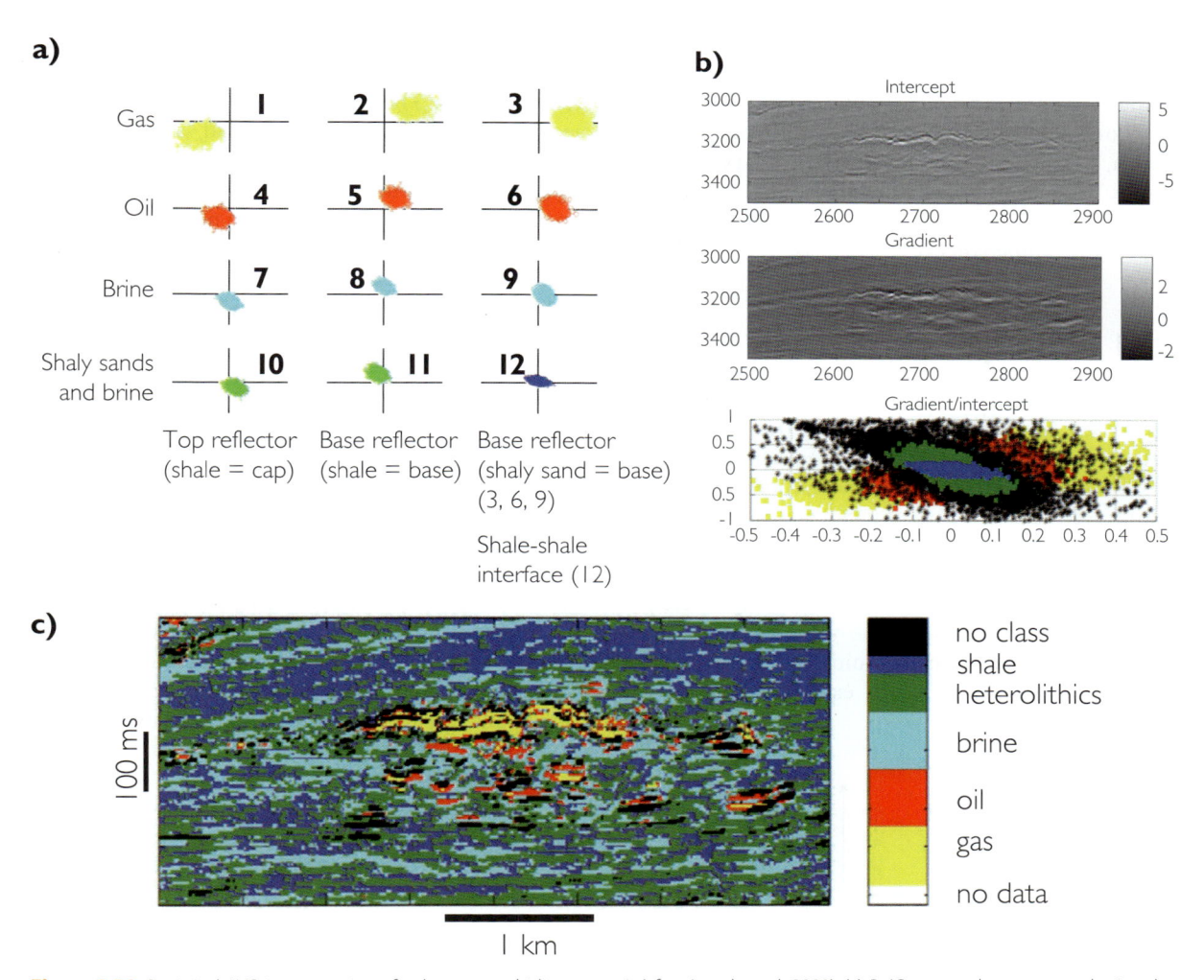

Figure 5.51 Statistical AVO interpretation of a deep sea turbidite reservoir (after Avseth et al., 2003): (a) R_0/G scatter plots generated using the statistical rock physics database and Monte Carlo simulation, (b) intercept and gradient sections and calibrated AVO plot (note that the 'background' trend is in black), (c) section showing predicted lithofacies.

some significant issues that the interpreter would need to address before accepting the results with confidence. Assuming that the lithofacies are optimally defined it is clear that the statistics need to be representative (and upscaled), with several wells at least needed to condition the model. Other issues include reflector interference and the effect of seismic noise, both on the AVO crossplot (see Section 5.6) and in AVO calibration (see Section 6.3.3).

5.5 Rock properties, AVO reflectivity and impedance

There are many ways of using AVO for the interpretation of fluid and rock content from seismic. Perhaps the most important, however, is the use of linearised (two-term) approximations (i.e. based on Shuey's equation) (Chapter 2). Linear combinations of intercept and gradient describe a continuum of rock and fluid effects, forming the basis for a set of readily useable tools that aid fluid and rock discrimination from seismic. In the 1980s and 1990s various workers combined intercept and gradient sections in different ways to create a variety of AVO attributes. Examples of these included the 'fluid factor' attribute (Smith and Gidlow, 1987), pseudo-Poisson's reflectivity (Verm and Hilterman, 1995) and R_p-R_s (Castagna and Smith, 1994). At the time it was thought that there may be a universal AVO indicator for fluid identification from seismic. However, over time it was realised that most of the 'hard wired' products can be understood as representing particular angle projections or coordinate rotations on the AVO crossplot (e.g. Smith, 2003) and that AVO is an adaptive technique. This section describes the mechanics of linearised AVO in terms of both reflectivity and angle dependent 'impedance'.

5.5.1 AVO projections, coordinate rotations and weighted stacks

Linearised AVO enables the construction of a seismic dataset at any angle using Shuey's equation $R = R_0 + G \sin^2 \theta$ (Fig. 5.52). For example, an intercept/gradient projection at an angle of incidence of 30° (i.e. $R = R_0 + G \sin^2 30°$) will look similar to an angle stack with an effective angle of 30°, although they will not be exactly the same owing to the different effects of noise in the two datasets. The value of projections, however, is in generating seismic displays beyond the acquired angles in order to exploit differences in AVO.

Figure 5.52 illustrates the concept. Three separate AVO responses are shown, typical of a Class IIp oil sand scenario. The amplitude difference of the projected responses for the oil sand and wet sand increases with increasing angle. It also shows how the difference diminishes with increasing negative angle (of course negative angles do not exist in reality; they are simply a graphical construct). A simple 2D model (Fig. 5.53) shows how, with this dataset, the fluid effect can be maximised at $\sin^2 \theta = 0.35$, whereas the fluid effect is removed with a projection at $\sin^2 \theta = -0.35$.

Figure 5.54 illustrates a model AVO plot in which the porosity is varied as well as the fluid content. At $\sin^2 \theta = 0.5$ (i.e. $\theta = 45°$) the various porosity cases for each fluid type converge (effectively presenting an angle at which fluid effects might be optimised). Porosity is optimised where the different fluid cases for each porosity converge at around $\sin^2 \theta = -0.5$. Note that the angle at which the convergence occurs is not the same for all porosity cases, but it is close enough for the idea of a lithological projection to be a practical possibility. Reality is often more complex than the simple models shown but they highlight how AVO is a continuum of

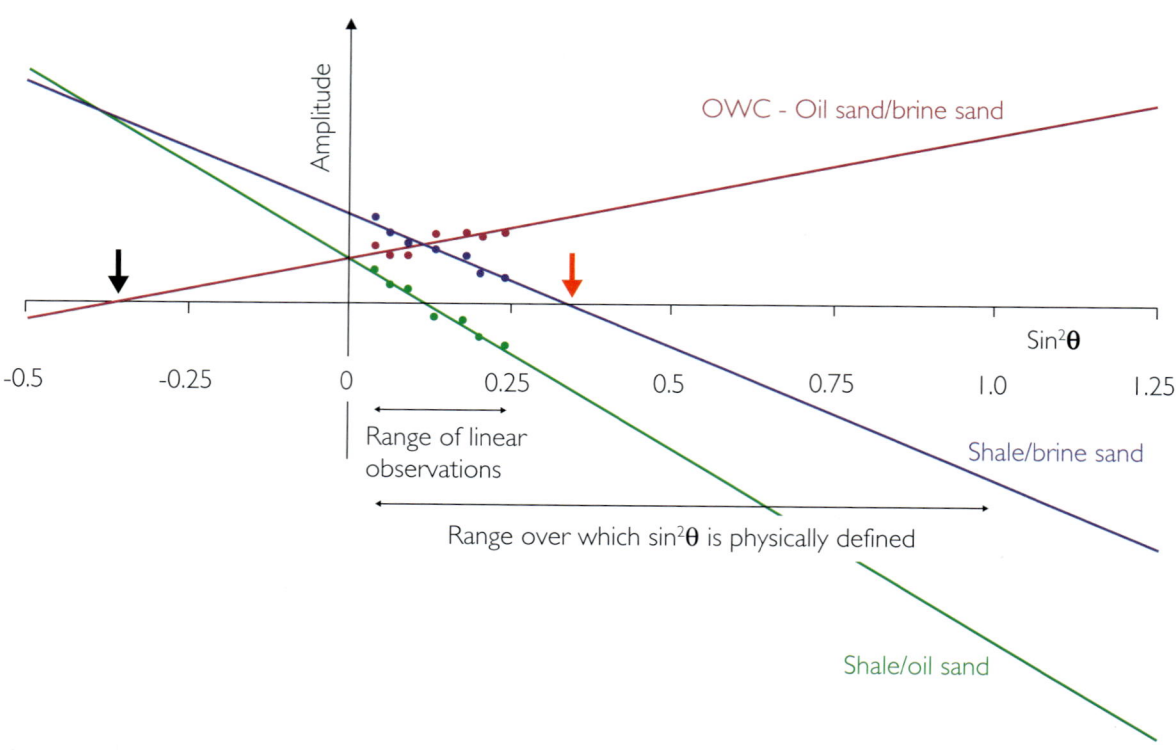

Figure 5.52 Linearised (two-term) AVO (modified after Whitcombe et al., 2002).

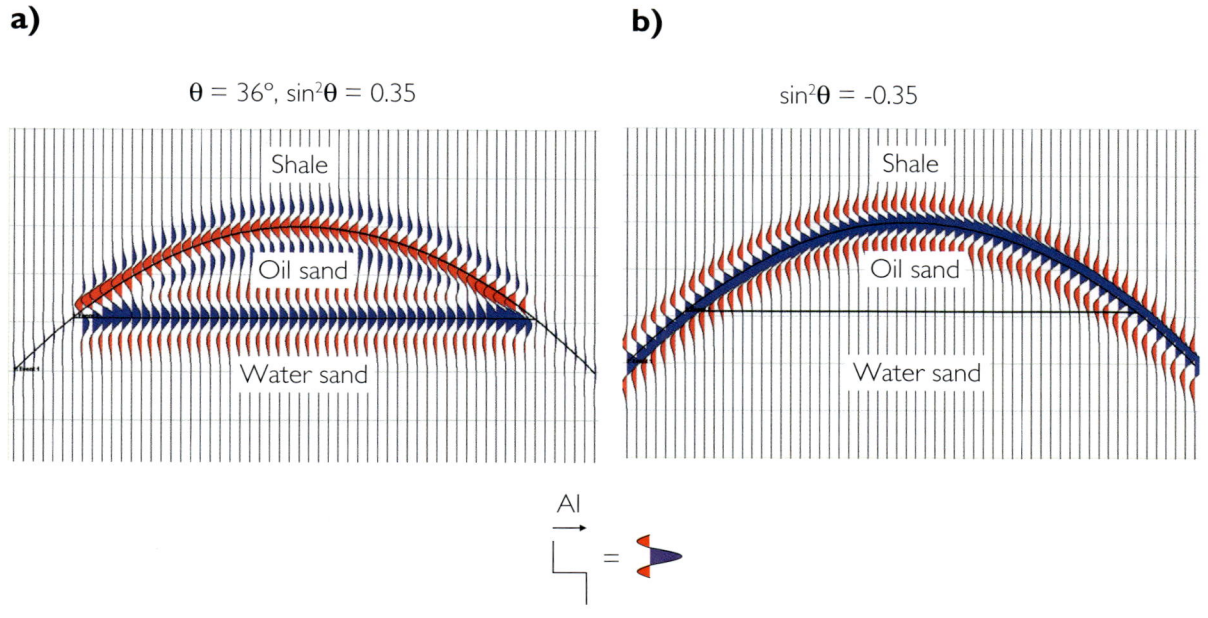

Rock properties, AVO reflectivity and impedance

a) $\theta = 36°$, $\sin^2\theta = 0.35$

b) $\sin^2\theta = -0.35$

Figure 5.53 The 2D model showing seismic projections at different values of $\sin^2\theta$; (a) $\sin^2\theta = 0.35$ ($\theta = 36°$) and (b) $\sin^2\theta = -0.35$.

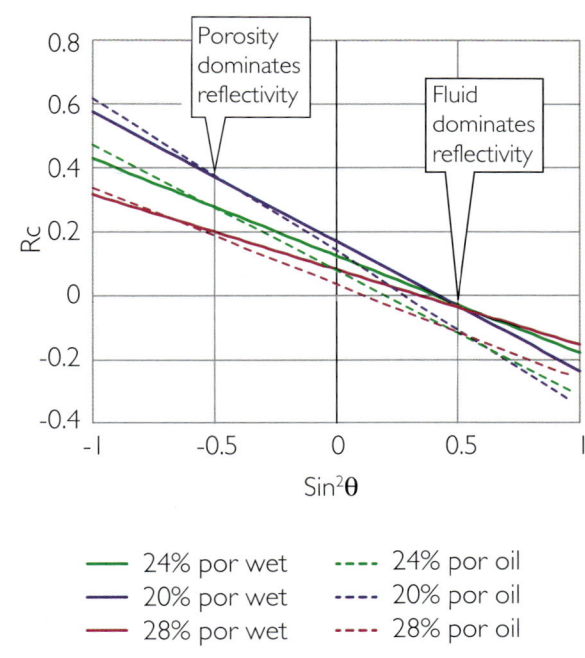

Figure 5.54 AVO plot showing Shuey reflectivity responses for different shale/sand and fluid fill combinations.

— 24% por wet   ---- 24% por oil
— 20% por wet   ---- 20% por oil
— 28% por wet   ---- 28% por oil

rock and fluid effects. Individual seismic stacks are effectively snapshots along this continuum.

Whilst the principle of AVO projections can be appreciated by using the AVO plots shown in Figs. 5.52 and 5.54, the AVO (intercept and gradient) crossplot is the best way to approach AVO analysis owing to the ability to identify different populations of responses. Figure 5.55 illustrates how projections work in R_0–G space using model data for a Class I brine sand/ Class IIp oil sand situation. It is useful to think of the x axis as the reflectivity projection axis onto which the data points are projected. So, an angle projection at zero degrees (i.e. $R = R_0 + G \sin^2 0°$) is simply a projection of the points along the y axis onto the intercept axis of the AVO crossplot (Figs. 5.55a, b). Increasing the angle of the projection involves rotating the axes in an anticlockwise direction (Figs. 5.55c, d), with the degree of rotation determined by the function that relates the angle of incidence (θ) to the AVO crossplot angle (χ):

$$\sin^2\theta = \tan\chi \text{ (Whitcombe et al., 2002).} \quad (5.10)$$

For example, angle projections at $\theta = 20°$ and $\theta = 30°$ are achieved by rotating the AVO crossplot axes anticlockwise by 7° and 14° respectively (Figs. 5.55c, d). Note how in the case of the $\chi = 14°$ projection the amplitude of the shale on water sand is effectively zero whereas the shale on oil sand is a soft (negative) response. This is what Hendrickson (1999) terms the 'magic of AVO', turning a dim spot on a full stack section into a bright spot on a projected stack.

Rock properties and AVO

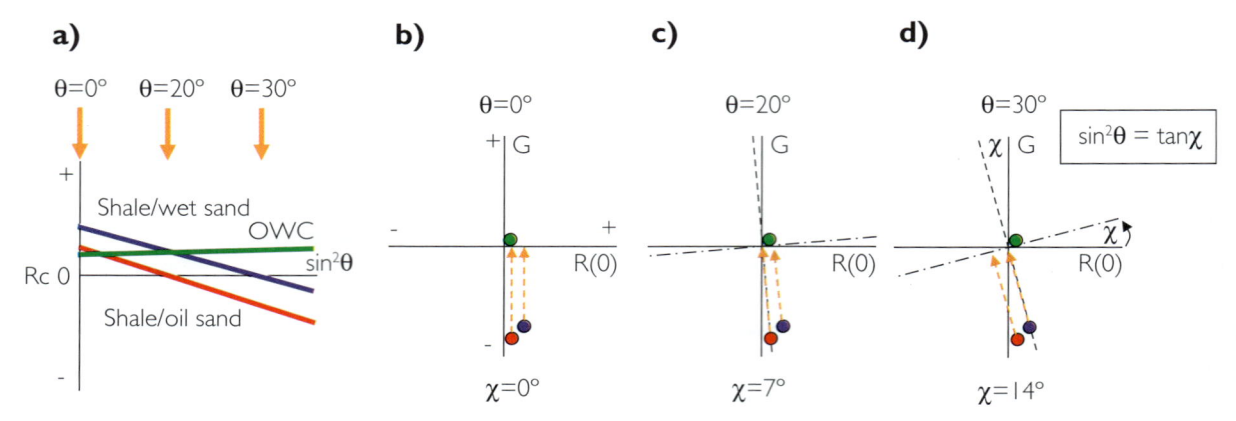

Figure 5.55 The principle of AVO projections; (a) AVO data plotted on the AVO plot, (b)–(d) AVO crossplots showing data projections for various angle rotations.

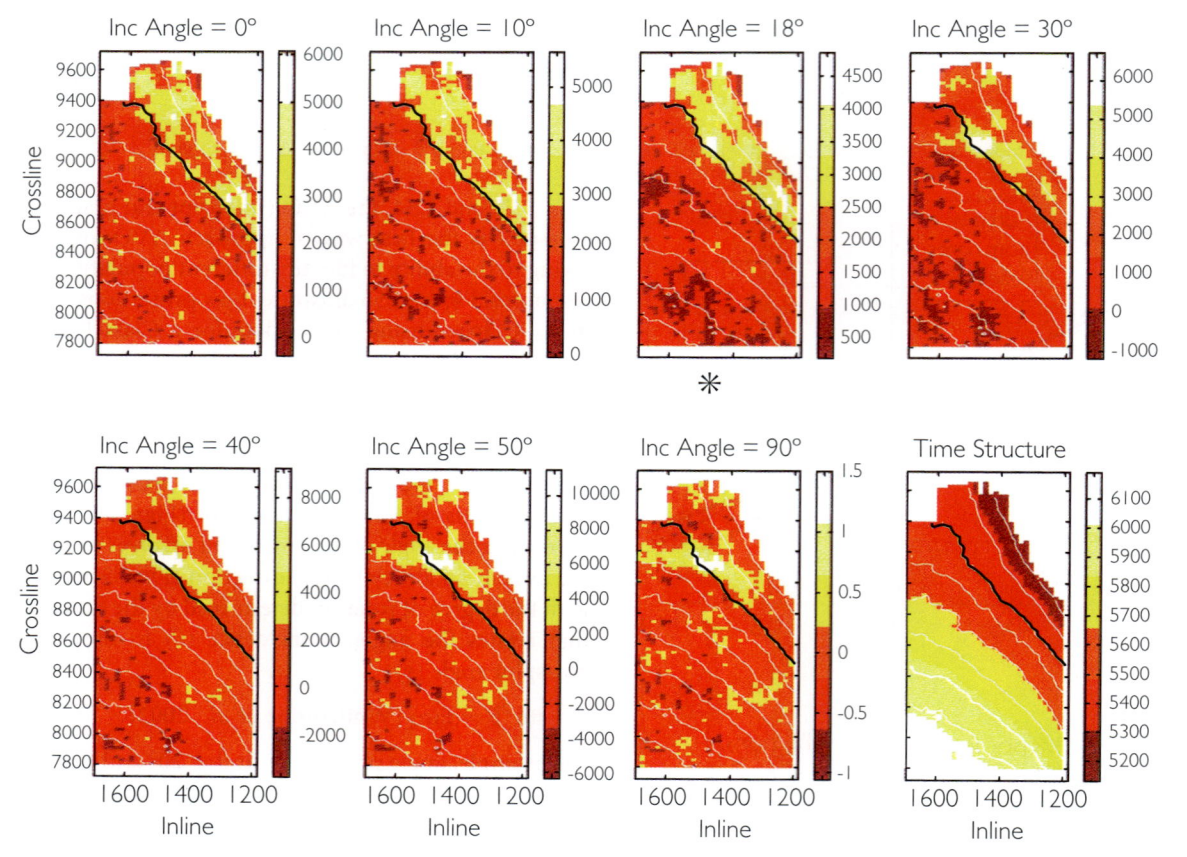

Figure 5.56 AVO projections generated using Shuey's equation, combining intercept and gradient maps with varying angle of incidence (θ) (after Hendrickson, 1999).

The idea of AVO projections has a fundamental impact on the practical approach to AVO analysis. Hendrickson (1999) realised that maps can be generated at any incidence angle and that a comparison of these maps is useful in appreciating the nature of the AVO response. Figure 5.56 illustrates an example from the Auger Field in the Gulf of Mexico. Note that the optimum angle of $\theta = 18°$ gives the most consistent anomaly which fits closely to the structural spill of the field (i.e. this is the optimum fluid angle). In contrast, $\theta = 30°$ shows a different map with the main amplitude anomaly oriented at an angle and extending down-dip beyond the closing contour. The change is interpreted to be due to the dominant influence of lithology rather than fluid and the amplitude is related to the orientation of turbidite channels within the field. Note that beyond this angle the maps do not change significantly. It is characteristic that there is usually a high degree of sensitivity to angle in the region of the fluid angle.

Notionally, it should be possible to generate projections at any AVO crossplot (χ) angle, but Shuey's equation has an angle limit of $\theta = 90°$ (i.e. $\chi = 45°$). It might be thought that $\sin^2\theta$ could simply be replaced with $\tan\chi$, but at large χ angles (beyond around 70°) the projections show a rapid rise in calculated values and the reflection coefficient may exceed unity. Whitcombe et al. (2002) showed that this problem was essentially overcome by scaling Shuey's equation in terms of $\cos\chi$, which gives:

$$R = R_0 \cos\chi + G\sin\chi. \tag{5.11}$$

This 'modified Shuey' equation allows for reflectivity calculations across the whole range of χ angles and gives results that maintain correct relative differences between AVO responses. As such it can be considered the generalised two-term AVO equation. To quote from Whitcombe et al., 2002, it is important that the interpreter understands that the aim of this manipulation is not to 'produce a model that replicates observed reflectivity beyond 30° and up to the critical angle' but to create a 'useful model that can be constructed for any real linear combination of A (intercept) and B (gradient), effectively extrapolating the observations along the $\sin^2\theta$ axis in either direction beyond the physically observed range'.

One of the reasons for modifying Shuey's equation was to accommodate the possibility of projections which sometimes emphasise lithological variations. Figure 5.57 shows AVO responses for two sands with different porosity, both brine and gas filled. A lithology optimised (fluid minimised) angle can be generated by rotating the AVO crossplot axes either by 55° in a clockwise (i.e. negative direction) or anticlockwise by 305°. The resulting equations are equivalent:

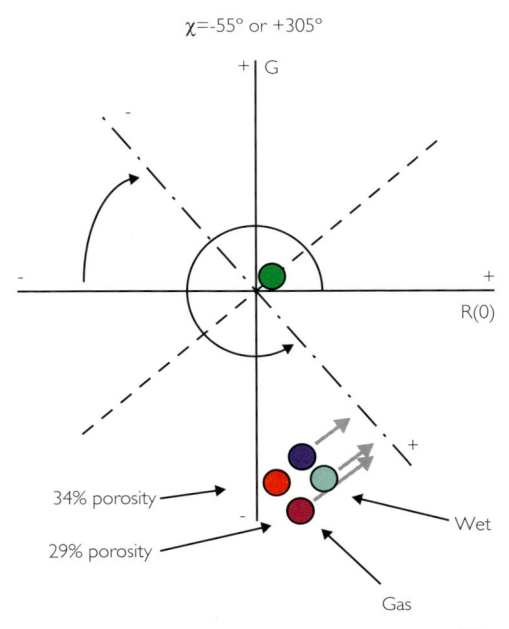

Figure 5.57 AVO projection maximising porosity differences and minimising the effect of fluid fill.

$$R = R_0 \cos 305° + G \sin 305°$$
$$R = R_0 \cos 55° - G \sin 55°. \tag{5.12}$$

An example of AVO projections applied to log data (in a Class I oil sand setting) is shown in Fig. 5.58. The figure shows the log curves with a synthetic gather and intercept and gradient traces. An AVO crossplot is shown in Fig. 5.58b, with the blue and red points representing wet sands plus shales and the oil zone respectively. An important feature is that the general reflectivity background (i.e. wet sands and shales) has a trend from top left to bottom right (i.e. decreasing amplitude with increasing angle). As expected, top oil sand reflections fall to the left of the general wet trend and oil water contact reflections fall in the upper right quadrant. Drawing a line through the wet data that intersects the origin effectively defines the fluid angle (Fig. 5.58c). The projection angle χ can be simply calculated (Fig. 5.58c):

$$\chi = \tan^{-1}(-y/z). \tag{5.13}$$

In this case $y = -0.08$ and $z = 0.2$, so $\chi = 22°$, or $\theta = 39°$.

To illustrate the potential impact of AVO projections in seismic interpretation the results of a 2D model are shown in Fig. 5.59. The 2D model was

Rock properties and AVO

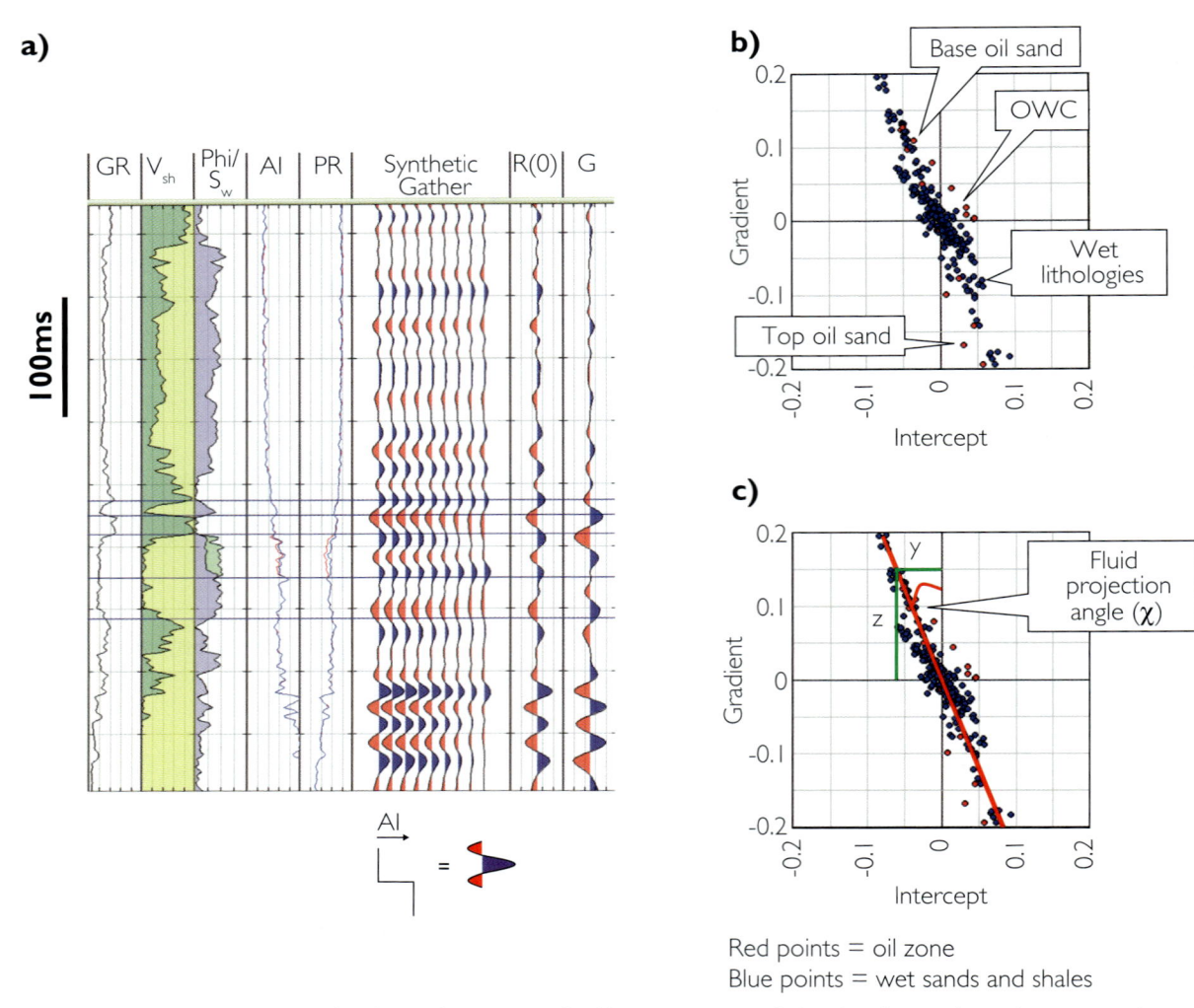

Figure 5.58 AVO projections using log data; (a) log section with oil bearing reservoir, calculated synthetic gather and intercept and gradient traces, (b) AVO crossplot showing the dominant trend of non-reservoir responses, (c) AVO crossplot with calculated fluid projection angle.

Red points = oil zone
Blue points = wet sands and shales

created using well data containing a water bearing sand with the data simply propagated parallel to the top sand horizon. Oil was substituted into the model after defining the position of the contact. Seismic sections have been generated using the Zoeppritz equations at angles of incidence of 10° and 30° (i.e. notional angles for near and far stacks) (Figs. 5.59a,b). Figures 5.59c, d show linear projections at fluid and lithology optimised angles respectively. The sections show that, while the far angle stack is quite good at showing the fluid effects, the projection is more effective at subduing background reflectivity and highlighting the oil sand. The lithology angle was empirically determined by inspection of a range of projection angles for the model.

Implementing AVO projections with partial stack data is fairly straightforward if it is assumed that the two-term approximation applies. So, for example, intercept and gradient might be derived from near and far stack data using the equations:

$$G = \frac{A_f - A_n}{\sin^2\theta_f - \sin^2\theta_n} \quad (5.14)$$

$$R_0 = A_n - G\sin^2\theta_n, \quad (5.15)$$

where A = amplitude, θ = incidence angle, f = far, n = near.

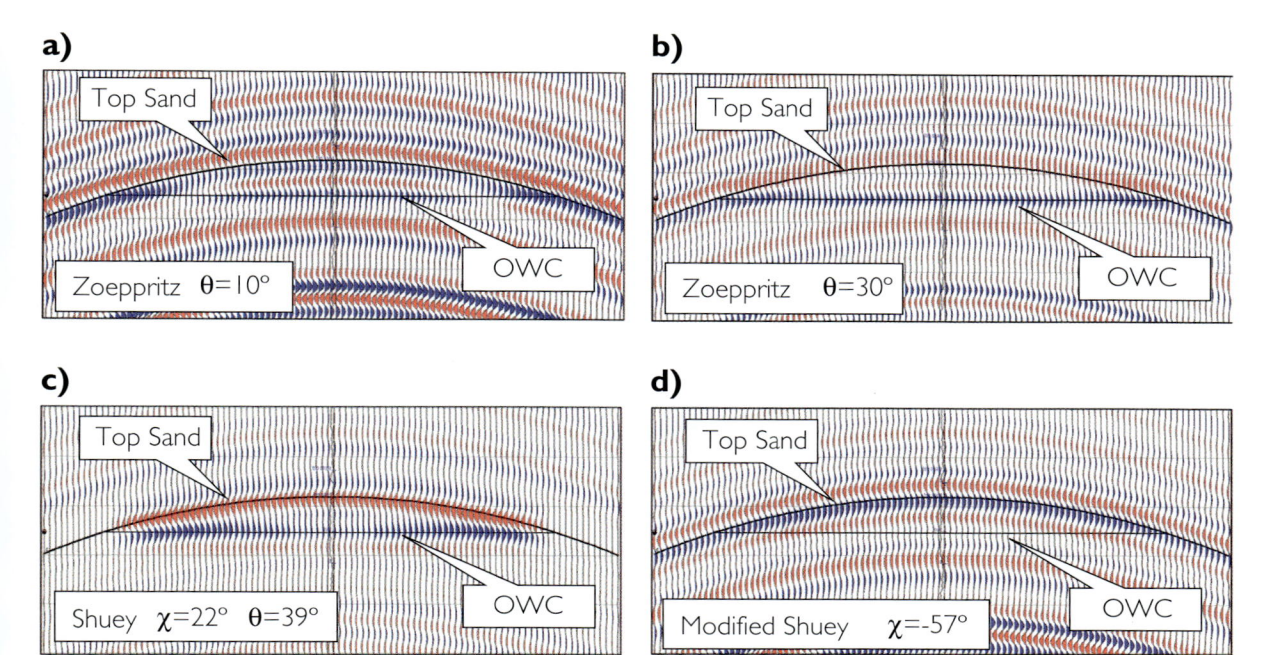

Rock properties, AVO reflectivity and impedance

Figure 5.59 Simple 2D models of a Class I oil sand; (a) reflectivity section calculated using Zoeppritz $\theta = 10°$, (b) reflectivity section calculated using Zoeppritz $\theta = 30°$, (c) section calculated at $\theta = 39°$, $\chi = 22°$ using Shuey's equation and showing optimum fluid response, (d) modified Shuey projection $\chi = -57°$ or $\chi = 303°$ highlighting the top reservoir as a continuous reflection.

Alternatively, AVO crossplot projection angles can be directly applied to near and far angle stack data (D. Whitcombe, personal communication):

$$Projection(\chi) = (Near * c) + (Far * k), \quad (5.16)$$

where

$$k = (\tan \chi - \tan \chi_{near})/\Delta \tan \chi, \quad (5.17)$$
$$c = 1 - k,$$

and

$$\Delta \tan \chi = (\tan \chi_{far} - \tan \chi_{near}). \quad (5.18)$$

One potential advantage in AVO analysis of partial stacks over intercept and gradient is that residual moveout effects (see Chapter 6) can be taken into account. So, for example, in the presence of residual moveout, nears and fars can be picked separately, adjusting the interpretation for differences in timing, and the AVO projections can be applied to the amplitude maps.

An alternative projection approach with partial stacks might be to generate the projection directly from near and far data with an angle of projection determined from the near vs far crossplot (Fig. 5.60). The angle in near/far space is not the same as the χ

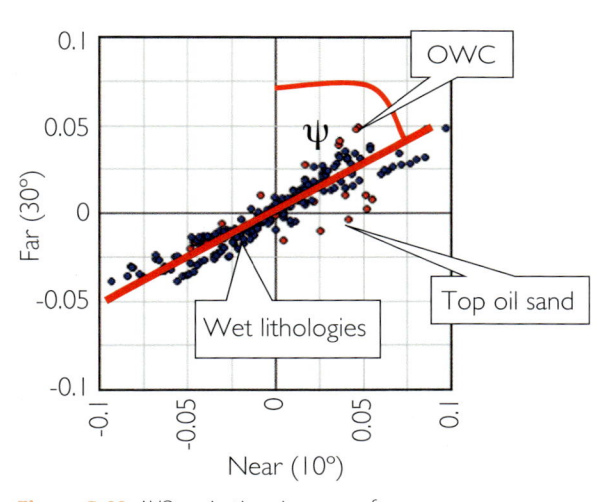

Figure 5.60 AVO projections in near vs far space.

angle discussed previously so it has been given the symbol ψ. The projection equation then becomes:

$$Projection(\psi) = (Near * \cos \psi) - (Far * \sin \psi). \quad (5.19)$$

With this formulation fluid projections tend to have quite large ψ angles whereas lithology projections tend to have smaller ψ angles.

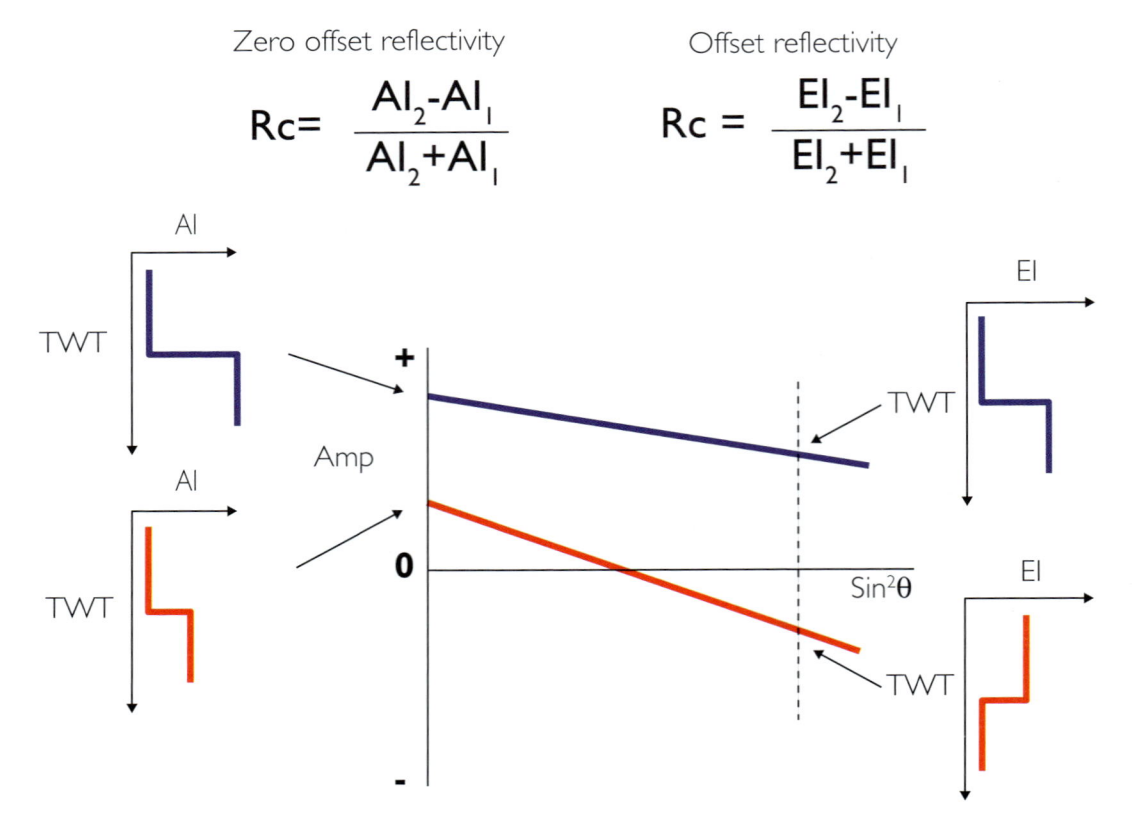

Figure 5.61 The concept of elastic impedance.

5.5.2 Angle-dependent impedance

It was commonplace prior to the late 1990s for seismic inversions of full stack seismic data to be calibrated using acoustic impedance. Given the developments in the understanding of AVO, however, it became clear that acoustic impedance is relevant only for zero offset (i.e. intercept) 'stacks'. Connolly (1999) re-wrote the Aki–Richards (1980) equations such that an *elastic impedance* could be calculated at any angle of incidence (Fig. 5.61). Not only did this enable greater precision in non-zero offset inversions but it also opened the way for evaluating the discriminating power of AVO using log data. Elastic impedance (*EI*) applies when reflection coefficients are small and in its generalised (two-term) form is defined as:

$$EI(\theta) = V_p^a V_s^b \rho^c, \quad (5.20)$$

where

$a = 1 + \sin^2 \theta$
$b = -8k \sin^2 \theta$
$c = (1 - 4k \sin^2 \theta)$ and
$k = \left[\frac{V_s}{V_p}\right]^2$.

For the three-term case:

$$a = 1 + \tan^2 \theta.$$

In a given area, the value of *k* is kept constant between different wells.

At $\quad \theta = 0, EI = AI. \quad (5.21)$

Owing to the fact that the general magnitude of elastic impedance as initially defined varies with angle, Whitcombe (2002) formulated it by normalising with average values of V_p, V_s and density. Thus visual comparisons can readily be made between *EI* logs calculated at different values. Normalised *EI* is defined as:

$$EI(\theta) = V_{p0}\rho_0 \left[\left(\frac{V_p}{V_{p0}}\right)^a \left(\frac{V_s}{V_{s0}}\right)^b \left(\frac{\rho}{\rho_0}\right)^c\right], \quad (5.22)$$

where

V_{p0} = average V_p,
V_{s0} = average V_s and
ρ_0 = average ρ.

An illustration of the relationship between *EI* and reflectivity is shown in Fig. 5.62. A synthetic gather comprising five traces in the range $\theta = 0$–$40°$ is plotted together with the corresponding two-term elastic impedance curves.

Whitcombe *et al.* (2002) subsequently approached angle-dependent impedance from an AVO perspective (i.e. by using χ for AVO projections rather than θ). Just as elastic impedance relates to two-term Shuey reflectivity, there is a corresponding impedance for the modification of Shuey's equation (Eq. (5.23)). Given that the modification was designed to extend the angular range of Shuey's equation this impedance parameter is called *extended elastic impedance* (*EEI*) and is defined as:

$$EEI(\chi) = V_{p0}\rho_0 \left[\left(\frac{V_p}{V_{p0}}\right)^p \left(\frac{V_s}{V_{s0}}\right)^q \left(\frac{\rho}{\rho_0}\right)^r \right] \quad (5.23)$$

where

$p = \cos \chi + \sin \chi$,
$q = -8k \sin \chi$ and
$r = \cos \chi - 4k \sin \chi$.

EEI effectively encompasses *EI*, so it can be considered the more general treatment for two-term angle-dependent impedance. Given that *EEI* is driven by the AVO crossplot angle it follows that *EEI* (χ) = $90°$ is the impedance related to gradient reflectivity and is appropriately called *gradient impedance* (*GI*). Given the logarithmic relationship of reflectivity and impedance (Eq. (2.2)) the AVO crossplot can effectively be re-drawn with ln(*AI*) replacing R_0 (zero incidence reflectivity) and ln(*GI*) (gradient impedance) replacing gradient reflectivity.

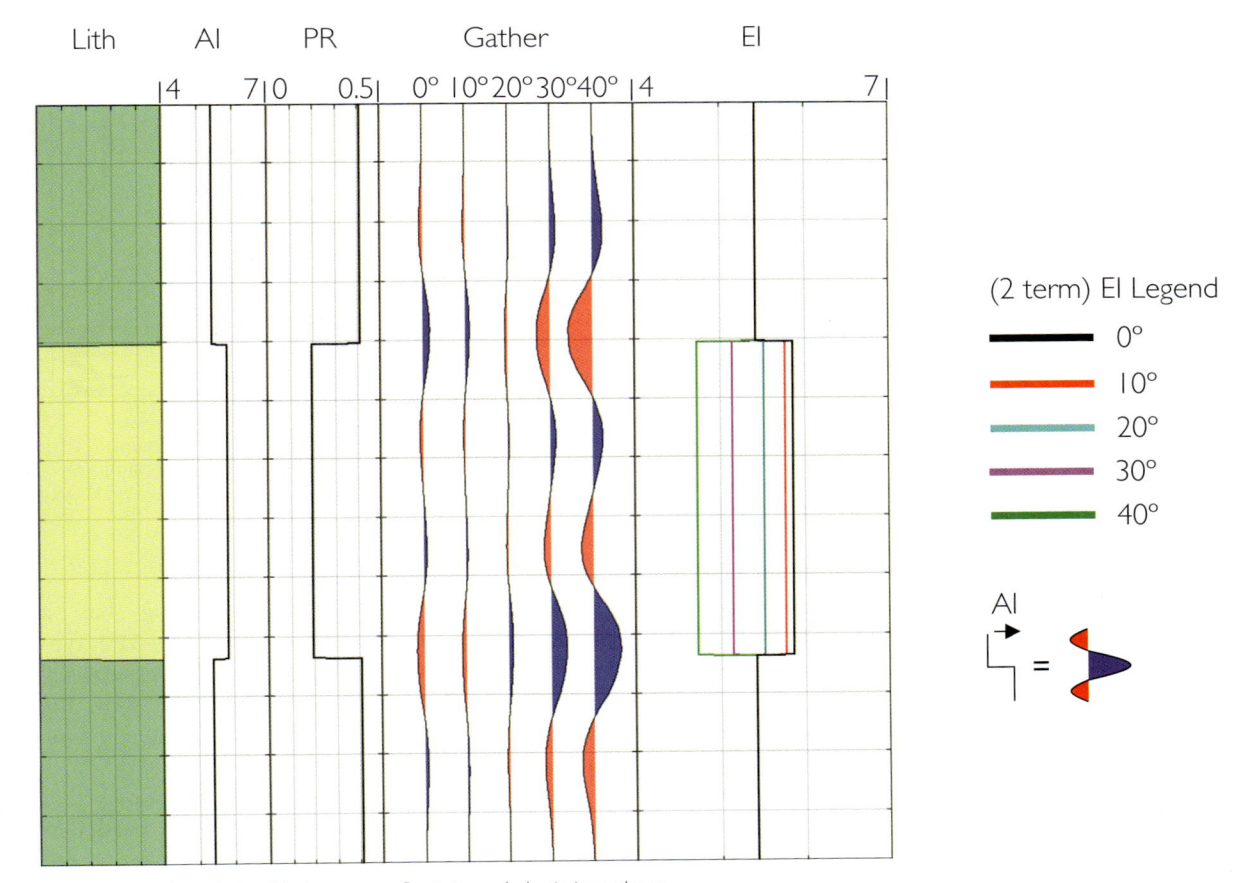

Figure 5.62 The relationship between reflectivity and elastic impedance.

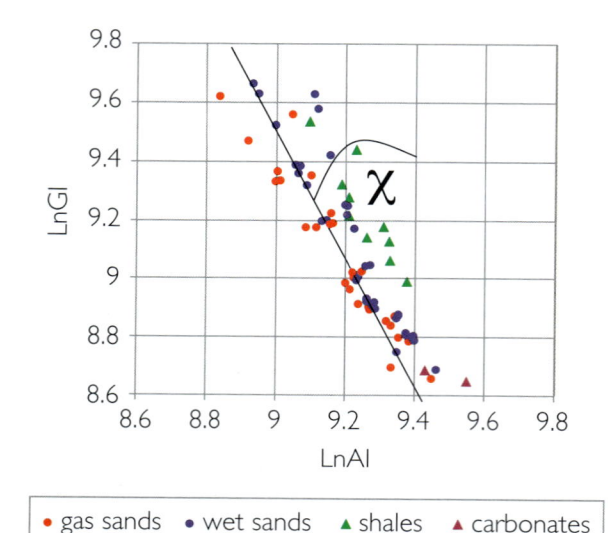

Figure 5.63 The AI/GI crossplot.

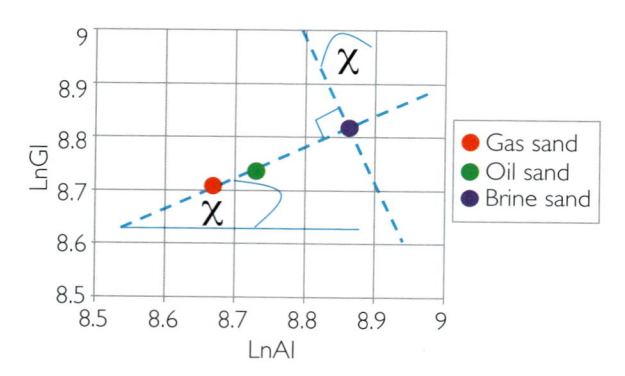

- gas sands
- wet sands
- shales
- carbonates

Figure 5.64 Determination of fluid χ angle from averaged log data using the method described by Whitcombe and Fletcher, 2001.

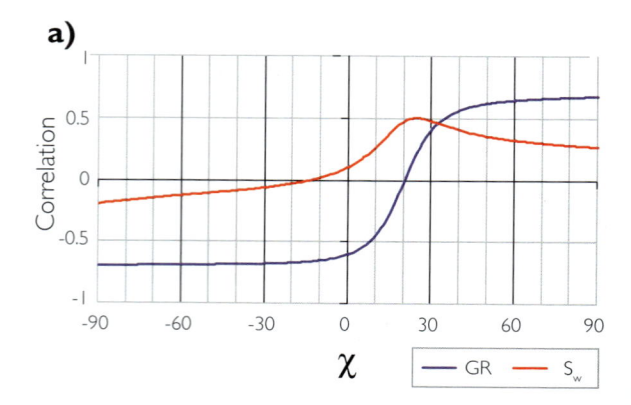

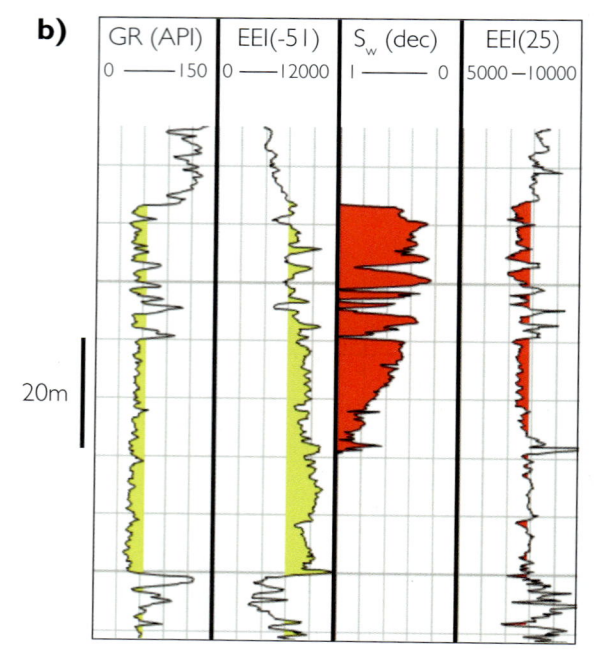

Figure 5.65 Correlation of *EEI* with gamma ray and water saturation logs; (a) correlation diagram showing maximum correlations at 25° and −51° for water saturation and gamma ray respectively, (b) log plot showing comparisons of *EEI* (χ = −51°) with gamma ray and *EEI* (χ = 25°) with water saturation.

Elegantly, the *AI* vs *GI* crossplot maintains the same angular relationships as the intercept vs gradient crossplot, and an example is shown in Fig. 5.63.

There are a number of ways of calculating χ angles from *EEI* data. Figure 5.64 shows the determination of the fluid angle from averaged log data from sands with various fluid fill (Whitcombe and Fletcher, 2001).

Note that χ angles can also be calculated by correlating *EEI* logs (calculated from +90° to −90°) with fluid and lithology logs. For example, Fig. 5.65 shows the correlation of a gamma ray and water saturation (S_w) log with *EEI*. The S_w log shows a maximum positive correlation with *EEI*(25°) whilst the gamma ray log has a maximum negative correlation with *EEI*(−51°), typical values for fluid and lithology angles respectively (Fig. 5.65a). The log plots in Fig. 5.65b show the sand defined on the *EEI*(−51°) log and the pay zone on the *EEI*(25°) log.

Similarly the correlation technique shows that certain χ angles correlate strongly with particular angle independent elastic properties (Fig. 5.66) (Whitcombe et al., 2002). Numerous authors

Rock properties, AVO reflectivity and impedance

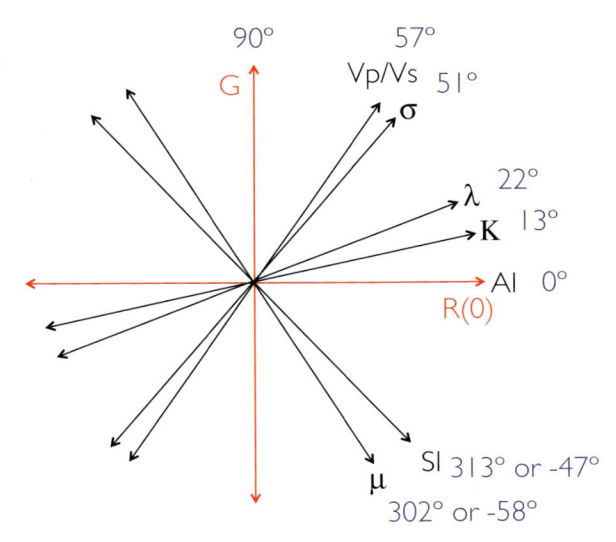

Figure 5.66 AVO crossplot showing projection axes which typically correlate strongly with particular elastic parameters (after Connolly, 2010).

(such as Dong, 1996) had observed that certain combinations of intercept and gradient relate to changes in particular elastic parameters but the idea of AVO projections and the *EEI* formulation is an elegant way of illustrating these relationships.

Perhaps a more geological approach is to analyse *EEI* in terms of lithological and fluid related facies. Figure 5.67 illustrates the idea with shales, water sands and oil sands defined from well logs. The degree of discrimination can be appreciated by using histogram displays with simple Gaussian fits and interactively changing χ angle.

It should be noted that there is not always a lithology angle at a high angle to the fluid projection. Data from three wells are shown in Fig. 5.68, with varying disposition of sands and shales. Both (a) and (b) have possible lithology trends at negative χ angles, whereas in (c) the fluid angle will also be the lithology angle.

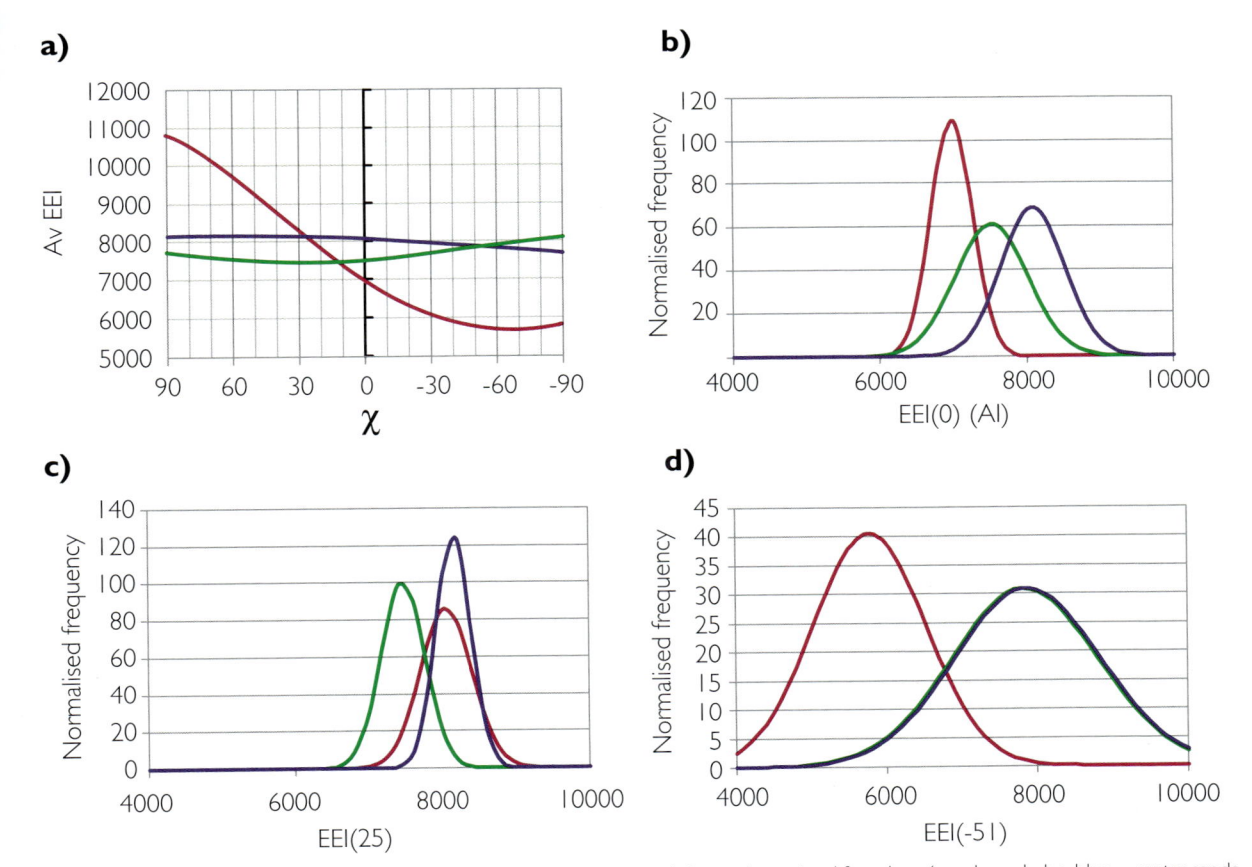

Figure 5.67 Theoretical fluid and lithological discrimination using *EEI* with facies determined from logs (purple = shales, blue = water sands and green = oil sands).

Rock properties and AVO

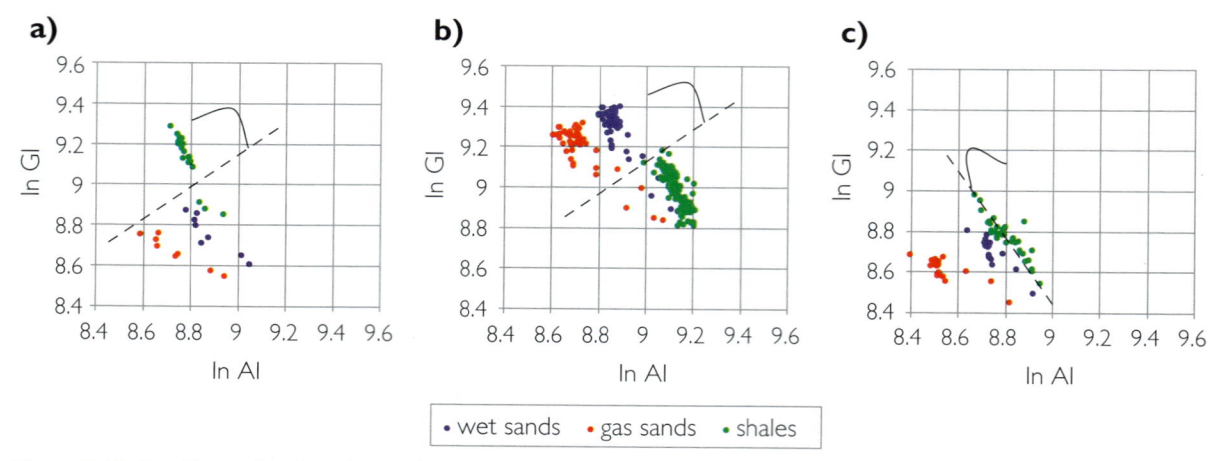

Figure 5.68 AI vs GI crossplots from three wells West of Shetlands showing variable lithology projection trends; (a) data from a well in which shales have high gradient values and can be separated from the sands with a negative χ angle, (b) data from a well where shales have low gradient values and sands and shales can similarly be differentiated using a negative χ angle, (c) data from a well where sands and shales have similar gradient values; thus the fluid χ angle will also serve to differentiate lithology.

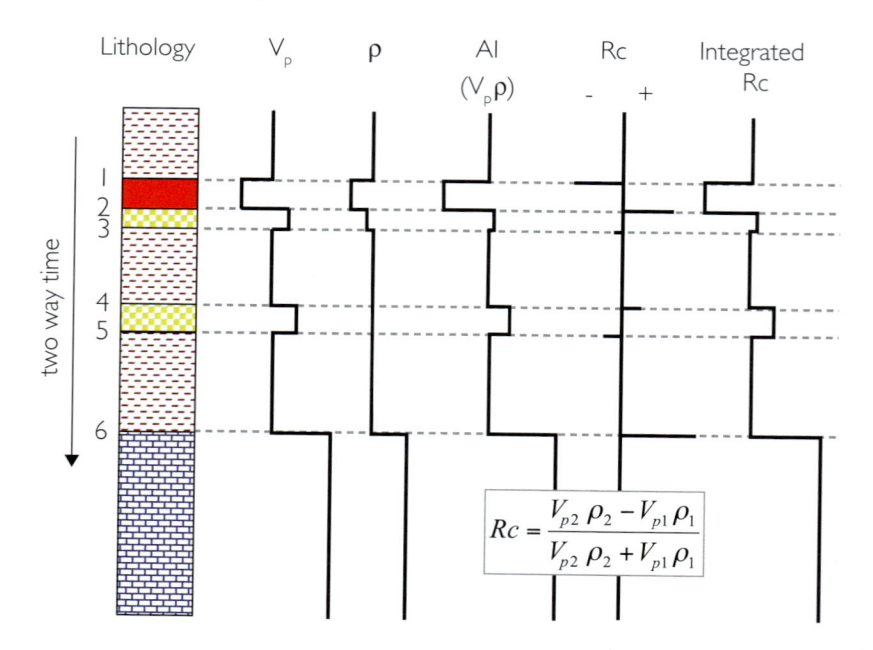

Figure 5.69 Integration of the reflection coefficient series gives the form of the impedance log (after Anstey, 1982).

$$Rc = \frac{V_{p2}\rho_2 - V_{p1}\rho_1}{V_{p2}\rho_2 + V_{p1}\rho_1}$$

5.5.3 Bandlimited impedance

The starting point for considering seismic inversion (in terms of the derivation of impedance from seismic traces) is the relationship between reflectivity and impedance. Integration of the reflection coefficient series gives a scaled version of the acoustic impedance log (Fig. 5.69). A practical problem for seismic inversion is that seismic reflectivity is bandlimited (Chapter 3), meaning that an integration of a seismic trace will give only a smoothed form of the impedance. This is equivalent then to the impedance log being convolved with a (zero phase) wavelet or, alternatively, preferentially removing low and high frequencies from the impedance log (Fig. 5.70).

Generation of bandlimited impedance from seismic effectively requires convolving the seismic with a bandlimited integration operator. Waters (1987)

Rock properties, AVO reflectivity and impedance

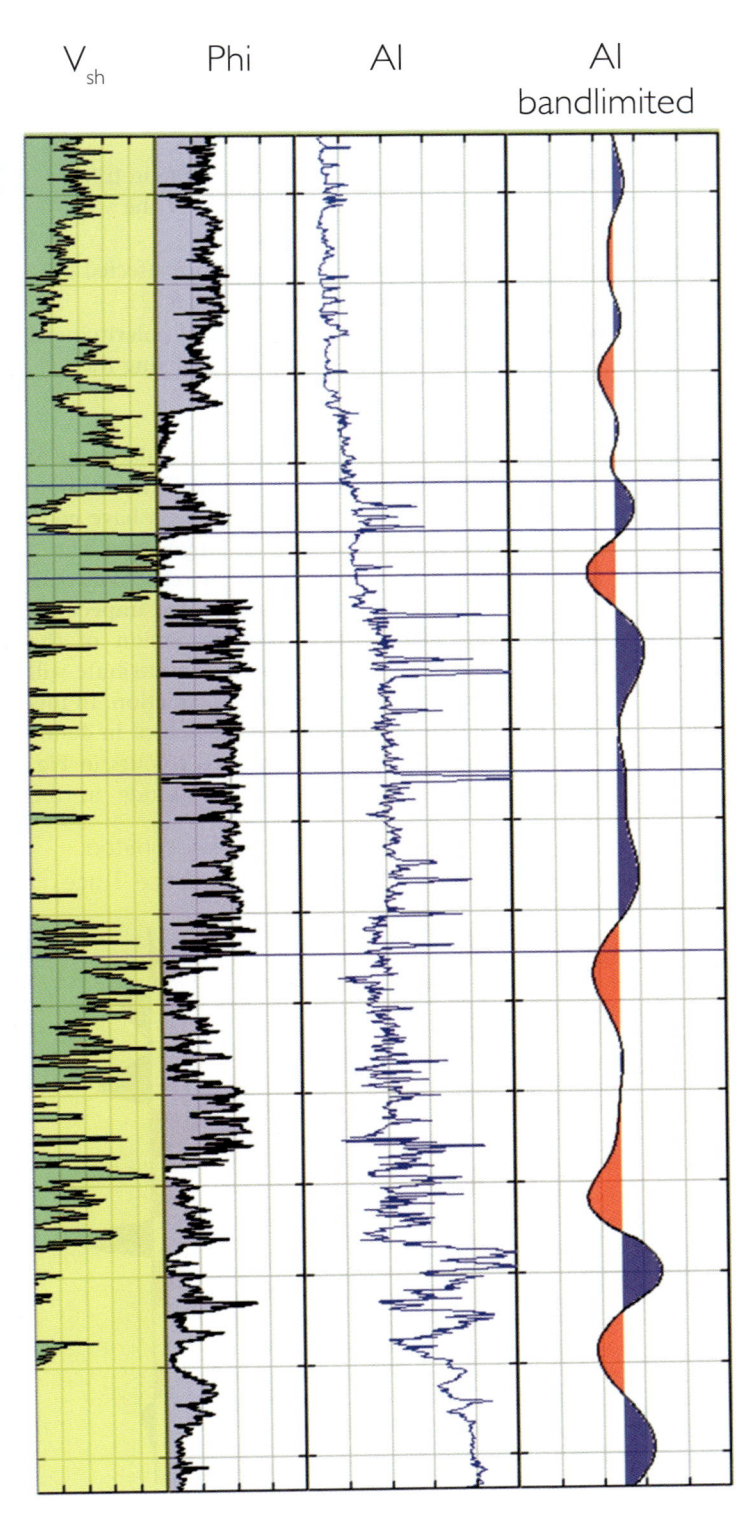

V_{sh} Phi AI AI bandlimited

Figure 5.70 A bandlimited impedance log. Low and high frequencies of the impedance log have been removed by convolution of the log with a seismic wavelet (vertical scale is 100 ft).

Rock properties and AVO

showed that this is equivalent to applying a −90° phase rotation and high cut filter of 6 dB/octave to a zero phase seismic dataset. This is called the *Seismic Approximate Impedance Log* (SAIL) and is similar to the 'pseudo log' described by Anstey (1982). Note that if, on inspection, it appears that the data were not initially zero phase, a phase rotation can be applied to the bandlimited impedance to tie geological boundaries in wells.

From an interpretation point of view, thinking in terms of bandlimited impedance can be very useful as it presents the data in layers rather than boundary reflections. On logs it is a useful tool to evaluate resolution (Fig. 5.70). In basin fill sequences it often clarifies stratigraphic relationships and emphasises hydrocarbon contacts. In particular geological situations (e.g. deep sea sedimentary environments) bandlimited impedance can be a good tool to evaluate net pay (see Chapter 10). Figure 5.71 shows a schematic model of how bandlimited impedance relates to seismic reflectivity. Several features can be noted:

- the horizon pick changes from a maximum or minimum on the reflectivity to a zero crossing on the bandlimited impedance data,
- the bandlimited impedance signature of the peak/trough doublet related to the topmost reflection is proportional to the magnitude of the real impedance contrast,
- where there is no signal (for example in thick homogenous units) the amplitude of the bandlimited impedance is zero,
- as the data are bandlimited they are affected by reflector interference (see also Fig. 5.72),
- seismic data with negative standard polarity need to be multiplied by −1 prior to applying the operator.

Figure 5.73 shows an example from the Palaeocene of the Central North Sea where a simple SAIL attribute works well. The seismic data in Fig 5.73a have good bandwidth and the bandlimited impedance has a lower frequency look compared to the reflectivity data (Fig. 5.73b). Note that if the bandlimited impedance simply looks like a phase rotated version of the reflectivity it is likely that the inversion attribute will not provide any additional useful information. In this case, the bandlimited impedance has helped in defining the presence of hydrocarbon; red colours in Fig. 5.73b represent low impedance oil sands in the

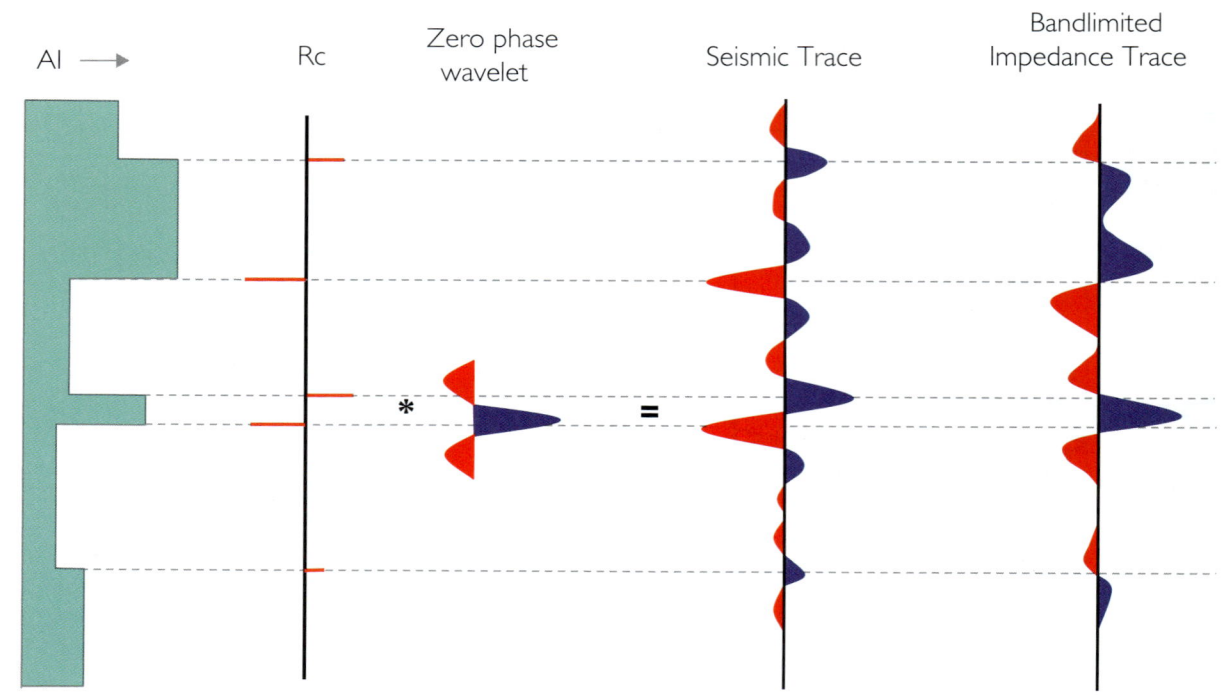

Figure 5.71 Schematic model showing relationship of seismic reflectivity and bandlimited impedance (after Kleyn, 1983); with kind permission from Springer Science+Business Media B.V.

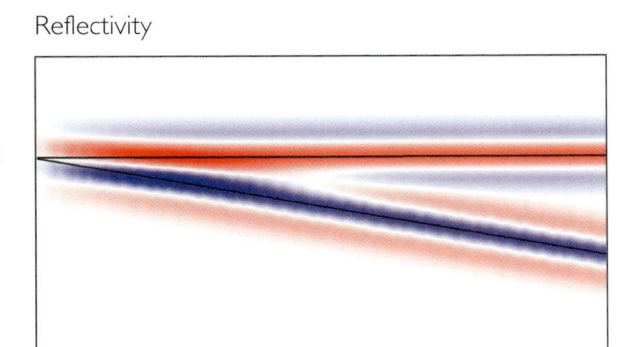

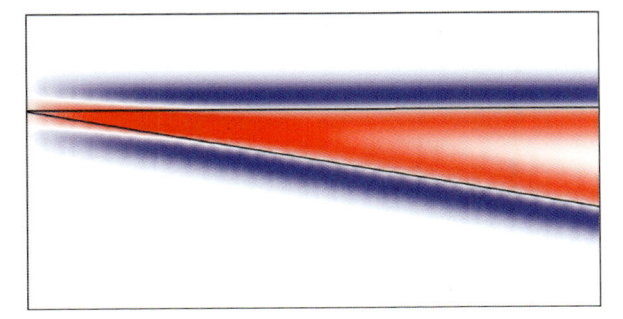

Rock properties, AVO reflectivity and impedance

Reflectivity

Bandlimited Impedance

Figure 5.72 Tuning wedge model contrasting reflectivity and bandlimited impedance.

structural culmination (Simm *et al.*, 1996; Ahmadi *et al.*, 2002).

An enhancement to the SAIL approach, known as *coloured inversion*, was proposed by Lancaster and Whitcombe (2000). The concept is that the bandlimited impedance from seismic is matched to the spectral characteristics of the impedances from log data (Fig. 5.74). A critical step in the method is the definition of the operator as it can involve a shaping of the spectrum. The coloured inversion operator effectively maps the frequency response from the seismic to that of the impedance logs (Fig. 5.74). This technique is simple and robust, requiring few assumptions except that the input data are zero phase, and has therefore been widely adopted as a first-pass approach to deriving impedance from reflectivity data.

Combining the concepts of AVO projections and coloured inversion has proved to be a powerful methodology (e.g. Connolly, 2010). Coloured inversions can be generated from reflectivity projections or can be applied to intercept and gradient prior to projection. Figure 5.75 shows amplitude maps

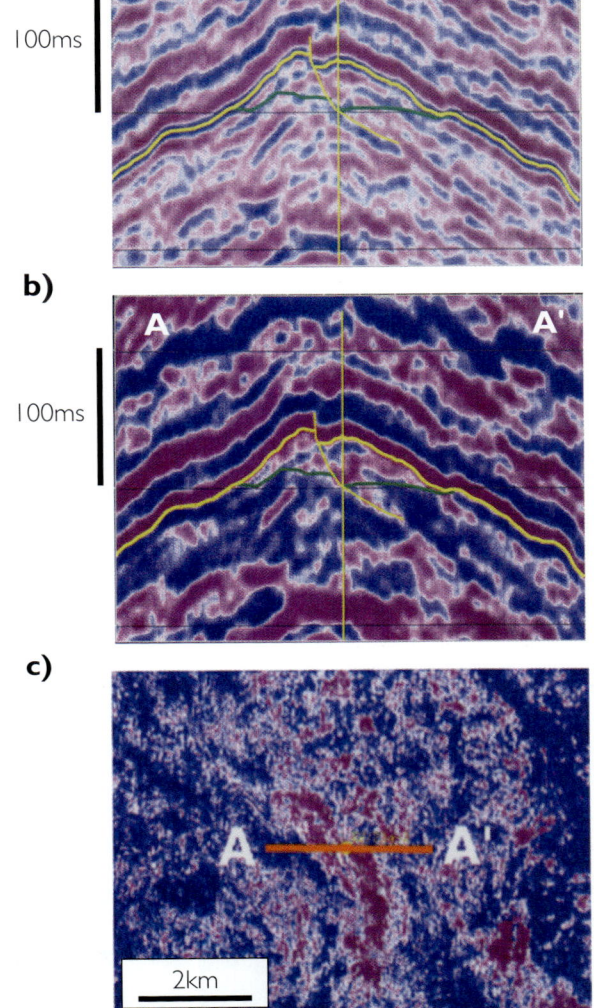

Figure 5.73 An example of bandlimited impedance (the SAIL attribute); (a) reflectivity section with top reservoir in yellow, (b) SAIL (bandlimited impedance) attribute, (c) slice through SAIL volume 10ms below top reservoir. Red = low impedance, blue = high impedance.

derived from coloured inversions generated at different angles across a deep-water Angola discovery (Connolly *et al.*, 2002). Fluid effects are optimised at an angle of $\chi = 25°$ (note the down-dip shut-off at the oil water contact), whereas the gradient impedance display ($\chi = 90°$) is emphasising sand continuity.

Rock properties and AVO

Figure 5.74 Coloured inversion.

SEISMIC DATA — Mean response (Amp vs Frequency)

CI OPERATOR — Amp vs Frequency; Time Series; Phase = −90°

AMPLITUDE SPECTRUM OF AI LOGS — Amp vs Log Frequency, f^α

Bandlimited Impedance

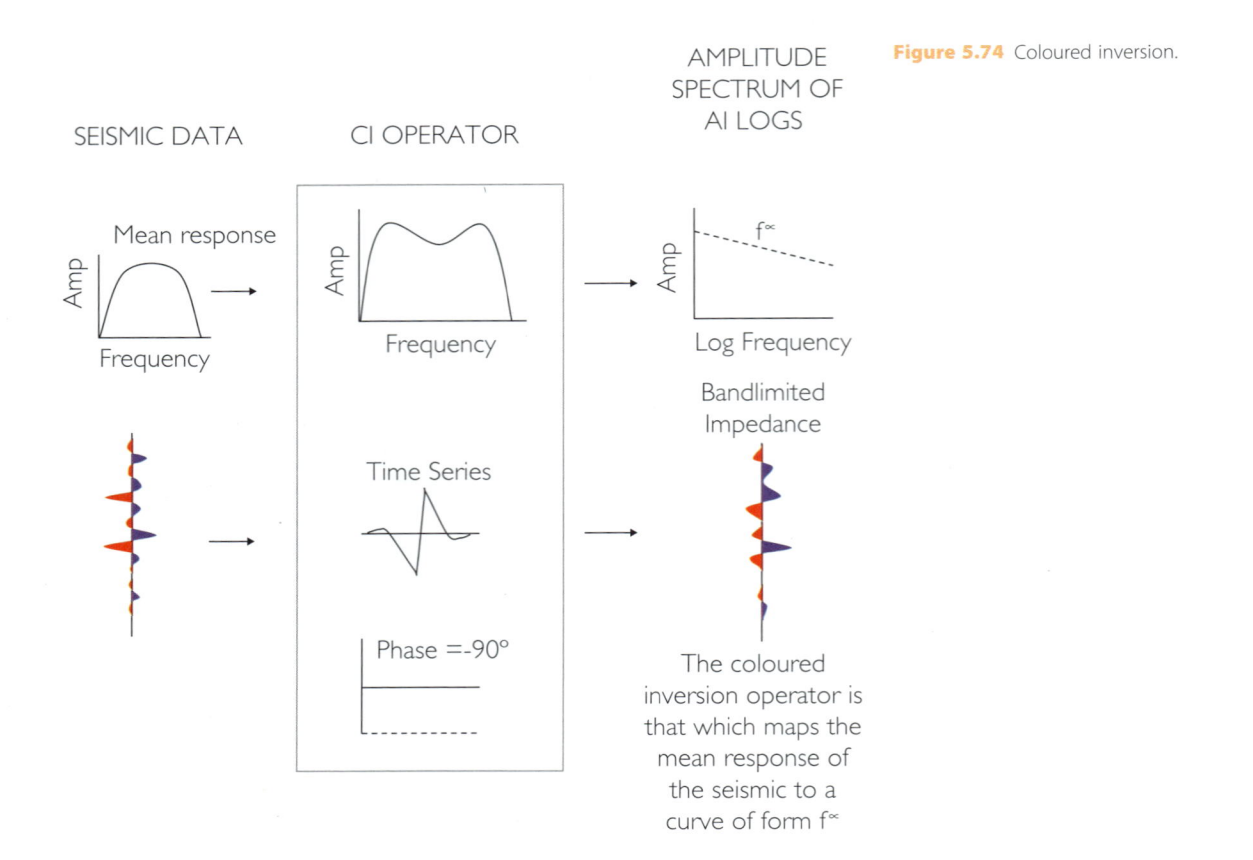

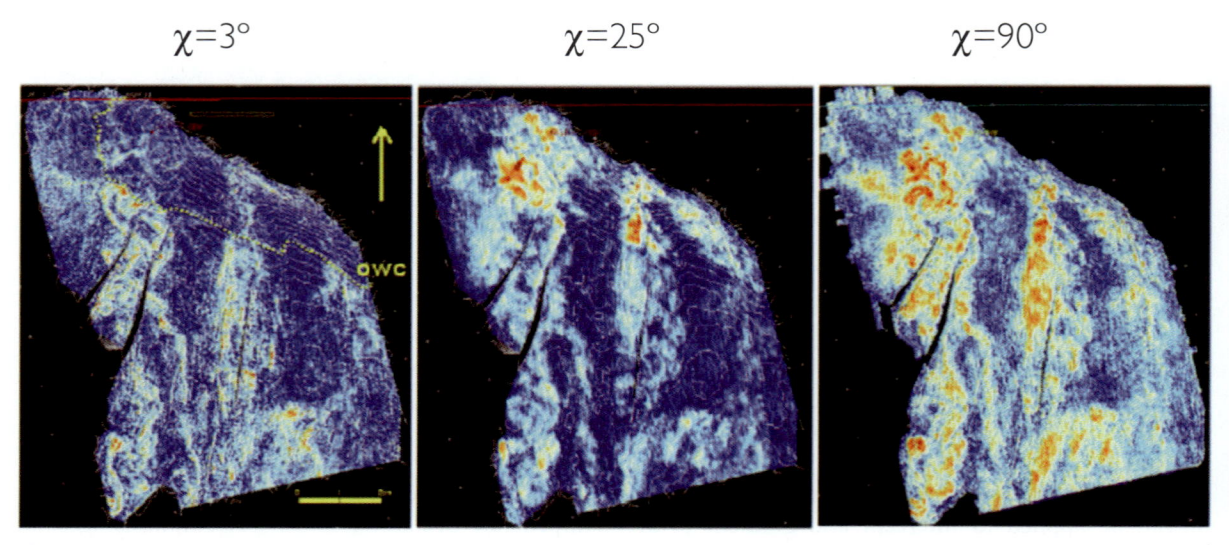

The coloured inversion operator is that which maps the mean response of the seismic to a curve of form f^α

$\chi=3°$ $\chi=25°$ $\chi=90°$

owc

Figure 5.75 Average coloured inversion impedance maps from offshore Angola (from Connolly et al., 2002).

5.6 Seismic noise and AVO

Even with the best processed seismic data there is always the issue of random noise to consider. Random noise has a systematic effect on intercept and gradient and this can have a number of implications for interpretation. The presence of random noise in move-out corrected seismic gathers effectively means that, for a given horizon, slightly different intercept and gradient fits will be made at each gather. The effect is more pronounced on the gradient compared to the intercept and the two parameters are highly anti-correlated (i.e. if the gradient becomes more negative the intercept will become (slightly) more positive) (Hendrickson, 1999; Simm et al., 2000). The angle of the noise trend on the AVO crossplot is dependent on the range of angles used in the parameter estimation and the greater the noise in the data, the greater is the extension of data in the gradient direction (Hendrickson, 1999). The suppression of noise is a critical aspect of any AVO processing workflow and this is discussed in more detail in Chapter 6.

An important effect of random noise is to rotate the optimum fluid angle. Thus, the fluid angle derived from seismic is usually lower than that derived from well based models. Figure 5.76 shows

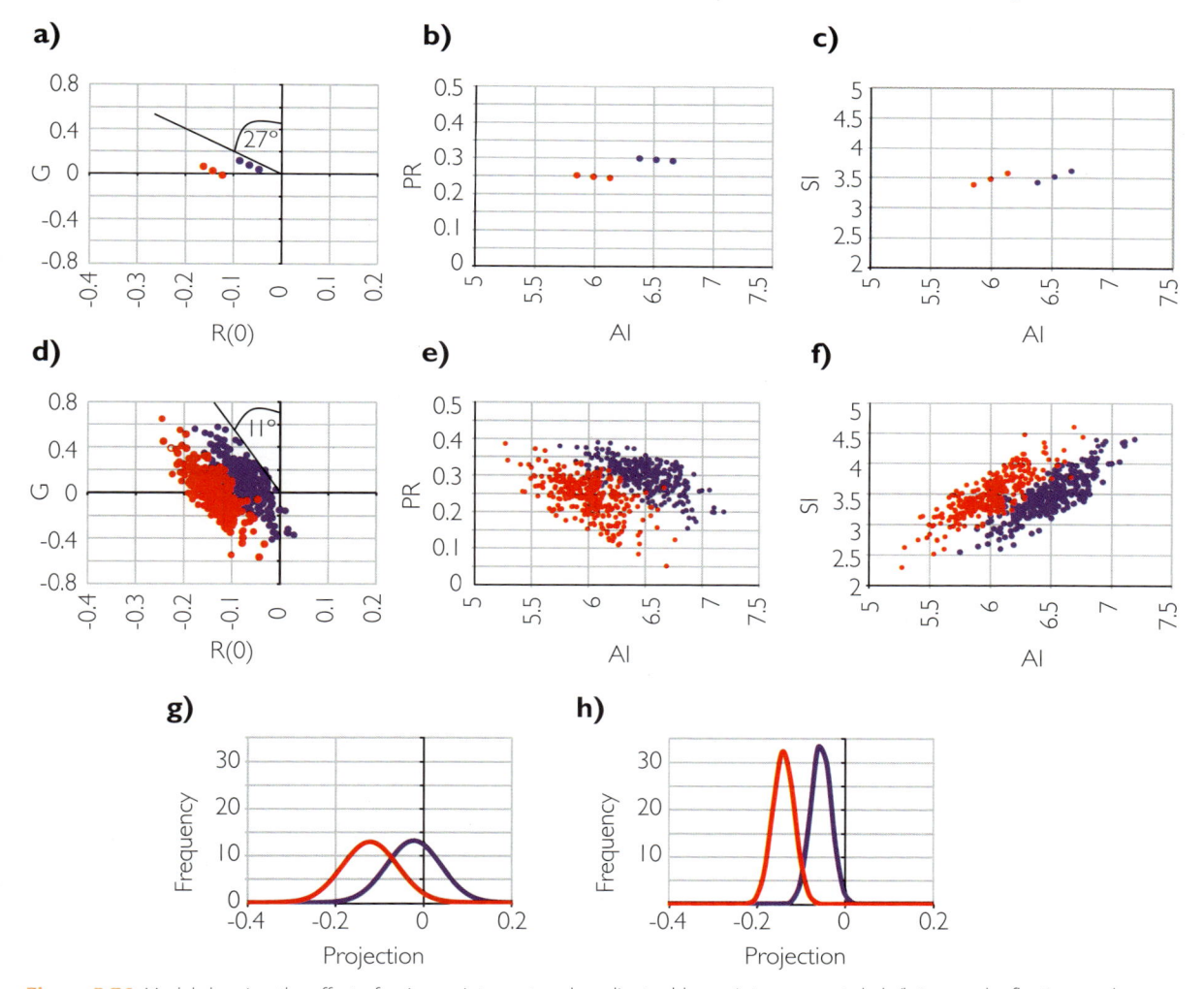

Figure 5.76 Model showing the effect of noise on intercept and gradient – blue points represent shale/brine sand reflections, red points represent shale/oil sand reflections; (a–c) = noise free model data (three single interface models representing different porosities and fluid fill), (d–f) = noise added to the models; (g) and (h) are projections at $\chi = 27°$ and $\chi = 11°$ respectively from the noisy data (data from Roy White, personal communication).

Rock properties and AVO

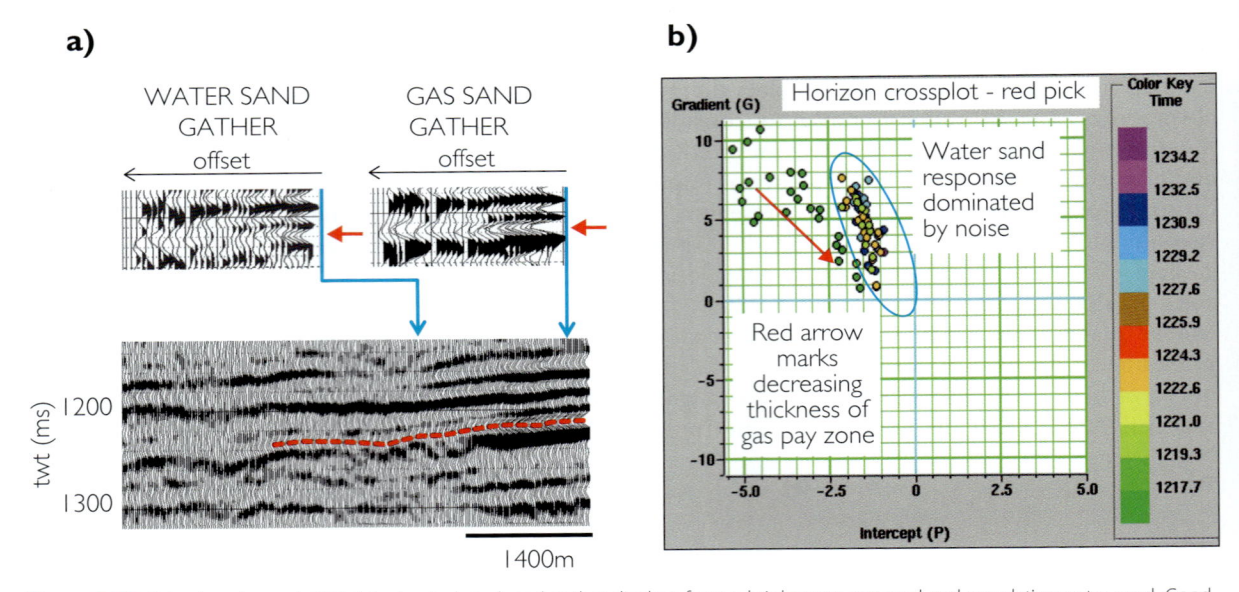

Figure 5.77 Seismic noise and AVO; (a) migrated stack and gather displays from a bright spot gas sand and correlative water sand. Good signal-to-noise associated with the gas sand and noisy data from the low amplitude water sand. (b) AVO crossplot. Note how the noise-prone water sand data forms a high angle 'noise ellipse' (modified after Simm et al., 2000).

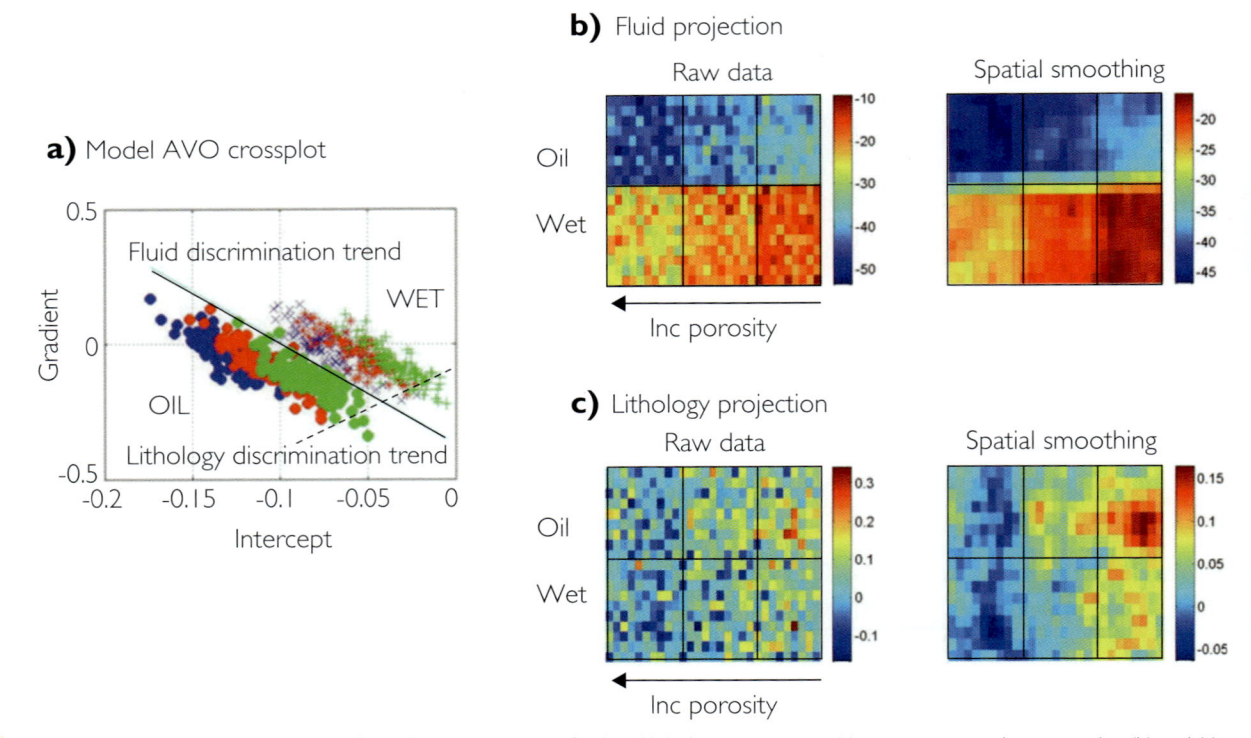

Figure 5.78 AVO model showing effect of random noise on fluid and lithology projections; (a) intercept vs gradient crossplot, (b) and (c) pseudo-maps generated from the model clusters. Note that a 3 × 3 filter has been applied for the spatial smoothing (Roy White, personal communication).

model data for a boundary comprising shale on sand with three different porosities and with water fill (blue) and hydrocarbon (red). The optimum fluid (χ) angle in the data with no random noise is 27° (Fig. 5.76a), whereas in the model containing a fairly typical level of random noise the angle is 11° (Fig. 5.76d). Figure 5.76 illustrates the effect of projecting the noise-prone data for each of these angles. There is less discrimination when the theoretical angle is used. Finding the optimum fluid angle in seismic data thus requires some trial and error.

Figure 5.76 also illustrates how noise might propagate into the derivation of elastic properties through seismic inversion (*SI*). *AI* and *SI* have been inverted from the reflectivity data in Fig. 5.76d (using equations described in Section 9.2.7.3) and *AI* vs *PR* and *AI* vs *SI* plots have been generated (Figs. 5.76e, f). On the *AI* vs *PR* plot it is clear that the effect of noise is to broaden the range of values but also to introduce a high angle trend. The shear estimation is determined effectively from the gradient; thus noise in the gradient is propagated into the Poisson's ratio parameter. In the *AI* vs *SI* domain the noise trend is roughly aligned with the porosity trend. These results point to the need for noise attenuating data conditioning prior to simultaneous inversion for absolute elastic parameters.

Figure 5.77 shows a seismic example of the effect of random noise. The crossplot shows data only from the top sand pick (red dash line in Fig. 5.77a). Note how the water sand data have a strong clustering at a low angle to the vertical (the *noise ellipse* described by Hendrickson (1999)). The water sand gathers are dominated by random noise and the extension in the gradient direction is an indication of the large amount of noise in the data. The gas sand data have much higher signal to noise ratio (S:N) and show a trend that is related to the decreasing thickness of the pay zone. With the eye of faith the noise trend can be seen superimposed on parts of the general thickness trend of the gas sand.

Random noise potentially has a different impact on fluid versus lithology projections. The model data in Fig. 5.78 are similar to those shown in Fig. 5.76 with three porosities and two fluid fills. Two projections are defined and the data clusters are arranged to give a pseudo-map display. The noise has little adverse effect in identifying a boundary (i.e. hydrocarbon contact) on the fluid projection, largely because the noise and signal are in similar directions. It is clear, however, that lithology projections at high angles to the noise trend are very noisy. Note that spatial smoothing (e.g. 3 × 3 point average smoothing operator) can be used to good effect to reduce the effect of the noise on the maps. The beauty of amplitude map displays is that they can contain significant amounts of noise whilst still revealing geological lineaments (Section 3.8).

Chapter 6

Seismic processing issues

6.1 Introduction

It has long been recognised that AVO analysis makes onerous demands of the seismic processing industry, well beyond those typical of the requirement for accurate imaging. What is needed are intercept and gradient data for conventional AVO analysis, or angle gathers and partial stacks for input into impedance inversion, of high enough quality that the results can be used for fluid and lithology discrimination or reservoir characterisation. A key requirement is that seismic amplitudes should be proportional to synthetic amplitudes that would be calculated from an accurate impedance model of the subsurface. Achieving this to within satisfactory tolerances is not always possible. To quote Cambois (2001), 'Although preserved amplitude processing is a clear requirement for AVO studies, such a processing sequence is not uniquely defined...One possible definition could be: Any sequence that makes the data compatible with Shuey's equation (if this is the model used for the AVO analysis). The simplicity of this definition should not hide the Herculean nature of this task'.

Some aspects of seismic processing that are important for AVO analysis have already been introduced in Chapter 2. In this chapter they will be discussed in more detail, considering first the steps in common processing sequences that are most important for AVO analysis, and then describing how gathers can be conditioned to improve the quality of AVO results. This leads on to a consideration of how AVO intercept and gradient can best be calculated. The chapter is not intended as a comprehensive discussion of seismic processing (for which see e.g. Yilmaz, 2001; Jones, 2010), but rather aims to highlight the main issues that the amplitude interpreter needs to recognise. From the point of view of amplitude interpretation, the key aims of seismic processing are:

- recovery of correct relative amplitudes
- improvement of signal/noise ratio (e.g. removal of multiples and other noise)
- data enhancement for improved interpretability (i.e. zero phasing, bandwidth improvement)
- imaging the data correctly (i.e. pre-stack time or depth migration)
- aligning reflections correctly across the gather (i.e. moveout and statics correction).

The critical issues for the interpreter include:

- data quality
  - signal/noise
    - recognising bad data
    - understanding the effect of noise on the AVO gradient
  - alignment
    - the effect of residual moveout on the gradient
    - differentiating residual moveout from Class IIp AVO effects
    - moveout stretch
- amplitude scaling
  - 'true' amplitude recovery
  - final adjustment to scaling with offset.

An appreciation of these issues requires that the interpreter is vigilant about the possible effects of various seismic processing steps on the AVO response. Sometimes it may be impossible to carry out meaningful AVO analysis because the AVO behaviour is severely distorted. If at all possible the interpreter needs to have access to the angle gathers to assess noise levels in relation to partial stacks. Residual multiples that may be easy to recognise on the gathers are usually impossible to see by inspection of partial stacks. Here and throughout the data analysis, well ties (Chapter 4) are very important. The steps taken to condition the data prior to AVO analysis should be checked against the well synthetics to make sure that they really are improving the data. It should be recognised that conditioning is not a mechanical process. Careful thought as to underlying data quality is required to assess the likelihood of genuine improvement. In some cases no amount of conditioning will establish the correct gradient. A useful discussion of data conditioning prior to pre-stack impedance inversion, which is also relevant to AVO analysis, has been published by Singleton (2009).

6.2 General processing issues

The ideal dataset for AVO analysis would have the following characteristics:

- zero phase data
- clear continuity of reflections across the gather
- adequate offset/angle sampling
- good multiple and noise attenuation
- flat gathers (correct moveout velocities and algorithm)
- correct relative amplitude across the gather
- consistent frequency content across the gather (no moveout stretch)
- consistent scaling of each trace (high source and receiver repeatability)
- no critical or post-critical energy
- correct pre-stack imaging
- high bandwidth.

Sometimes the evaluation of data quality is straightforward simply from inspection of stacked data. For example, Fig. 6.1 illustrates a full stack comparison between 3D surface streamer and ocean bottom cable (OBC) data where it is clear that the data quality of the OBC data is superior. Often, however, data effects are subtle and require pre-stack data displays for their evaluation. Figure 6.2 shows an example of residual moveout that can only effectively be appreciated in the pre-stack domain. The presence of this type of residual moveout can adversely impact the AVO gradient estimation.

6.2.1 Initial amplitude corrections

6.2.1.1 Divergence correction

One of the first processes applied to seismic data, which is fundamental to the general behaviour of amplitude with offset, is a scaling step to correct for the 'spherical' divergence phenomenon, described in Chapter 2. The effect is quite large, and is corrected for using a simple function of two-way time (TWT) and stacking velocity. This correction is based on an approximation and is unlikely to be entirely accurate, particularly if there is anisotropy in the overburden. Anisotropy can cause significant modification of geometric spreading with angle (Stovas and Ursin, 2009) and inadequate correction will distort the amplitude variation with offset. When using AVO for quantitative interpretation this effect has to be corrected prior to AVO analysis by applying some form of offset-related scaling (see Section 6.3.3).

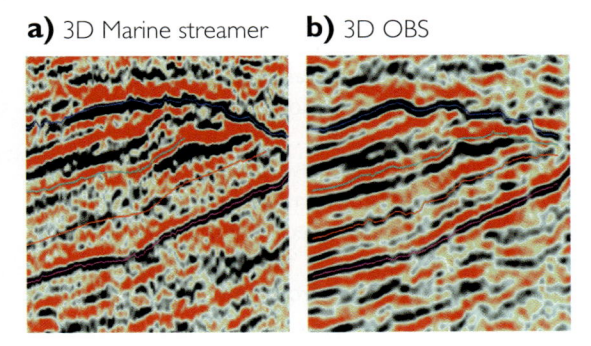

Figure 6.1 Comparison of marine surface towed streamer data with ocean bottom seismic (OBS) data (after Thompson et al., 2007).

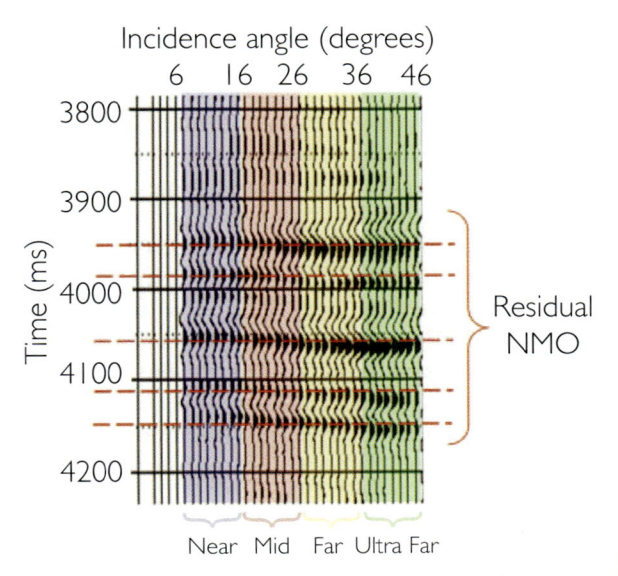

a) 3D Marine streamer **b)** 3D OBS

Figure 6.2 Angle gather showing residual moveout (after Contreras et al., 2007).

6.2.1.2 Surface consistent corrections

AVO analysis is performed on a gather of traces that have a common reflection point but are acquired from different shot–receiver pairs. Clearly, the effect of shot-to-shot variation in signal strength and variations in receiver sensitivity needs to be removed. In the marine case, these variations are likely to be fairly small. Air guns emit highly reproducible signals and receivers are manufactured to close tolerances, and both source and receiver are surrounded by seawater, the properties of which do not usually vary laterally. In the land case, the problem is much more serious. Sources and receivers are still manufactured to give uniform performance, but coupling to the subsurface is strongly affected by the nature of the near-surface layering. Sometimes it is possible to achieve fairly uniform conditions, for example by burying geophones or by sweeping loose material away so that a Vibroseis source rests on solid rock, but often this is not practical and corrections need to be applied in processing. In general, seismic amplitudes are more reliable for marine than for land data.

The near-surface effects are dealt with by surface-consistent corrections (Taner and Koehler, 1981). The idea is that near-surface effects associated with a particular surface position remain constant regardless of the raypath involved. Thus, source strength will affect all of the traces recorded from that source, and geophone coupling will be the same for all traces recorded at a particular receiver station from different source positions. Unwanted surface-consistent factors can be split into three categories: source response, receiver response, and an offset-related response, which takes account of directivity in source or receiver arrays. Owing to the possibility of modifying first order offset amplitude changes related to geology it is seldom advised to apply corrections for energy directivity. In most instances corrections are applied to source and receivers only. Since there are traces for many source–receiver pairs, the source and receiver responses can be calculated in a least-squares sense. The process is analogous to calculating shot and receiver statics and, in a similar way, is quite effective at dealing with short-wavelength variations.

6.2.2 Long-wavelength overburden effects

A key issue for the interpreter to appreciate prior to amplitude interpretation is variation in amplitudes at the level of interest that may be the result of propagation effects from shallower rock layers. For example, Fig. 6.3 shows a section from a gas field offshore Indonesia (Rudiana *et al.*, 2008), where a shallow gas accumulation strongly attenuates seismic amplitudes below it, giving rise to an amplitude 'shadow'. High impedance layers such as basalts and thick carbonate layers can have similar effects. The extent of amplitude shadows will vary with offset, depending on how the source and receiver paths are affected. Important factors are source–receiver distance as well as the depth and lateral extent of the attenuating layer. The effects on measured AVO at a target horizon can be complicated; near traces will almost certainly be affected but it is possible that at long offsets the raypaths may fully or partially undershoot the feature. This means that in the affected area it will not be possible to relate AVO on a deep reflector to properties of the interface concerned without carrying out some form of data normalisation.

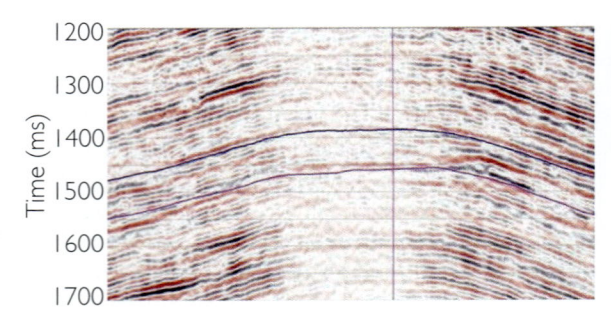

Figure 6.3 Example of amplitude shadow caused by shallow gas (after Rudiana *et al.*, 2008).

For simple amplitude mapping, a possible approach is to equalise amplitudes over a long gate by means of *automatic gain control* (AGC). This aims to scale the seismic data so that the average RMS amplitude in a window is the same across all traces, and assumes that the average reflectivity is invariant. This is dangerous if applied over a short window, potentially destroying lateral amplitude variations of interest. If applied over a long window (e.g. 1.5–2 s TWT), however, it can remove much of the shadowing effect of the overburden without damaging lateral amplitude variations in a target reflector. This solution can be applied to partial offset stacks, but it will obviously affect the intercept vs gradient relationships determined from the stacks. Long gate AGC is dubious for angle stacks because the offsets (and the degree of undershooting) contributing to any one trace will vary with TWT.

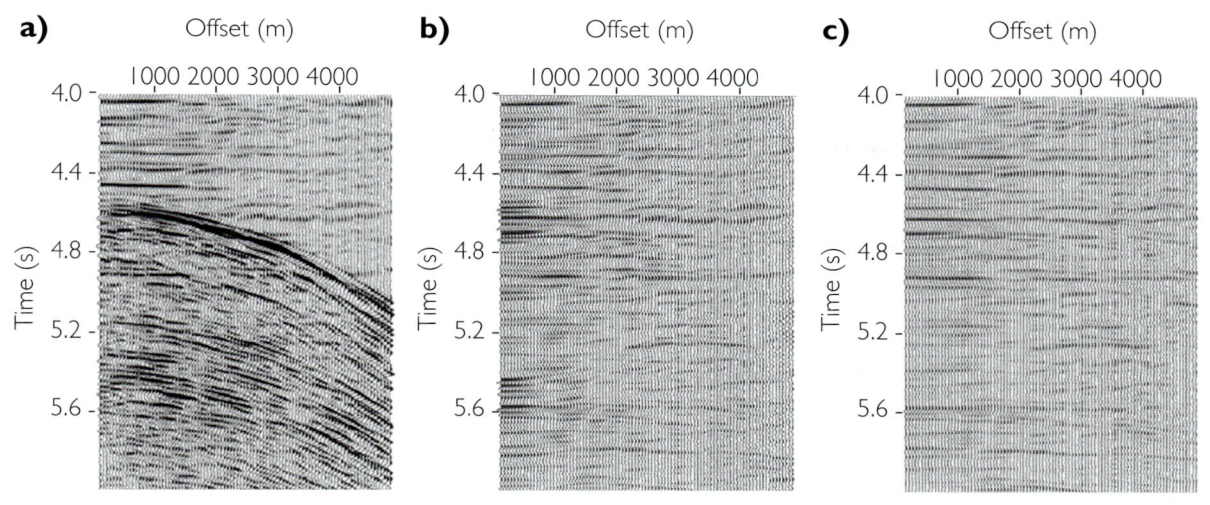

Figure 6.4 (a) CMP gather after moveout correction, with strong multiples; (b) result after standard Radon demultiple: although the multiples have been largely removed there is residual multiple energy at near offsets which will affect the calculated AVO; (c) as (b) but now using the high-resolution Radon transform: the residual multiple energy at near offsets is much reduced (redrawn from Verschuur, 2007, with permission from the CSEG Recorder).

A different solution would be to normalise the amplitude map at target level by dividing it by the amplitude at a reflector above or below the target that is not expected, on geological grounds, to show lateral variability. The reference horizon needs to be fairly close in TWT to the target so that raypaths through the attenuating layer are the same for both. None of these corrections can be applied with any confidence when shadowing is severe. In such a case it might be better to recognise that it is simply not possible to make a reliable interpretation of amplitude data.

6.2.3 Multiple removal

Most seismic datasets require some processing to remove multiples. A common method is Radon demultiple, first introduced by Thorson and Claerbout (1985), which discriminates between primaries and multiples on the basis of moveout behaviour (i.e. velocity differences arising from differences in raypaths). After applying moveout corrections to a CDP gather, the primaries will be flat and the multiples will in general show residual moveout. The Radon transform maps the data into a time-curvature domain, in which the primaries will have near-zero curvature compared to multiples which will have larger curvature (e.g. Kabir and Marfurt, 1999). Unfortunately, owing to the limited offset range available, the change in frequency content and amplitudes varying with offset, there is often partial overlap between primaries and multiples in the transformed data. This can lead to residual multiple energy being left in the data, especially on the near offsets, after muting the multiples. A partial solution is provided by the high-resolution Radon transform which aims to find the sparsest representation of the input in the Radon domain (Fig. 6.4), leading to less smear and thus less of a problem with the near offsets. Residual multiple energy on the near offsets is potentially a problem for the offset scaling comparisons that will be described shortly; it would not be correct to scale down the amplitude of the near traces if their higher than expected amplitude was due solely to residual multiple energy.

Alternative techniques of multiple elimination are available, such as 'surface-related multiple elimination' (SRME) (e.g. Dragoset et al., 2010). This technique predicts surface-related multiples by a convolution process and does not require a velocity field for its application. Once predicted, the multiples are removed by a filtering process called '*adaptive subtraction*', which attempts to match the modelled multiples to the data (Guitton and Verschuur, 2004). Unfortunately, the sparse distribution of sources and receivers available in real datasets does not generally conform to the dense surface sampling required by the algorithm to predict multiples accurately, but good results can be obtained by various interpolation schemes, removing multiples without distorting relative amplitude variation across the offset range. Well synthetics, generated with and without inclusion of multiples, are a useful tool to check for the presence of residual multiples or for damage to primary reflections by over-aggressive application of demultiple techniques. In order to model seabed multiples a water layer would need to be inserted into the model.

6.2.4 Migration

In all but the simplest subsurface geometry (i.e. horizontal reflectors at all levels), pre-stack migration is needed for detailed AVO work. Without migration, the traces of a common midpoint (CMP) gather will contain data reflected from different subsurface locations and these will not in general be positioned directly below the common midpoint. The purpose of migration is to process the signal collected from many surface locations so that the gathers contain traces with reflection points vertically below the source–receiver midpoint, focusing energy and increasing lateral resolution. There are many different algorithms in use for pre-stack migration, and the choice of method depends on the complexity of the structure of both target and overburden. Jones (2010) provides a useful review and discussion. For AVO, an issue is the amplitude fidelity in the migrated gather. This is partly determined by the type of algorithm used, but also by its detailed implementation; where possible, this needs to be discussed with the seismic processor in the context of the intended use of the gathers.

6.2.5 Moveout correction

Given the importance for AVO analysis of time registration of reflectors across the gather, accurate moveout correction is necessary. For an isolated interface, this means that the reflector should be flat across the gather enabling, for example, the calculation of a reasonably accurate AVO gradient at the same TWT on every trace.

6.2.5.1 Problems in residual moveout identification

Whilst flat gathers are generally a requirement for AVO, evaluating the 'flatness' of gathers may not always be straightforward. For example, if there are multiple interfaces in a stack of thin beds, then interference effects (tuning) will change across the gather

as the individual interface reflection coefficients change with angle, and individual peaks and troughs will not necessarily line up exactly horizontally. Class IIp AVO effects, with a polarity flip at some intermediate offset, will similarly not give rise to a neat alignment across the offset range. In these cases, the apparent residual moveout should be the same as that seen in a well synthetic. In practice, the synthetic gives a general idea of whether apparent residual moveout is an issue, but does not give much assistance in calculating a detailed residual moveout correction (Fig. 6.5).

6.2.5.2 Higher order moveout corrections

In some cases, trying to flatten reflections across the gather using a hyperbolic equation may distort the moveout correction, particularly at mid and far offsets. In such cases the first refinement to be considered is a more accurate allowance for vertical velocity heterogeneity by including higher order terms in the moveout equation. The simple hyperbolic moveout approximation is

$$T_{X,n}^2 = T_0^2 + \frac{X^2}{V_{rms}^2}, \quad (6.1)$$

where T_0 is the travel time at zero offset and V_{rms} is the RMS average velocity between the surface and the base of the nth layer. This is derived from a horizontally layered earth model, in which the travel time for a reflection from the base of the nth layer at a source–receiver offset x is given by (Taner and Koehler, 1969):

$$T_{X,n}^2 = c_1 + c_2 X^2 + c_3 X^4 + c_4 X^6 + \cdots, \quad (6.2)$$

where the coefficients c depend on the layer thicknesses and the velocities in each layer.

Thus, a more accurate approximation is obtained in most cases by retaining the term in X^4.

6.2.5.3 Anisotropy

Another complication for gather flattening is anisotropy. When fine scale layering (i.e. transverse isotropy, see Chapter 5) becomes important it is not possible to flatten gathers beyond an incidence angle of about 30° with standard moveout corrections. Alkhalifa (1997) addressed the effect of transverse isotropy (VTI) on the seismic travel time. For a horizontal interface in a homogeneous VTI medium then the travel time can be described by:

General processing issues

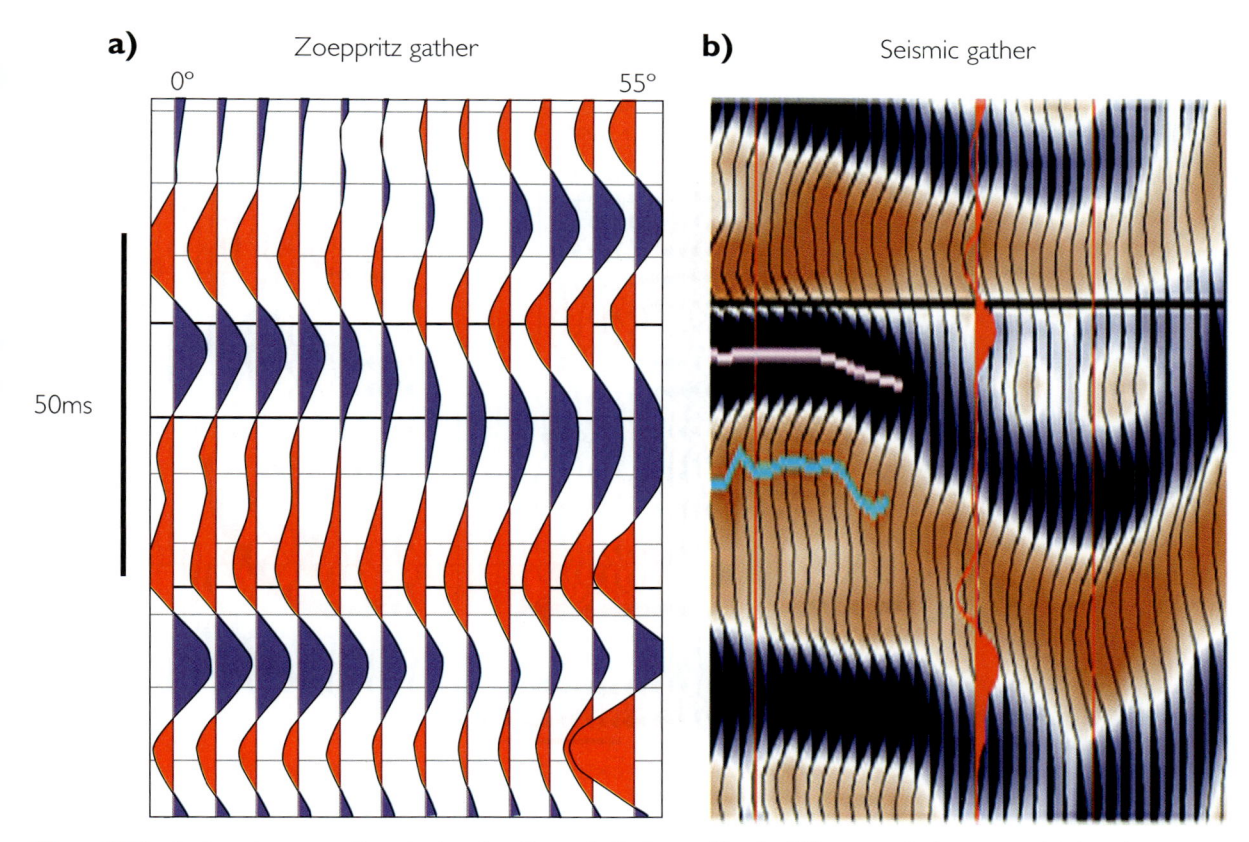

Figure 6.5 Synthetic and corresponding seismic gather. The synthetic shows a Class IIp AVO response in the upper part, in broad agreement with the real data; in the lower part the reflections are nearly horizontal in the synthetic but show residual moveout in the real data.

$$T_x^2 = T_0^2 + \frac{X^2}{V_{nmo}^2} - \frac{2\eta X^4}{V_{nmo}^2 \left[T_0^2 V_{nmo}^2 + (1+2\eta)X^2\right]}, \quad (6.3)$$

where η is given in terms of Thomsen's anisotropy parameters (see Chapter 5) by

$$\eta = \frac{\varepsilon - \delta}{1+2\delta} \quad (6.4)$$

and

$$V_{nmo} = V_{P0}\sqrt{1+2\delta}, \quad (6.5)$$

where V_{p0} is the P wave vertical velocity.

In time processing η is usually derived from maximising stack semblance after initial velocity estimation (e.g. Toldi *et al.*, 1999). Figure 6.6 shows an example where the AVO response of gas sands is seen at offsets large enough to require anisotropic moveout corrections.

In principle, a combination of time and depth migration can provide information on the anisotropic parameters δ and ε that can modify the AVO response (Chapter 5). For example, δ can be estimated iteratively from the misties between a depth migrated image and the true depth as measured in a well (Jones, 2010), and ε can then be estimated from the combination of η and δ. In practice, the ε and δ functions derived from seismic are usually quite generalised and not usually applicable to the seismic amplitude modelling of specific target horizons. There may also be ambiguity regarding how much of the η estimation is simply accommodating the effects of vertical velocity gradients. The reader is referred to an excellent treatment on velocity issues in migration by Jones (2010).

6.2.6 Final scaling

The application of modern pre-stack migration and demultiple techniques usually results in pre-stack gathers that have reasonable amplitude scaling with offset. In some instances, however, particularly in land data, an evaluation of the moveout corrected gathers reveals gross scaling differences between sets of traces

Seismic processing issues

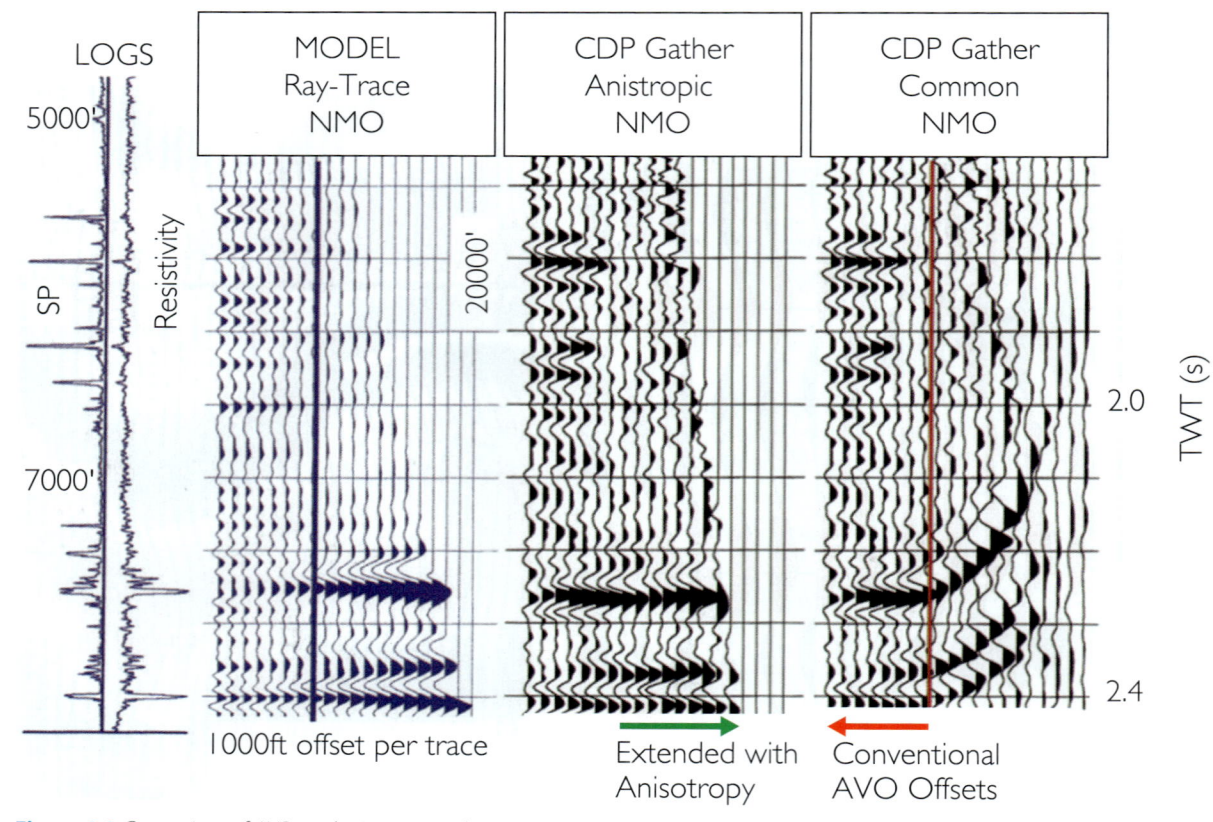

Figure 6.6 Comparison of AVO synthetic generated using anisotropic ray theory and migrated CDP gathers with both anisotropic and isotropic moveout correction (from Hilterman et al., 1998).

across the gather. Figure 6.7a shows an example from a gas reservoir where the nearest traces have amplitudes higher than the rest of the gather. The modelled response from wells at the top of the gas sand is Class III but the amplitude imbalance in the gather is distorting the AVO, giving it a Class IV appearance. In this case a simple application of automatic gain control (AGC) over a large window (Fig. 6.7b) served to establish the Class III response. Note how reducing the AGC window can have the effect of reducing the AVO gradient (Fig. 6.7c). This use of AGC should rarely be necessary for modern data, but it is still likely that some adjustments to the gain across the offset range will be needed to ensure consistency of AVO with well synthetics (see Section 6.3.3).

6.2.7 Angle gathers and angle stacks

If possible, it is usually best to carry out an AVO study starting from angle gathers. This gives total flexibility in choosing what angle range to use. In addition, it is easier to assess the presence of multiples and other types of coherent noise by looking at the gathers rather than at angle stacks. For interpretation, angle stacks are commonly generated as they benefit from the S:N improvement of stacking whilst retaining the main features of the amplitude vs angle behaviour.

The choice of angle range in each angle stack is determined by a balance between several factors:

- it is difficult for the interpreter to work with more than three angle stacks, for lack of space on the screen to display them side by side;
- if the number of angle stacks is small, a fairly large number of pre-stack traces contribute to each stack trace, with useful noise reduction;
- on the other hand, a small number of stack volumes implies summing over quite a large angle range, with consequent smear in the AVO response.

In deciding which angle ranges to stack, the first step is to look at the gathers and define the useful

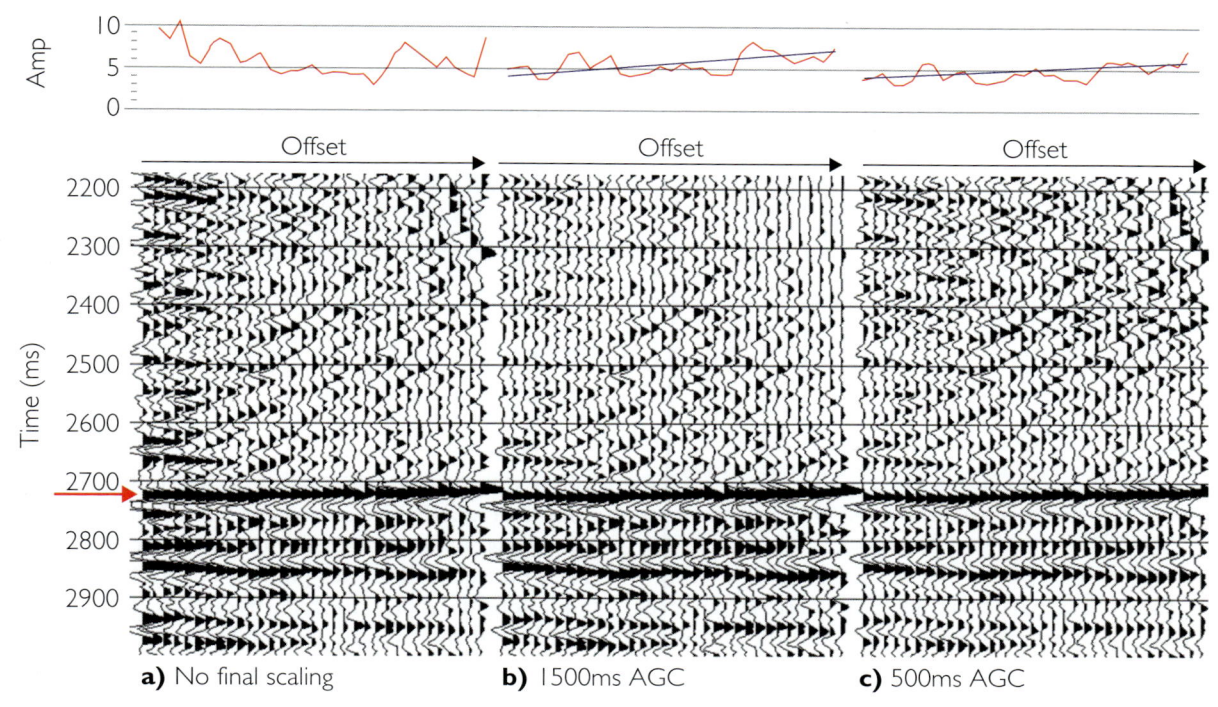

Figure 6.7 Example of a final scaling step using long gate AGC. The AVO gradient on the arrowed event is distorted by the high amplitude of the near traces if no scaling is applied, but as the AGC window narrows the AVO gradient flattens.

angle range available. Usually it is possible to make use of near angle data down to the lowest angle available at the level of interest (which is usually a few degrees as the shortest source–receiver offset available is not zero). In many instances, angle ranges of angle stacks generally conform with the angle range over which the change of amplitude with $\sin^2\theta$ is linear, so that two-term (e.g. intercept/gradient) interpretation techniques may be applied to the data. Ideally the processor would check the linearity of the amplitude response (in general linearity holds to an incidence angle of around 30°). Certainly the interpreter needs to be careful that direct arrival and refraction energy has been omitted from far stacks. Sometimes there are considered to be useful data beyond the angles at which amplitudes vary linearly with $\sin^2\theta$ and these data may be used to create an 'ultra-far' stack. The interpretation of ultra-far stacks can be fraught with uncertainty owing to the potential for interplay of imaging and geological effects, including anisotropy.

Having defined the useful range a usual approach would be to create near and far or near, mid and far stacks for general interpretation use by dividing the angle range into equal parts. So, for example, three angle stacks to be made from data with a range of 3° to 39° would each cover a range of 12° in incidence angle. For the purposes of AVO inversion a smaller angle range is often considered ideal. For example, around 7° might be more appropriate, suggesting that in this case five angle stacks should be generated. Since narrow-range stacks will tend to be noisy, it is worth considering application of a noise-reduction process such as structure-oriented filtering (Höcker and Fehmers, 2002; Helmore et al., 2007).

6.3 Data conditioning for AVO analysis

In summary, the steps that may be needed to condition seismic data for AVO analysis are:

- conversion to angle gathers
- spectral equalisation
- residual moveout removal
- residual scaling of amplitude with offset (*offset balancing*).

The first step is to convert the offset gathers to angle gathers, so that each trace represents the seismic

Seismic processing issues

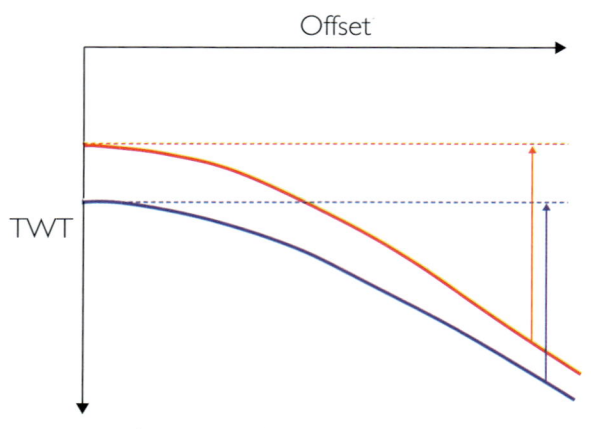

Figure 6.8 Schematic normal moveout curves for two reflections in a CDP gather, showing convergence of the curves as offset increases. Moveout correction moves the reflections onto the horizontal lines, which implies that the trace will be stretched by an amount that increases with offset.

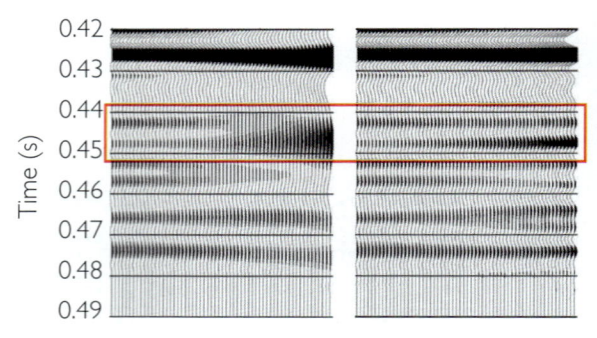

Figure 6.9 Synthetic angle gather before (left) and after (right) correction for moveout stretch, after Roy et al. (2005). Gradients calculated from these data (across the full offset range) would be different after stretch correction.

response at a particular angle or narrow range of angles (see Section 2.3.5). Given that the processor uses a smoothed velocity function to calculate the offset vs angle relationship, it is important to compare the results to a model generated from well velocity data. Some depth migration algorithms are able to output angle gathers directly, although the angles predicted will depend on the accuracy of the velocity model inherent in the depth migration. Again, it is advised to check the calculations at well locations.

6.3.1 Spectral equalisation

The frequency of seismic data generally decreases with increasing offset with the effect primarily related to the application of moveout correction to hyperbolic reflections. Given that moveout hyperbolae are not parallel, moveout correction leads to a stretching of the signal at larger offsets to match the time separation at near offsets and this inevitably causes a shift to lower frequency across the gather (Fig. 6.8). These effects may also be reinforced by a dip-dependent shift to lower frequencies with increasing incidence angle due to the effect of the migration process (Tygel et al., 1994) and by absorption effects caused by the longer raypath for far offset traces.

A key problem for AVO analysis in the presence of a large spectral imbalance across the gather is that data on near traces may not necessarily relate properly to the data on the far traces, giving rise to erroneous gradients. In addition, frequency changes with offset will cause variations in thin-bed tuning behaviour that might obscure true AVO effects. Various proposals for 'stretch-free' moveout correction have been put forward, but as pointed out by Williamson and Robein (2006) the high-frequency information was lost at the time of acquisition and no moveout correction process can recover it. Traditionally, data with stretch of more than 2:1 is muted, with the remaining data having a spectral balancing process applied. One approach is to apply a deterministic correction for moveout stretch and absorption (Lazaratos and Finn, 2004). Alternatively, frequency spectra at various offsets can be compared and operators derived to match them to a reference trace. The reference trace might for example be the full stack. Using a near stack will tend to magnify noise on the far traces, and using a far stack will in all likelihood remove valid information on the nears. Correction for spectral variation across gathers can have a noticeable effect on the AVO gradient (Fig. 6.9). In seismic inversion (Chapter 9) pre-stack spectral variations are taken into account by specifying a separate wavelet for each angle stack.

6.3.2 Residual moveout removal

There are often residual moveout effects visible on the gathers, perhaps as a result of inaccuracies in picking velocities or residual anisotropic effects (Fig. 6.10). If the reflections in the corrected gather are not flat the calculation of AVO gradient can be strongly affected. Figure 6.11 shows an example of the effect of residual moveout on near and far stacks. The horizon picks were made on the near stack and transferred to the far stack. The red pick shows an apparent phase reversal

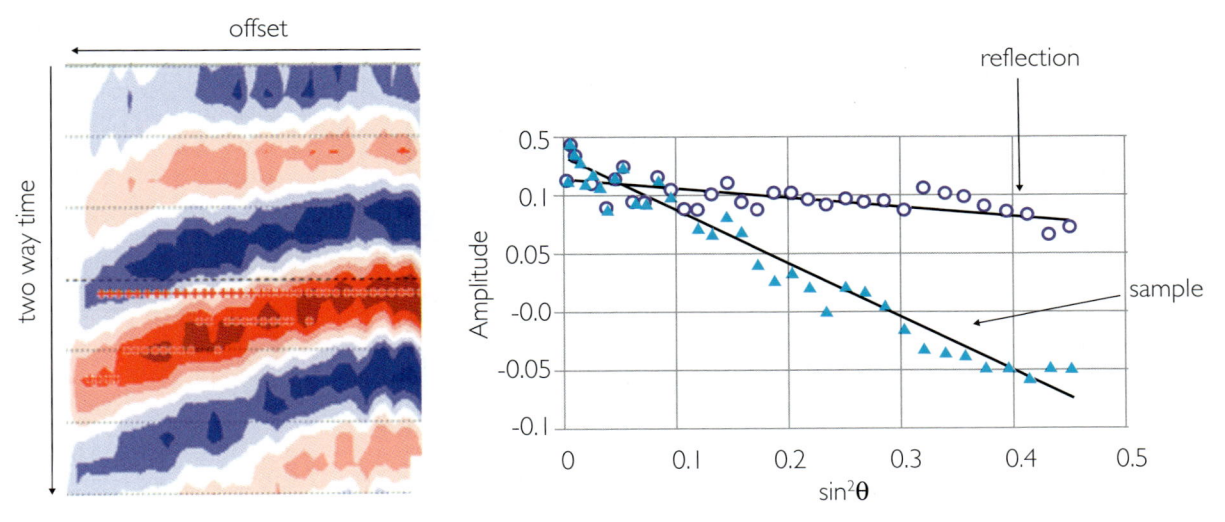

Figure 6.10 The effect of residual moveout on the AVO intercept and gradient. The gradient is far more sensitive to the effect than the intercept (after Bacon *et al.*, 2003).

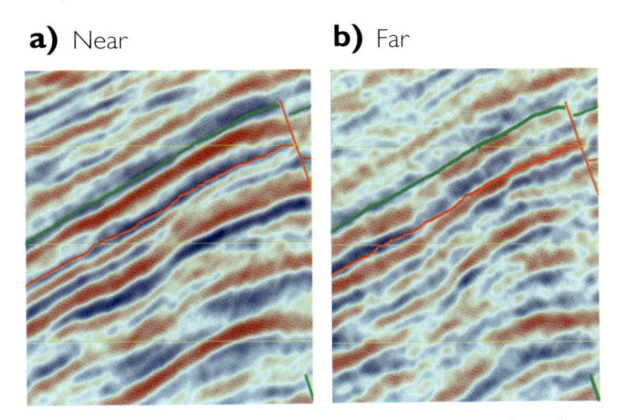

a) Near **b)** Far

Figure 6.11 Partial stacks showing residual moveout. The red picked horizon shows an apparent phase reversal from near to far in the updip part of the structure.

from near to far in the right-hand part of the section, but this is in fact simply an effect of residual moveout. This would be much easier to recognise in the gathers but interpreters often do not have quick access to gathers during interpretation. A way of handling residual moveout on stacked data is to make separate horizon picks on the near and far stacks, then measure the amplitudes and perform AVO analysis (such as described in Chapter 5) on the horizon amplitude maps rather than in a volume.

If the cause of the misalignment is an inaccurate velocity model, the proper solution is to correct the velocities. Indeed, the sensitivity of AVO to velocity errors can be exploited as a way to obtain more accurate velocities in an automated procedure (Swan, 2001). If the AVO study proceeds from raw gathers with processing software available then applying such a process is relatively straightforward. However, in many cases there is not enough time available before completion of an AVO study for re-interpretation of velocities to be a realistic option.

A possible approach would be to improve the alignment by applying a trim static time-shift, calculated as the shift required to give the best alignment between the trace and a reference trace (e.g. the full stack trace). This requires some care in the size of the gate used for correlation and the maximum time-shift permitted; it works best when there are some high-amplitude events of distinctive character, to avoid miscorrelations due to noise. Class IIp AVO effects are a particular problem because of the polarity flip between near and far traces, creating the possibility of incorrectly correlating a peak (for example) on the near trace with a peak on the far trace, whereas the correct correlation should be to a trough. Comparison with an offset synthetic may be needed to determine whether residual moveout is present. Also, wavelet phase changes across the offset range should be removed prior to applying trim statics, preferably by making all the data zero phase.

Figures 6.12 and 6.13 show a North Sea example where gather flattening has led to an improvement in the AVO gradient estimation (Gulunay *et al.*, 2007).

Seismic processing issues

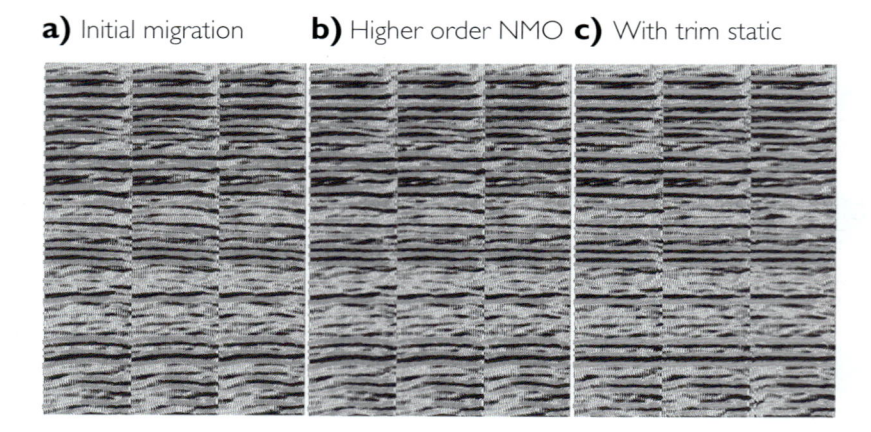

Figure 6.12 Three adjacent gathers after (a) initial migration, (b) after additional higher order residual moveout correction and (c) after additional trim static flattening. Gather flatness has been significantly improved (from Gulunay *et al.*, 2007).

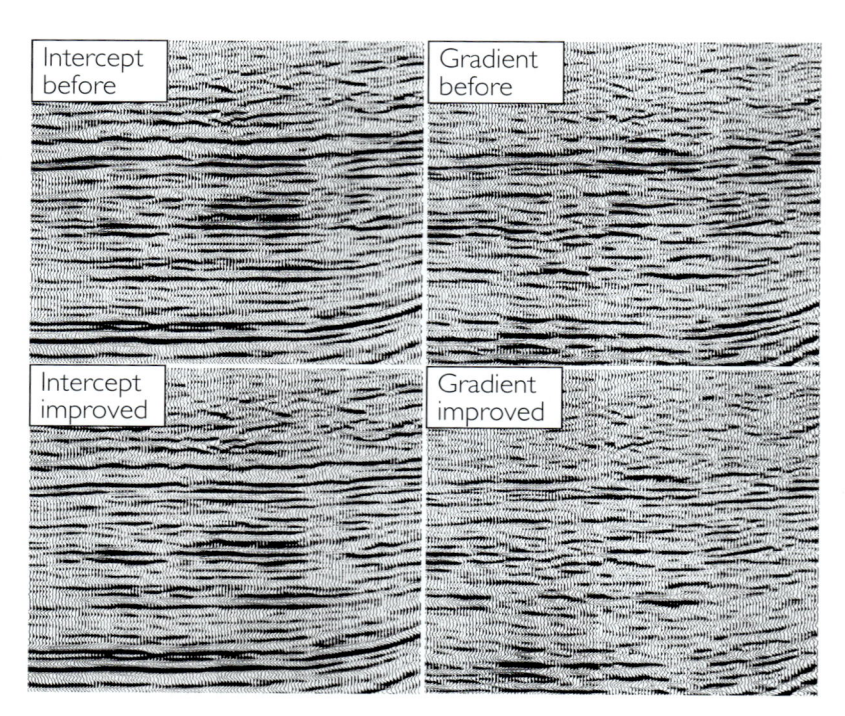

Figure 6.13 Intercept and gradient sections calculated from gathers before (top) and after (bottom) trim static flattening. A significant amount of residual energy has been removed from the gradient (Gulunay *et al.*, 2007).

Of course, the trim static can be calculated in multiple gates down the traces, so as to align events at different TWTs. It should also ideally be applied in a multi-trace methodology to avoid introducing steps in reflectors where different static shifts have been calculated on adjacent sets of traces.

6.3.3 Amplitude scaling with offset

Despite the best efforts of the seismic processor, it is often the case that the amplitude variation with offset in the final CDP gathers does not faithfully represent the amplitude variation that would be expected purely from AVO effects at the individual reflectors. This often leads to uncertainties in how to scale the offset component of the seismic (i.e. AVO calibration). Whilst it may not be necessary to correct the scaling if the purpose of a study is restricted to identifying anomalous AVO behaviour, the issue can become important in situations where some quantitative interpretation of the AVO signature is to be made (Ross and Beale, 1994).

Perhaps the most robust approach to offset scaling is provided by well ties (see Section 4.6.5). If the

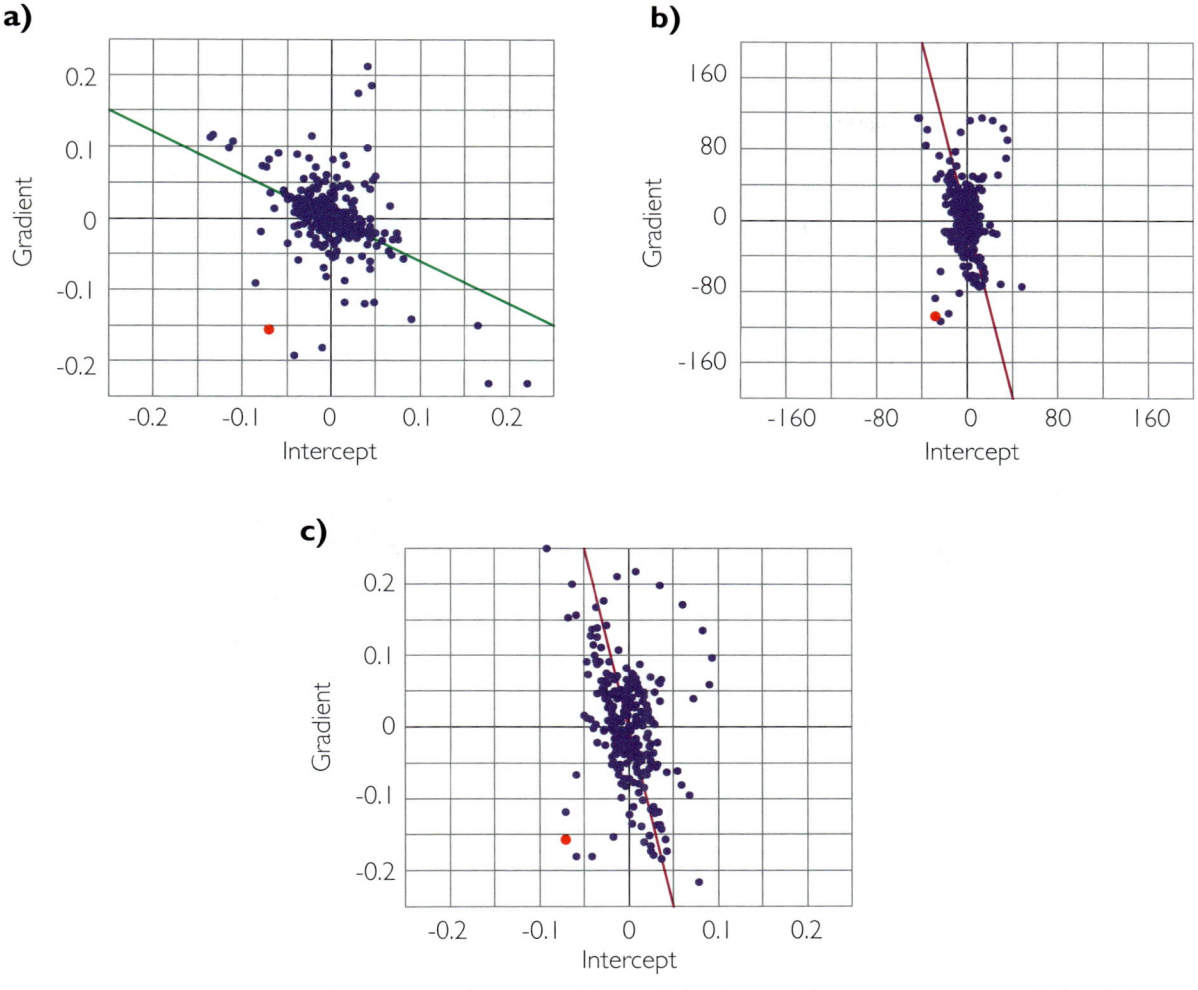

Figure 6.14 AVO calibration at a well; (a) AVO crossplot generated from well data (note red dot represents top reservoir), (b) seismic AVO crossplot determined from the same time segment as the well crossplot, (c) seismic AVO crossplot calibrated by scaling the reflectivities at top reservoir and re-calculating intercept and gradient. Note how the seismic background trend does not change significantly with calibration as it is contaminated by noise.

scaling of the seismic has the same relative change as the synthetic (i.e. the seismic scaling is 'correct') then the amplitude of the extracted wavelets on nears, mids and fars should essentially be the same, although the frequency content and phase may change across the offset range. Corrections can be derived simply from the relative scaling of the extracted wavelets. Clearly, if there are big scaling differences from one well to another then further analysis is needed. If well ties are poor then comparison can be made using RMS amplitudes in a window which contains several strong reflectors to avoid the effects of noise or bias from a single dominant reflection.

It is paramount in offset scaling that the effects of noise should be avoided. Figure 6.14 illustrates the problem in the context of the AVO crossplot, where the effects of noise are most apparent. The example is fairly extreme as it is based on an unconsolidated sand target and data with fairly poor signal-to-noise ratio. A crossplot derived from the well synthetic (Fig. 6.14a) shows a background trend at an angle of $\chi = 53°$. Compared to the well crossplot (Fig. 6.14a), the seismic crossplot (Fig. 6.14b) has a much lower angle of trend (around 11°). It might be supposed that the AVO calibration step could involve a rotation of this trend to fit the well data. Unfortunately, this is not

Seismic processing issues

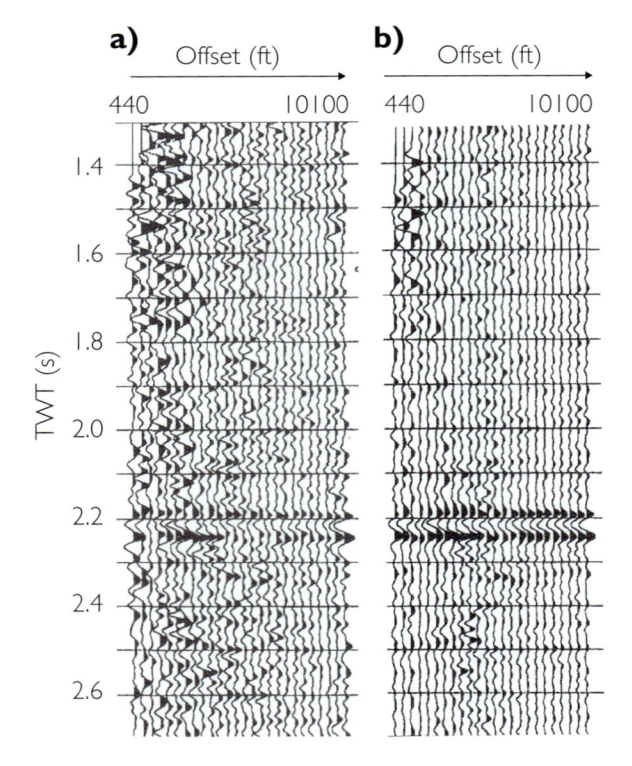

Figure 6.15 An example of gathers and supergathers (after Allen and Peddy, 1993); (a) processed gather, (b) supergather generated by stacking offsets across a number of cdps. Noise levels are significantly reduced throughout the traces in the supergather, but lateral resolution is compromised to some extent.

the case as the seismic trend is not simply a function of geology but a combination of the geology and noise (e.g. Hendrickson, 1999; Simm et al., 2000). In fact, most of the crossplot data in Fig. 6.14 are heavily contaminated with noise. In such a case the calibration should be done on the reflectivity associated with the target sand (shown as the red point on the plots). Effectively, the calibration is done by deriving different scalers for near and far angles, then re-calculating the intercept and gradient. Given the noise contamination, the background trend on the calibrated crossplot (Fig. 6.14c) in this case is no different from the original seismic.

6.3.4 Supergathers

A way of enhancing signal-to-noise ratio is by forming a supergather. This is done by collecting traces from adjacent CMPs and stacking traces with similar offsets or angles. Of course, the lateral stacking element of supergathers means that a certain amount of lateral resolution is traded for improved signal-to-noise ratio. An example is shown in Fig. 6.15.

A practical issue is the amount of fold (or number of traces) available to perform the supergather operation. In 3D datasets the fold varies across the offset range; typically, fold is low at near and far offsets, and traces with intermediate offsets dominate the overall fold. Xu and Chopra (2007) have described how using an adaptive approach, in which traces are borrowed from adjacent CMP locations to create an even fold distribution with offset, provides

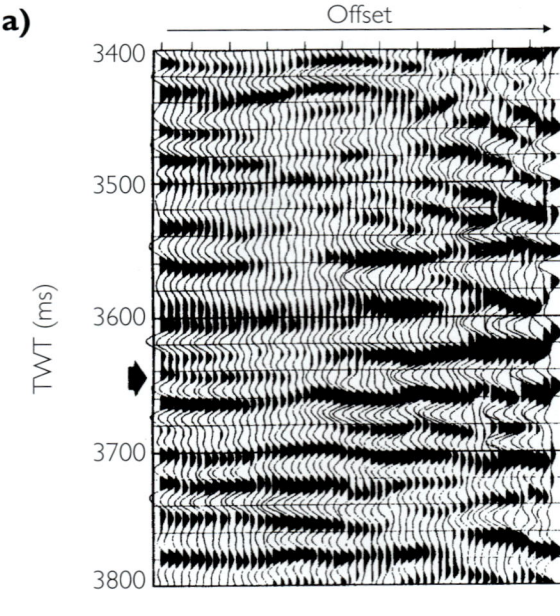

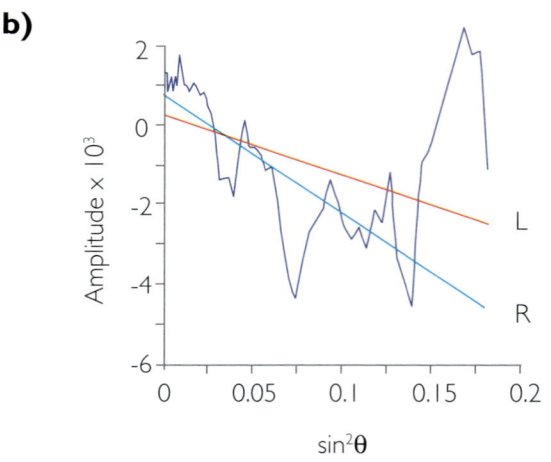

Figure 6.16 Comparison of AVO intercept and gradient calculation by least squares (L) and by robust fit (R) (after Walden, 1991).

extra stability in generating supergathers with attendant benefits in the extraction of AVO attributes.

6.3.5 Gradient estimation and noise reduction

Notionally, it is a simple matter to calculate a volume of intercept R_0 and of gradient G traces by making a least-squares fit of the amplitude R (for each TWT sample of each angle gather) to the equation $R = R_0 + G \sin^2\theta$, where θ is the angle of incidence. It is important that the angle range used to calculate the gradient has good signal-to-noise ratio and that the linearity assumption is valid. This can be tested by analysing the angle behaviour of amplitudes in the gathers and comparing it to well synthetics. As has been discussed in previous sections a key issue is the role of noise in the variability of the gradient (e.g. Hendrickson, 1999; Cambois, 1998) and various methods have been proposed to improve the robustness of the gradient estimate in the presence of noise.

6.3.5.1 Walden's robust fitting

Walden (1991) proposed a two-stage approach as follows. In the first step the data are divided into two groups: nears and fars, separated at the median value of $\sin^2\theta$. For each group, the median amplitude and value of $\sin^2\theta$ are calculated. A straight line drawn between the near-group point and the far-group point gives an estimate of the gradient that is robust against the noise, because the median is insensitive to outliers. If this estimate for the gradient is β, then the residuals calculated by subtracting $\beta \sin^2\theta$ from each starting amplitude can be subjected to the same median fit calculation. This process could be iterated further, but instead Walden suggests making a maximum-likelihood fit to the residuals at this stage. This differs from a standard least-squares fit by giving less weight to points with high residuals, and zero weight to points with residuals beyond a threshold. An example is shown in Fig. 6.16, where the robust fit has been effective in removing the influence of the high amplitudes on the last few traces.

6.3.5.2 Whitcombe's noise reduction

Another method for reducing noise in the gradient has been proposed by Whitcombe et al. (2004). It makes use of the fact that the noise is well imaged as a coordinate rotation on the intercept vs gradient crossplot. Filtering of the rotated data is carried out by a process rooted in the hodogram concept (Keho et al., 2001). First, a coordinate rotation is applied to the data in the intercept–gradient domain to align the axes parallel and perpendicular to the noise trend (Fig. 6.17a). Rotation of the axes is achieved by:

$$R'_0 = R_0 \cos \chi + G \sin \chi, \quad (6.6)$$

$$G' = -R_0 \sin \chi + G \cos \chi. \quad (6.7)$$

Within a sliding window the rotated values of intercept and gradient for adjacent datapoints have a regression line fit through them. The filtered output for the central datapoint is determined directly from

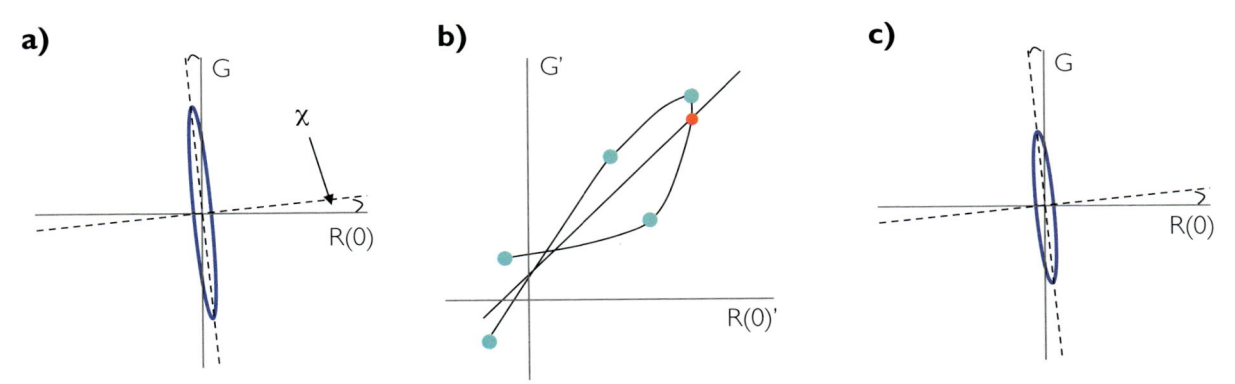

Figure 6.17 Filtering process to improve AVO gradient estimation, after Whitcombe et al. (2004); (a) intercept vs gradient crossplot showing the noise ellipse, (b) determining new values for the rotated gradient based on a regression of rotated intercept and rotated gradient points in a sliding window (five samples in this case), (c) noise ellipse reduced in magnitude after rotation back into intercept gradient space (after Whitcombe et al., 2004).

Seismic processing issues

the regression assuming that the rotated intercept value is correct (red dot in Fig. 6.17b). This is repeated for all samples by sliding the window down one sample at a time. Data are then transformed back to the intercept–gradient domain (Fig. 6.17c). The result is that scatter in the noise trend direction has been reduced, while there has been much less change in the perpendicular direction. To minimise smoothing through geology, a short time window is chosen (e.g. five samples) but samples are also included from adjacent inlines and crosslines, in a direction parallel to the local dip.

Chapter 7

Amplitude and AVO interpretation

7.1 Introduction

The purpose of seismic amplitude interpretation and AVO analysis is to explain changes in seismic signatures in terms of fluid and rock variations. In hydrocarbon exploration, the interpreter looks particularly for seismic amplitude changes that may be related to a change in fluid (i.e. from water to hydrocarbon). These seismic effects are often called *direct hydrocarbon indicators* ('DHIs').

The search for DHIs and the analysis of AVO data go hand in hand to develop a view of the prospectivity of an area. There are a variety of interpretation techniques that can be used to exploit AVO phenomena and aid interpretation. Commonly used techniques include:

- partial offset and angle stacks (e.g. near, mid, far) (Chapter 2),
- AVO crossplot colour coding (e.g. Verm and Hilterman, 1995),
- AVO hodograms (e.g. Keho et al., 2001),
- AVO projections (e.g. Whitcombe et al., 2002; and Chapter 5),
- target/reference 'relative AVO' method (e.g. Chiburis, 1993).

It should be remembered that confidence in amplitude interpretation is to a large extent determined by the geological and spatial context. DHIs are most commonly, but not exclusively, found in relatively shallow unconsolidated and partially consolidated sand/shale sequences, where fluid compressibility can have a significant effect on the whole rock compressibility. In well consolidated sandstone and carbonate situations the fluid tends to make a smaller contribution to whole rock compressibility (Chapters 5 and 8) and consequently DHI and AVO effects are usually more subtle.

Some of the key factors for the interpreter to consider are:

- reservoir porosity, mineralogy and stiffness characteristics,
- relative compressibility of hydrocarbon and water,
- non-reservoir properties,
- seismic bandwidth (resolution) and data quality,
- AVO angle,
- reservoir architecture
  . layering characteristics (e.g. thick or thin isolated reservoir units vs thick sequences of thin-bedded reservoirs)
  . lateral extent and morphology.

Clearly, well calibration is invaluable for understanding the seismic response. The preceding chapters have emphasised how seismic modelling with well data is a critical tool in guiding the interpreter in what to expect from local rock and fluid effects. In areas without well calibration detailed interpretations may not be possible but AVO might be used as a scanning procedure to highlight anomalies and rank different areas.

The following discussion focusses on the problem of hydrocarbon detection and examples will be presented of how various techniques might be used in a variety of different geological settings. Given the large variety of expressions of AVO behaviour this discussion is intended to get the interpreter thinking rather than as an exhaustive account of all published approaches to AVO analysis.

7.2 AVO and amplitude scenarios

The following discussion describes common amplitude and AVO scenarios encountered in siliciclastic sands and shales, carbonates and fractured reservoirs, highlighting the use of various interpretation techniques and illustrating potential variability. A useful starting point is the AVO classification presented in Chapter 5, particularly for the description of tops and

Amplitude and AVO interpretation

bases of sands and hydrocarbon contacts in thick-bedded sand settings. The classification becomes less useful, however, when considering reservoir sections with multiple thin beds.

7.2.1 Class II/III hydrocarbon sands and Class I water sands

Figure 7.1 illustrates a full stack section from a gas field in an ideal setting to observe hydrocarbon effects. The hydrocarbon sands have Class II or III responses whilst the water bearing sands have a Class I response. Effectively this means that on full stack sections there will be a phase reversal at the reflection from the top of the sand as the water (red loop) gives way to gas (black loop). Bright soft responses are characteristic of hydrocarbon bearing sands in this scenario. Note that the gas water contact reflection in Fig. 7.1 is not flat. It shows 'push-down' owing to the very low velocities associated with the gas sand compared to the surrounding rocks.

The simple AVO model shown in Fig 7.1, generated by using average values taken from logs and input to the Aki–Richards equations, illustrates the general trends of the amplitude changes with increasing angle. Note that the model indicates decreasing water sand reflectivity with increasing angle, whereas the gas sand effectively brightens with increasing angle. Thus, far stacks may be best for observing the hydrocarbon signatures.

In this setting the identification of increasing amplitude with angle (sometimes referred to as 'rising AVO' or 'positive AVO'; Chapter 5) may well be characteristic of the presence of hydrocarbon. However, care needs to be taken not to apply this as a general rule. Given that the optimal angle for imaging fluids (i.e. the 'fluid angle') is effectively within the range of seismically acquired angles (Chapter 5) projection techniques may be of little benefit.

A popular AVO attribute combining intercept and gradient that effectively enhances 'rising AVO' (i.e. Class III reflections and positive reflections with

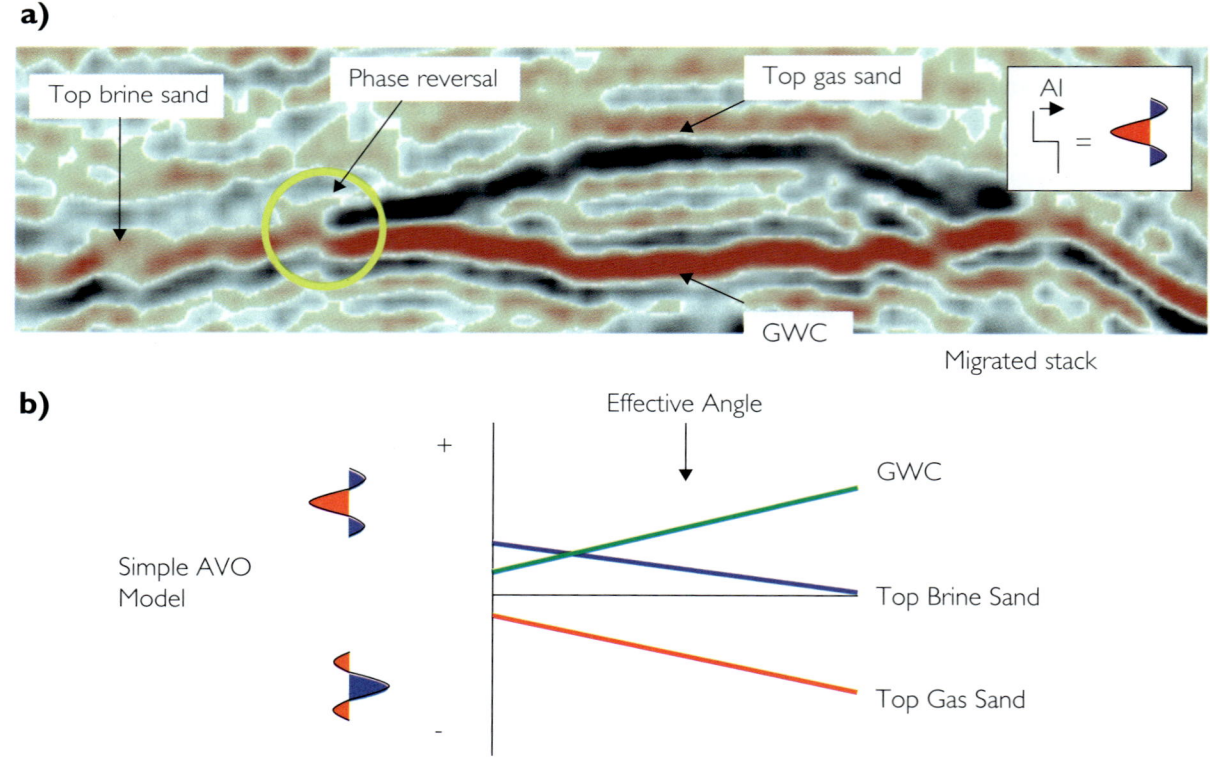

Figure 7.1 A bright spot on migrated stack display related to the presence of gas; (a) migrated stack display (European polarity), courtesy Rashid Petroleum Company, (b) schematic rock physics model.

AVO and amplitude scenarios

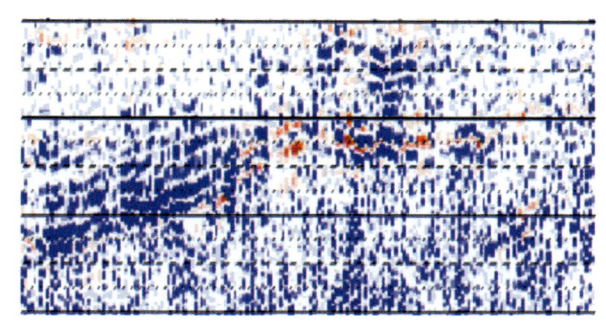

Figure 7.2 The AVO product indicator (intercept × gradient) (or 'AB' or 'PG' stack) highlighting Class III gas sands with increasing amplitude with offset (red) in contrast to Class I water sands with decreasing amplitude with offset (blue) (after Bacon et al., 2003).

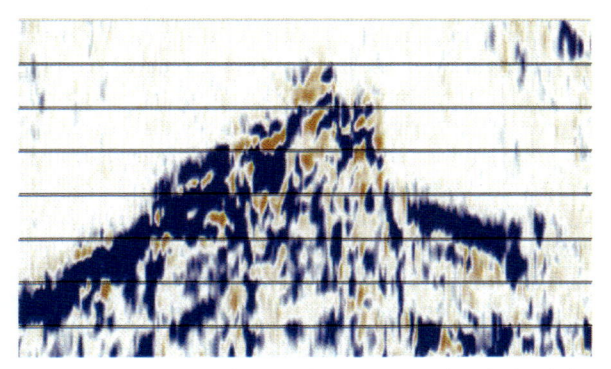

Figure 7.3 Gas accumulation with Class III AVO identified with the ERG (enhanced restricted gradient) attribute. Red colours represent positive AVO and blue colours represent negative AVO.

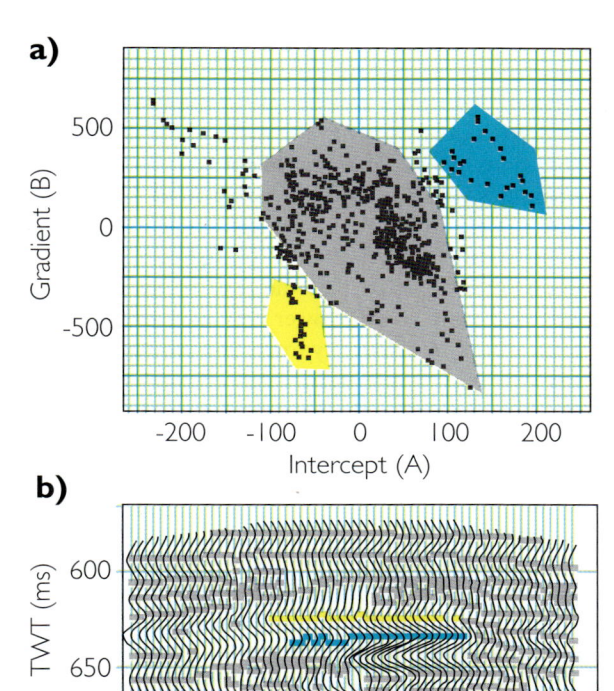

Figure 7.4 An example of colour coded crossplots highlighting 'anomalous' Class III signatures associated with the presence of gas: (a) AVO crossplot, (b) intercept section (wiggle) with data from yellow and blue zones highlighted. Courtesy Hampson-Russell.

positive gradients) is the intercept × gradient attribute (also called the 'PG' or 'AB' attribute). An AB attribute section showing two thin gas sands is displayed in Fig. 7.2. The red signature indicates 'positive AVO', in this case associated with thin gas sands, whereas the blue signature indicates 'negative AVO' associated with water bearing sands.

A useful reconnaissance attribute that tends to highlight Class III (positive AVO) responses and which is relatively insensitive to residual NMO is the enhanced restricted gradient (ERG) (Barton and Gullette, 1996). It is defined on the basis of the amplitude envelope complex attribute:

$ERG =$ (far envelope $-$ near envelope)$\times$far envelope.

Figure 7.3 shows a gas accumulation with the ERG attribute. The hydrocarbon is identified by the mottled red/blue signature (red values represent positive AVO) whereas the water sands have a broad blue signature typical of a Class I AVO signature.

7.2.2 Class III hydrocarbon and water sands

A more difficult situation for interpretation is where both hydrocarbon sands and water bearing sands have Class III AVO signatures. In this instance a soft response with a negative gradient is not necessarily diagnostic of hydrocarbons. Both hydrocarbon sands and water sands have soft responses but hydrocarbon sands will have higher amplitudes than water bearing sands (assuming they have similar porosities).

In this situation the interpreter needs to assess whether a response is bright enough to be due to the presence of hydrocarbon. In some basins the relative amplitude difference between hydrocarbon and water bearing sands is exploited by calculating the AB ratio (not to be confused with the AB AVO attribute). The AB ratio is the ratio of the target amplitude to selected 'background' amplitudes. High values of the AB ratio (e.g. 5–6) would tend to indicate hydrocarbon presence. In basins where the AB ratios are much smaller (e.g. 2–3) and the reservoir units are relatively thin it may not be possible to confidently discriminate wet sands from hydrocarbon sands.

Figure 7.4 shows an example of the use of colour coded crossplots from an area with Class III gas and water sands. Intercept and gradient have been plotted for samples within a time window around the zone of interest. Various clusters of samples have been highlighted in different colours and posted on the seismic section. It is expected that the tops of hydrocarbon sands will plot to the lower left and the base of hydrocarbon sands or hydrocarbon contact reflections will plot in the upper right (Chapter 5). The points highlighted with the yellow and blue colours are therefore of interest. On the seismic section it is evident that these points fit with the structural culmination on the seismic section, supporting the interpretation of a hydrocarbon sand.

A common pitfall in this AVO scenario is misinterpreting 'rising AVO' from water sands as a hydrocarbon effect. Figure 7.5 shows an example. The AVO effect is dramatic on the partial stacks (Fig. 7.5a) and the seismic gathers (Fig. 7.5b) but it is caused by a thin shaley water bearing sand. In this instance, the amplitude effect shows no obvious consistency with structural closure, nor does it have the characteristics of a geologically plausible trap (Fig. 7.5c). Without a connection of the amplitudes to a trap, the interpreter should be wary of assigning a DHI interpretation.

The anomaly shown in Fig 7.6 is also accentuated by tuning effects (Chapter 5), a common pitfall in scenarios where brine sands give soft seismic responses. In these types of situations it may be impossible to distinguish between a productive pay interval and a water bearing sand. Fig 7.6 shows an example where a productive gas zone and a thin water wet sand produce similar AVO responses. Such observations are useful in the risking process (Chapter 10).

AVO exploration is most successful in areas where the background geology is laterally invariant and the hydrocarbon signatures stand out from the background. Variations in lithology can give rise to ambiguities in the interpretation of AVO signatures. Figure 7.7 illustrates an example in which there are both variations in fluids (water and hydrocarbons) and a dominant facies change from clean sand to shaley sand. The coloured segments from the schematic section (Fig. 7.7a) represent zones with different AVO signatures (Fig. 7.7b). The clean sands in the

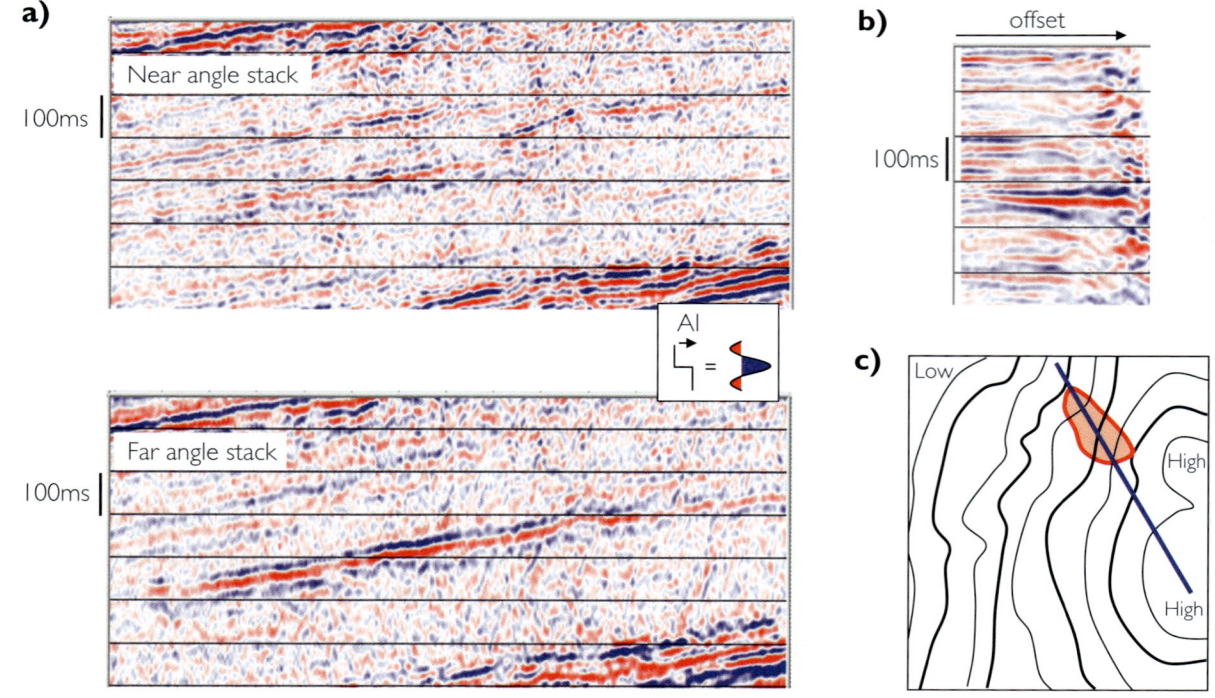

Figure 7.5 A Class III AVO response caused by a water bearing shaley sand; (a) near and far angle stacks, (b) gather example, (c) map showing monocline structure and zone of high amplitude with no obvious consistency with structure. Tuning is contributing to the high amplitude.

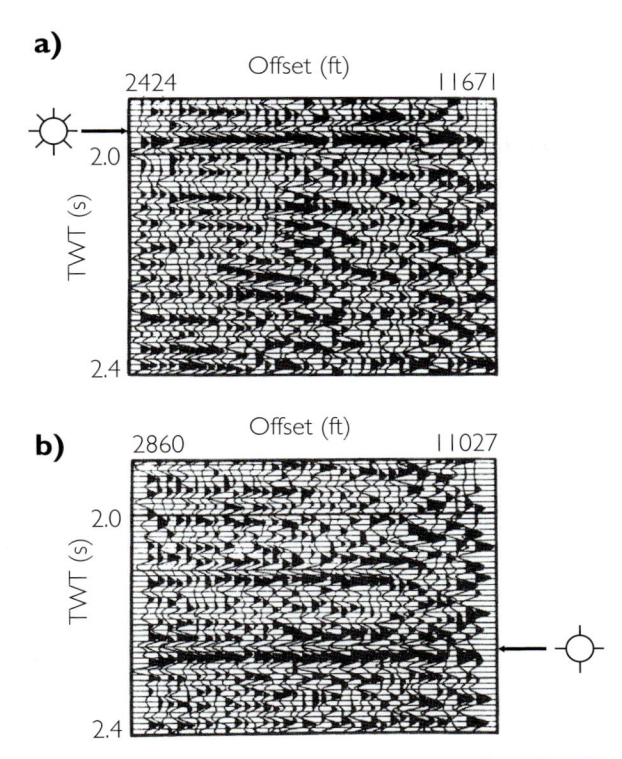

Figure 7.6 Tuning and AVO; (a) gather showing a Class II/III AVO response from a productive gas zone, (b) gather showing similar response from a tuned water wet sand (after Allen and Peddy, 1993).

south show Class III responses with hydrocarbon sands having amplitudes about three times brighter than the wet sands. The differential AVO effect between water and hydrocarbon fill in the shaley sands is much more subtle and the shaley hydrocarbon bearing sands fall in the same position as the clean water sands. Figure 7.7c illustrates how these variations might impact AVO mapping. It is probable that the AVO change will be consistent with structure in the south but the northern boundary of the southern culmination will be blurred owing to the facies change. There is likely to be an AVO anomaly over the northern culmination but far less dramatic than in the south.

Another pitfall arises when low gas saturation produces soft amplitudes and AVO signatures similar to commercial saturations (the so-called 'fizz gas' problem) (Chapter 5). Figure 7.8 shows an example from the Green Canyon area of the Gulf of Mexico published by O'Brien (2004). Gas was found at the King Kong prospect with the two prominent high-amplitude zones representing the pay zones. All exploration indicators (e.g. AB ratios) for the Lisa Anne prospect appeared to show gas but although the well found the target sandstones it did not find commercial pay. The explanation given was that low saturations of gas were present within the sandstones. A single interface model based on the field data (Fig. 7.8b) illustrates how the low gas saturation can give rise to AVO similar to that of commercial gas. Although low gas saturation is clearly an issue for interpreting amplitudes it can be difficult to petrophysically determine low gas saturations from well logs. Thus, it can be hard to be definitive about a 'fizz gas' interpretation. It should be noted that the 'fizz gas' problem only occurs at relatively low temperatures and pressures (Han and Batzle, 2002; Section 8.2.4.2).

7.2.3 Class IV hydrocarbon and water sands

Although relatively uncommon, it is possible that the tops of hydrocarbon sands and water bearing sands both have Class IV responses. Typically this occurs where the lithology overlying the sand has a shear velocity significantly greater than the shear velocity of the sand (Chapter 5). This scenario is similar to the previous scenario, in that it may be difficult to assign a hydrocarbon interpretation simply by looking at a full stack (i.e. both water sands and hydrocarbon sands have soft responses). However, unlike the previous scenario, AVO projections may be useful. Figure 7.9 (based on data from a gas field) shows how projections can highlight the hydrocarbon reservoir and the contact. In this scenario the shale/water sand interface has a steeper positive gradient than the hydrocarbons (Fig. 7.9a). Thus, linearly projecting the AVO responses will give an angle at which the brine sand reflectivity is minimised. In this model this approximates to an angle of $\theta = 90°$ or $\chi = 45°$ (Fig. 7.9b). Figure 7.9c shows how typical far angle stacks will show soft signatures in both the water and the hydrocarbon zone, but the reservoir will be best discriminated at the 'fluid angle'.

Class IV signatures are more commonly associated with hard units, such as silts and limestones, overlying reservoir sands. The presence or absence of a hard unit overlying porous sands can have a dramatic effect on the AVO reflectivity. Figure 7.10a shows a model example from a well with a hard silt above a sand with 30% porosity. The wet sand has a

Amplitude and AVO interpretation

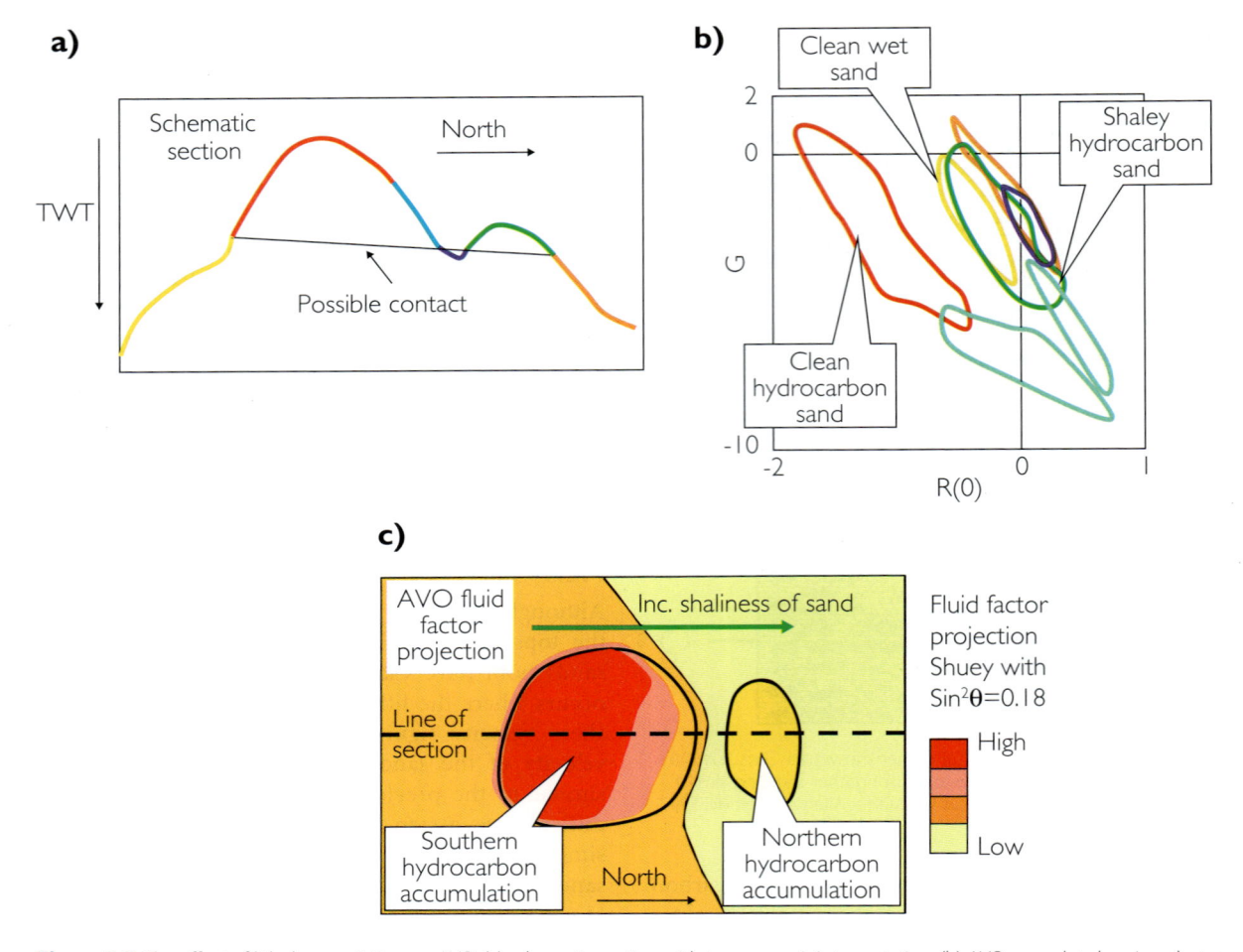

Figure 7.7 The effect of lithology variation on AVO; (a) schematic section with top reservoir interpretation, (b) AVO crossplot showing cluster shapes from segments shown in (a), (c) schematic fluid factor map. (Re-drawn and modified after Sams, 1998.)

soft signature owing to the presence of the silt and the hydrocarbon (oil) effect is simply to increase the amplitude (i.e. the differential AVO effect between the hydrocarbon and water cases is minimal).

When the silt is removed (Fig. 7.10b) the top sand reflection becomes a weak Class I response when the sand is wet and a Class II response when oil bearing. The key signature effect for hydrocarbons in this case is a soft response at the top of the sands on far stacks. AVO crossplots for the two cases illustrate how the presence of the hard unit can affect the optimal fluid projection angle. A high angle, $\chi \sim 35°$ ($\theta = 57°$), is required to identify the Class IV response at the top of the sand when the silt is present (Fig. 7.10c), whereas a lower angle of $\chi \sim 10°$ ($\theta = 25°$) is required for the non-silt case.

The presence or absence of these hard layers could create ambiguities for AVO interpretation. For example, if a low angle projection (in Fig. 7.10d) is applied when silt layers are present above sands (Fig. 7.10c), water wet responses will appear anomalous as well as the hydrocarbon sands.

7.2.4 Class IIp hydrocarbon sands, Class I water sands

Perhaps the most subtle and potentially ambiguous AVO scenario occurs when hydrocarbon sands have a Class IIp signature and water sands have a Class I signature. A good example from Rutherford and Williams (1989) is shown in Fig. 7.11a. The feature has been tested by a well which found gas.

AVO and amplitude scenarios

a)

King Kong (Gas) GC473   Lisa Ane (Fizz) GC474

TWT (s)

W   E

b)

Effective Angle

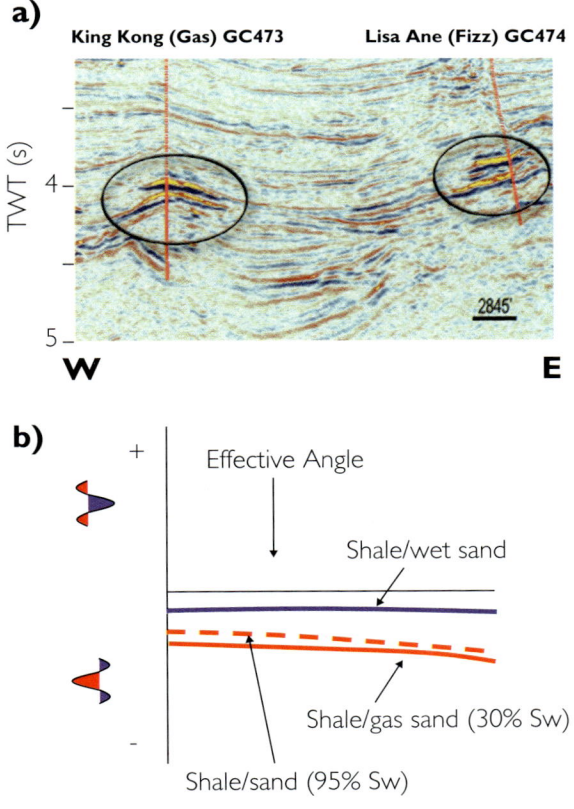

Shale/wet sand

Shale/gas sand (30% Sw)

Shale/sand (95% Sw)

Figure 7.8 Soft amplitudes related to variable saturations of gas; (a) seismic sections (after O'Brien, 2004) showing no significant differences between a discovery well and a well with low gas saturation, (b) top sand AVO plot generated from published log data.

The clearest feature on the full stack migrated section is a bright peak reflection related to the response from the gas water contact whilst the top of the reservoir does not have a distinct reflection. A notional AVO model consistent with these observations is shown in Fig. 7.11b. The presence of a hydrocarbon accumulation is demonstrated by the apparent thickening of the isochron between the bright peak and the peak above. The time thickness of the pay zone can be estimated by the extra thickening observed. Of course in exploration recognising these effects requires careful observation of the seismic signatures combined with a rock physics model developed from appropriate well information. In all likelihood the correct interpretation would be clearer on the far stack where (as indicated by the rock physics model in Fig. 7.11b) the base reflection would be even brighter but would be matched by a soft response at the top of the gas sand. The interpreter should also check for a dim spot effect on the near stack to further justify the hydrocarbon interpretation.

Figure 7.12 shows an example which illustrates the value of the far stack for hydrocarbon identification in this AVO scenario. The top of the hydrocarbon-bearing oil sand reservoir is clear on the far stack (Fig. 7.12b), being the high-amplitude soft black loop. On the nears (Fig 7.12a) it is difficult to see the subtle dimming of the hard red/orange loop that would be consistent with the presence of oil. The phase reversal shown on the gather (Fig. 7.11c) emphasises the need for accurate velocity analysis to correct for normal moveout.

The philosophy of using simple models to aid interpretation is illustrated very well by the example shown in Fig. 7.13, which led to an oil discovery in the Gulf of Mexico (Ross, 1995). In this case, the rock physics analysis done as part of the exploration effort suggested that oil sands would have Class IIp responses whilst wet sands would have Class I responses. It was realised that if the near stack was crossplotted against the far stack the oil sand reflectivity (top and base) would fall in particular areas of the plot (Fig. 7.13a). Colour coding the data on the seismic sections (Fig. 7.13b) enabled the identification of a major Class IIp anomaly that could be mapped over a large area and which had good geological justification.

AVO projections can be used to good effect in this AVO scenario. Whitcombe et al. (2002) illustrate the mapping of fluid and lithology effects using a modification of the Shuey equation (Chapter 5). The lithology and fluid maps shown in Fig. 7.14 were generated by using intercept and gradient bandlimited impedance volumes that were combined at AVO (χ) angles of 308.7° (or −51.3°) and 12.4° respectively. The oil water contact is shown as the blue outline. Note how the lithology volume effectively cancels the fluid effect in the data revealing the channel shapes which crosscut the contact, whereas the red areas on the fluid map reveal the presence of oil sands. The misfit of the red colours with the oil water contact in the west is due to poor data quality in this part of the field. Figure 7.15 shows some examples of these types of fluid and lithology cubes, with schematic AVO crossplots illustrating the mechanics behind their generation (see Section 5.5).

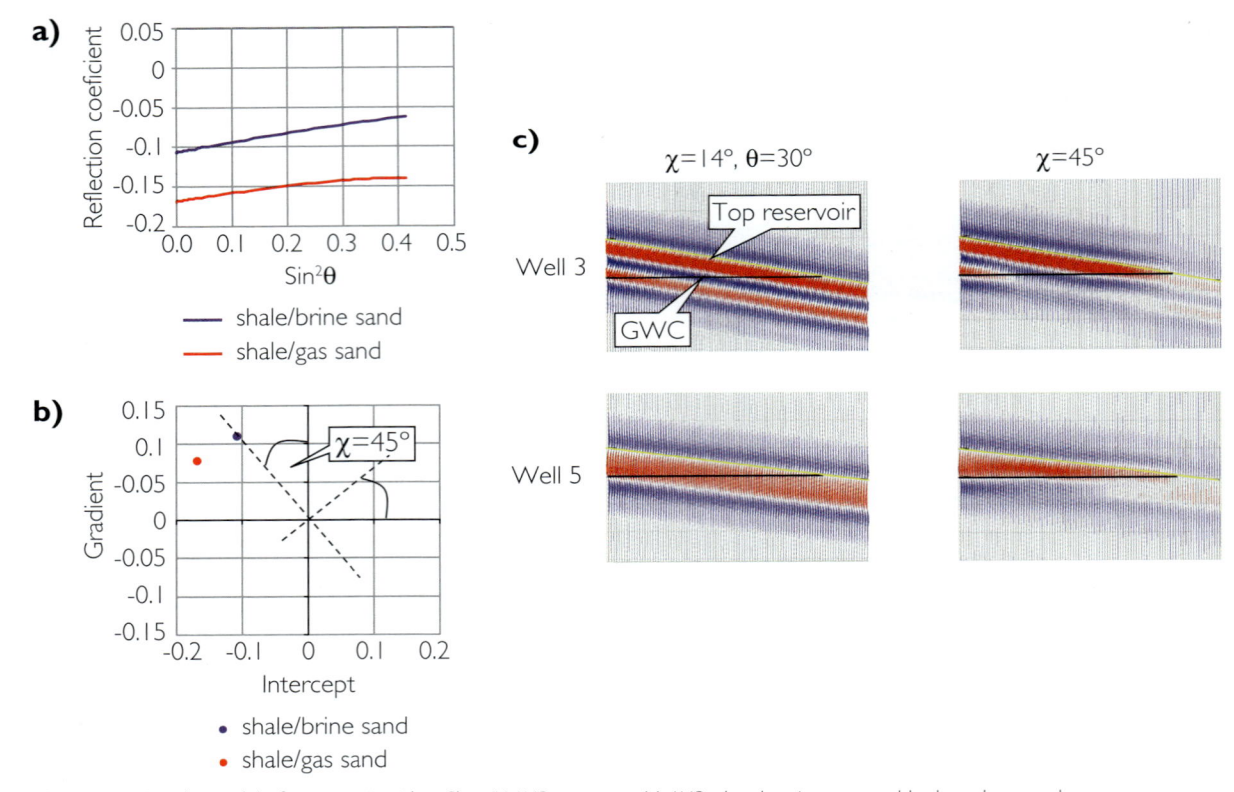

Figure 7.9 Simple model of a reservoir with a Class IV AVO response; (a) AVO plot showing wet and hydrocarbon sand responses, (b) AVO crossplot, (c) seismic models from two pseudo-wells, showing the use of AVO projections and bandlimited impedance in highlighting the gas water contact (after Simm, 2009).

7.2.5 Class I hydrocarbon sands, Class I water sands

When both water sands and hydrocarbon sands have Class I responses the effect of hydrocarbon is to dim the hard response at the top of the sand. On full stack sections the effect can be quite subtle, as is shown by Fig. 7.16. In some cases there may be a slight phase reversal at the top hydrocarbon sand that may be diagnostic.

There is likely to be interpretation ambiguity if only full stack sections are used in this scenario. For example, it may not be possible to differentiate a dimming due to an increase in porosity from a hydrocarbon effect. Figure 7.17 shows a model based on well log data in which a high-porosity sand is overlain by a low impedance shale. The near angle model (Fig. 7.17a) shows a subtle dimming of the top sand response, whilst the far angle model (Fig. 7.17b) shows a more dramatic change with a dim spot at the top of the hydrocarbon sand. Note that picking from the top water sand across the hydrocarbon contact is a possibility. The termination of the red trough on the far stack associated with the base of the sand is a useful diagnostic of the presence of a contact.

AVO projections (Chapter 5) may be useful for discriminating hydrocarbon sand in Class I situations. Figure 7.17 shows a modelled example. Amplitude changes are subtle on near and far angles (Figs. 7.17a,b); however, a fluid projection (Figs. 7.17c,d) emphasises the presence of hydrocarbon. With real data the result may not be so clear owing to the limited differential AVO effect and the noise-prone nature of the gradient estimation, but it is a technique worth trying.

One pitfall in Class I AVO analysis is the misinterpretation of critical angle energy (Chapter 2). The processor needs to be careful to remove all traces of critical angle effects in order that angle stacks and gradient estimations are valid for interpretation.

AVO and amplitude scenarios

a) Synthetic: silt above sand

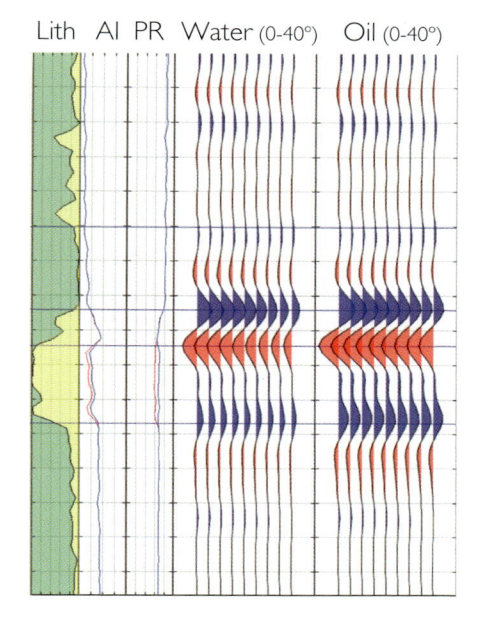

b) Synthetic: no silt above sand

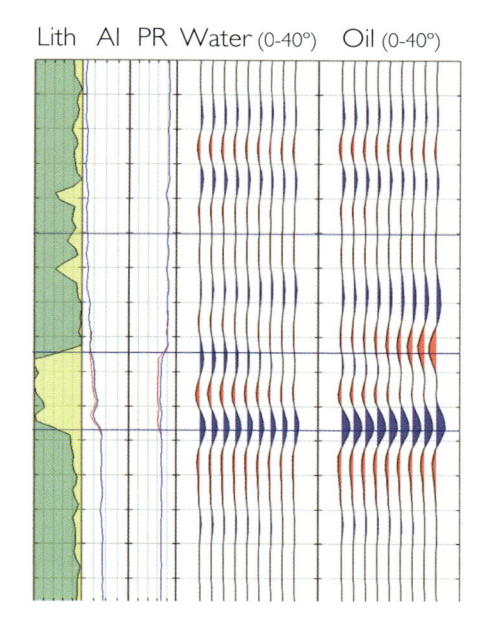

c) AVO Crossplot: silt above sand

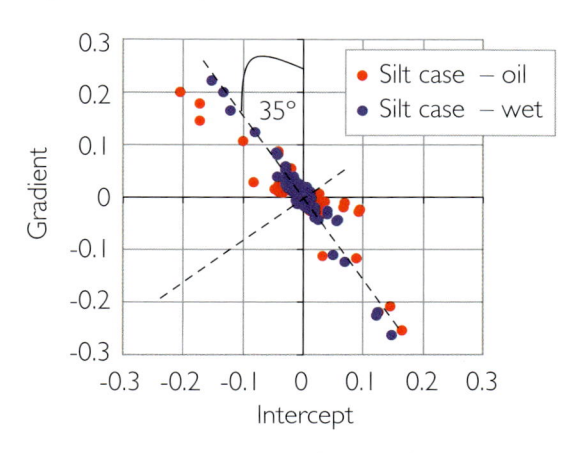

d) AVO Crossplot: no silt above sand

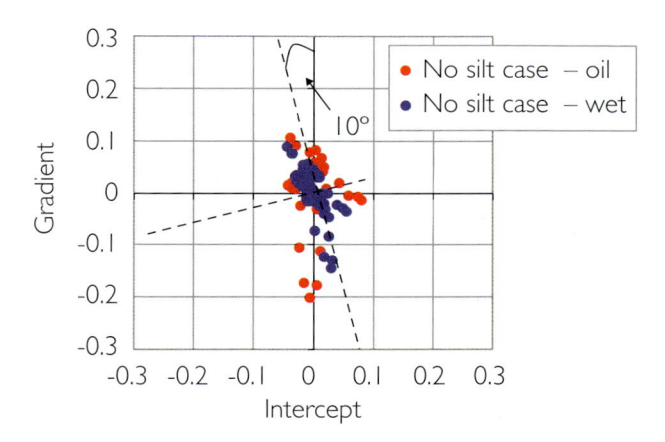

Figure 7.10 A model example of a top sand AVO signature that varies due to the presence or absence of a high impedance silt unit above sand; (a) synthetic with silt above sand, (b) synthetic with silt removed, (c) top sand AVO crossplot of data in (a), (d) top sand AVO crossplot of data in (b).

7.2.6 Multi-layered reservoirs

There are some thick reservoir sequences in which the differential AVO effects are very limited but owing to the multi-layered contrasts a general brightening of the section is a good hydrocarbon indicator. Figure 7.18 shows an example of a deltaic sequence of fairly low porosity, thin-bedded sands and shales with lateral thickness variations. It is difficult to reliably interpret hydrocarbon effects from brightening or dimming at the top sequence reflection as has been described in the previous AVO scenarios, mainly because it is an unconformity with variable

Amplitude and AVO interpretation

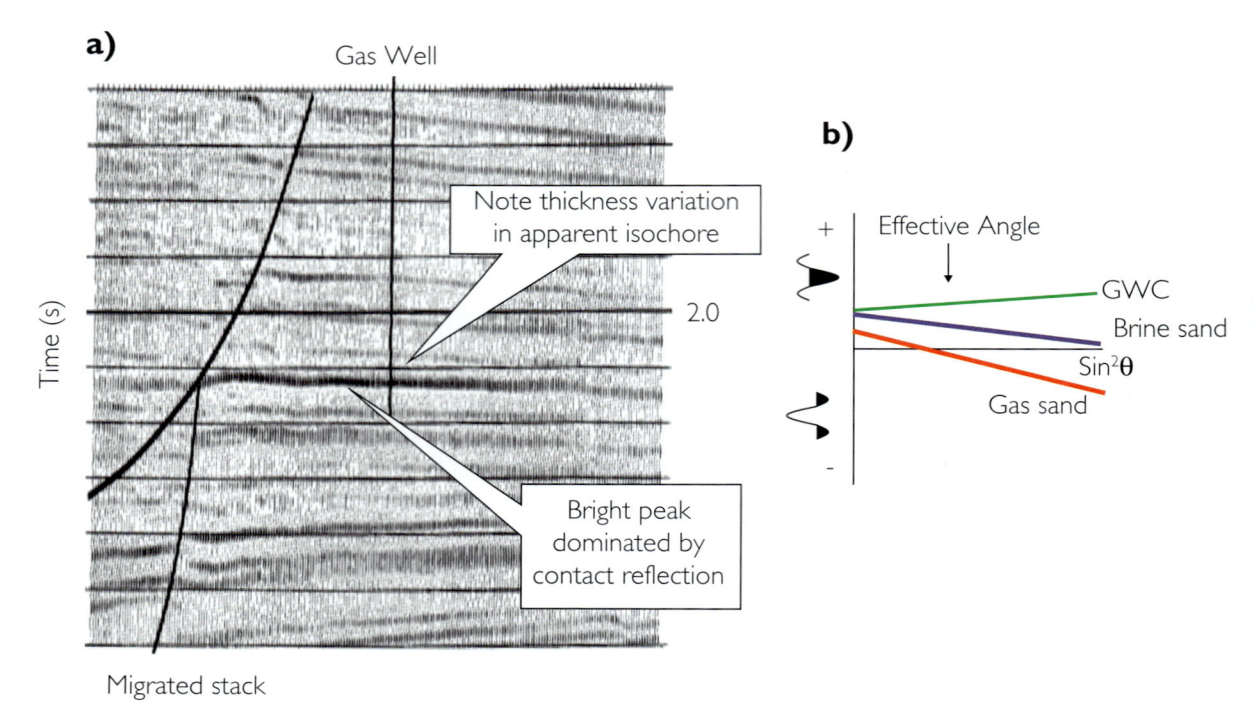

Figure 7.11 A Class IIp gas sand; (a) migrated full stack section (after Rutherford and Williams, 1989), (b) notional AVO model explaining the bright peak associated with the gas water contact and the lack of a reflection at the top of the sand.

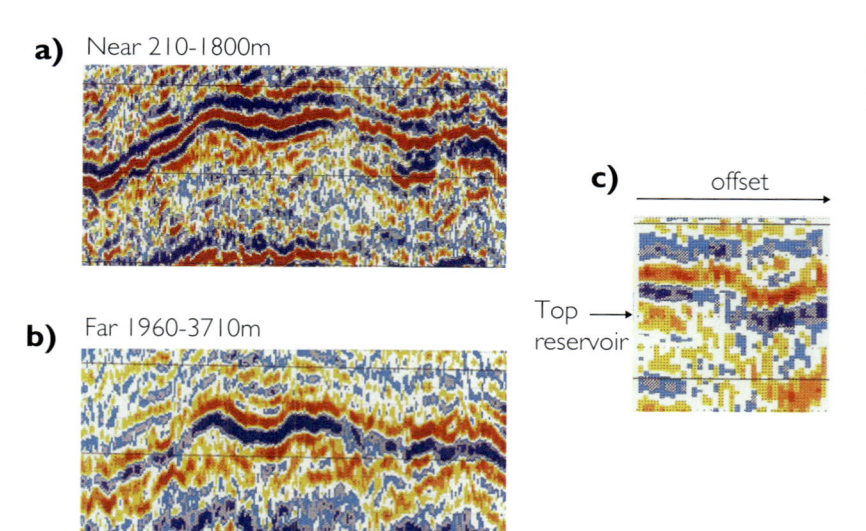

Figure 7.12 A Class IIp oil sand; (a) near stack section, (b) far stack section, (c) gather showing the phase reversal at the top of the reservoir.

sub-cropping lithologies. However, the reservoir section can be seen to generally brighten in response to the lowering of sand impedance. In these types of multi-layered scenarios the brightening is typically associated with thin-bed couplet reflections, sand base/top or top/base depending on the relative thickness and impedances of porous and non-porous units. Figure 7.19 shows an example from an aeolian

AVO and amplitude scenarios

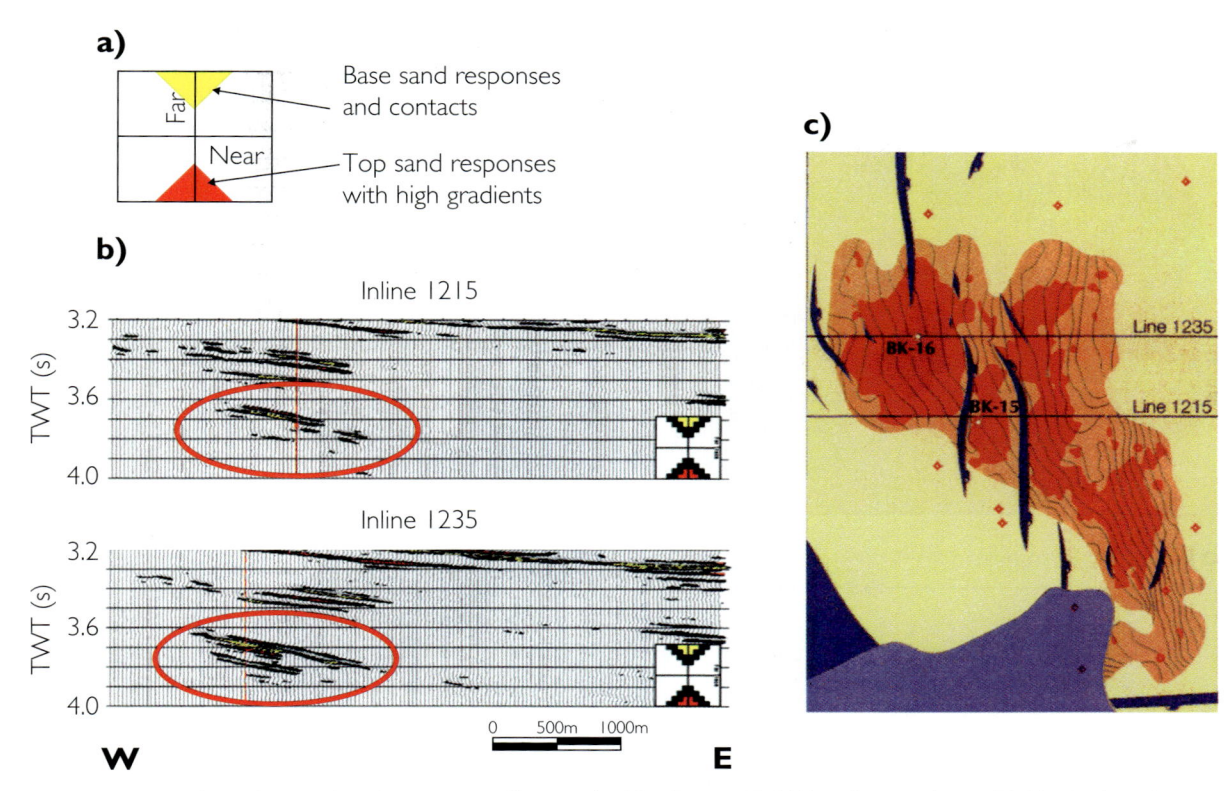

Figure 7.13 Describing Class IIp oil sands using a near/far crossplot (after Ross, 1995). (a) Near–far crossplot model; (b) seismic sections showing responses in red and yellow zones; (c) map of far amplitude strength.

reservoir in which variations in porosity (related to changes in the sedimentary environment) control the brightening. The brightening hydrocarbon effect of multi-layered reservoirs may be visible on vertical seismic sections but can be enhanced by techniques such as optical stacking (Fig. 7.20).

7.2.7 Hydrocarbon contacts

In favourable circumstances hydrocarbon contacts may give rise to *flat events* or *flat spots* on seismic and the recognition of such features can have a significant impact on the perceived chance of exploration success (Chapter 10). From a general point of view the presence of hydrocarbons in any prospect displaying a flat event is considered very likely if the flat event shows the following characteristics (Blom and Bacon, 2009):

- displays the correct phase and polarity,
- crosses the reservoir layering,
- shows a difference in dip with respect to possible multiples,
- is truncated at top and base reservoir or by a fault,
- shows AVO behaviour as expected for a fluid transition (i.e. higher amplitude on far offsets than near offsets),
- is located at the base of a structurally conformable soft reflection amplitude anomaly,
- corresponds in depth to a mapped structural spill point,
- is consistent with other direct hydrocarbon indicators, such as
  - dimming or phase reversals at the top of the reservoir,
  - lowering of dominant frequency beneath the contact (usually when gas is present).

These observations, of course, should be most evident on fluid optimised stacks. If the reservoir is fairly continuous, lithology stacks should reveal no obvious changes at the level of the contact. The interpreter should be mindful of the likely relative amplitudes of top and base reservoir and contact reflections based on modelling. Not all contact-related seismic

Amplitude and AVO interpretation

a) Fluid Projection

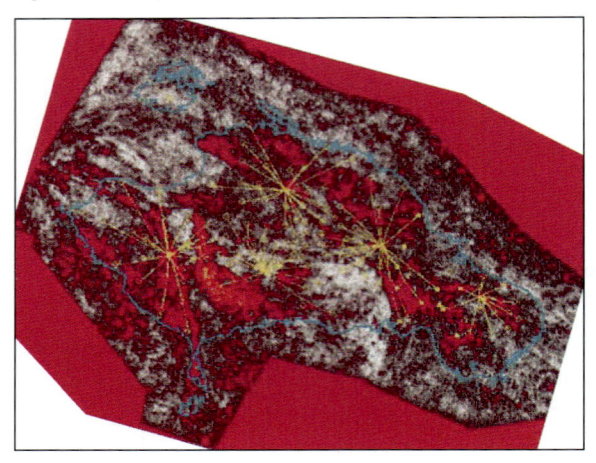

b) Lithology Projection

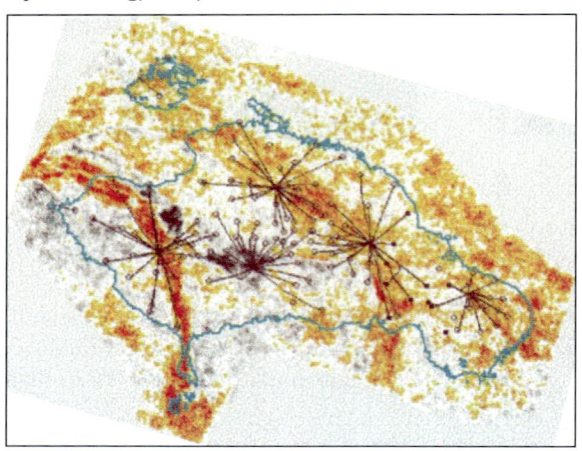

Figure 7.14 Fluid and lithology projections in a Class IIp oil sand scenario (after Whitcombe et al., 2002). Note how the fluid related AVO signature (red in (a)) follows the field oil water contact (blue line) (except for in the far west where there is a problem with data quality).

flat spot, discordant relationship to the fault block stratigraphy, clear lateral terminations at the fault and top reservoir consistent with a fluid change in a dipping fault block, as well as dimming (and possible phase change) at top reservoir and lowering of frequency beneath the flat spot.

A good example of amplitudes conforming to structure is shown in Fig. 7.22. Such conformance often increases confidence in a hydrocarbon interpretation. However, there may be good geological reasons why the seismic representation of the contact is not flat in two-way time. This may be due to the effects of overburden velocities varying laterally (e.g. Blow and Hardman, 1997). On the other hand, it is quite possible that the contacts are not actually flat in depth but inclined in response to pressure gradients within the sedimentary basin (Fig. 7.23).

Unfortunately, a flat spot related to the presence of hydrocarbon may not always be a guarantee of reservoir quality or productivity. For example, gas water contacts have been observed in shale and silt dominated sequences. Figure 7.24 is an example of a bright spot and flat spot related to oil in laminated sands where the permeability of the sands is such that the formation is not producible.

The character of hydrocarbon contacts depends on a number of factors including

- formation thicknesses,
- stratigraphic dip,
- dry rock frame stiffness of the reservoir rocks,
- difference between hydrocarbon and water elastic moduli,
- effective angle of the seismic stack.

When the reservoir comprises a series of thin-bedded sequences it is usual to see contacts as a series of separate segments each of which is related to interference effects between the fluid contact and the tops and bases of the sands (Fig. 7.25).

In carbonates and relatively stiff sandstone rocks interpretable contact effects are relatively rare, owing to the low reflection coefficients associated with the contacts and the presence of strong lithological contrasts. In these situations, contact effects are usually related to subtle tuning effects between top and bases of porous units and the fluid contact. Figure 7.26 shows a modelled effect in a relatively homogenous carbonate grainstone reservoir with porosity varying between 15% and 20%.

signatures show all these effects so it is important that seismic observations are integrated with all available geological information. Non-hydrocarbon explanations for the flat event should be investigated, for example presence of strong multiple energy or diagenetic changes such as temperature-related quartz transformations.

A dramatic *flat spot* in Jurassic sediments from the Troll gas field in offshore Norway is shown in Fig. 7.21. This example shows a number of DHI effects, including the clear positive reflection of the

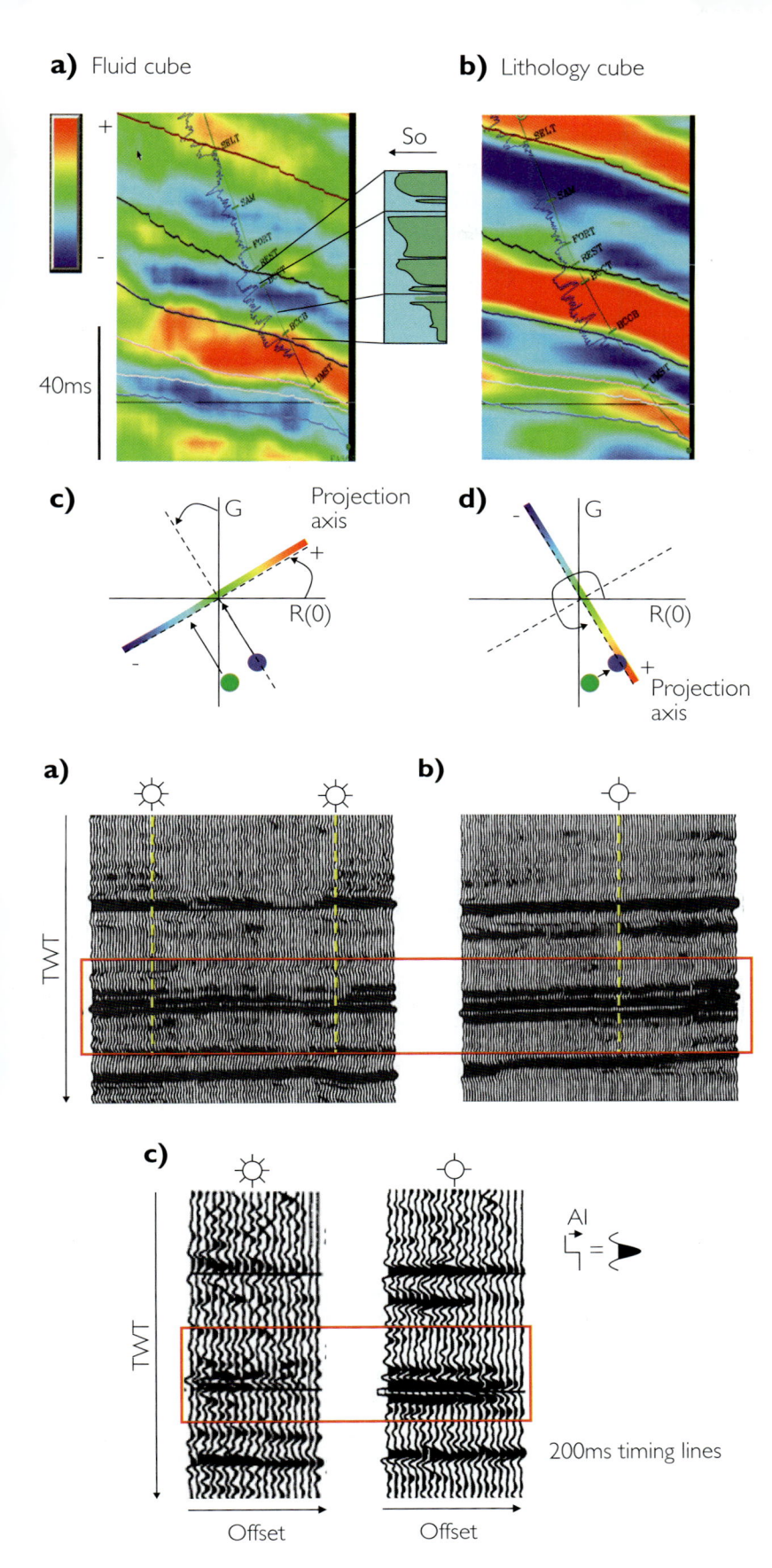

Figure 7.15 Examples of (a) fluid and (b) lithology cubes together with corresponding interpretation crossplots (c) and (d). Log curve shown in (a) and (b) is gamma ray (sections courtesy Apache Corporation).

a) Fluid cube

b) Lithology cube

c)

d)

Figure 7.16 Woodbine Sands, East Texas, Class I gas sand scenario (after Peddy et al., 1995); (a) migrated full stack section showing dimming of Woodbine sand reflectors when gas filled, (b) section showing bright water wet reflectors, (c) gathers contrasting gas and water wet AVO signatures.

200ms timing lines

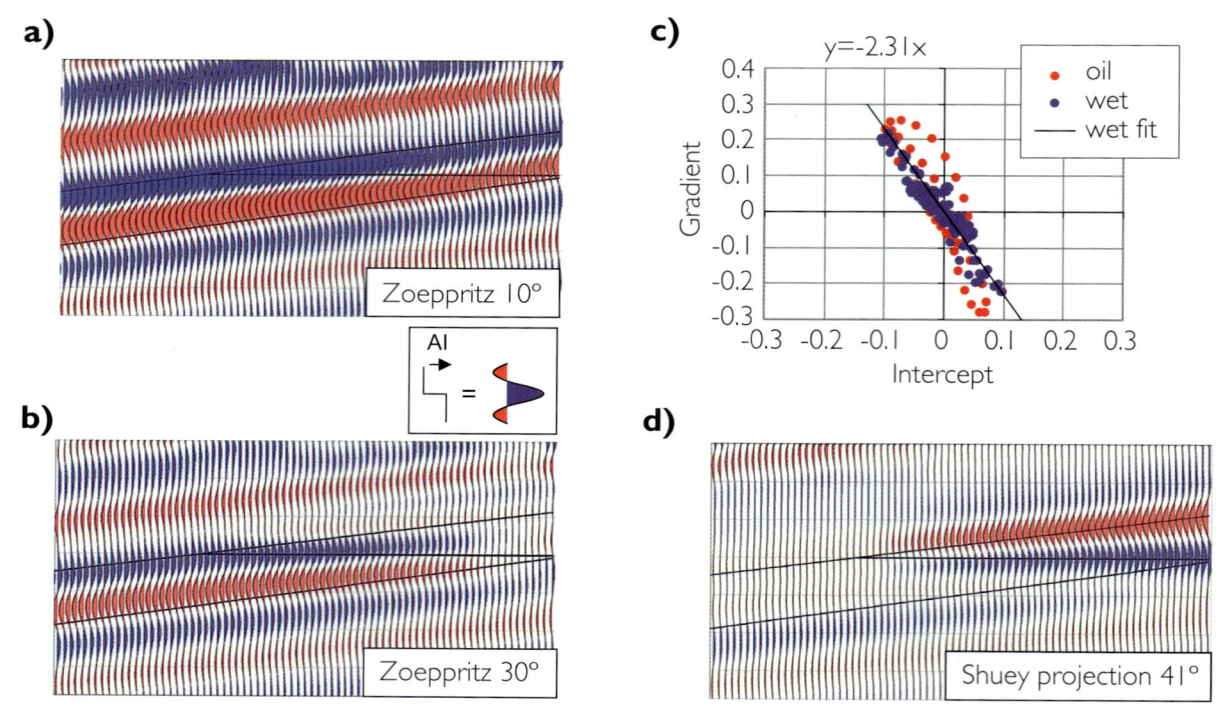

Figure 7.17 Class I oil sand model; (a) near angle section showing dimming related to oil presence, (b) far angle model showing little reflectivity at the top reservoir and trough termination at the contact, (c) AVO crossplot, (d) projection section (Shuey's equation used with $\theta = 41°$).

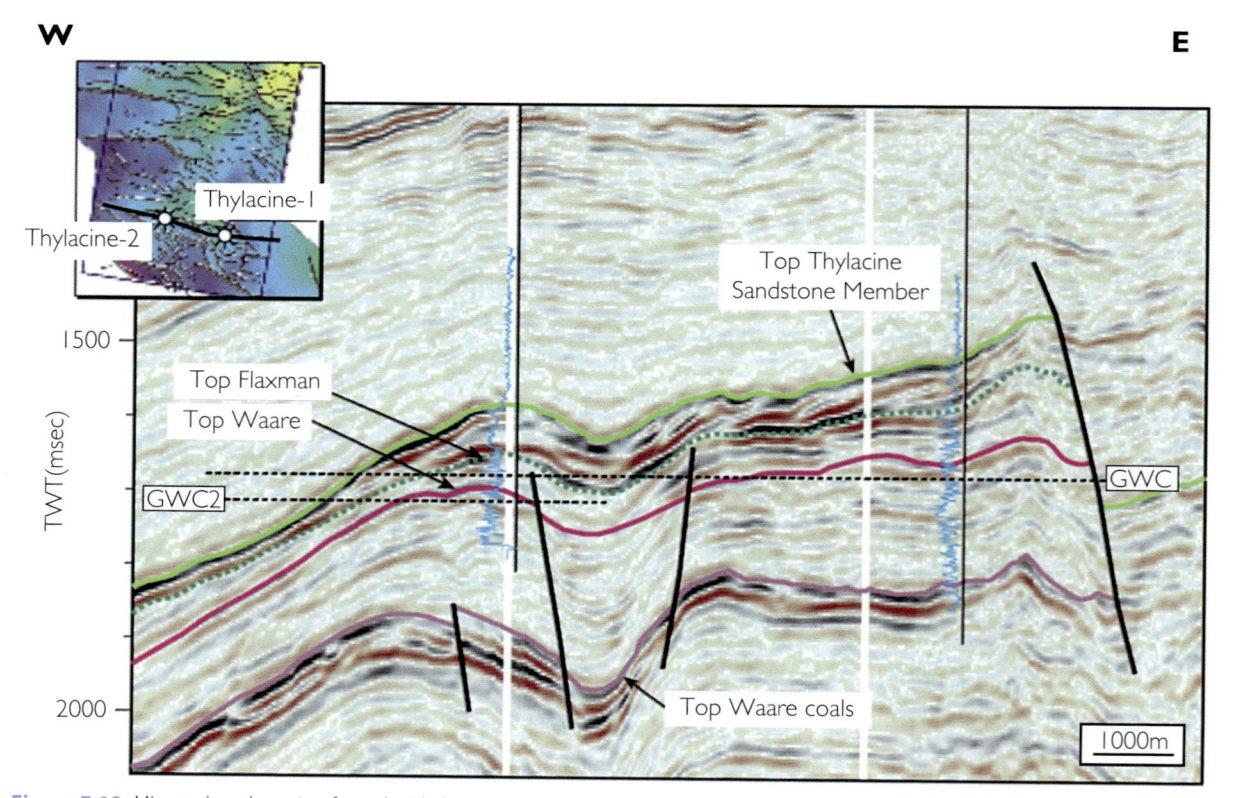

Figure 7.18 Migrated stack section from the Thylacine gas discovery in the Otway basin, offshore Australia (after Cliff *et al.*, 2004).

AVO and amplitude scenarios

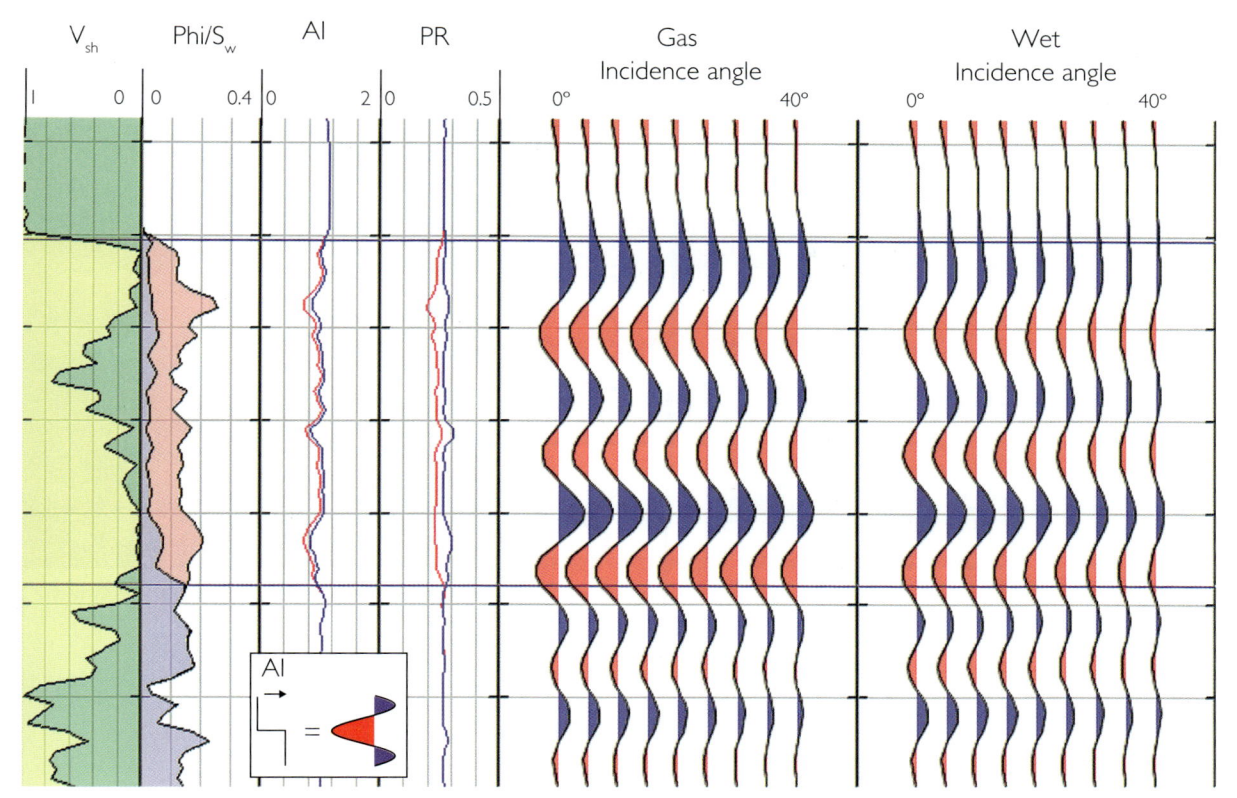

Figure 7.19 Modelled synthetic gathers from a sandstone reservoir with variable porosity layering. The hydrocarbon diagnostic is simply the brightening of the reflectivity associated with the contrasts in porosity. Note timing lines have 20 ms separation.

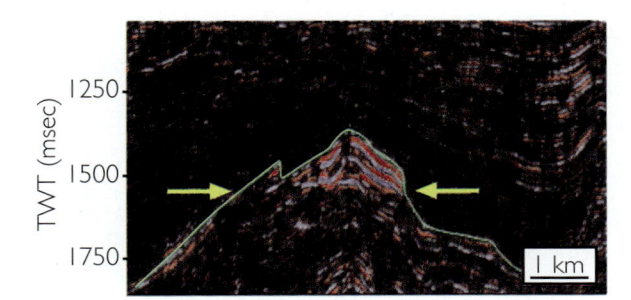

Figure 7.20 Enhancing the hydrocarbon effect in thin bedded reservoirs using optical stacking (after Cliff et al., 2004). In optical stacking opacity is adjusted to highlight the bright amplitudes then several adjacent sections are summed (in this case four sections are summed).

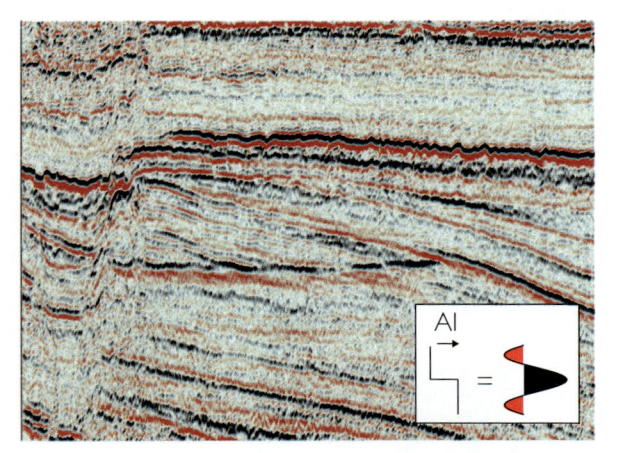

Figure 7.21 Troll gas field flat spot related to a gas water contact (positive standard polarity with black representing a peak). Note also the change in reflectivity at the top of the fault block (dimming of black peak) as well as the apparent change in dominant frequency beneath the flat spot.

Amplitude and AVO interpretation

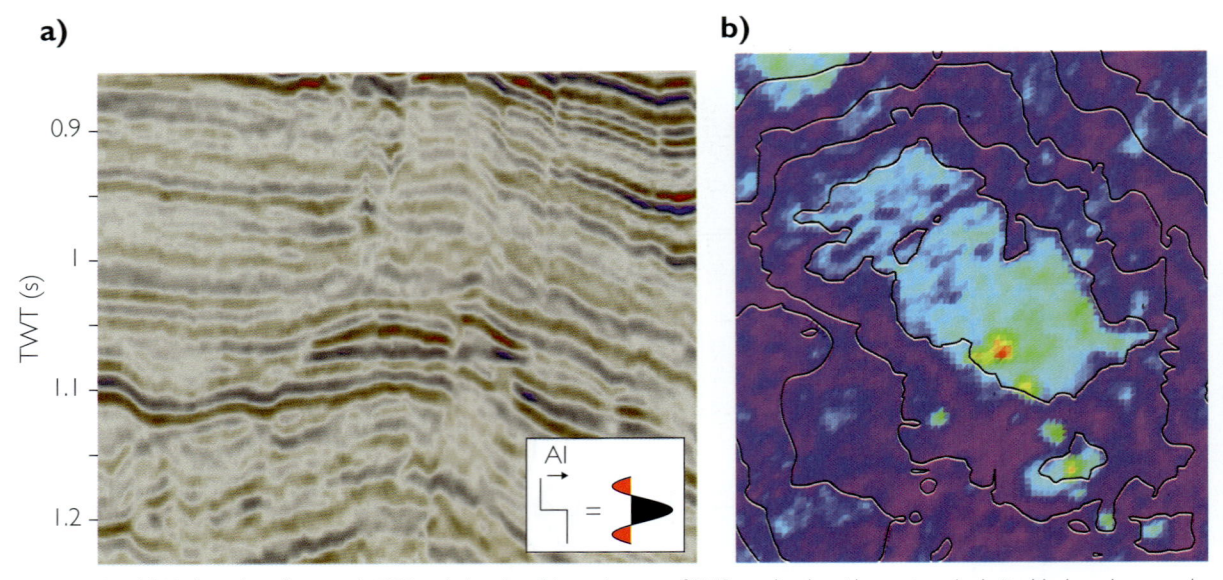

Figure 7.22 (a) Hydrocarbon flat spot (~1070 ms) showing (b) consistency of RMS amplitude with structure (polarity: black = downward increase in impedance) (courtesy WesternGeco).

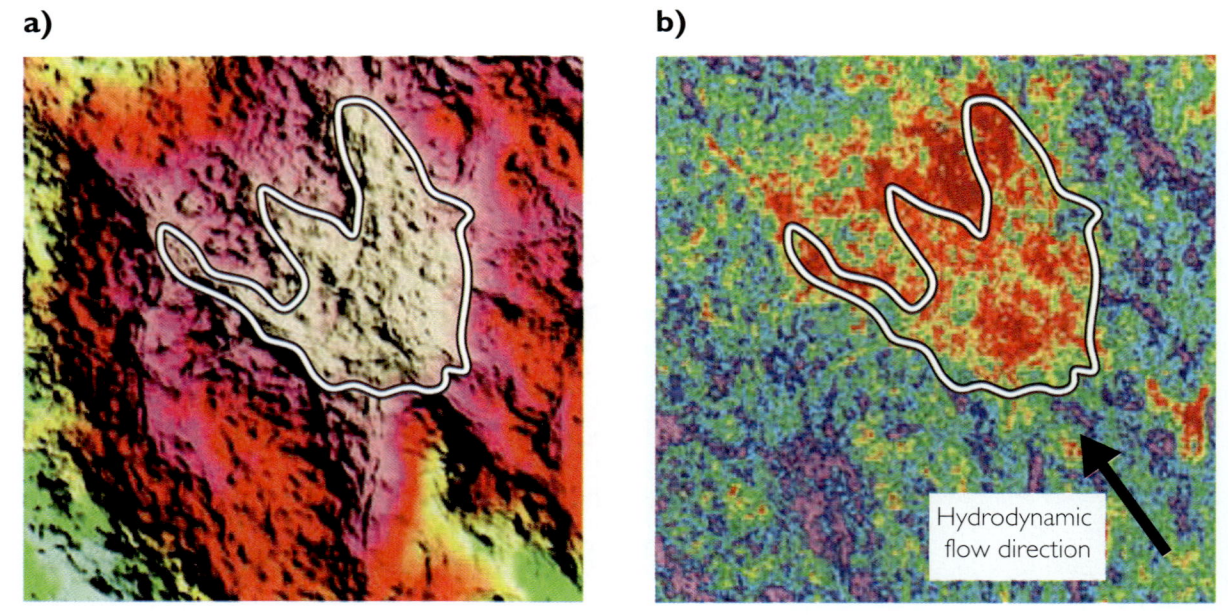

Figure 7.23 Structure, seismic amplitude and a tilted contact; (a) 3D-shaded structural relief map with the dip-closure outlined in white; (b) bandlimited impedance extracted close to the top reservoir (red colours indicate relatively low impedance related to oil fill in channelised sandstones oriented NW–SE). Note that the amplitude outline is shifted north-westwards from the structural closure owing to the effect of a hydrodynamic gradient (modified after Dennis et al., 2005).

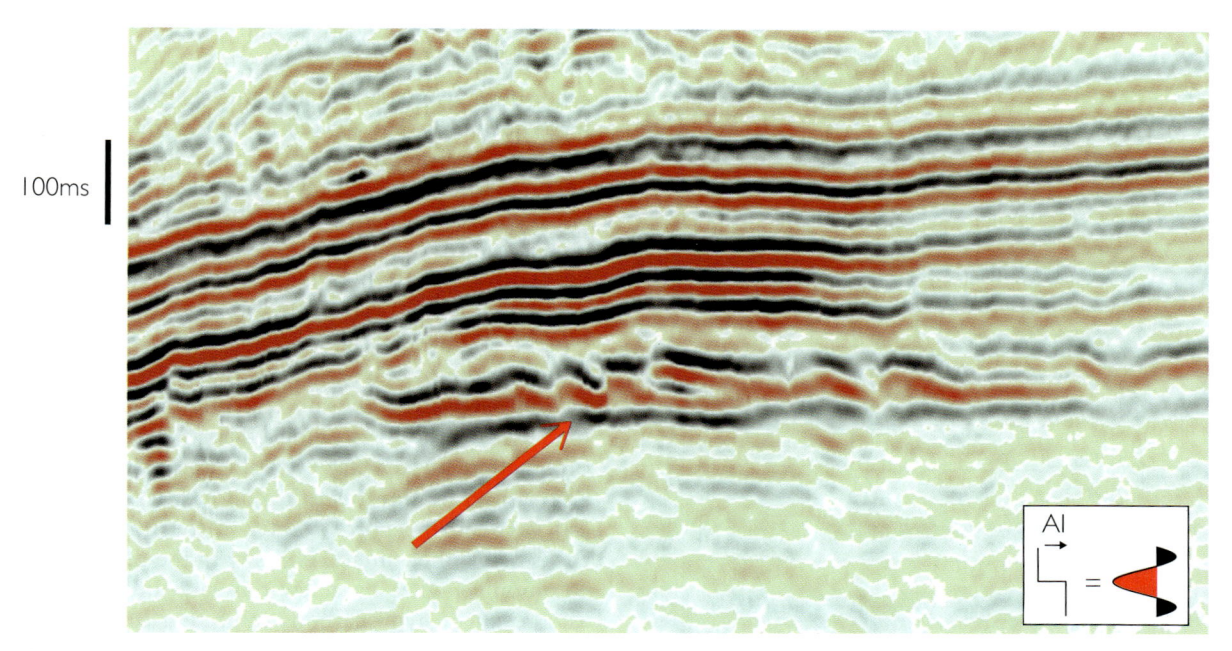

Figure 7.24 A DHI associated with an oil-bearing reservoir. Unfortunately, the reservoir has very low permeability. The presence of a DHI is no guarantee of producibility. Data are provided courtesy of WesternGeco.

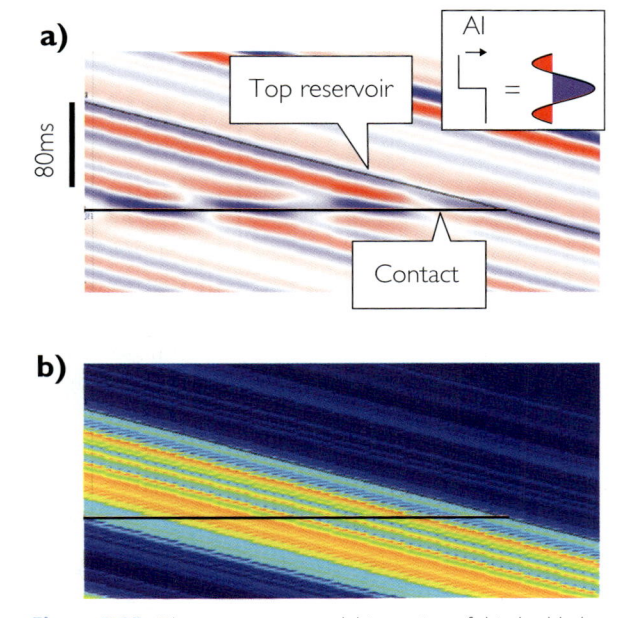

Figure 7.25 Oil water contact model in section of thin-bedded sands and shales; (a) angle stack ($\theta = 30°$) showing contact as a series of reflection segments with slight inclination, (b) porosity section – orange colours indicate sand layers with greater than 20% porosity.

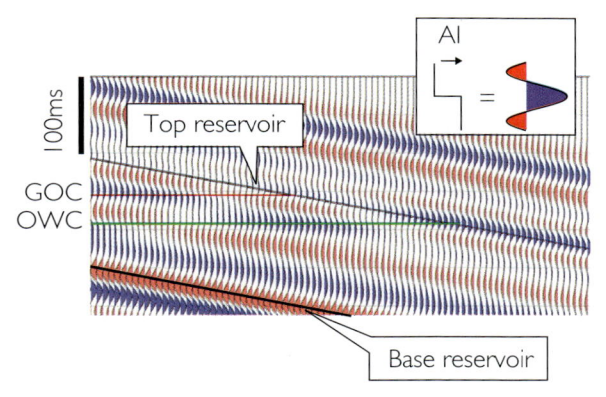

Figure 7.26 Hydrocarbon contact model in a grainstone carbonate reservoir. Contact effects are relatively subtle, with the main DHI being a dimming of the top reservoir reflector.

It is usually the case that seismic models predict that a hydrocarbon contact will have increasing amplitude with offset. This may not always be the case in real data, particularly if there are problems in imaging the far offsets. Figure 7.27 illustrates a comparison between a real gather (Fig. 7.27a) and a modelled

Amplitude and AVO interpretation

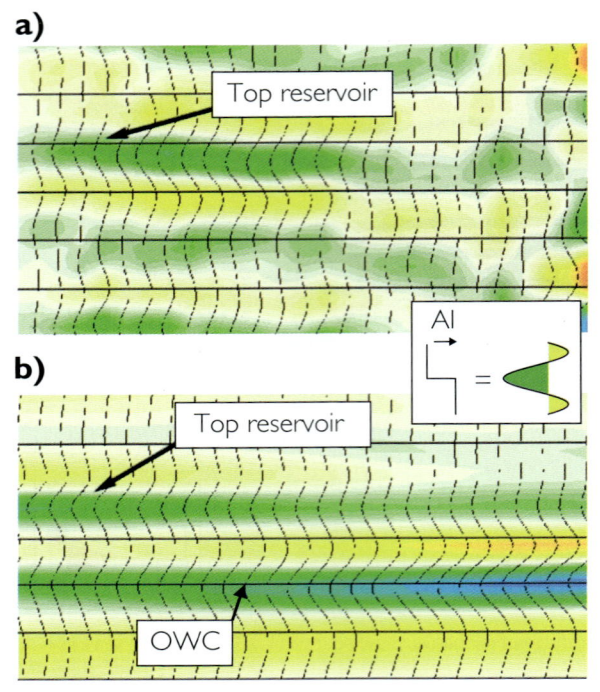

Figure 7.27 Real vs modelled seismic gather. The far angle brightening predicted by the model (b) is not replicated by the real gather (a) (G. Drivenes, personal communication).

gather (Fig. 7.27b) from an oil field discovery. The model is fairly sophisticated, including the effects of acquisition geometry and applying the same processing sequence as that applied to the seismic. The seismic shows a brightening of the contact at mid offsets but the strong modelled AVO effect at far angles is not replicated on the real gather probably because of poor imaging.

Flat-lying reflectors that cross-cut stratigraphy are not always related to the presence of hydrocarbon. In some cases the flat-lying event is related to the position of a previous contact with the trap having been filled but subsequently breached, with the hydrocarbon escaping. The most common explanation for the flat event in these situations is a porosity variation related to differences in diagenesis between the porous units in the trap and the rocks below the contact. Porosity in the relict trap is commonly higher by a few porosity units perhaps as a result of hydrocarbons inhibiting cementation (e.g. Gluyas et al., 1993) or other diagenetic processes such as bituminisation around the level of the relict contact (e.g. Yaliz and McKim, 2003). Clearly, it is important for the interpreter to be aware of the interplay between the reservoir

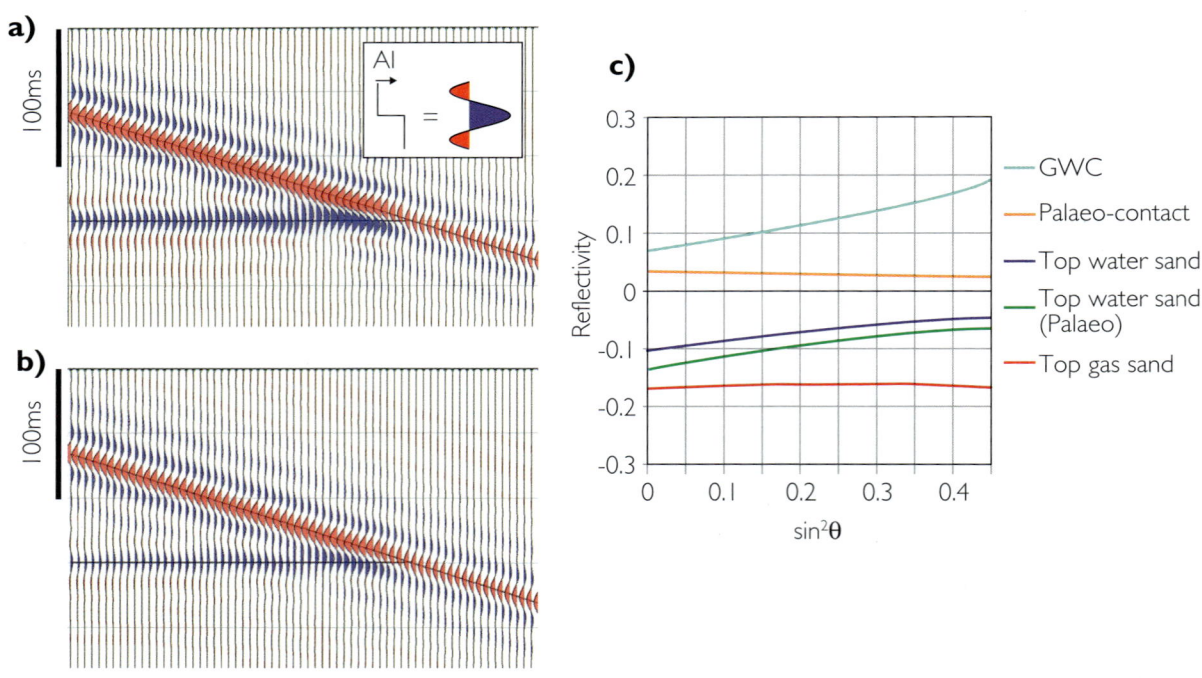

Figure 7.28 Hydrocarbon and relict contact models; (a) full stack hydrocarbon (gas) contact model, (b) full stack relict contact model (difference in porosity across the relict contact is 2%–3%), (c) AVO responses of different reflecting boundaries (based on data from Francis et al., 1997).

AVO and amplitude scenarios

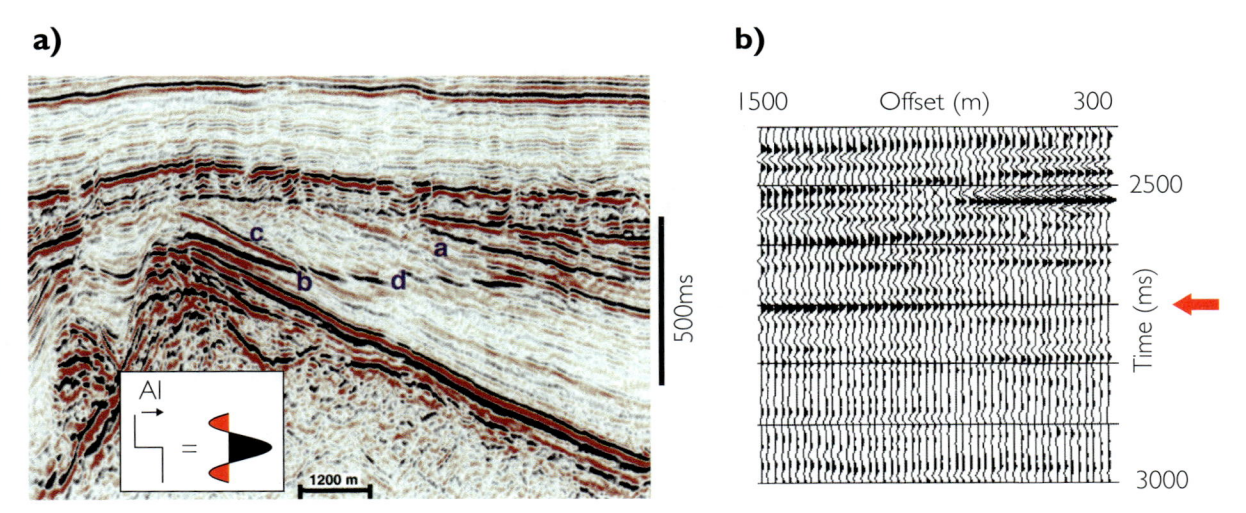

Figure 7.29 The Fylla flat spot – related to an opal/cristobalite transition; (a) full stack section ('flat spot' is marked as 'd', (b) moveout corrected gather showing positive AVO associated with the 'flat spot' (after Isaacson and Neff, 1999).

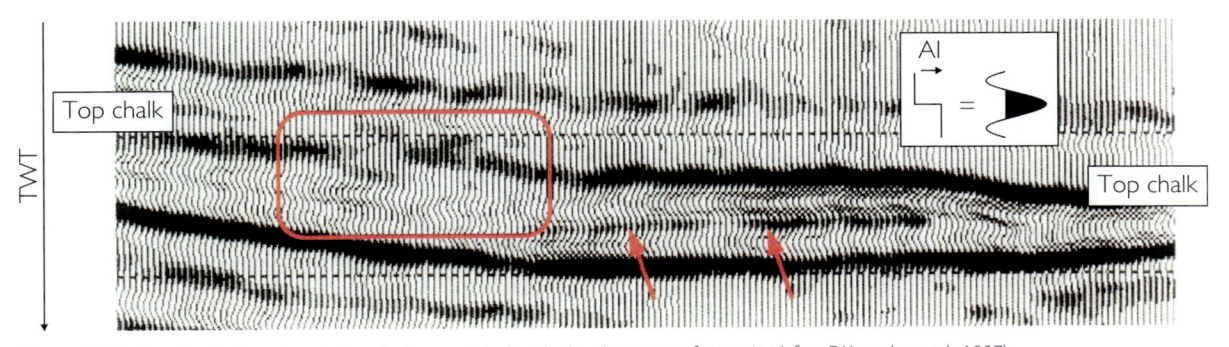

Figure 7.30 Amplitude dimming at top chalk associated with development of porosity (after D'Angelo et al., 1997).

types, burial history, timing of geological structuration and fluid migration.

Francis et al. (1997) have described a dramatic relict flat spot in the Triassic Sherwood sandstone in the East Irish Sea basin. Fig 7.28 shows models based on the published data. On full stack models the gas case is slightly higher amplitude than the palaeo-contact case (Fig. 7.28a,b). However, in principle, AVO analysis could help in differentiating the two cases (Fig. 7.28c). In terms of the top response the hydrocarbon case has zero gradient whereas the top water sand has a strong Class IV response. At the contact, hydrocarbon gives rise to a strong positive gradient but the palaeo-contact has little gradient expression.

Other examples of non-hydrocarbon related flat events include temperature-controlled mineral transformations. A well-known example is the Fylla flat spot described by Isaacson and Neff (1999) and Aram (1999) (Fig. 7.29). A hydrocarbon interpretation was given to the 'flat spot' as it appears to be restricted to a distinct stratigraphic unit (Fig. 7.29a) and shows increasing amplitude with offset (Fig. 7.29b). A well drilled into the feature encountered only mudstones at this level. X-ray diffraction analysis of sidewall cores suggests that it may be related to a phase change of amorphous opal (opal–A) converting to cristobalite and tridymite (opal–CT) (Christiansen et al., 2001). The reflection signature is poorly understood. Christiansen et al. (2001) postulate that there may be a component of anisotropy, although this has not been verified. Although the opal–CT transition gives rise to a

Amplitude and AVO interpretation

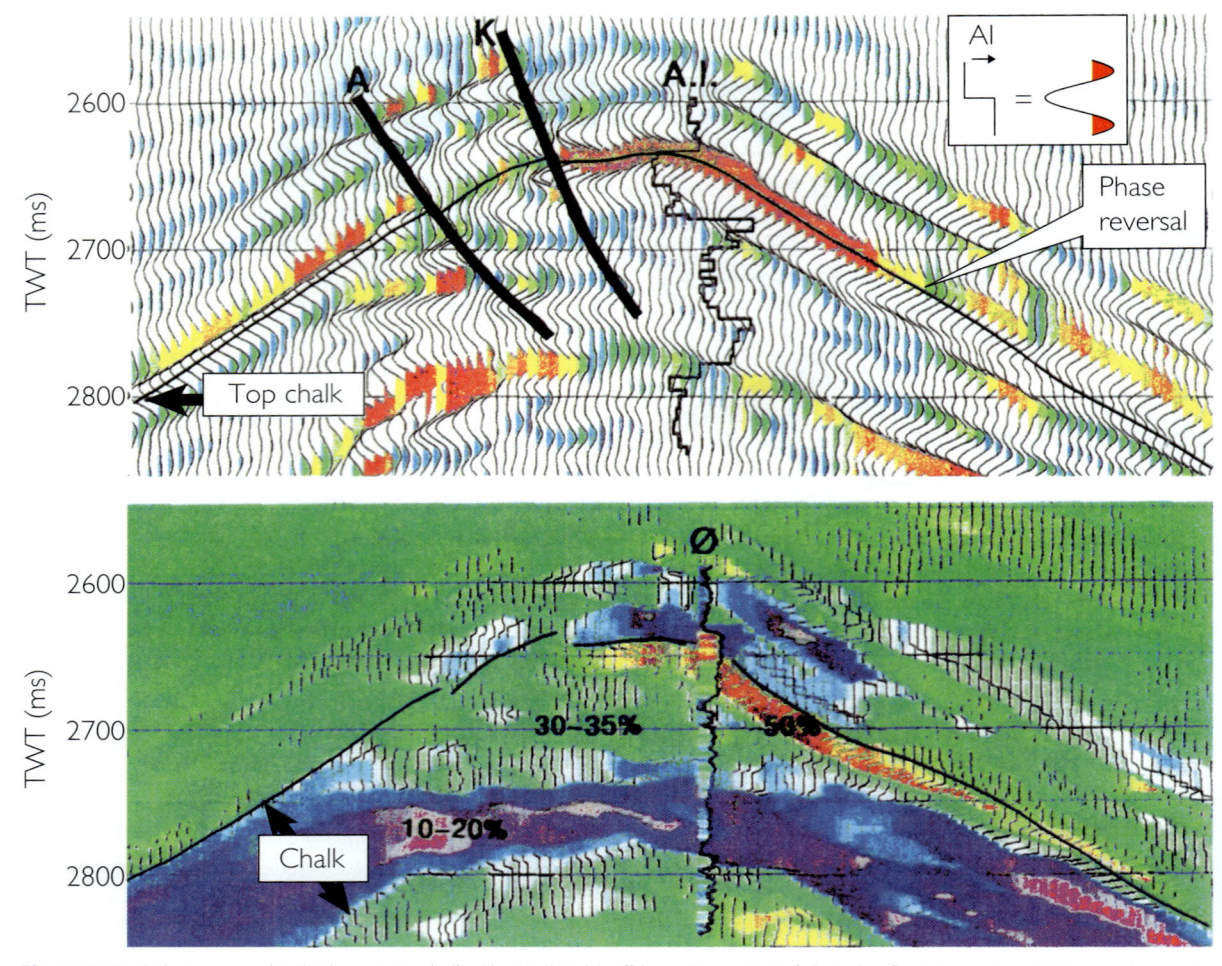

Figure 7.31 Polarity reversal in high porosity chalks (the Hod Field, offshore Norway); (a) full stack reflectivity section, (b) inverted acoustic impedance section scaled to porosity (after Pearse and Ozdemir, 1994).

stiffening of the rock and an increase in density, the transition is also associated with an increase in clay content (Mikada et al., 2002). Thus, it is possible that the positive AVO is at least in part generated by an increase in V_p/V_s downward across the transition boundary.

7.2.8 Carbonates

Amplitude interpretation in carbonate rocks is generally held to be more challenging than in siliciclastic environments (e.g. Chopra et al., 2005; Dong et al., 2003). Some reasons for this include the following.

- Significant potential for rapid lateral and vertical facies variations which may be difficult to predict.
- Owing to the fact that calcite and dolomite have higher velocities than silica-based minerals, velocities in carbonate rocks tend to be more dominated by mineral properties than clastics. Changes in porosity tend to have larger effects than changes in fluid fill. This means that fluid characterisation is generally more difficult than in siliciclastic environments (e.g. Anderson, 1999).
- Given the complexities of post-depositional processes that can influence carbonate rock fabric and porosity distributions, the rock physics of carbonates tends to be more complex than clastics (Chapter 8). This can lead to significant uncertainty in the construction and application of rock physics models in seismic interpretation.

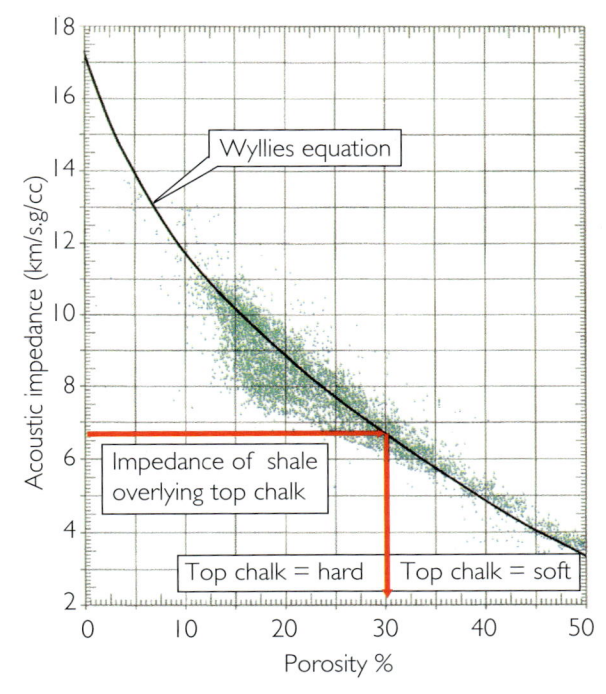

Figure 7.32 Relationship between porosity and acoustic impedance in East Hod Field (after Campbell and Gravdal, 1995).

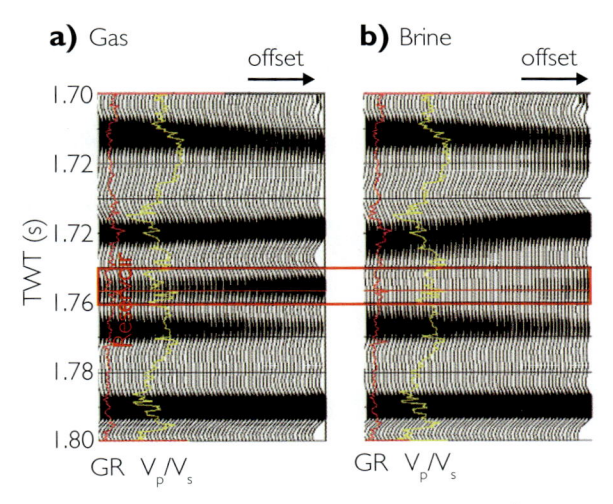

Figure 7.33 Model gathers showing fluid substitution effect in dolomite (14%–20% porosity) encased in tight limestone, based on data from the Western Canada sedimentary basin (after Li et al., 2003).

- In terms of seismic data quality, high velocities in carbonates generally mean that seismic resolution is low. In addition, carbonates are commonly encountered on land or in shallow water, and characterised by relatively poor data.

Typical amplitude interpretation scenarios in carbonate environments are associated with the development of porosity either at the top of the carbonate section or in zones, owing to carbonate facies variations or differences in diagenesis. Figure 7.30 illustrates a dim spot on a full stack seismic section associated with increased porosity at the top of a North Sea Chalk sequence. Dramatic changes in porosity give rise to phase reversals (Fig. 7.31). Similar phase reversal effects have also been described in reef carbonate settings (e.g. Chacko, 1989). The strong relationship between porosity and acoustic impedance (Fig. 7.32 shows an example) has been the basis for using seismic inversion techniques in carbonates (Fig. 7.31 and Chapter 9).

Amplitudes and AVO analysis may also help in the detection of porous dolomite within tight limestone sequences in platform carbonate settings (e.g. Li et al., 2003; Eissa et al., 2003). A key amplitude diagnostic for recognising porous dolomite with gas may be the strong positive AVO associated with the base of the dolomite (Fig. 7.33), although it is possible that large variations in porosity of brine bearing dolomite will give rise to ambiguity in the fluid interpretation. The presence of gas in fractures can have a beneficial effect on detection, lowering the Poisson's ratio and accentuating the AVO effect (e.g. Harvey, 1993). AVO crossplot techniques might be effectively used to discriminate dolomites and tight limestones (Fig. 7.34).

The 'relative AVO' method of Chiburis (1987, 1993) (Chapter 5) is also appropriate for analysing the sub-parallel reflectivity of carbonate platform sequences and it is possible to use the technique for recognising hydrocarbon bearing zones (Fig. 7.35). The technique comprises the following steps.

(1) Extract amplitudes on gathers (or partial angle stacks) at target and reference picks.
(2) Calculate the normalised ratio (NR):

$NR = (T/T_{avg})/(R/R_{avg})$,

where T = target amplitude, T_{avg} = target average amplitude, R = reference amplitude and R_{avg} = reference average amplitude.

(3) Constrain range of values such that if $NR > 1$ then $NR\_edit = 2 - (1/NR)$.

Amplitude and AVO interpretation

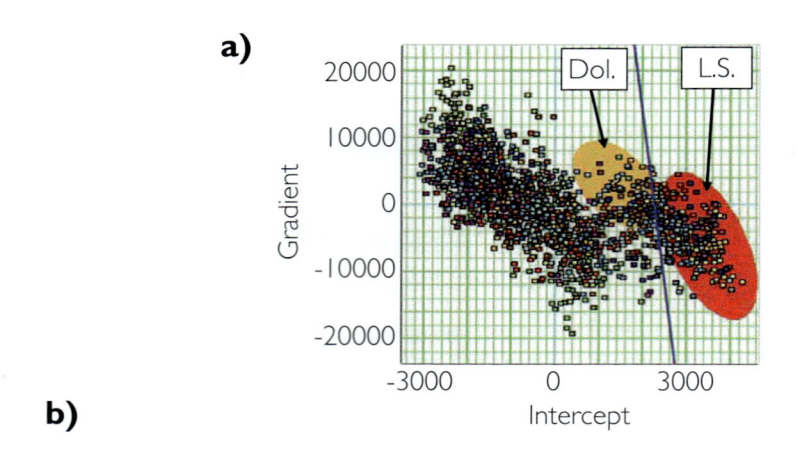

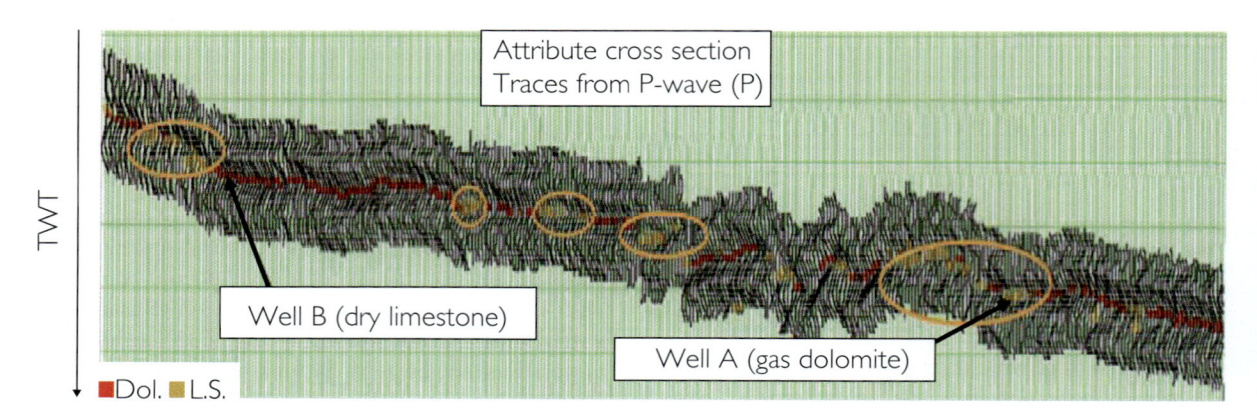

Figure 7.34 Detection of porous dolomite zones using AVO analysis; (a) intercept/gradient crossplot from small seismic window around the top Black River Carbonate reflector, (b) red (tight limestone) and orange (porous dolomite) zones highlighted on intercept stack (after Eissa et al., 2003).

(4) Fit a trend to the edited normalised ratios. For example fit $= A + Bx^2$, where A is the zero angle intercept and B is a rate of change coefficient and x is the offset distance.

(5) Derive the 'AVO difference' attribute $= B/A*x^2$ so that the AVO character can be mapped. Effectively, increasing values of normalised ratio give rise to positive AVO difference values.

In a comparison of AVO techniques for locating zones of fractured dolomite within non-reservoir limestone, Ho et al. (1992) determined that Chiburis' technique was the most robust. An appealing aspect of the technique is that the ratio procedure effectively accounts for most of the causes of amplitude distortion associated with land acquisition. Unfortunately confidence in the technique may be reduced when the carbonates show strong lateral lithological change. More recently several authors have reported success in using simultaneous inversion techniques (Chapter 9) for discrimination of porosity and fluid type in carbonates (e.g. Pelletier and Gunderson, 2005; Li et al., 2003; Ishiyama et al., 2010).

7.2.9 Fractured reservoirs

The exploration of fractured reservoirs using geophysics generally comprises azimuthal P wave AVO and shear wave birefringence techniques. There are numerous examples of fractures detected with

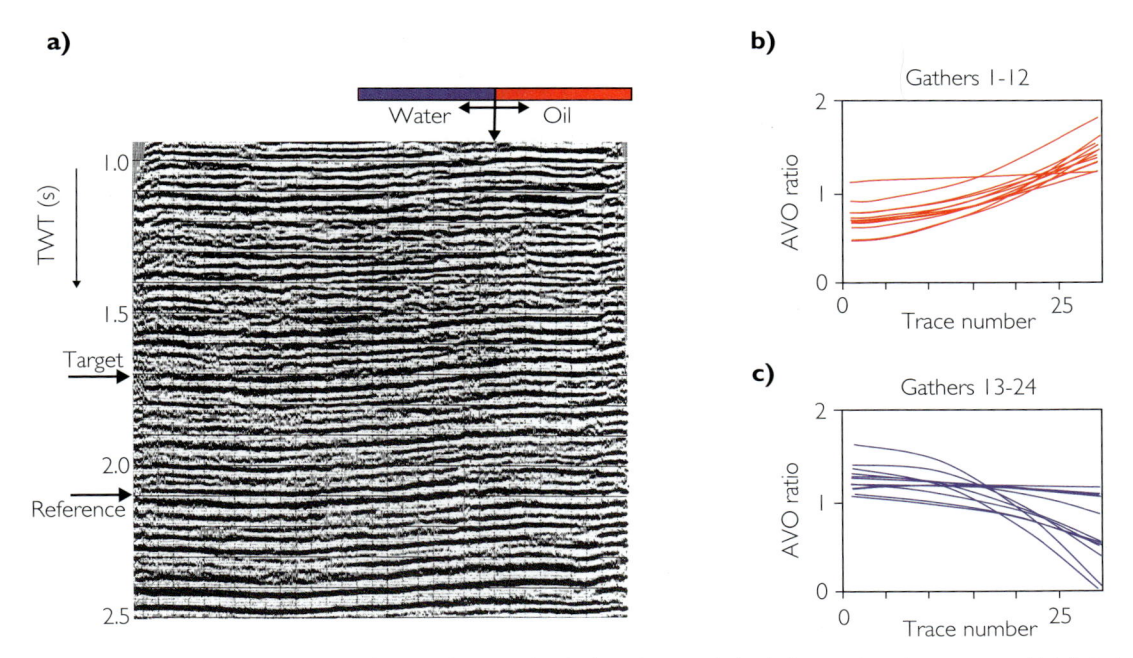

Figure 7.35 Chiburis' 'relative AVO' technique discriminating hydrocarbon sands in platform carbonate sequences; (a) full stack showing target and reference reflectors, (b) target/reference AVO ratio showing positive AVO associated with oil in calcarenites, (c) target/reference AVO ratio showing negative AVO in wet calcarenites. Increasing AVO ratio with offset defines the presence of the hydrocarbon (after Chiburis, 1993).

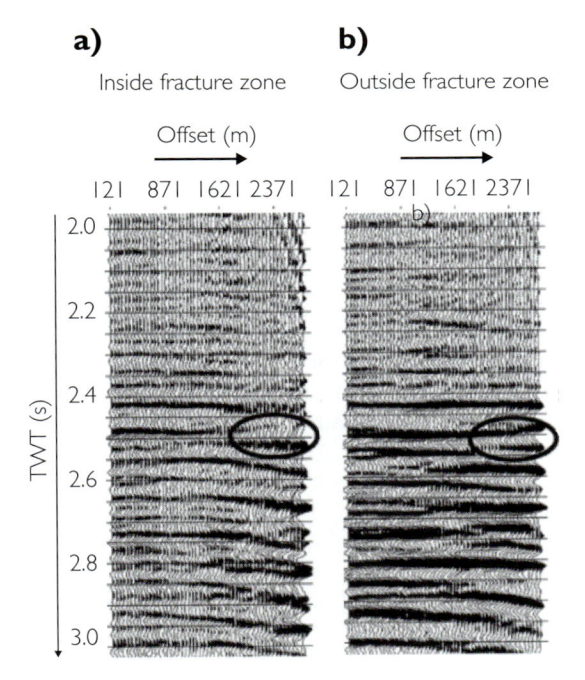

Figure 7.36 Gathers from a 2D line parallel to a fracture swarm in the Chalk of the Fife Field, UK North Sea (after Macbeth et al., 1999); (a) gather from inside the fracture zone showing dimming at far offsets at the Base Chalk, (b) gather from outside the fracture zone showing high-amplitude reflectivity at far offsets. Gather (b) is typical of gathers from seismic lines perpendicular to fractures and from areas of no fractures.

azimuthal AVO (e.g. Macbeth et al., 1999; Gray et al., 2002; Neves et al., 2003; Johns et al., 2008). Typical amplitude effects are shown in Fig. 7.36, where Macbeth et al. (1999) model and interpret the presence of fractures on the basis of Base Chalk dimming on seismic lines that are parallel to the strike of the fractures. As noted in Chapter 5, other workers have interpreted similar far offset dimming to be related to the direction normal to fracture strike.

Increasingly, workers are attempting to make quantitative predictions based on anisotropic parameter inversion (effectively deriving the coefficients in Rüger's (1998) equations (Chapter 5)) and combining amplitude measures with other geophysical attributes such as coherence (e.g. Hunt et al., 2010). The AVO response through fractured zones is a complex function of fracture density, azimuth, porosity, aspect ratio and target impedance contrast, so fracture description from seismic is clearly a significant challenge. Most studies focus on integrating azimuthal amplitude observations with other seismic attributes and incorporating all relevant geological knowledge to arrive at an interpretation.

Fracture detection on land can make use of shear wave sources and the phenomenon of shear wave

Amplitude and AVO interpretation

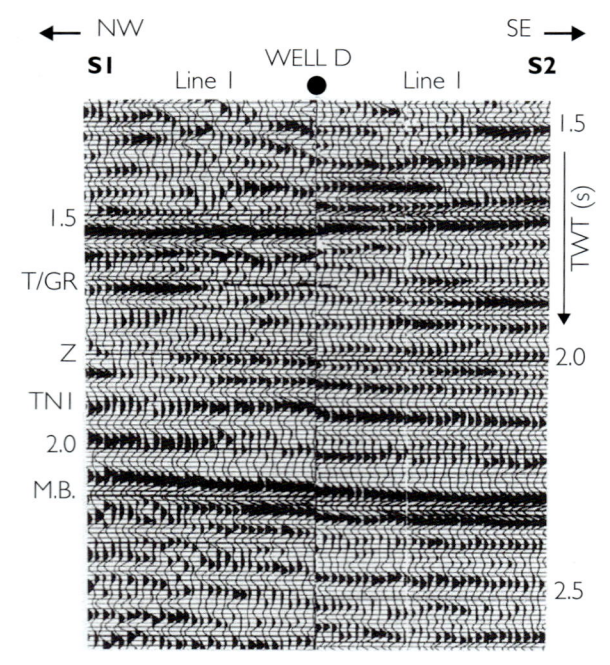

Figure 7.37 Travel time differences of fast (S1) shear wave (left hand section) and slow (S2) shear wave (right hand section) are evidence for the presence of fractures in the Green River Formation, N E Utah (after Lynn et al., 1995). Note the top of the fractured unit is close to the Z marker on the sections.

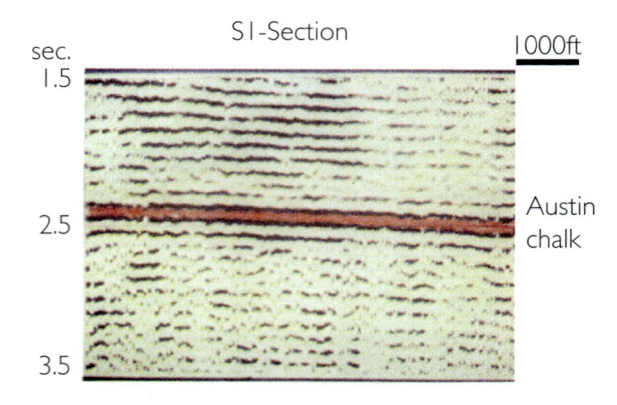

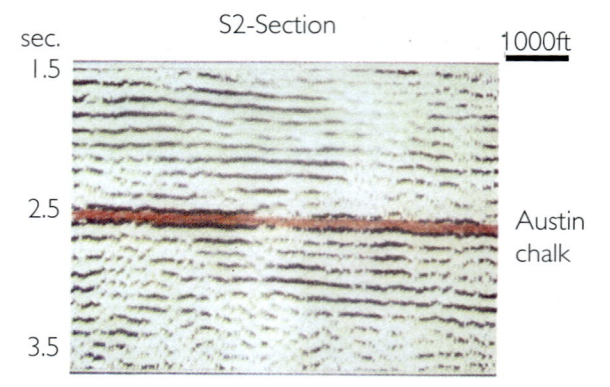

Figure 7.38 Fast (S1) and slow (S2) shear wave sections in the Austin Chalk (after Mueller, 1992). Note the zone of decreased amplitude on the slow shear wave section associated with the presence of fractures.

birefringence (Chapter 5). In the presence of fractures, shear waves are polarised in two directions, parallel to the fractures (S1 or fast) and perpendicular to the fractures (S2 or slow) (Chapter 5). Figure 7.37 shows an example of the differential timing of fast and slow shear wave sections derived from multi-component seismic data through a well which intersects fractures in the Green River formation of NE Utah. Fast (S1) shear waves carry information about the rock matrix, whereas slow shear waves carry information on relative fracture density and other fracture parameters (Lynn et al., 1995). Thus, the amplitudes of fast shear waves at the top of the fractured unit tend to be insensitive to the presence of fractures whereas the slow shear wave will show amplitude dimming. An example from the Austin Chalk is shown in Fig. 7.38.

Chapter 8

Rock physics for seismic modelling

8.1 Introduction

Seismic amplitudes are interpreted using models generated from well data and supported by a range of knowledge collectively described as 'rock physics'. This chapter describes the practical rock physics relevant for the interpreter to begin constructing suitable models for interpreting seismic signatures. The key objectives in this use of rock physics (Fig. 8.1) are:

(1) log preparation and conditioning,
(2) rock characterisation from logs,
(3) seismic modelling.

Conditioning of log data involves editing log curves and making predictions for erroneous or missing sections as well as making corrections for the adverse effects of drilling mud on log measurements (i.e. correcting for invasion). In deviated holes, corrections for anisotropic effects may also be required. Conditioned logs are the basis for well ties (Chapter 4).

Rock characterisation (introduced in Chapter 5) aims to describe seismic litho-facies and analyse the relationships between rock and acoustic properties. Mineralogy, porosity, pore types, fluid type and effective pressure are all factors to be evaluated from the log data (e.g. Castagna, 1993). In addition, any preferred vertical ordering of lithologies should be established as this information can be used in generation of various geological scenarios. The presence of distinct depth or sedimentological trends may in certain situations enable predictions outside the measured range (Avseth *et al.*, 2003).

As described in Chapters 2 and 5, seismic models can take various forms, such as single interface AVO models, 1D angle synthetics, and various 2D models

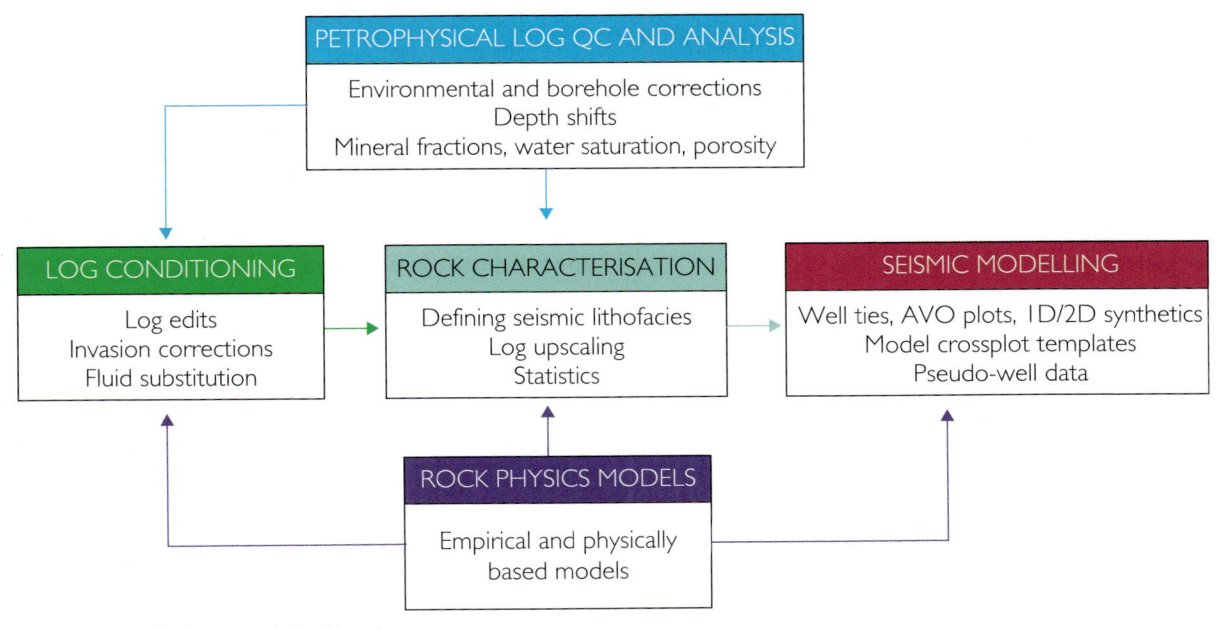

Figure 8.1 Rock physics analysis of log data.

including simple interference models. Importantly, rock physics data can be used in conjunction with relevant geological knowledge to generate modelled scenarios, often termed 'pseudo-wells', which can contribute to the understanding of the relationship between seismic amplitude attributes and reservoir properties (Chapter 10) as well as the attendant uncertainty in predictions.

Fundamental to each of these steps are good quality velocity and density log data, together with an accurate petrophysical interpretation (i.e. mineral content, porosity and water saturation determined from various log suites). Another major component is the use of rock physics models, algorithms or empirical relationships, which can be fit to measured velocity and density log responses and then used predictively. In this way rock physics models can be used to predict missing log sections or to highlight potentially erroneous log sections. Perhaps the most important rock physics model is Gassmann's (1951) equation, used for determining the effect of substituting fluids on the velocities of rocks with intergranular porosity.

8.2 Rock physics models and relations

There are a large number of rock physics models and relations that provide tools for data QC, characterisation and generation of model scenarios. Not all are covered here for the sake of brevity; the reader is referred to Mavko et al. (1998) and Avseth et al. (2005) for a thorough treatment.

The most commonly employed models can be categorised as follows:

- theoretical bounds,
- empirical models,
- Gassmann's equation,
- contact models,
- inclusion models.

8.2.1 Theoretical bounds

Theoretical bounds establish the physical limits of the properties of mixtures of minerals and fluids. The idea of bounds was introduced in Chapter 5 in terms of the effective limits of porosity–velocity behaviour of sandstones, with the lower bound determined by the Reuss average (describing a suspension of mineral and fluid) and the upper bound by a modified Voigt bound.

The Voigt and Reuss bounds are defined, for example with a mix of quartz and water, as follows.
Voigt modulus:

$$K_{sat\_voigt} = (K_{qtz} Vol_{qtz}) + (K_w Vol_w), \quad (8.1)$$

where

K_{qtz} = modulus of quartz,
K_w = modulus of water,
Vol_{qtz} = volume fraction of quartz,
Vol_w = volume fraction of water.

Reuss modulus (describing a lower bound for mineral/fluid suspensions):

$$K_{sat\_Reuss} = \frac{1}{\frac{Vol_{qtz}}{K_{qtz}} + \frac{Vol_w}{K_w}}. \quad (8.2)$$

The modified Voigt or critical porosity model (Nur et al., 1998) provides a more realistic upper bound for sandstones and is defined by:

$$K_{sat\_mod\_voigt} = \left(1 - \frac{\phi}{\phi_c}\right)(K_{qtz} - K_{\phi c}) + K_{\phi c}, \quad (8.3)$$

where $K_{\phi c}$ is the Reuss modulus at critical porosity.

Another commonly used set of bounds was published by Hashin and Shtrikman (1963). These bounds in general provide a slightly tighter constraint than the Voigt–Reuss values, although the equations are more complex. For a mixture of two materials the Hashin–Shtrikman upper (HS$^+$) and lower (HS$^-$) bounds are defined by:

$$K^{HS\pm} = K_1 + \frac{f_2}{(K_2 - K_1)^{-1} + f_1(K_1 + \tfrac{4}{3}\mu_1)^{-1}}$$

$$\mu^{HS\pm} = \mu_1 + \frac{f_2}{(\mu_2 - \mu_1)^{-1} + \frac{2f_1(K_1 + 2\mu_1)}{5\mu_1(K_1 + \tfrac{4}{3}\mu_1)}}. \quad (8.4)$$

where K, μ and f represent the bulk moduli, shear moduli and volume fractions of the individual phases 1 and 2. If the stiffest material is denoted by 1 and the softest material is denoted by 2 the equations calculate the upper bound. To calculate the lower bound requires that the stiffest material is denoted by 2 and the softest material is denoted by 1. Berryman (1995) has provided a general form that can applied to more than 2 phases (see also Mavko et al., 1998). Figure 8.2 illustrates the bounds in terms of a mix of quartz and water. Note that the lower HS bound is identical to the Reuss bound because the shear modulus of brine is zero.

Whilst theoretical bounds are not especially predictive for the purposes of describing the velocities of

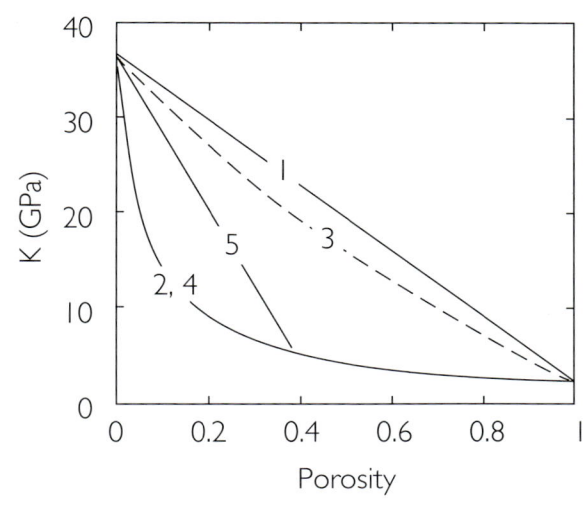

Figure 8.2 Bounds on bulk modulus of a brine-filled sandstone, after Nur et al. (1998). In this case the constituents are quartz and brine. The curves shown are (1) Voigt average, (2) Reuss average, (3) upper Hashin–Shtrikman, (4) lower Hashin–Shtrikman and (5) modified Voigt bound.

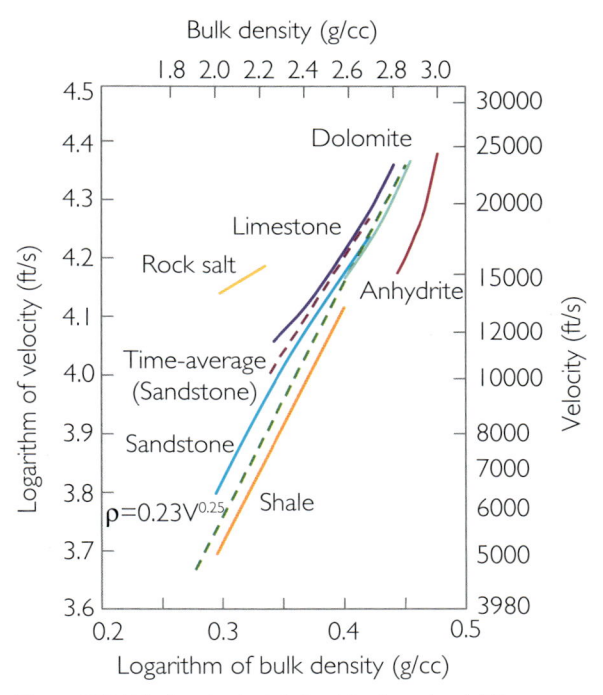

Figure 8.3 Velocity–density relationships in rocks of different lithology (re-drawn after Gardner et al., 1974).

brine filled sandstones, given the fact that their form roughly coincides with sorting and diagenetic trends (Chapter 5) it is possible to generate predictive rock physics models based on modification of the bounds. Avseth et al. (2005) describe modifications to the Hashin–Shtrikman bounds in order to mimic the effect of cementing and sorting trends.

Theoretical bounds are commonly used in predicting the effective moduli of mineral or fluid mixtures. For example, a common approach to calculating the effective mineral and fluid moduli for input to Gassmann's equation is to use the average of the Voigt and Reuss bounds (called the Voigt–Reuss–Hill average (Hill, 1952)) for mineral mixing and the Reuss average for fluid mixing. A Reuss average effectively assumes that the fractions (or saturations) of different fluids is the same in each pore within the rock (i.e. a homogenous fluid mix). Alternative approaches to fluid mixing may be required in flushed or swept reservoir settings (see Section 8.4.4).

8.2.2 Empirical models

Empirical rock physics models are derived from fits made to experimental results. Numerous models have been published. These are often simple regression-type relationships involving two or three parameters. Experience has shown that despite often having little physical justification they can be very useful. Presented below are some of the most commonly used.

- Gardner's relations – compressional velocity and density.
- Wyllie's equation – compressional velocity and porosity.
- Han's relations – compressional velocity, porosity and clay content.
- Greenberg–Castagna – compressional and shear velocity.
- Faust's relation – resistivity and compressional velocity.

8.2.2.1 Gardner's relations

In many rocks, compressional velocity and bulk density have a positive relationship, so that as velocity increases so does density (Fig. 8.3).

Gardner et al. (1974) developed a series of (brine-saturated) lithology dependent relations of the form:

$$\rho_b = dV_p^f, \tag{8.5}$$

where ρ_b is bulk density in g/cc, V_p is the P wave velocity in km/s, and d and f are constants. The various lithology dependent coefficients are shown in Table 8.1.

Rock physics for seismic modelling

Table 8.1. Coefficients for Gardner's relations. Note that, based on practical experience, the coeficients initially published for limestone have been amended from $d = 1.50$, $f = 0.225$ to $d = 1.55$, $f = 0.3$ (H. Morris, personal communication).

Lithology	d	f
Ss/sh avg	1.741	0.25
Shale	1.75	0.265
Sandstone	1.66	0.261
Limestone	1.55	0.3
Dolomite	1.74	0.252
Anhydrite	2.19	0.16

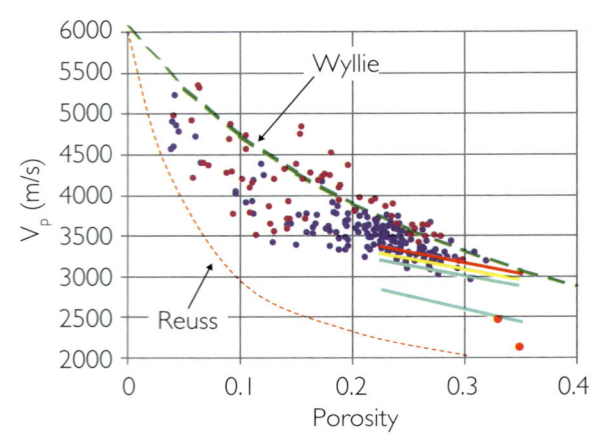

Figure 8.4 Crossplot of porosity vs compressional velocity for various sandstone datasets: purple points – data from Han et al. (1986), dark blue points – Tertiary sandstone dataset from North Sea, dashed red line – Reuss mix of water and quartz, red and yellow lines – trends from Oseberg high-porosity data (Dvorkin and Nur, 1996), blue lines – trends from Troll high-porosity data (Dvorkin and Nur, 1996) with upper line 20 MPa effective pressure and lower line 5MPa effective pressure, red points – selected unconsolidated sand data.

The Gardner relations may be used in transforming sonic or density logs for the purposes of replacing missing sections or in constraining the results of inversions for P and S reflectivity (e.g. White, 2000). Owing to the lack of universal applicability it is advised that area-specific density – velocity relations are derived from the available data. In some cases it may be more appropriate to derive density from a transform based on shear velocity rather than compressional velocity (e.g. Potter and Stewart, 1998).

8.2.2.2 Wyllie's ('time average') equation

Wyllie et al. (1958) derived a relationship between velocity and porosity that fits data from well-consolidated sandstones and limestones. The relation is essentially intuitive rather than based on physical principles. Wyllie's equation relates the sonic interval transit time (i.e. the reciprocal of the seismic velocity, usually measured in μs/ft in sonic logs) to the weighted addition of interval transit times through the pore fluid and the mineral matrix. It is often referred to as the 'time-average' equation:

$$t = \phi t_{fl} + (1 - \phi) t_0, \qquad (8.6)$$

where ϕ is the porosity (as a fraction) and t, t_{fl} and t_0 are the interval transit times in the rock, the fluid and the mineral matrix. Figure 8.4 illustrates the form of Wyllie's equation in the context of a selection of different (brine-bearing) sandstone data. It is evident that Wyllie's equation falls at the top of the datapoints, indicating that it is a model appropriate for stiff well consolidated sands. It is often observed to fit well with data from carbonate rocks (see Chapter 7) where the mineralogy has a key role in establishing a relatively stiff rock frame. Figure 8.5 illustrates some carbonate data from Eberli et al. (2003). The best fit line drawn on the graph is close to a prediction using Wyllie's equation. Intriguingly, high permeability carbonates with moldic and intra-frame porosity plot above the trend and low permeability carbonates with microporosity and interparticle porosity plot below the trend (Eberli et al., 2003). In general, to quote Weger et al. (2009), 'carbonates with a large amount of microporosity, a complex pore structure (high specific surface), and small pores generally show low acoustic velocity at a given porosity. Samples with a simple pore structure (low specific surface) and large pores show high acoustic velocity for their porosity'.

When traditional porosity logs are not available, the Wyllie equation is sometimes used by petrophysicists as a way to calculate porosity:

$$\phi = (t - t_0)/(t_f - t_0) \qquad (8.7)$$

Clearly the porosity estimate will be erroneous for rocks which fall below the Wyllie trend. An empirical correction to this porosity estimate that extends it to relatively unconsolidated rocks is to multiply it by $100/t_{sh}$, where t_{sh} is the shale interval transit time at or near the depth of interest, in μs/ft.

An improved version of the Wyllie equation was published by Raymer et al. (1980):

$$V = (1 - \phi)^2 V_0 + \phi V_{fl}, \qquad (8.8)$$

Rock physics models and relations

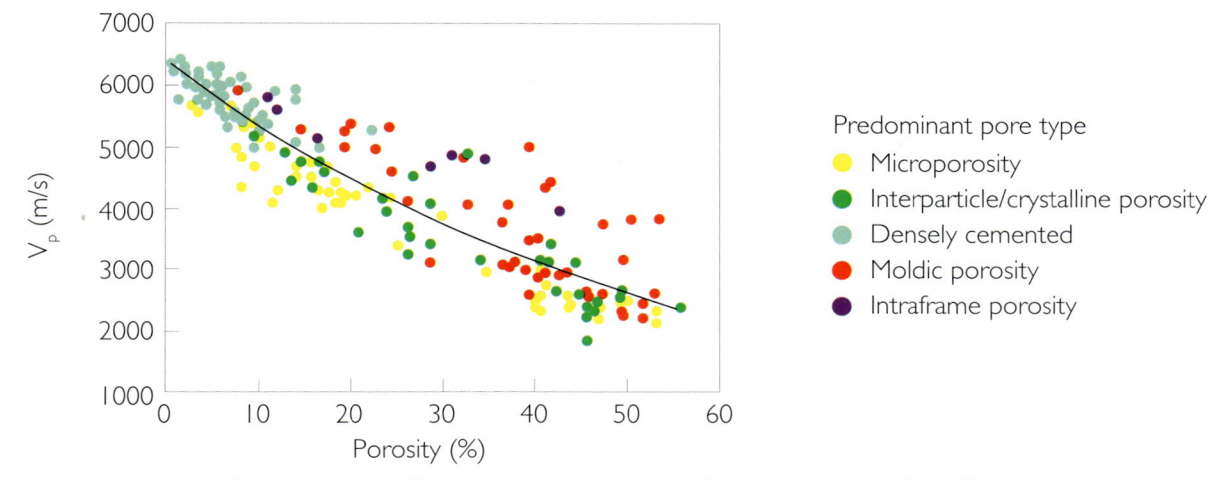

Figure 8.5 Crossplot of velocity (at 8 MPa effective pressure) versus porosity of various pore types of (brine filled) carbonates with an exponential best fit curve through the data for reference (re-drawn after Eberli et al., 2003).

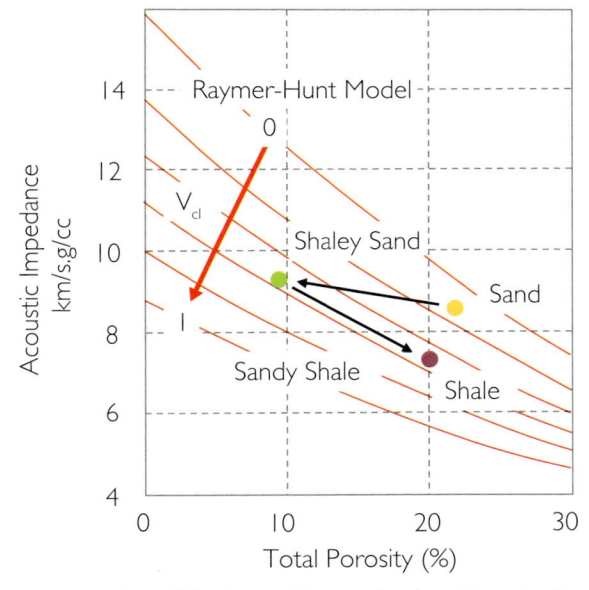

Figure 8.6 Use of the Raymer–Hunt empirical model to describe changes in porosity and clay content (Dvorkin et al., 2004).

where ϕ is the fractional porosity and V, V_0 and V_{fl} are the velocities of the rock, the mineral matrix and the fluid.

The Raymer–Hunt–Gardner model is a flexible model that can be readily calibrated to measured data. An example of the application of the model in describing dispersed shale in fluvial sandstones is shown in Fig. 8.6. Of course the application of a model in this way requires accurate estimates of mineralogy fractions and porosity.

Following the work of Raymer et al. (1980), Raiga-Clemenceau et al. (1988) provided an improvement in the prediction of porosity from sonic by including lithology-specific dependence:

$$\phi = 1 - \left(\frac{t_0}{t}\right)^{\frac{1}{x}}, \quad (8.9)$$

where t_0 = matrix slowness, t = sonic slowness and x is a constant dependent on lithology.

Common values for matrix slowness and the fitting coefficient 'x' are shown in Table 8.2.

Interestingly, this sonic transform is based on the formulation:

$$\frac{t}{t_0} = (1 - \phi)^{-x}, \quad (8.10)$$

which is of a similar form to the relationship of resistivity and porosity (Archie, 1942):

$$\frac{R}{R_w} = a\phi^{-m} = F \quad (8.11)$$

where

R = true formation resistivity,
m = cementation exponent (see Table 8.3 for typical values),
a = constant, sometimes referred to as the tortuosity factor,
F = formation factor.

The relationship of sonic and resistivity measurements is discussed in Section 8.2.2.5.

Spikes and Dvorkin (2005) have noted that the Wyllie time average equation is not consistent with

Table 8.2. Mineral slowness values and coefficient values for Eq. (8.9) (after Raiga-Clemenceau et al., 1988).

Matrix	$T_{ma}(\mu s/ft)$	x
Silica	55.5	1.6
Calcite	47.6	1.76
Dolomite	43.5	2

Table 8.3. Typical values for Archie coefficients 'a' and 'm' (from Hacikoylu et al, 2006).

Lithology	a	m
Consolidated sandstone	0.81	2
Unconsolidated sandstone	0.62	2.15
Average sands	1.45	1.54
Shaly sands	1.65	1.33
Clean granular rock	1.00	2.05–3

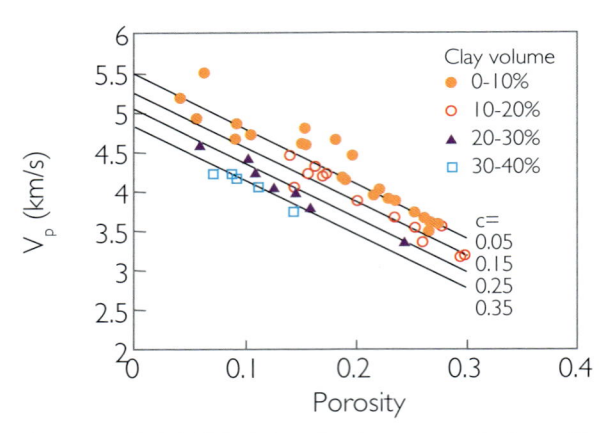

Figure 8.7 Han's (1986) relations showing constant clay lines, with a subset of Han's (1986) data (after Avseth et al., 2005).

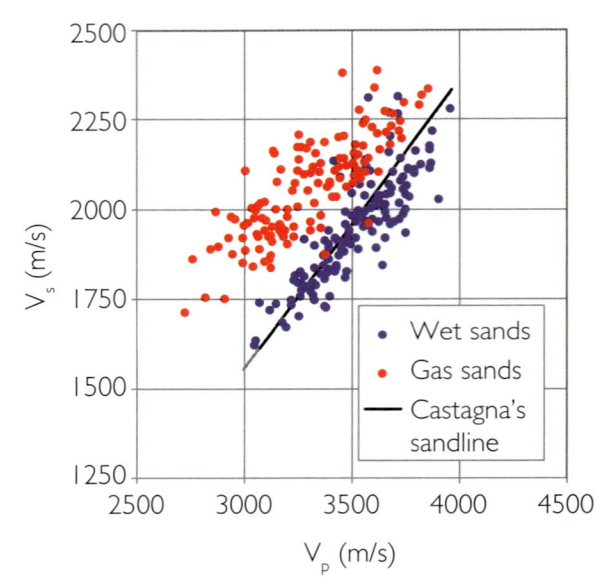

Figure 8.8 Comparison of a North Sea shaley sandstone dataset and Castagna's sandline, an empirical V_p–V_s relation based on sands from Gulf of Mexico and onshore United States.

the Gassmann (1951) fluid substitution equation. Therefore, the time average equation should only be used with brine as the fluid. Substituting to another fluid would require the use of the Gassmann formulation. The Raymer–Hunt–Gardner equation on the other hand is broadly consistent with Gassmann, suggesting that any pore fluid may be used with this model (Spikes and Dvorkin, 2005).

8.2.2.3 Han's relations

The work of Han (1986) and Han et al. (1986) identified relationships between porosity, compressional velocity and clay content in laboratory experiments on well consolidated sandstones. It was realised that the effect of increasing clay content is to effectively soften the rock and reduce the compressional velocity. The key relationships from the high pressure (40 MPa) measurements are:

$$V_p = 5.59 - 6.93\phi - 2.18C \quad (8.12)$$

$$V_s = 3.52 - 4.91\phi - 1.89C, \quad (8.13)$$

where V_p is in km/s, C is clay volume fraction and ϕ is porosity.

Avseth et al. (2005) describe how if the relations are re-written slightly the equations can be used to determine a series of parallel lines on the porosity – velocity crossplot, representing the effect of changing clay content (Fig. 8.7).

8.2.2.4 Greenberg–Castagna relations

Given the high fidelity of modern shear logging tools, the ideal shear velocity input to rock physics analysis should be based on a log measurement. It is often the case, particularly with old wells, however, that there is no shear log and the shear velocity needs to be predicted. Fortunately there is usually a strong lithology dependent, but largely pressure independent, positive correlation between compressional

Rock physics models and relations

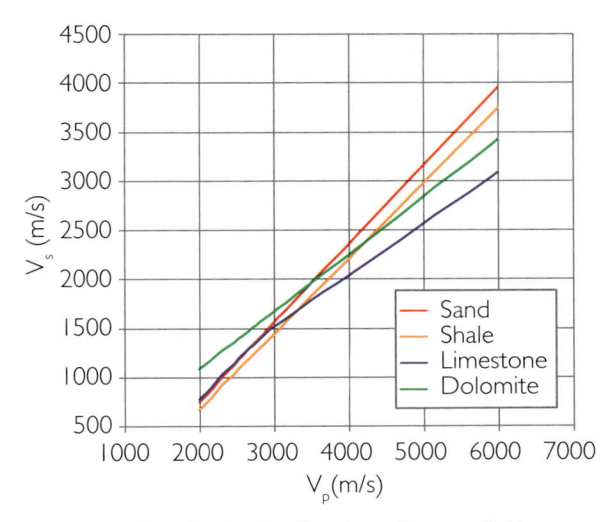

Figure 8.9 Crossplot showing Greenberg–Castagna V_p–V_s relations.

and shear velocity (Castagna *et al.*, 1985). Figure 8.8 shows an example of a North Sea sandstone dataset compared to the sandstone relation published by Castagna *et al.* (1985), based on a variety of data types from the Gulf Coast and onshore United States. It is evident that the sandstone relation is only valid for the brine fill case. The effect of replacing water with gas moves the points up and to the left so that they lie above the Castagna sandstone trend (Fig. 8.8).

Greenberg and Castagna (1992) defined four trends for commonly occurring (brine bearing) lithologies (Fig. 8.9):

- sandstone: $V_s = 0.8042 V_p - 0.8559$,
- limestones: $V_s = 0.0551 V_p^2 + 1.016 V_p - 1.0305$,
- dolomite $V_s = 0.58321 V_p - 0.07775$,
- shale $V_s = 0.7697 V_p - 0.86735$,

where V is in km/s.

Combining these trends to predict V_s for mixed lithologies effectively follows the Voigt–Reuss–Hill average approach of averaging the arithmetic and harmonic averages (Greenberg and Castagna, 1992).

Other useful data on V_p–V_s relations include the following.

- Gas sands: $V_p/V_s \sim 1.4$–1.8 (an average V_p/V_s ratio of 1.6 for gas sands is a good rule of thumb).
- Halite: $V_p/V_s = 1.74$ (Simmons and Wang, 1971).
- Anhydrite: $V_p/V_s = 1.8$ (Simmons and Wang, 1971).

- Coal: V_p/V_s can vary between 1.9 (typical of anthracite) to 2.1 (typical of bituminous coals). Morcote *et al.* (2010) have published a V_p–V_s relation appropriate for effective pressures above 5MPa:

$$V_s = 0.4811 V_p - 0.0038.$$

- Volcaniclastics: relatively high V_p/V_s, often lying on the Greenberg–Castagna limestone trend (Klarner and Klarner, 2012).
- High-velocity volcanic rocks: $V_s = 0.51 V_p + 0.148$ (Klarner and Klarner, 2012).

Using empirical relations for V_s prediction has been shown to be as accurate if not better than using effective medium models (such as the Xu–White (1995) model) for V_s prediction (Avseth *et al.*, 2005; Jorstad *et al.*, 1999). However, care should be taken in using the Greenberg–Castagna V_s prediction model without local validation from shear sonic logs. Experience has shown that there can be significant variations away from these trends.

- Unconsolidated and partially consolidated mudrocks can have a slightly different trend to the shale relation. Castagna's mudrock relation is defined as:

$$V_s = 0.862 V_p - 1.172 \text{ (Castagna et al., 1985)}.$$

- Various sandstone lithologies can have V_s higher than that predicted by the sandline (e.g. Smith, 2011), including clean (quartz prone) sands and glauconitic sands. A clean sand line (referred to by the authors as 'the quartz line') can be derived from the data presented by Murphy *et al.* (1993):

$$V_s = 0.802 V_p - 0.75.$$

Hossain *et al.* (2012) describe a V_p–V_s relation for glauconitic greensand that gives slightly higher V_s than Castagna's sandline:

$$V_s = 0.86 V_p - 0.96.$$

- In some areas, brine sands may fall below Castagna's sandline. Figure 8.10 shows shear log data in which the sands fall significantly below Castagna's sandline and a single V_p–V_s trend would be a reasonable model for predicting V_s for both brine bearing sands and shales.
- In the shallowest, most unconsolidated, part of the section it is possible that the V_p/V_s of sands is

Rock physics for seismic modelling

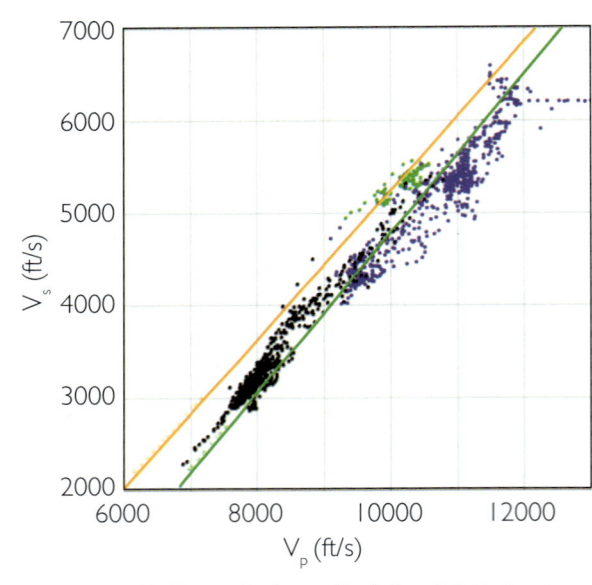

Figure 8.10 V_p–V_s crossplot from a North Sea oil field, showing significant differences between log data and the empirical trends of Castagna et al. (1985). Green points – oil sands, blue points – brine sands and black points – shale.

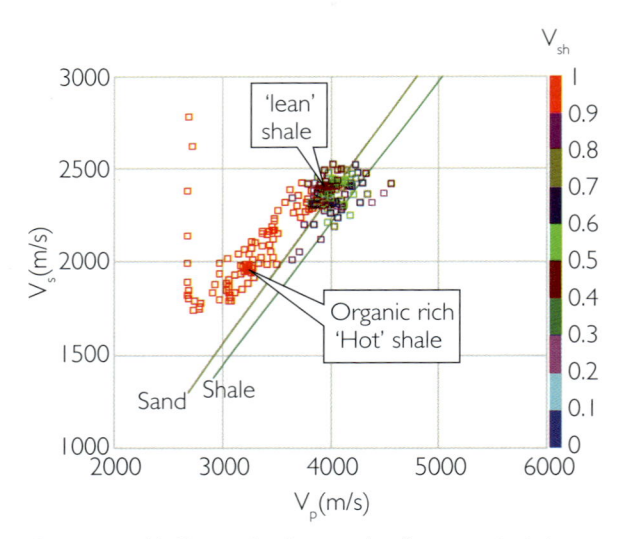

Figure 8.11 V_p–V_s crossplot showing data from organic shales plotting above the sandstone empirical trend of Castagna et al. (1985).

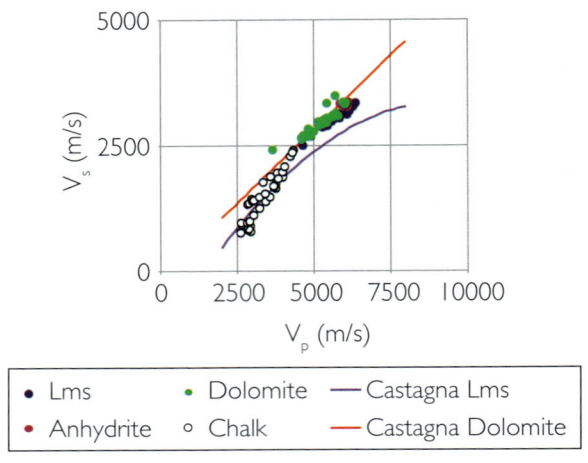

Figure 8.12 An example of V_p–V_s trends in carbonates (data from Rafavich et al. (1984) and Chalk data from the North Sea).

significantly higher than that predicted by Castagna's sandline. As a rule of thumb this can apply to sands where the compressional velocity is less than about 2300 m/s. Unfortunately, owing to the general lack of appropriate measurements in this setting (for example due to the difficulty of acquiring sonic logs in large boreholes) the relationship for the velocity ratio may be uncertain.

- Organic shales may have fairly low V_p/V_s ratios (e.g. Vernik and Milovac, 2011), plotting above the sandline in a similar region to hydrocarbon bearing sands (Fig. 8.11). For example, Bailey and Dutton (2012) present a V_p–V_s relation for the Kimmeridge Clay formation in the Central North Sea:

$$V_s = 0.75V_p - 0.5625.$$

Figure 8.12 illustrates variability in the V_p/V_s behaviour of selected carbonate data. Data presented for limestones, anhydrite and dolomite from Oklahoma by Rafavich et al. (1984) are consistent with Castagna's dolomite trend (roughly $V_p/V_s = 1.8$) whereas the data from some North Sea Chalks show a variation between Castagna's limestone and dolomite trends.

Owing to the variability in the relationships of V_p and V_s for different lithologies it is imperative for shear log data (and possibly laboratory measurements on core samples) to be acquired. Empirical V_p–V_s relations can also have other uses, such as providing V_p/V_s constraints in elastic inversion (Chapter 9) and also in log analysis. Williams (1990) describes a log attribute called the 'acoustic hydrocarbon indicator log' which is applied to sandstone data and essentially reveals the deviation of the logged V_p/V_s curve away from the Castagna brine-saturated trends (Fig. 8.13). The workflow Williams (1990) describes is as follows.

shared dependence of resistivity and sonic measurements on porosity. Faust (1953) was effectively the first worker to propose a usable relation:

$$V_p = \gamma(ZF)^{\frac{1}{6}}, \qquad (8.14)$$

where V_p is in ft/s with $\gamma = 1948$ and Z in feet, or V_p is in km/s with $\gamma = 2.2888$ and Z in km, and $F =$ resistivity formation factor (R_t/R_w), $R_t =$ deep formation resistivity and $R_w =$ resistivity of water, and resistivity units are ohm-metres.

A useful way of looking at Faust's relation is that it is effectively a scaling function of the resistivity combined with a low-frequency trend, in this case determined by depth. In the absence of a significant depth-related trend, scale functions are readily derived and applied. A number of different resistivity–sonic relations have been proposed:

$$t = a + bR^{-1/c} \qquad (8.15)$$

(where t is in μs/ft, and a,b and c are determined from the data) is known as the 'Kim–Rudman' equation, after Kim (1964) and Rudman et al. (1975);

$$t = aR^b, \qquad (8.16)$$

where $a = 94.2$, $b = 0.15$, t is in μs/ft and resistivity in ohm-metres. This is generally known as Smith's equation (Adcock, 1993).

It is important that these types of scaling functions are verified from wells with both sonic and resistivity and that due attention is paid to stratigraphy, lithology and pore pressure as well as the presence of erroneous log responses due to bad hole or other factors (see Section 8.4.1). It is also important to recognise that the functions will fail in hydrocarbon zones. Which resistivity log to use is essentially a data dependent decision. The resistivity log needs to have a similar character to the sonic log for scaling functions to be successful of course. In the Gulf Coast region the SFL log (shallow investigating Spherically Focussed Log) appears to be more appropriate for a one step resistivity–sonic transform than the deep induction ILD log (Smith, 2007) (Fig. 8.14). However, in other areas borehole effects such as mud filtrate invasion may be an important issue to address in the choice of resistivity log.

A more petrophysical approach, effectively based on Wyllie's relation, has been described by Fischer and Good (1985):

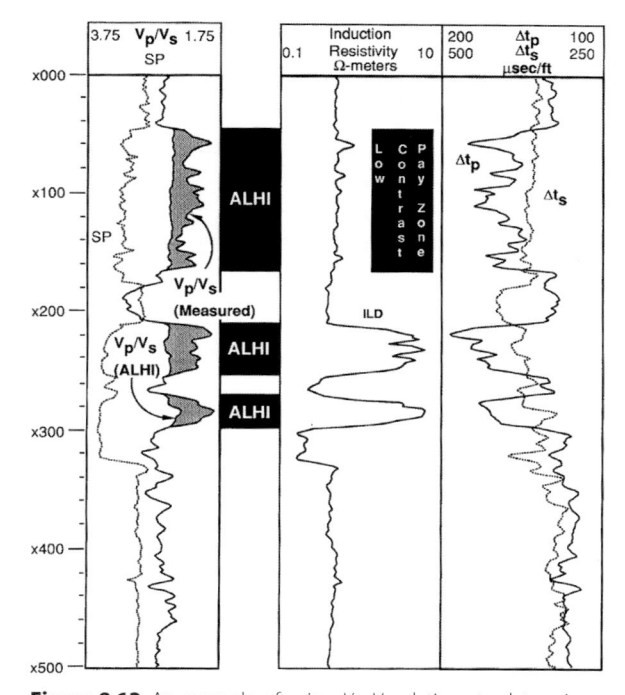

Figure 8.13 An example of using V_p–V_s relations to determine pay zones (after Williams, 1990). Column 1 – differences between the ALHI attribute and V_p/V_s (measured) are shaded grey and represent zones that the ALHI technique would predict to be hydrocarbon-bearing. Note that the shaded areas in the low-contrast pay zone are as definitive of the hydrocarbon-bearing interval as they are in the cleaner zones deeper in the well. Column 2 – resistivity logs showing low resistivity zone around × 100. Column 3 – compressional (Δt_p) and shear (Δt_s) sonic logs.

(1) Calculate sandline $V_p/V_s = 1.182 + 0.00422$ dts (where dts = shear sonic slowness in μs/ft).
(2) Calculate shaleline $V_p/V_s = 1.276 + 0.00374$ dts.
(3) Calculate water-bearing minimum $V_p/V_s = $ min $[V_p/V_s$ (sand)$, V_p/V_s$ (shale))$] - 0.09$.
(4) Calculate $ALHI =$ water-bearing min $V_p/V_s -$ measured V_p/V_s. Whilst the $ALHI$ attribute is not especially sensitive to saturation (especially with gas), it can be particularly useful in indicating the presence of pay in low resistivity situations.

8.2.2.5 Faust's relation

In wells that do not have a sonic log it is common practice to transform resistivity to sonic for the purposes of seismic horizon ties, particularly in areas where the geology is dominantly a passive fill sequence. The basis for the transformation is the

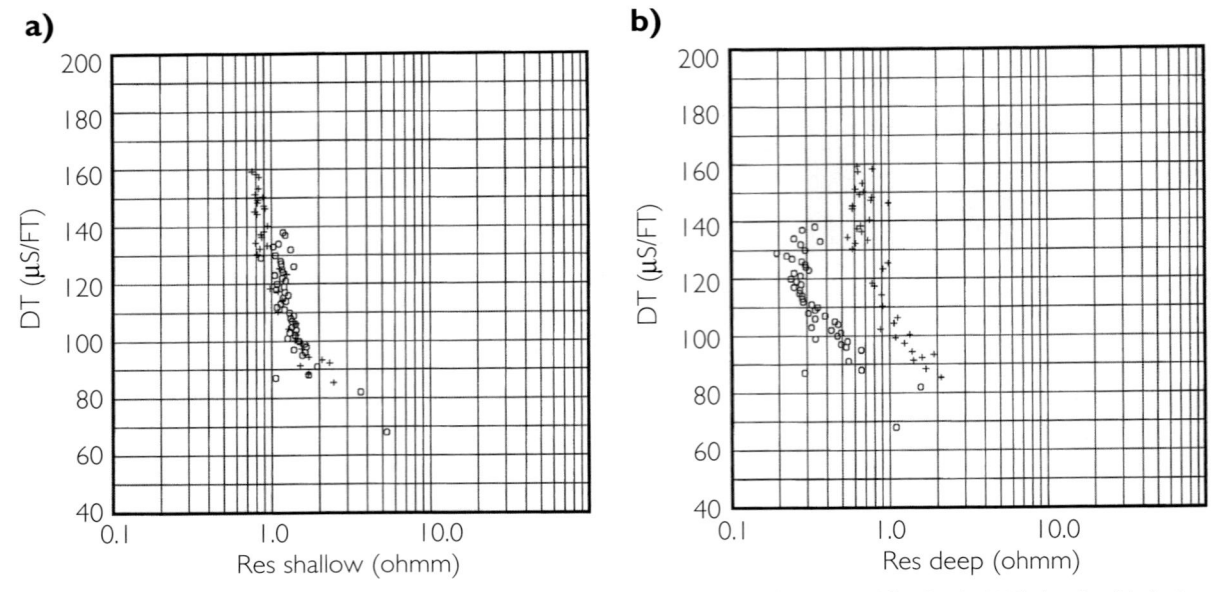

Figure 8.14 Resistivity–sonic crossplots from 1000 ft of Miocene sand/shale section onshore Texas (after Smith, 2007) showing lithologic differences between sonic and shallow SFL (a) and deep ILD (b) resistivity logs. Circles represent sands and crosses represent shales. Republished by permission of the Gulf Coast Association of Geological Societies, whose permission is required for further publication use.

$$t = t_0 + c\left(\frac{R_r}{R}\right)^{0.5}, \quad (8.17)$$

where c is a constant dependent on lithology and R_r is the resistivity of water for a water-bearing sand/shale mixture:

$$R_r = \frac{1}{\left(\left(\frac{V_{sh}}{R_{w\_shale}}\right) + \left(\frac{(1-V_{sh})}{R_{w\_sand}}\right)\right)}. \quad (8.18)$$

It assumes that R_w is constant and the resistivity is measuring water-filled rocks. Hacikoylu et al. (2006) concluded that Faust's relation was appropriate mainly for consolidated rocks. For unconsolidated rocks they propose a relation that includes the Archie formation factor:

$$V_p = \frac{F}{(0.9 + cF)}, \quad (8.19)$$

where c is a coefficient which ranges from 0.27 to 0.32.

R_w is a critical factor in these types of resistivity–sonic transforms. Prediction of R_w requires knowledge of water salinity and temperature variations with depth. It can be directly obtained from laboratory measurements on formation water samples or from spontaneous potential logs which measure at least one clean and permeable zone.

The effect of depth needs to be carefully considered when applying resistivity–sonic transforms. Burch (2002) describes an approach that utilises available sonic data in an area to estimate the first order depth trend, which is then combined with the scaled and filtered resistivity data to generate pseudo-sonic curves in wells without sonic curves.

Other workers have approached the issue of sonic log prediction from resistivity from the perspective of combining resistivity and velocity models. For example, Dos Santos et al. (1988) utilised the resistivity–porosity model of Bussian (1983) and Wyllie et al.'s (1958) equation to generate a sonic–resistivity transform. The practical limitation of this approach is that Wyllie's equation does not account for variations in pore geometry, for example associated with the variation of shale content in sandstones (e.g. Xu and White, 1995). A consideration of these issues led to the development of the Xu–White model described in Section 8.2.7.

8.2.3 Gassmann's equation

A key concept in interpreting seismic amplitudes is an understanding of how rock properties are affected by a change in fluid fill, for example from brine to hydrocarbons (e.g. Smith et al., 2003). The calculation is straightforward for density, where the rock density

Rock physics models and relations

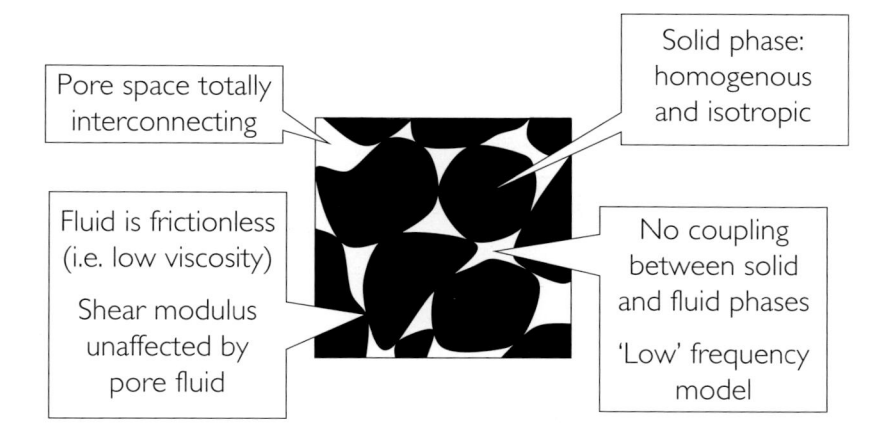

Figure 8.15 Schematic illustration of the key assumptions in Gassmann's equation.

- Pore space totally interconnecting
- Solid phase: homogenous and isotropic
- Fluid is frictionless (i.e. low viscosity)
- Shear modulus unaffected by pore fluid
- No coupling between solid and fluid phases
- 'Low' frequency model

is simply the arithmetic average of the various solid and fluid components of the rock, weighted according to their volume fractions. The effect of fluids on velocity is more complicated.

The usual starting point for modelling fluid substitution effects at seismic frequencies is Gassmann's equation (Gassmann, 1951; Geertsma and Smit, 1961). It describes rocks in terms of the bulk moduli of a two-phase medium (fluid and mineral matrix). The reader is referred to Chapter 2 for an introduction to elastic moduli. Gassmann's equation is applicable to rocks with intergranular porosity and more or less uniform grain size. It can be written as:

$$\frac{K_{sat}}{K_0 - K_{sat}} = \frac{K_d}{K_0 - K_d} + \frac{K_{fl}}{\phi(K_0 - K_{fl})} \quad (8.20)$$
$$\mu_{sat} = \mu_d,$$

where K_{sat} is the bulk modulus of the fluid-saturated rock, K_0 is the bulk modulus of the matrix material, K_d is the bulk modulus of the dry rock frame, K_{fl} is the bulk modulus of the pore fluid, and ϕ is the (fractional) porosity, μ_{sat} is the shear modulus of the fluid-saturated rock and μ_d is the shear modulus of the dry rock frame. The second part of Eqs. (8.20) simply states that the shear modulus is not affected by pore fill, effectively because shear waves do not travel through fluids. The equations introduce the concept of the moduli of the dry rock frame, which is the rock frame with all fluid (liquid or gas) removed. It is possible to measure the dry rock modulus in the laboratory, but in most Gassmann applications using wireline log data this modulus is inverted from the other inputs. As will be discussed, the dry rock modulus is an important parameter for quality control of Gassmann results.

Some key assumptions in the Gassmann model (Wang and Nur, 1992; Fig. 8.15) are that:

- the solid is homogeneous and isotropic,
- all the pore space is in communication,
- wave-induced pressure changes throughout the pore space have time to equilibrate during a seismic period (the low-frequency assumption),
- the fluid that fills the pore space is frictionless (i.e. low viscosity),
- no coupling between solid and fluid phases.

Given these constraints Gassmann's model in the strictest sense is likely to apply only to 'clean' sandstones with moderate to high porosity at low (seismic) frequency (Figs. 8.16 and 8.17). In this scenario there is sufficient time for fluid movement between pores during the passage of the wave. At high frequency (i.e. in the laboratory) this process is limited, leading to increased pore stiffness and higher velocity. To account for this dispersion effect, Gassmann needs to be extended using additional models, such as those proposed by Biot (1962) and Mavko *et al.* (1998). It is worth noting that laboratory measurements on dry rocks are independent of frequency and can be used directly in Gassmann relations. Sonic frequencies are typically around the transition from low to high frequency as defined by Biot (1956) and it is generally assumed that Gassmann can be applied to porous rocks at these frequencies. Tight sands and shaley sands are scenarios in which it is possible that Gassmann's assumptions are violated. This will be discussed in Section 8.5.

It is generally held that Gassmann is appropriate for carbonate rocks with relatively homogenous pore systems (e.g. Wang and Nur, 1992; Wang

Rock physics for seismic modelling

```
  Seismic              Sonic              Ultrasonic
 |_____|          |_____|           |_____|
 0    100Hz      1kHz    10kHz   0.1MHz        1MHz

         500Hz                 100kHz
       |____|_____|____|
       Low frequency     ?     High frequency
```

High porosity/permeability rocks
with low viscosity fluids

Gassmann ───────────────→   ←── Biot-Gassmann ──→

Low/moderate porosity/permeability
sands with high viscosity fluids

                          Biot-Gassmann
Gassmann ?   ←──────────────── + ────────────────→
                           Squirt flow

Sandstones and carbonates with moderate to high porosity	Gassmann OK	Considered that sonic log responses in these logs are essentially low frequency and Gassmann is applicable
Tight sand	Gassmann application requires care	**At logged frequencies:** Velocities may be dispersive Patchy saturation may characterise low saturation gas scenarios
		At seismic frequencies: Potential for uncertainty in fluid and mineral moduli
Shaley sands and laminated sands		Common Gassmann pitfall of exaggerated fluid substitution effect Gassmann needs to be adjusted to achieve intuitive result
Rocks with fractures or dual porosity systems	Other models required	

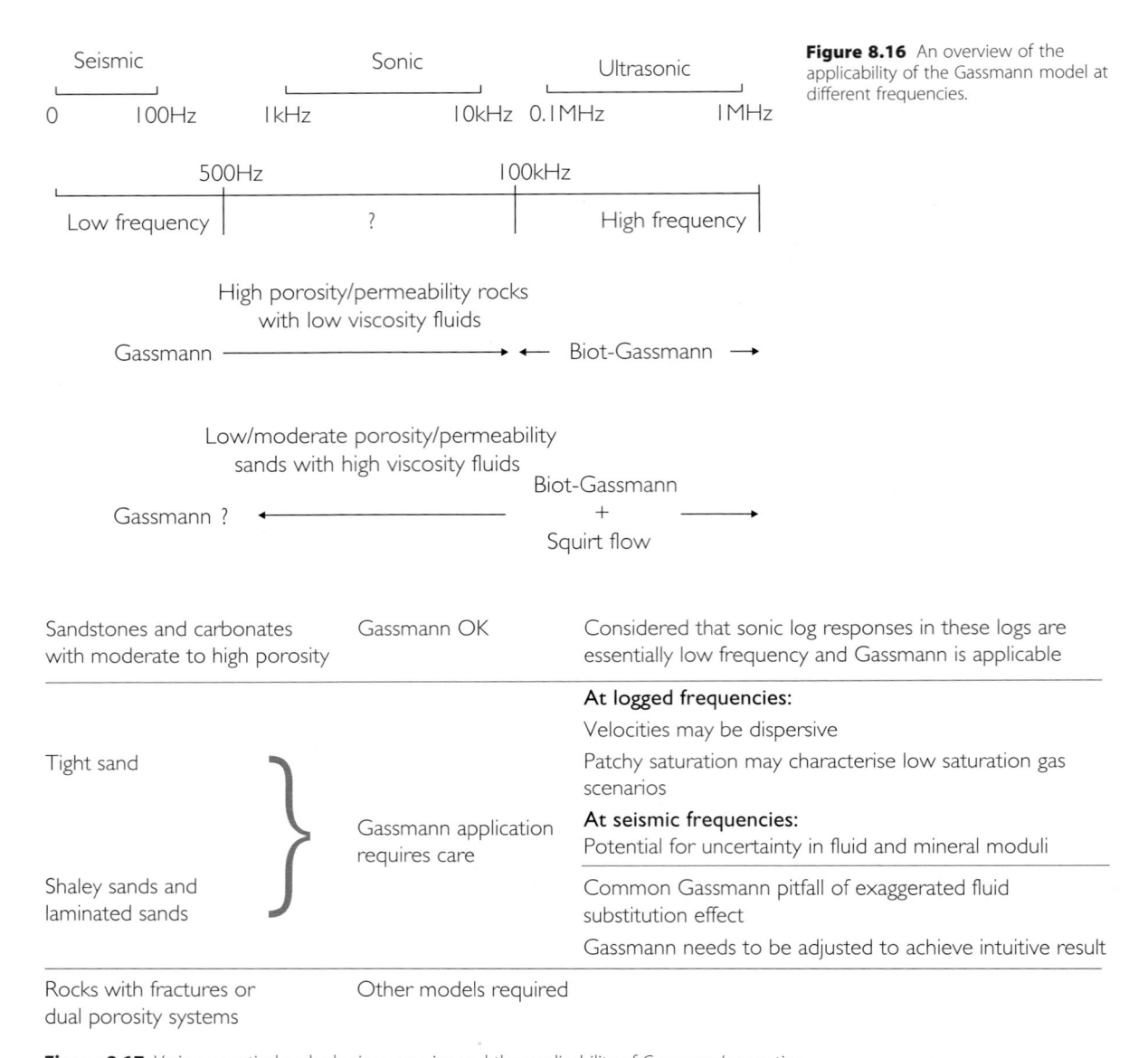

Figure 8.16 An overview of the applicability of the Gassmann model at different frequencies.

Figure 8.17 Various practical rock physics scenarios and the applicability of Gassmann's equation.

et al., 2001) but recent work suggests that the relevance of Gassmann may depend on the nature of the pore types and distributions (e.g. Xu and Payne, 2009). There is also a potential issue in the interaction between fluid and carbonate matrix. For example, Baechle et al. (2007) have described how the introduction of brine into a dry carbonate may soften or harden the rock. Certainly, if fractures or dual porosity systems are present then Gassmann will be inappropriate for the purposes of modelling fluid substitution effects.

Figure 8.18 presents a series of equations that enable the practical implementation of Gassmann's relations with log data. It describes how the saturated bulk modulus is defined by four components, namely mineral, fluid, porosity and the dry rock frame. The stiffness characteristics of the rock introduced in Chapter 5, an important element in the magnitude of fluid substitution effects on compressional velocity, are controlled essentially by a combination of the mineral, porosity and the dry rock frame modulus, referred to as the pore space stiffness (Mavko and Mukerji, 1995).

Rock physics models and relations

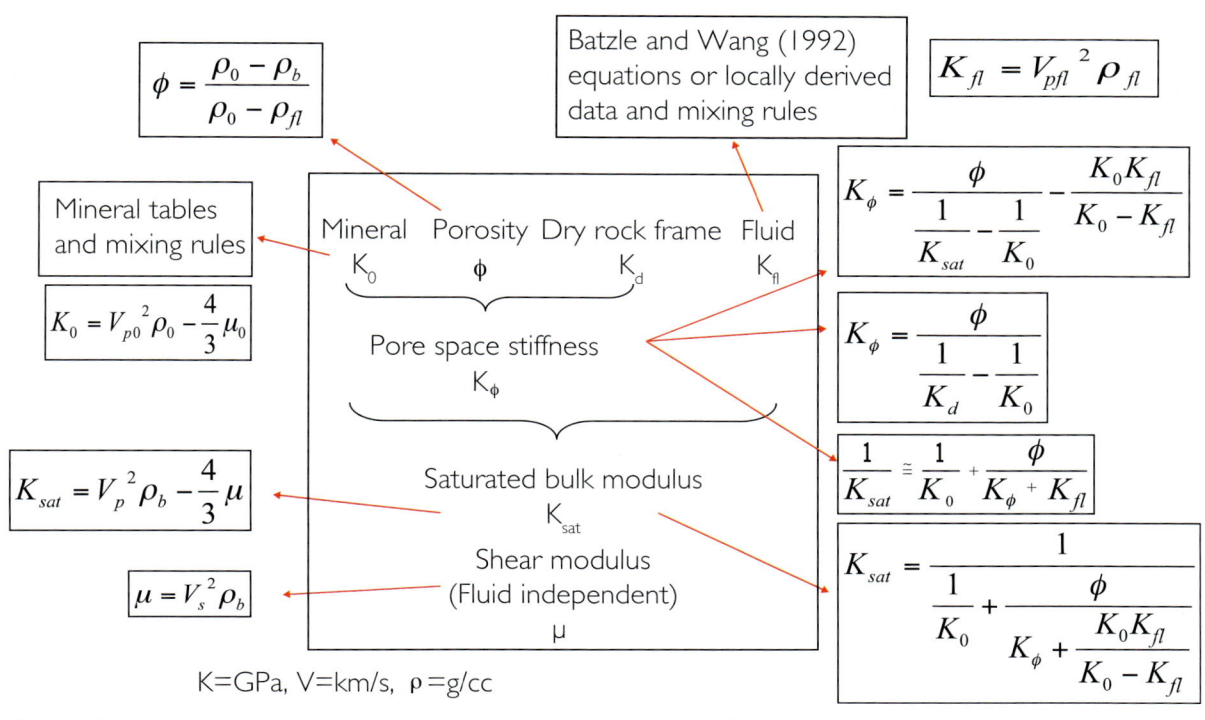

Figure 8.18 Practical equations for the application of Gassmann's relations to log data.

Top-left: $\phi = \dfrac{\rho_0 - \rho_b}{\rho_0 - \rho_{fl}}$

Top-center: Batzle and Wang (1992) equations or locally derived data and mixing rules

Top-right: $K_{fl} = V_{pfl}^2 \rho_{fl}$

Left side: Mineral tables and mixing rules

$K_0 = V_{p0}^2 \rho_0 - \dfrac{4}{3}\mu_0$

$K_{sat} = V_p^2 \rho_b - \dfrac{4}{3}\mu$

$\mu = V_s^2 \rho_b$

Center box: Mineral K_0 | Porosity ϕ | Dry rock frame K_d | Fluid K_{fl}
Pore space stiffness K_ϕ
Saturated bulk modulus K_{sat}
Shear modulus (Fluid independent) μ

Right side:

$K_\phi = \dfrac{\phi}{\dfrac{1}{K_{sat}} - \dfrac{1}{K_0}} - \dfrac{K_0 K_{fl}}{K_0 - K_{fl}}$

$K_\phi = \dfrac{\phi}{\dfrac{1}{K_d} - \dfrac{1}{K_0}}$

$\dfrac{1}{K_{sat}} \cong \dfrac{1}{K_0} + \dfrac{\phi}{K_\phi + K_{fl}}$

$K_{sat} = \dfrac{1}{\dfrac{1}{K_0} + \dfrac{\phi}{K_\phi + \dfrac{K_0 K_{fl}}{K_0 - K_{fl}}}}$

K=GPa, V=km/s, ρ=g/cc

An implication of Fig. 8.18 is that the dry rock parameters that are inverted from the various inputs are also those used in the fluid substitution step. This data-driven approach is appropriate only for clean blocky sands of moderate to high porosity and with good quality log data. In practice, there is a QC step to evaluate the dry rock moduli before proceeding. There are a number of reasons why the dry rock moduli may be in error such as inconsistencies in the various log curves, either in terms of depth registration or the fact that they do not all investigate the same volume of rock. It is recommended therefore that the fluid substitution step is conditioned by a dry rock model. Following Simm (2007) it is proposed that the dry rock model is conditioned in ϕ vs normalised modulus space (Mavko and Mukerji, 1995; Avseth et al., 2005) and the use of this plot will be explained in the following sections. A practical workflow for Gassmann is presented below. For clarity, the fluid substitution step has been divided into three steps (labelled 4, 5 and 6), namely density substitution, shear velocity substitution and compressional velocity substitution; the equation for fluid substitution of the compressional velocity is taken from Downton and Gunderson (2005).

(1) Moduli calculation

$$\mu = V_s^2 \rho_b, \quad K_{sat} = V_p^2 \rho_b - \dfrac{4}{3}\mu.$$

(2) Dry rock inversion

$$K_\phi = \dfrac{\phi}{\dfrac{1}{K_{sat}} - \dfrac{1}{K_0}} - \dfrac{K_0 K_{fl}}{K_0 - K_{fl}}, \quad K_d = \dfrac{1}{\dfrac{\phi}{K_\phi} + \dfrac{1}{K_0}}.$$

(3) Dry rock modelling

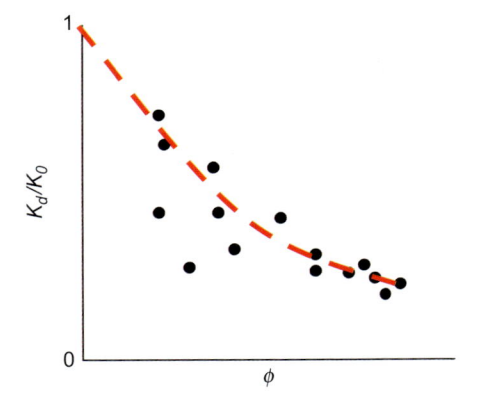

[Plot of K_d/K_0 vs ϕ, ranging 0 to 1, with scattered data points and a dashed red trend curve decreasing from upper left to lower right.]

(4) Fluid substitution – density

$$\rho_{b2} = \rho_{b1} - \left((\phi\rho_{fl1}) - (\phi\rho_{fl2})\right)$$

$$\rho_{fl2} = (\rho_w S_w) + (1 - S_w)\rho_h$$

$$\rho_{b2} = (\rho_{fl2}\phi) + (1 - \phi)\rho_0.$$

(5) Fluid substitution – shear velocity

$$V_{s2} = \sqrt{\frac{\mu}{\rho_{b2}}}.$$

(6) Fluid substitution – compressional velocity

$$V_{p2} = \sqrt{\frac{\rho_{b1}}{\rho_{b1} + (\rho_{fl2} - \rho_{fl1})\phi} V_{p1}^2 + \beta^2 \frac{N_2 - N_1}{\rho_{b1} + (\rho_{fl2} - \rho_{fl1})\phi}},$$

where

$$\beta = 1 - \frac{K_d}{K_0} \quad N_1 = \frac{1}{\frac{\phi}{K_{fl1}} + \frac{\beta - \phi}{K_0}} \quad N_2 = \frac{1}{\frac{\phi}{K_{fl2}} + \frac{\beta - \phi}{K_0}}.$$

(8.21)

The model that is fit to the dry rock data can take various forms depending on the dominant trends in the data. The following are examples of functions that might be used:

Linear fits – for porosity variations controlled by diagenesis (i.e. the critical porosity model of Nur, 1992):

$$\frac{K_d}{K_0} = -\frac{\phi}{\phi_c} + 1. \quad (8.22)$$

Iso-stiffness fits representing porosity variations controlled by sorting:

$$\frac{K_d}{K_0} = \frac{1}{1 + c\phi}. \quad (8.23)$$

Exponential fits – for intermediate trends:

$$\frac{K_d}{K_0} = \exp^{-c\phi}. \quad (8.24)$$

Models such as that proposed by Krief et al. (1990) might also be used:

$$\frac{K_d}{K_0} = (1 - \phi)^{\frac{x}{1-\phi}}, \quad (8.25)$$

where x is a fitting coefficient.

In Chapter 5 it was shown that sandstones with the same porosity but different stiffness can show different magnitudes of fluid substitution effect on the compressional velocity. This is explained in terms of the effect of pore space compressibility in Fig. 8.19. Data from two (unrelated) sands with 30% porosity are shown for both water and gas fill. The change in compressional velocity in the stiffer cemented sand is far less than in the softer uncemented sand (Figs. 8.19a,b).

The effect of stiffness on fluid substitution is elegantly explained by considering the crossplot of porosity vs normalised bulk modulus (K_{sat}/K_0) (Mavko and Mukerji, 1995; Mavko et al., 1998; Avseth et al., 2005). Given the Gassmann relation:

$$\frac{1}{K_d} = \frac{1}{K_0} + \frac{\phi}{K_\phi} \quad (8.26)$$

it is possible to draw lines of equal normalised pore space modulus (K_ϕ/K_0) on the plot. Wide spacing of the lines indicates relatively soft rock, whereas closely spaced lines indicate relatively stiff rocks. Given that:

$$\frac{1}{K_{sat}} \approx \frac{1}{K_0} + \frac{\phi}{K_\phi + K_{fl}} \quad (8.27)$$

the new saturated bulk modulus (K_{sat}) can be estimated by using the change in normalised fluid modulus to move between the normalised pore modulus lines. Thus, if the change in normalised fluid modulus is 0.07 (as in Fig. 8.19) and the normalised pore modulus lines have 0.02 spacing, the new K_{sat} can be read directly by moving 3.5 lines from the K_{sat} starting value. For a given change in the fluid modulus there will be a relatively large change in K_{sat} for soft rocks but a small change for stiff rocks.

8.2.4 Minerals, fluids and porosity

The parameterisation of Gassmann's equation requires values for mineral and fluid moduli as well as densities and porosity. In the application of Gassmann to well log data an accurate petrophysical analysis of mineral fractions, porosity and fluid saturations is a necessary starting point.

8.2.4.1 Mineral parameters

Values for the matrix mineral density (ρ_0) and bulk modulus (K_0) are generally based on published tabulations (Table 8.4). Information is therefore required on the various minerals present in the rock. Often the petrophysicist's shale volume calculation is used as

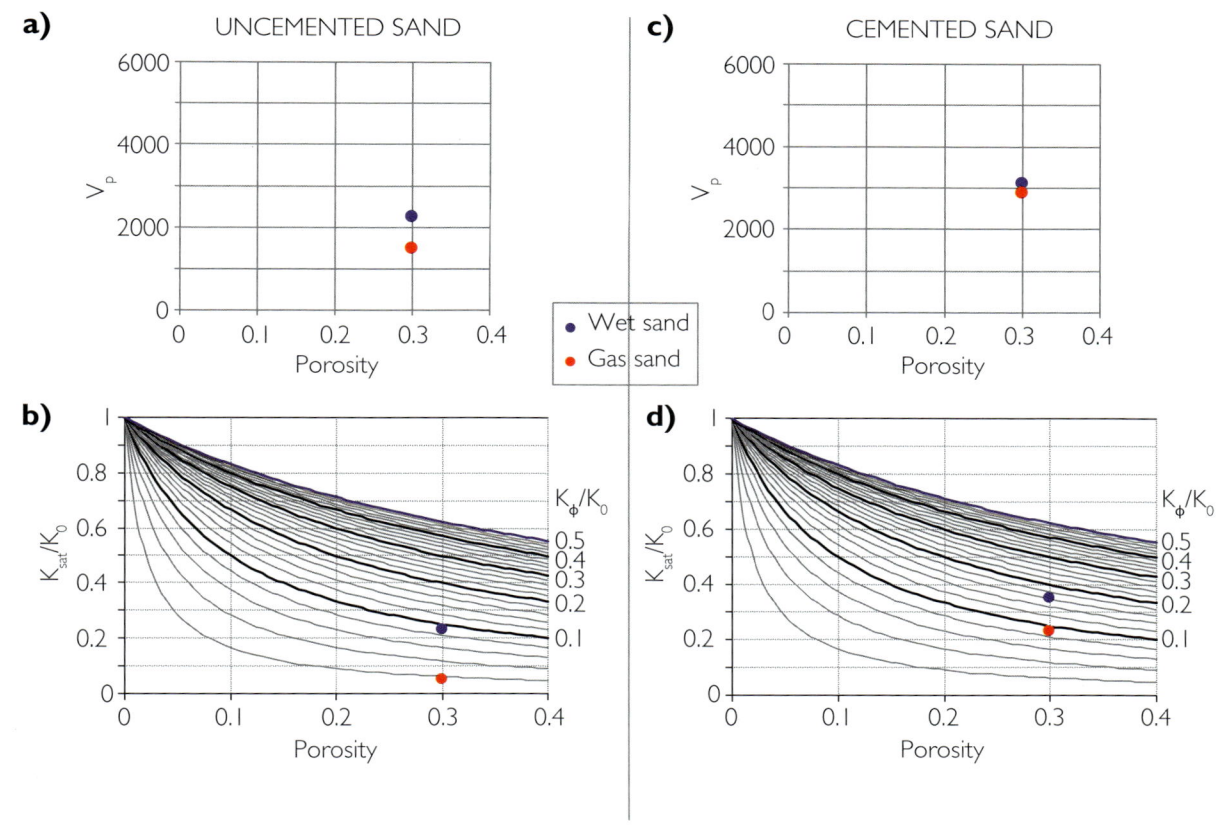

Figure 8.19 Fluid substitution and stiffness in sandstones. Two sands with 30% porosity are shown, soft uncemented sands in (a) and (b) and stiffer cemented sands in (c) and (d). For the same porosity softer sands show greater fluid substitution effects in terms of compressional velocity.

the basis for the relative volume of sand and shale. Shale properties are variable and therefore not shown in Table 8.4, but in any particular case it will usually be straightforward to read off shale values from wireline logs for shales within or adjacent to the interval of interest. In the absence of other information it is often assumed that the sand mineral is quartz. However, there may be some uncertainty in the effective mineral modulus if feldspar is present (Smith, 2011). Ideally the type of feldspar needs to be known, given the range of modulus values for different types of feldspar. Sometimes, detailed lithology logs will be available, where logs have been inverted for volume fractions of a range of minerals. Mineralogical reports can be useful if there are exotic minerals present with properties very different from those of quartz and shale.

Another potential issue for the interpreter in parameterising Gassmann's equation is the uncertainty in the modulus of dry clay. The parameters of dry clays vary dramatically (Fig. 8.20 and Table 8.5). Wang et al. (2001) has shown that dry clay densities can vary between 2.2 g/cc to 2.84 g/cc and the bulk moduli (K) can vary between 10 GPa and 70 GPa, while the shear modulus is roughly equal to $0.47K$. Often, especially in exploration settings, the types of clays are unknown.

Single effective mineral densities are calculated as the arithmetic average of the constituent minerals, weighted by volume fraction. To calculate the effective bulk modulus (K_0), the moduli of the constituent minerals must be combined according to a mixing scheme. There are several possibilities as discussed in Section 8.2.1, of which the Voigt–Reuss–Hill (VRH) average (Hill, 1952) and the Hashin-Shtrikman (1963) are the most commonly used.

8.2.4.2 Fluid parameters

If there is more than one fluid present, the properties of a fluid mixture need to be calculated. Similar to mineral mixing, the density of a fluid mixture is

Rock physics for seismic modelling

Table 8.4. Examples of solid mineral elastic properties (after Simmons and Wang, 1971).

	Bulk modulus (GPa)	Density (g/cc)	Shear modulus (GPa)	Young's modulus (GPa)	V_p (km/s)	V_s (km/s)	Poisson's ratio
Quartz	36.6	2.65	45.0	95.8	6.038	4.121	0.064
Chert	26.0	2.35	32.0	67.7	5.400	3.700	0.058
Calcite	76.8	2.71	32.0	84.3	6.640	3.436	0.317
Dolomite	94.9	2.87	45.0	116.6	7.347	3.960	0.295
Aragonite	47.0	2.94	39.0	92.6	5.800	3.600	0.187
Magnesite	114.0	3.01	68.0	169.7	8.200	4.750	0.247
Na-Felspar	55.0	2.62	28.0	71.3	5.900	3.300	0.272
K-Felspar	48.0	2.56	24.0	61.9	5.600	3.050	0.289
Ca-Felspar	85.0	2.73	38.0	99.0	7.050	3.750	0.303
Clays (approx)	41.0	2.68	17.0	45.0	4.900	2.500	0.324
Muscovite	52.0	2.82	31.5	78.7	5.800	3.350	0.250
Biotite	50.0	3.00	27.5	70.2	5.400	3.000	0.277
Halite	25.2	2.16	15.3	38.3	4.600	2.650	0.252
Anhydrite	66.5	3.00	34.0	87.0	6.150	3.400	0.280
Gypsum	58.0	2.31	30.0	77.1	6.750	3.700	0.285
Pyrite	158.0	5.02	149.0	338.7	8.400	5.450	0.137

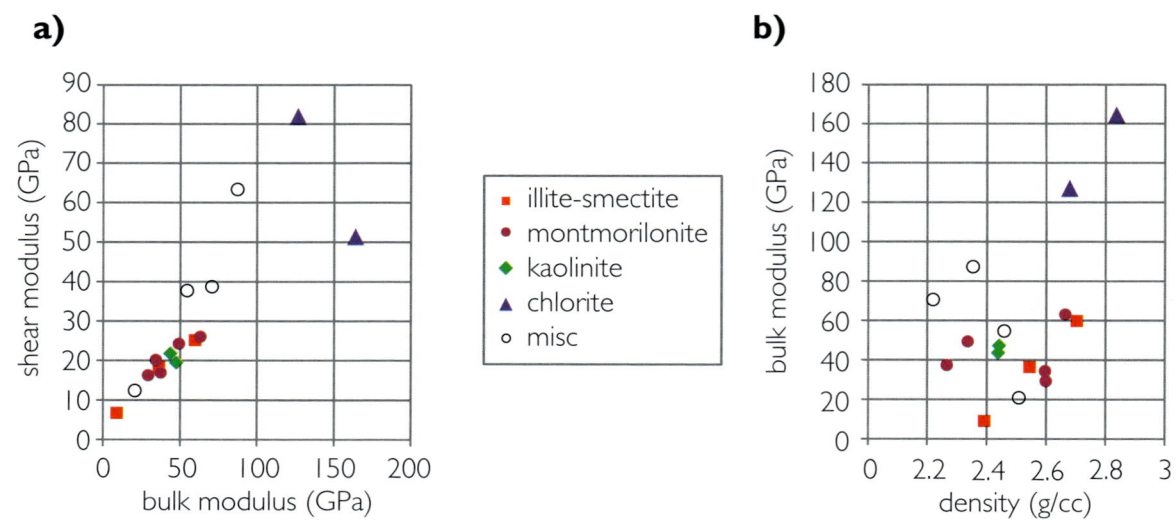

Figure 8.20 Dry clay elastic parameters (after Wang et al., 2001).

Table 8.5. Examples of dry clay elastic parameters (after Avseth et al., 2005).

Clay mineral	K (GPa)	μ (GPa)
Smectite	17.5	7.5
Illite	39.4	11.7
Kaolinite	1.5	1.4
Kaolinite	37.9	14.8
Kaolinite	12	6
Chlorite	95.3	11.4

simply the weighted addition of the densities of the various components:

$$\rho_{fl} = S_w \rho_w + (1 - S_w)\rho_h, \quad (8.28)$$

where S_w = water saturation, ρ_w = brine density and ρ_h = hydrocarbon density.

For the bulk modulus mix, it is generally assumed that Reuss' (also called Wood's) law applies:

$$\frac{1}{K_{fl}} = \frac{S_w}{K_w} + \frac{1 - S_w}{K_h}, \quad (8.29)$$

where K_{fl} is the bulk modulus of a mixture consisting of volume fraction S_w of brine, with bulk modulus K_w, and hydrocarbon with bulk modulus K_h. This equation describes homogeneous mixing, where the relative proportion of each fluid is the same in every pore space. This is usually valid for modelling at seismic frequencies in rocks that satisfy the Gassmann assumptions, but other mixing schemes may be needed to account for inhomogeneous ('patchy') saturation in some cases (Section 8.4.4).

The 'virgin' fluid is simply the fluid present in the formation, undisturbed by any invasion effect caused by the drilling of the well. The properties of brine and hydrocarbons at reservoir temperature and pressure can be estimated using the Batzle and Wang (1992) equations, which are generally satisfactory for gases and oils in the range 15–40 API. For the sake of brevity a detailed presentation of the equations will be omitted here.

There is usually some uncertainty in the application of these equations particularly with regard to heavy oils and condensates. When the viscosity is high, oils may support shear waves (Batzle et al., 2004) and there is the possibility of coupling between rock frame and fluid. For very light hydrocarbons and condensates it may be preferable to use direct laboratory measurements of compressibility if they are available. However, laboratory PVT measurements are acquired under isothermal rather than the adiabatic conditions that characterise seismic wave propagation in the petroleum reservoir. In some cases the difference between the two bulk moduli may be as much as a factor of 2 (Batzle and Wang, 1992). When estimating fluid compressibilities of condensates the interpreter should consult with the petroleum engineer.

The inputs required for the Batzle and Wang (1992) equations are

- reservoir temperature and pressure,
- brine salinity,
- dissolved gas index for brine (from 0 = no dissolved gas to 1 = gas-saturated),
- density of gas relative to air,
- oil *API* gravity
- oil gas-oil-ratio (*GOR*).

Subsequent work by Han and Batzle (2000a,b) suggests refinements to the original Batzle and Wang (1992) equations. In particular it was determined that the original Batzle and Wang (1992) equations overestimated the effect of GOR on the elastic properties of oils.

As an illustration, Fig. 8.21 shows some fluid properties based on North Sea data. It shows that gases are more compressible than oil and brine and also that gas responds to increasing pressure and temperature with an increase in density and fluid modulus, whereas oil and brine show a slight decrease in density and fluid modulus. It is important that the interpreter determines appropriate basin-specific constraints for the modelling of fluid parameters. It is common for example to find relationships between oil API and depth. Figure 8.22a shows an example from the Central North Sea where oil *API* generally increases with depth, principally as a result of biodegradation at shallow levels. Figure 8.22b shows how the maximum amount of dissolved gas (a parameter calculated as part of the Batzle and Wang (1992) equations) increases with *API* and pressure and temperature.

The effect of a small amount of gas on the acoustic properties of sandstones was discussed in Section 5.3.3, potentially giving rise to bright spots that cannot be distinguished from those related to commercial gas (Chapters 5 and 7). Figure 8.23 illustrates that this phenomenon is related principally to the

Rock physics for seismic modelling

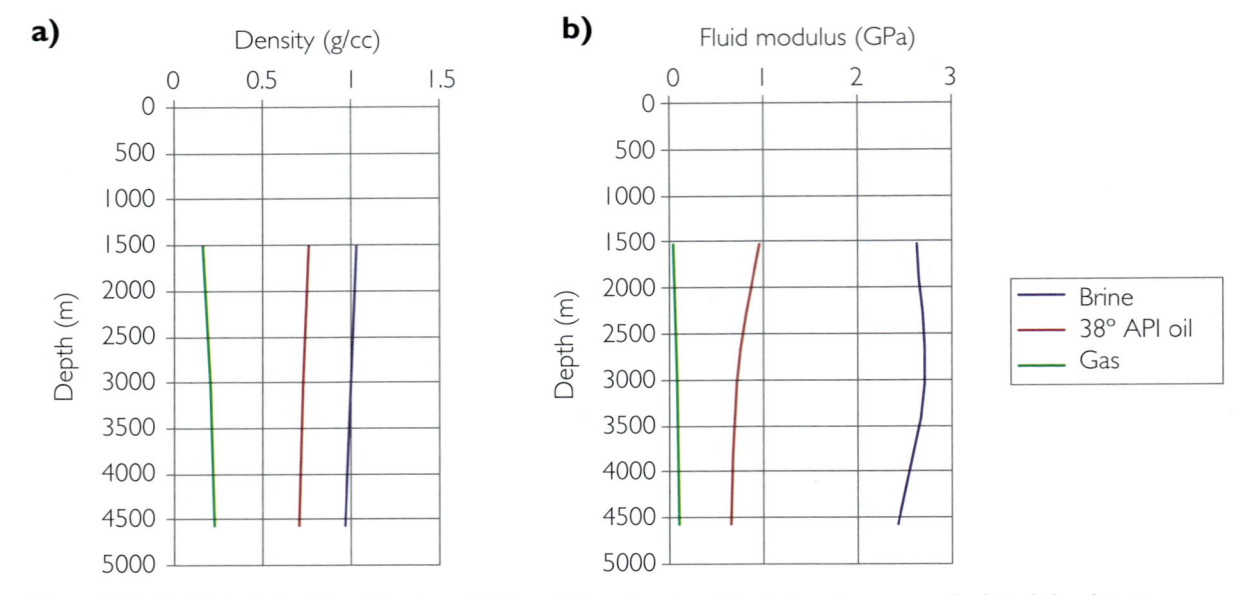

Figure 8.21 Modelled relationships of density and fluid moduli as a function of depth, based on a generalised North Sea dataset.

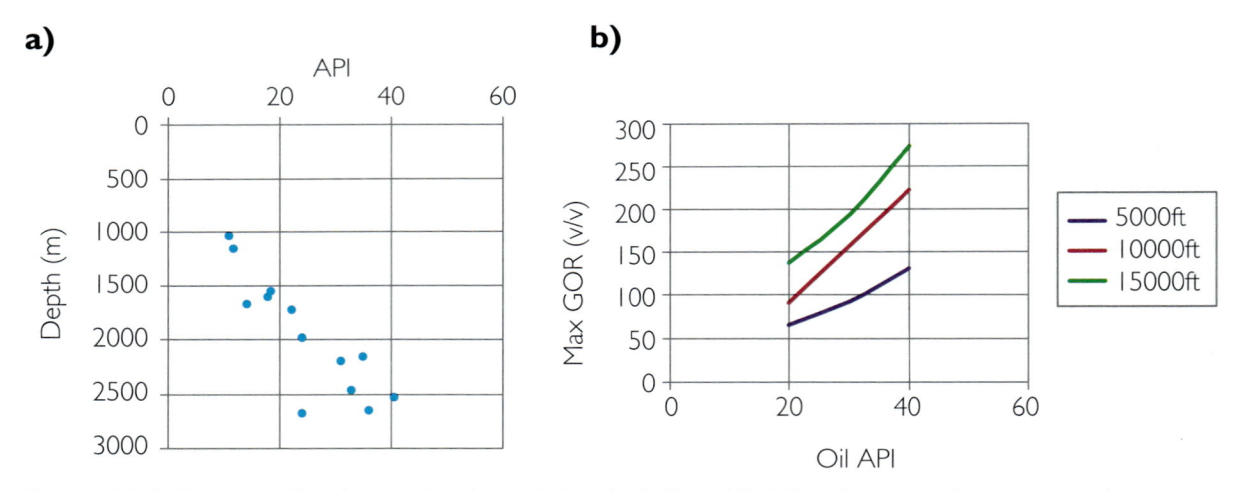

Figure 8.22 Fluid parameters from the Central North Sea; (a) API vs depth, (b) modelled effect of pressure and temperature on the maximum GOR for oils with different API.

homogenous mixing of water and gas at low temperature and pressure. The effect is less evident as the temperature and pressure increase.

8.2.4.3 Porosity

At first sight, porosity is a straightforward concept, being the ratio of the pore space volume to the volume of the rock, expressed as a decimal or percentage. It can thus be derived from fluid, mineral and bulk densities:

$$\phi = \frac{\rho_o - \rho_b}{\rho_o - \rho_{fl}}, \qquad (8.30)$$

where ρ_o is the matrix density, ρ_{fl} is the fluid density and ρ_b is the measured bulk density. This is straightforward if the mineral and fluid parameters are known and the density log is trustworthy. In real rocks, however, only part of the fluid in the pore space can move freely in and out of the rock. Thus, the porosity can be split into 'effective porosity' (i.e. the connected porosity in which fluids can move)

Rock physics models and relations

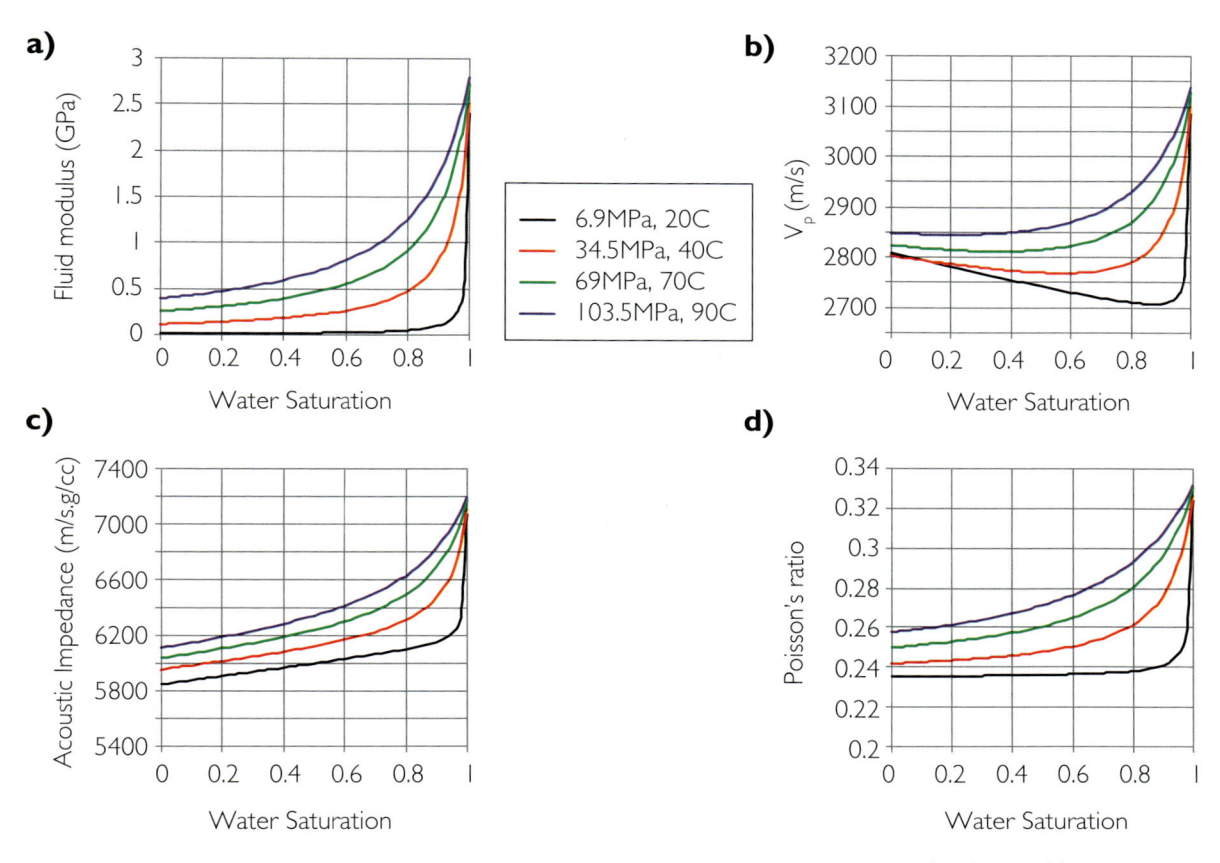

Figure 8.23 The effect of temperature and pressure on a gas-bearing sand. Graphs show the variation of (a) fluid modulus (using Reuss mixing), (b) compressional velocity, (c) acoustic impedance and (d) Poisson's ratio with changing water saturation. Constants used in models: $K_d = 8.8$ GPa, $\mu = 5.7$ GPa, porosity = 0.22, gas gravity = 0.65.

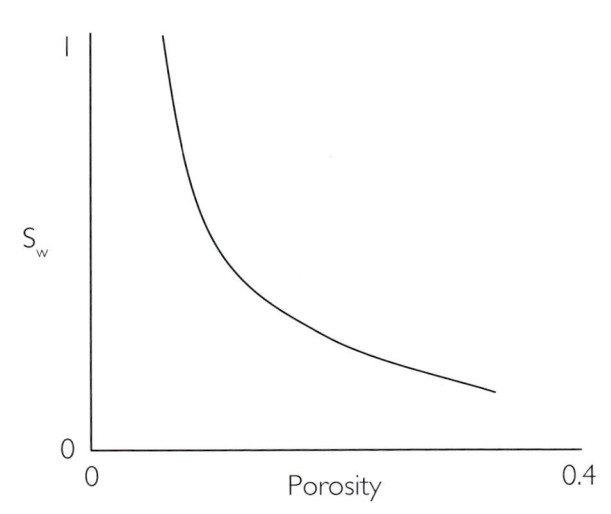

Figure 8.24 Typical form of the relationship between porosity and water saturation in an unproduced sand reservoir. The curve defines the irreducible water saturation. Note that with finer-grained sandstones the trend shifts to the right.

and the 'ineffective' porosity. In clean sandstones, the effective porosity is largely determined by grain size and sorting; for example, it is possible for fine-grained sandstone to have a total porosity of 15% but no effective porosity owing to the way that the grains are packed. The rock is effectively impermeable and it wouldn't be appropriate to perform fluid substitution on such a rock. In order to carry out fluid substitution correctly it is important to consider the relationship between porosity and saturation, particularly the irreducible water saturation characteristic of the reservoir (Fig. 8.24).

The effective porosity concept is also important in shaley sands (Chapter 5). Most sandstones have some shale as a component of the rock, where shale is defined as the combination of clay minerals, silt and chemically bound water (Fig. 8.25). Petrophysicists will often work with effective porosity and effective water saturation in shaley sands, owing to the ease

Rock physics for seismic modelling

with which shale parameters can be estimated from the log data and the relevance of the effective parameters to the estimation of moveable hydrocarbon. In terms of the porosity equation given above, the effective porosity of shaley sandstones is calculated when the mineral density is derived for a mixture of quartz and shale. Calculation of total porosity is often difficult because of the uncertainty in dry clay density. The following equations provide a link between effective and total porosity measurements:

$$(1 - S_{we})\phi_e = (1 - S_{wt})\phi_t, \quad (8.31)$$

$$S_{wt} = 1 - \frac{(1 - S_{we})\phi_e}{\phi_t}, \quad (8.32)$$

$$\phi_t = \phi_e + V_{sh}(1 - \phi_e)\left(\frac{\rho_{cl} - \rho_{sh}}{\rho_{cl} - \rho_w}\right), \quad (8.33)$$

where S_{we} and S_{wt} are effective and total water saturation, ϕ_e and ϕ_t are effective and total porosity, V_{sh} is the volume fraction of shale, and ρ_{cl}, ρ_{sh} and ρ_w are the densities of clay, shale and formation water. An example of water saturation relationship to porosity and shale content is shown in Fig. 8.26. The water saturation increases as the shale content increases and the porosity decreases.

From the perspective of using rock physics models to predict velocities (e.g. such as the Raymer–Hunt model or any other velocity model) it is the 'total' porosity that is used. However, in terms of fluid substitution, and owing to a number of practical issues (see Section 8.5), it can be argued that Gassmann might be parameterised either in terms of total or effective porosity (Section 8.5).

8.2.5 Dry rock relations

The dry rock values inverted from Gassmann's equation are important parameters for quality control. They are also important in rock characterisation studies.

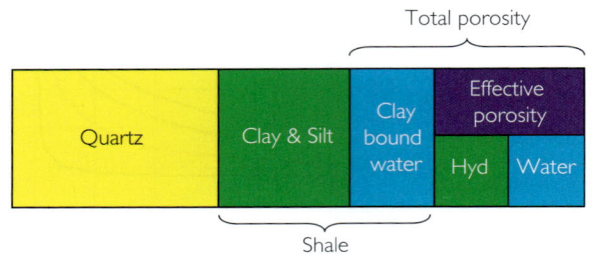

Figure 8.25 Schematic representation of the components of a shaley sandstone.

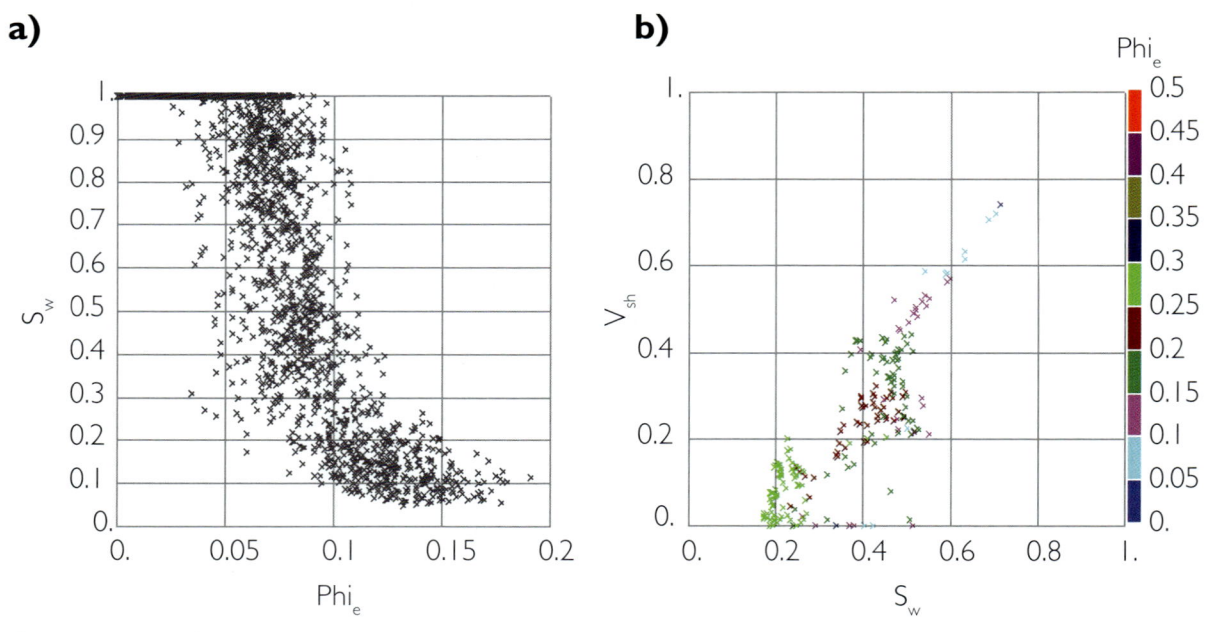

Figure 8.26 Results from a petrophysical analysis of an oil-bearing shaley sandstone; (a) effective porosity vs water saturation, (b) water saturation vs shale content.

Rock physics models and relations

a)

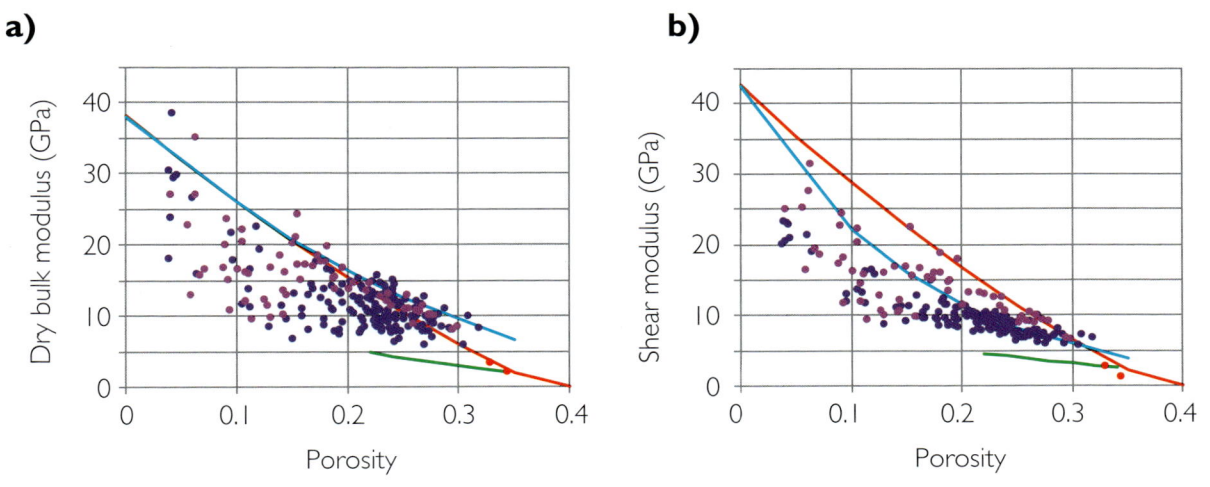

b)

Figure 8.27 Examples of dry rock measurements in sandstones; (a) porosity vs dry rock bulk modulus, (b) porosity vs shear modulus. Purple points – data from Han et al. (1986), dark blue points – Tertiary sandstone dataset from North Sea, red line – functions derived by Murphy et al. (1993) for clean sandstone: $K_d = 38.18 (1 - 3.39\phi + 1.95\phi^2)$ and $\mu = 42.65(1 - 3.48\phi + 2.19\phi^2)$, blue line – generalised trend for Tertiary sandstones in the North Sea, green line –Troll Field data at an effective pressure of 5MPa (Dvorkin and Nur, 1996), red points – selected unconsolidated sand data.

8.2.5.1 Dry rock characteristics of sandstones and carbonates

Typical values of dry rock parameters for sandstones are shown in Fig. 8.27. The plot looks similar to the porosity vs V_p crossplot for brine-saturated data in Fig. 8.4, with a variety of trends related to variations in rock stiffness and porosity reduction mechanisms. It is typical that for low porosity sandstones (i.e. below ~15%) the shear modulus is slightly higher or of similar magnitude to the dry bulk modulus. At higher porosities the dry bulk modulus is usually higher than the shear modulus.

An example of a normalised bulk modulus plot for dry rocks based on selected laboratory data from sandstones is shown in Fig. 8.28. The sorting and diagenesis trends are evident but it is also clear that K_ϕ / K_0 values are greater than 0.06 for porosities less than 32%. A combination of the dry bulk modulus and shear modulus defines another parameter which is useful in Gassmann QC, namely the dry rock Poisson's ratio (σ_d):

$$\sigma_d = \frac{3K_d - 2\mu}{2\mu + 6K_d}. \tag{8.34}$$

Typical values of dry rock Poisson's ratio tend to be between 0.1 and 0.3 for consolidated sandstones (e.g. Spencer et al., 1994) with higher values generally associated with increased clay content (Figs. 8.28, and 8.29).

An example of dry rock parameters for carbonates is shown in Fig. 8.30. It is typical that the bulk modulus is higher than the shear modulus across all porosities owing to a dominant effect of the mineral moduli. It is also characteristic that the dry rock Poisson's ratio shows a large scatter as a function of porosity. Whilst much progress has been made in linking velocity and moduli to pore geometry factors in carbonates (e.g. Weger et al., 2009; Verwer et al., 2010), the interpretation of velocity ratios in carbonates is as yet not fully explained (R. Weger, personal communication).

8.2.5.2 Dry rock model for shear velocity prediction

An alternative approach to the empirical regression methodology for shear velocity prediction described in section 8.2.2.4 is 'the dry rock Poisson's ratio' or 'Gregory–Pickett' approach (Gregory, 1977; Pickett, 1963). This method essentially uses Gassmann's equation to predict shear velocity with one of the key inputs being the dry rock Poisson's ratio. The method first calculates the dry bulk modulus from V_p, density (ρ), porosity (ϕ), fluid and mineral moduli (K_{fl} and K_0) and dry rock Poisson's ratio (σ_d):

$$K_d = (1 - y)K_0,$$

Rock physics for seismic modelling

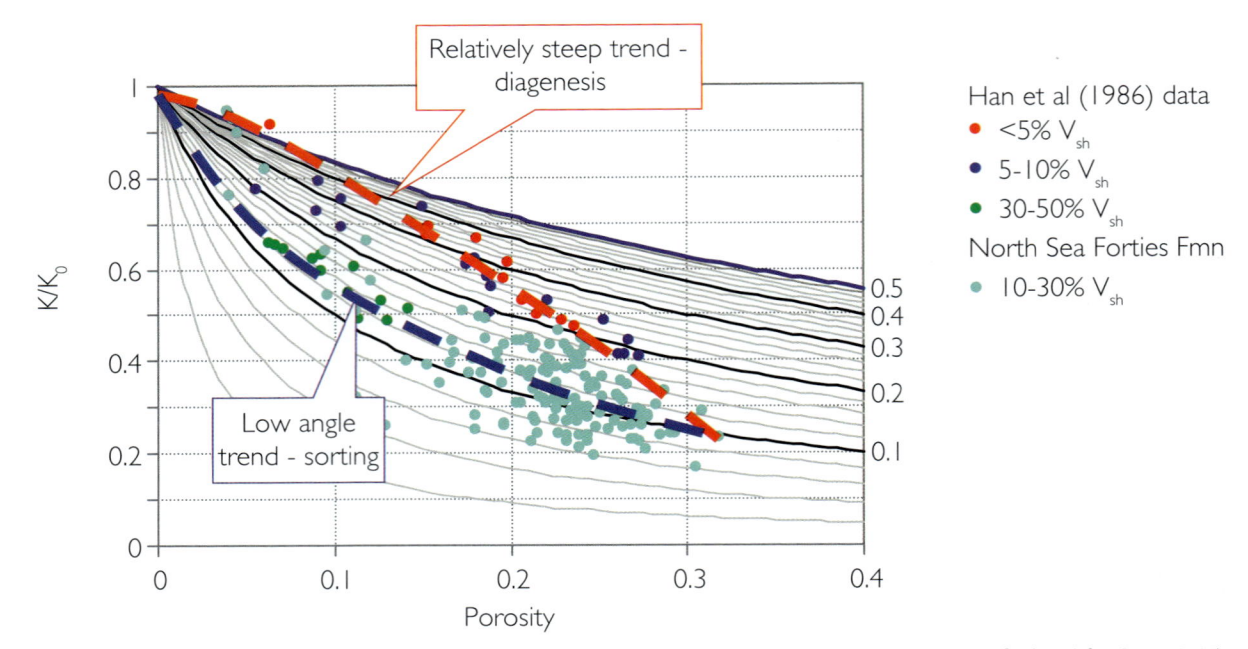

Figure 8.28 Normalised bulk modulus crossplot for selected sandstone data showing typical trends and ranges of values (after Simm, 2007).

where

$$y = \frac{-b + (b^2 - 4ac)^{0.5}}{2a}$$

$$a = S - 1$$

$$b = \phi S \left(\frac{K_0}{K_{fl}} - 1\right) - S + \frac{M}{K_0}$$

$$c = -\phi \left(S - \frac{M}{K_0}\right)\left(\frac{M}{K_{fl}} - 1\right)$$

$$S = \frac{3(1 - \sigma_d)}{1 + \sigma_d}$$

$$M = V_p^2 \rho. \qquad (8.35)$$

Once K_d is derived then the shear velocity can be calculated:

$$V_s = \sqrt{\frac{\mu}{\rho}}, \qquad (8.36)$$

where

$$\mu = \frac{3K_d(1 - 2\sigma_d)}{2(1 + \sigma_d)}. \qquad (8.37)$$

Hilterman (1990) used this technique with log data and modelled the dry rock Poisson's ratio as a function of volume of shale (Fig. 8.31). Owing to the fact that

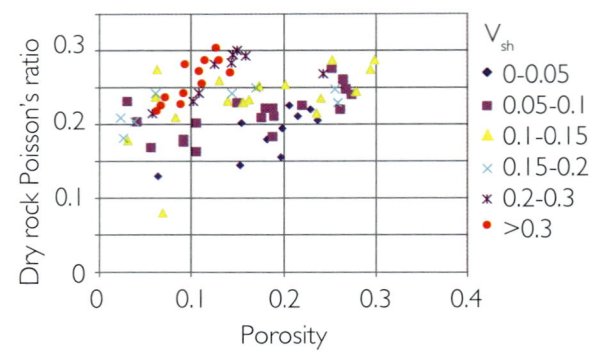

Figure 8.29 Dry rock Poisson's ratio as a function of porosity and clay content in consolidated sandstones (data from Han, 1986).

parameters such as porosity are included in the V_s prediction it can be used effectively to establish a consistency of the inputs to Gassmann's equation. It can work well in sandstone settings where there is a fairly narrow range of dry rock Poisson's ratio. However, in practice there may be limitations to the technique as it requires total porosity and can be difficult to stabilise, particularly for the prediction of shear velocity in shales.

8.2.5.3 A simple model for porosity change

In the context of trying to understand seismic amplitudes by modelling fluid effects using Gassmann, the question often arises: what would the effect of a

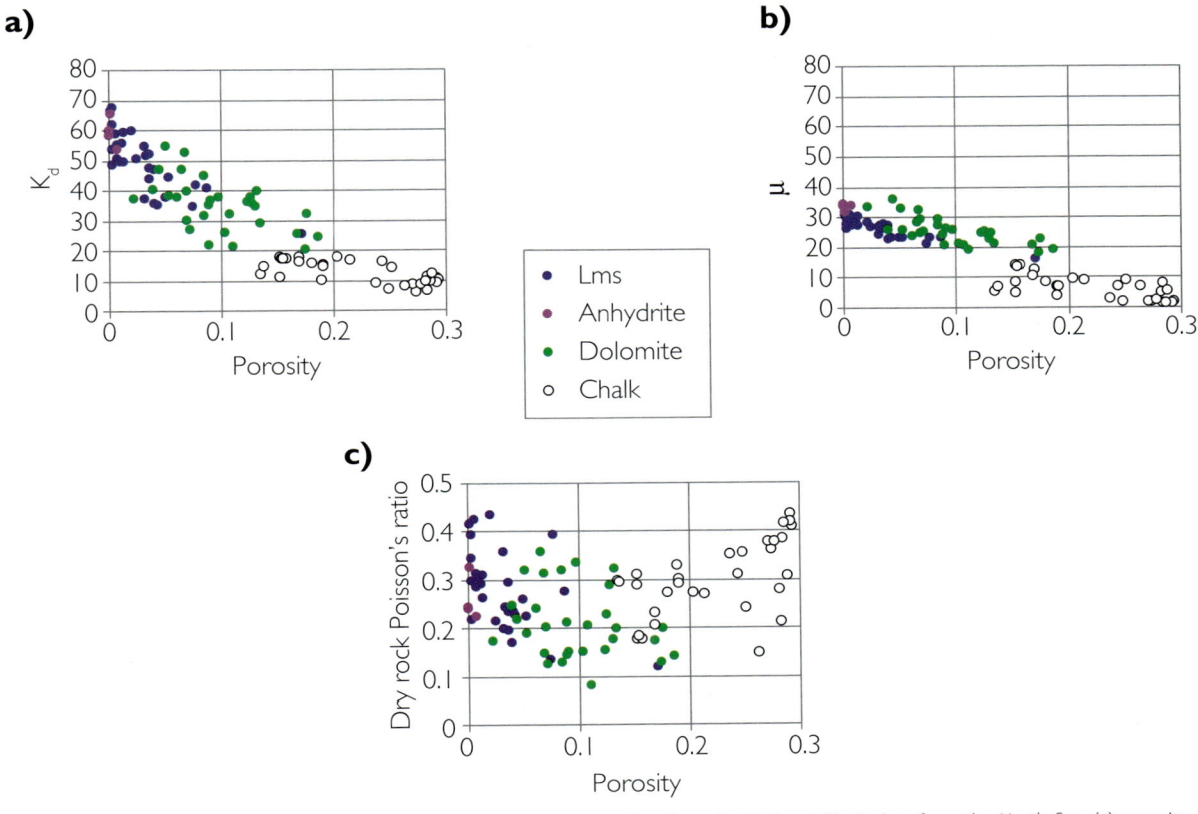

Figure 8.30 Dry rock parameters for selected carbonates (data from Rafavich et al. 1984) and Chalk data from the North Sea; (a) porosity vs dry bulk modulus, (b) porosity vs shear modulus, (c) porosity vs dry rock Poisson's ratio.

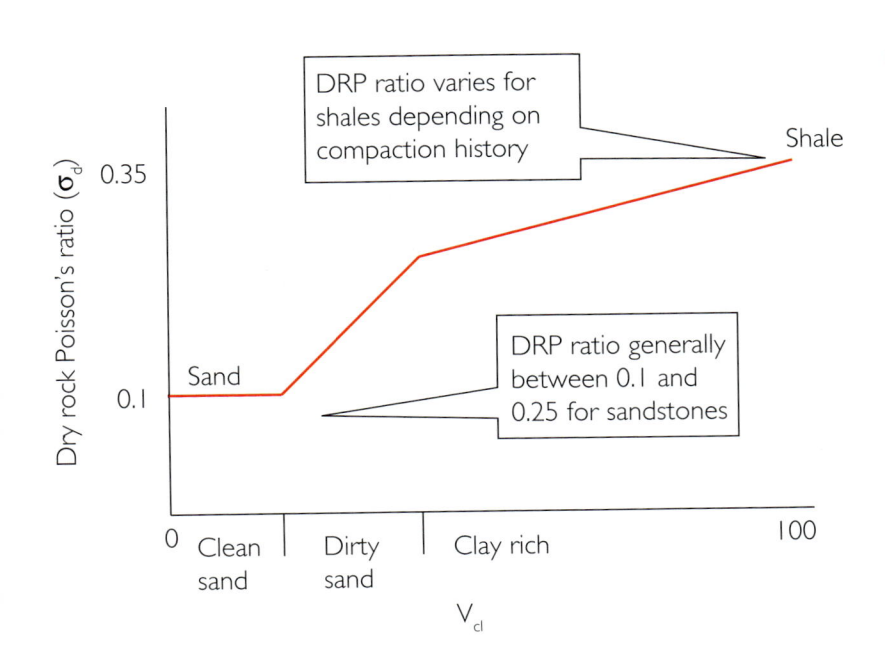

Figure 8.31 A model of dry rock Poisson's ratio as a function of volume of shale (modified after Hilterman, 1990).

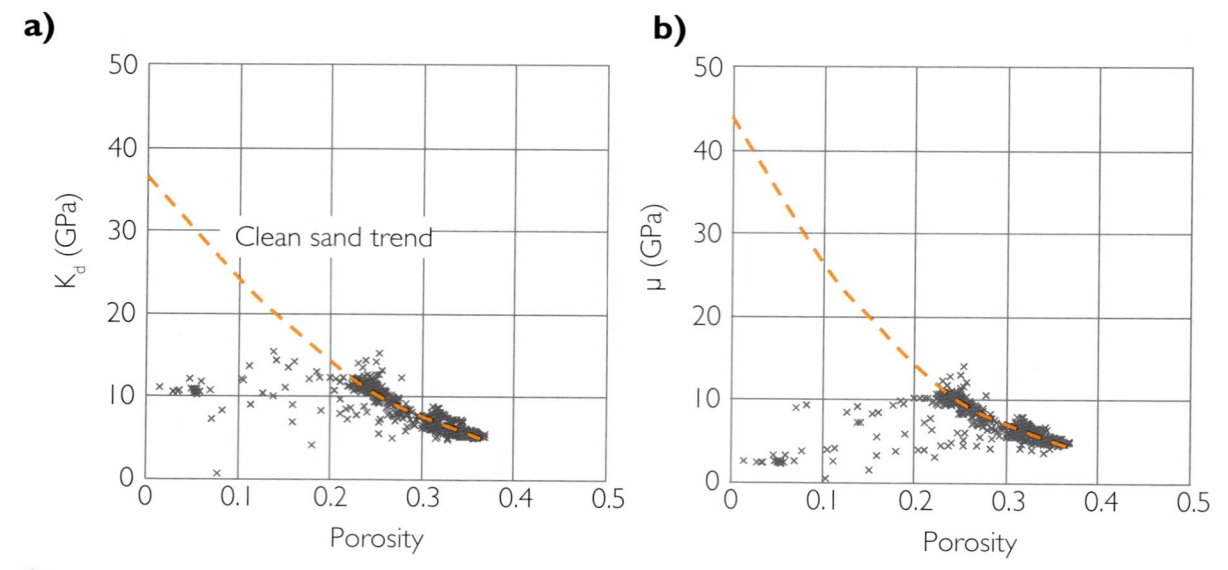

Figure 8.32 Dry rock parameters inverted from log data using Gassmann's equation; (a) porosity vs K_d, (b) porosity vs μ. Functions have been derived (dashed orange curves) based on fitting the clean high-porosity data to the mineral point. For more discussion on the interpretation of the data scatter see Section 8.5.

change in porosity be? This is not a trivial question, essentially requiring a calibrated rock physics model to describe the porosity changes.

For small porosity changes a reasonable approach is to assume that the pore space modulus K_ϕ and dry rock Poisson's ratio σ_d remain unchanged. Given these two constraints the new dry rock bulk and shear moduli for a change in porosity are given by Eqs. (8.26) and (8.37). The assumptions are generally valid for porosity changes in clean sand of up to about 10 porosity units (i.e. percentage points). Modelling changes greater than this requires an understanding of how the rock fabric changes with porosity. One approach for monomineralic rocks is to make use of trends for the moduli against porosity generated from Gassmann inversion to dry rock parameters (Fig. 8.32). Modelling porosity effects related to changing shale or clay content will require a specific rock physics model, such as Raymer–Hunt (Section 8.2.2) or inclusion type models (Section 8.2.7).

8.2.5.4 Dry rock pressure sensitivity

The sensitivity of reservoirs to changing stress is a critical factor in understanding reservoir performance through seismic monitoring. The effect of pore pressure on the fluid is easily calculated, for example through the equations of Batzle and Wang (1992). In clastics porosity reduction or density variations are very small although these need to be taken into account with rocks such as chalks that show significant volume changes during production (MacBeth, 2004). Dry rock moduli can vary significantly with changing effective pressure but unfortunately there are no simple guiding principles to help the interpreter.

The increase of velocity or dry bulk modulus with increasing effective pressure typically has an exponential profile (as shown in Chapter 5), with an initial high rate of change, followed by a more gradual change as the rock compressibility becomes less sensitive to effective pressure. MacBeth (2004) has shown that the variation of dry rock properties with effective pressure, as measured on core samples in the laboratory, is not simply dependent on a parameter such as porosity. Different sandstones show widely different characteristics (Fig. 8.33), related to variations in rock fabric and consolidation history. Given the lack of definitive rock physics relations to guide the modelling of effective pressure and dry rock properties, time lapse feasibility modelling requires laboratory measurements of core velocities at different effective pressures, or at least a good field analogue.

Several authors have published models that can be used to fit laboratory data and guide modelling studies. MacBeth (2004) presented relations in terms of the dry rock bulk modulus and the shear modulus:

Rock physics models and relations

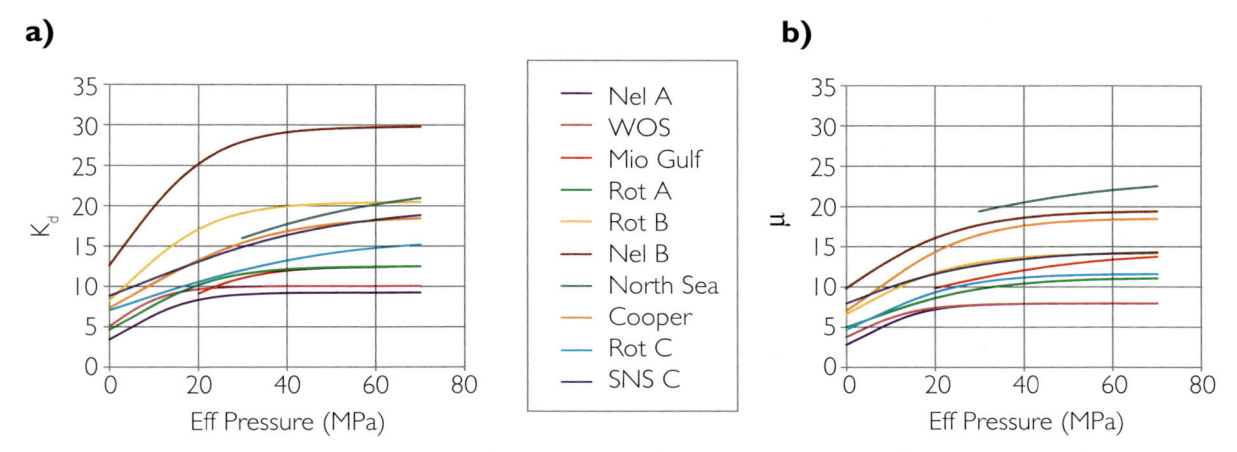

Figure 8.33 Dry rock parameters of sandstones and their sensitivity to effective pressure (trend data from MacBeth (2004).)

$$\kappa(P_e) = \frac{\kappa_\infty}{1+E_\kappa \exp^{\frac{-P_e}{P_\kappa}}}$$

$$\mu(P_e) = \frac{\mu_\infty}{1+E_\mu \exp^{\frac{-P_e}{P_\mu}}}, \quad (8.38)$$

where κ and μ are the bulk and shear modulus at effective pressure P_e, κ_∞ and μ_∞ are the asymptotic high-pressure values, and P_κ and P_μ are constants that characterise the rollover point beyond which the rock frame becomes relatively insensitive to pressure increase.

Vernik and Hamman (2009) published similar models for clean sandstones, parameterised in terms of dry velocities;

$$V_{pd} = V_{p0} + b_p\left[P_e - c.\exp^{(-dP_e)}\right]$$

$$V_{sd} = V_{s0} + b_s\left[P_e - c.\exp^{(-dP_e)}\right], \quad (8.39)$$

where b, c and d are fitting parameters with physical meaning as follows. c correlates with microcrack density; when pressure is high enough to close the microcracks in the rock then c is zero and the equations become linear, with slopes b that are inversely related to the level of consolidation and cementation. Note that d is related to the dominant aspect ratio of the microcracks.

In general it appears that rocks which are overpressured and are relaxed through pressure drawdown will show relatively large dry rock changes consistent with laboratory measurements. However, if a normally pressured reservoir is drawn down then the effect may well be less than that shown by laboratory measurements (C. MacBeth, personal communication). Drawdown in the overpressured situation results in the dry rock frame taking more of the weight of the overburden and thus the frame stiffens. In a normally pressured situation the effect is minimised as the rock frame already supports the overburden. In contrast to this, where pore pressures increase during hydrocarbon production (e.g. around water injectors), large changes in seismic amplitude have been observed (e.g. Sayers, 2006).

Whilst laboratory data are important in assessing the dry rock sensitivity to effective pressure variations it should be noted that even when these data are available there is always some doubt as to the relevance of high-frequency measurements on small samples (micro-fabric measurement) to a field-wide response under the influence of subsurface stress regimes. For example, the response in the low effective pressure regime may be dominated by microcracks which may be natural or generated through pressure release of core material. The reader is referred to McCann and Southcott (1992) for a review of the issues in laboratory measurements.

8.2.6 Contact models

Contact models are based on mathematical principles of the interaction of granular materials and are applicable to sandstones. They are generally constructed by determining high- and low-porosity dry rock end members which are then interpolated, often using modified Hashin–Shtrikman (1964) bounds.

Saturated elastic moduli are then calculated using Gassmann's (1951) equation. These models effectively address the issues of effective pressure on unconsolidated grain packs and the effect of cement in stiffening the rock. When fit to log data it is possible to infer whether the sand is unconsolidated or cemented. Not only is this useful for seismic lithofacies characterisation but it may also be useful in sanding assessment for reservoir production.

There are two commonly used contact models, the 'friable sand' model and the 'cemented sand' model. These two models were derived by Dvorkin and Nur (1996) from a study of two sets of laboratory data on sandstones from the North Sea, one from the Oseberg Field and one from the Troll Field. Thin-section images showed slight quartz cementation of the grains in the Oseberg samples, whereas cementation was absent in the Troll samples.

With the 'friable sand' or 'uncemented sand' model, a high-porosity dry rock end member is determined for a random pack of identical spherical grains at critical porosity. Hertz–Mindlin theory (Mindlin, 1949) is used. The lower Hashin–Shtrikman bound then interpolates between this point and the mineral moduli at zero porosity. This interpolation represents the deterioration in porosity due to decrease in sorting. The model inputs are the solid phase bulk and shear moduli, critical porosity, effective pressure and the average number of contacts per grain (the 'coordination number'). Empirical models relate the coordination number to porosity. The 'friable sand' model can be effectively used to derive a constant clay model for sandy shales by substituting a much larger critical porosity (60%–70%) and a Reuss mix of clay and silt components (assuming the silt is suspended in the clay) to derive the effective solid moduli (Avseth et al., 2005).

The 'cemented sand' or 'contact cement' model assumes that the porosity reduces owing to the uniform deposition of cement on the surface of the sand grains. Only a small amount of cement deposited at grain contacts is required to significantly increase the stiffness of the rock. The mathematical solution for this scenario was determined by Dvorkin et al. (1994). The model inputs are solid phase and cement phase bulk and shear moduli, and coordination number (representing the radius of contact of the cement layer). In the Oseberg samples clay cement gave slightly lower velocities than quartz cement. There is no effective pressure dependency built into the model.

Figure 8.34 illustrates examples of friable and cemented sand trends in data from the North Sea and Gulf of Mexico. The similarity of the trends in the two basins is striking. However, it should be remembered that there a large number of variables (i.e. sorting, shaliness, effective pressure, and amount of cement) that may give rise to similar values of velocity for a given porosity. The friable sand and cemented sand models are not necessarily mutually exclusive. For example, a combination of the friable and cemented sand models can potentially describe the situation where a sorting trend has a constant value of cement (i.e. the constant cement model) (Avseth, 2000; Avseth et al., 2005) (Fig. 8.35). It is recommended that the reader consults Avseth et al. (2005) for a thorough discussion of a variety of models describing specific sand/shale scenarios, including models to describe the dispersed clay and laminated shale effects described in Chapter 5.

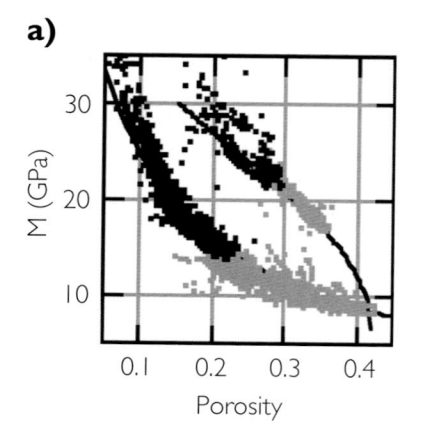

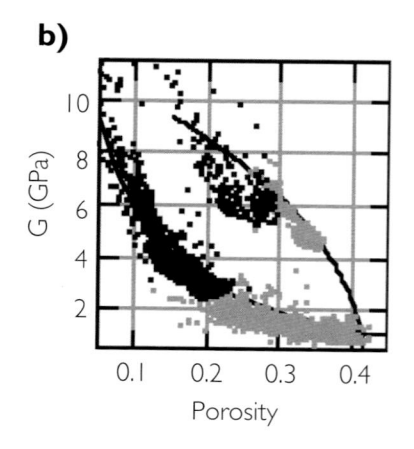

Figure 8.34 Modulus versus porosity crossplots for North Sea (grey) and Gulf Coast (black) wells; (a) porosity vs M (ρV_p^2), (b) porosity vs G (ρV_s^2). The curves are from the theoretical contact cement model (upper branch) and uncemented model (lower branch), after Dvorkin et al., 2002.

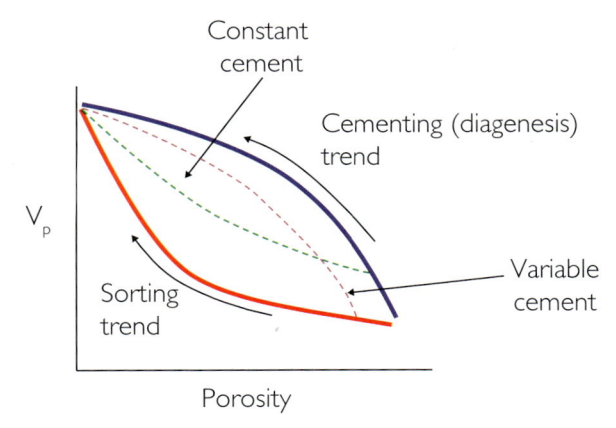

Figure 8.35 Schematic illustration of sorting and cementing trends in clean sandstones.

8.2.7 Inclusion models

Inclusion models approximate the rock as an elastic solid containing inclusions (i.e. pores). The Xu–White (1995, 1996) model is an excellent example of an inclusion model that is readily applied to log data. The model depends on theory developed by Kuster and Toksöz (1974) for a two-phase medium, relating porosity and pore aspect ratio to P and S velocities. The reader is recommended to consult the work of Mavko *et al.* (1998) for a detailed discussion of the Kuster–Toksöz model and its context within the broad range of rock physics models. The key aspect of the Xu–White pore model is that pores are split into clay-related and sand-related pores and each has a different pore aspect ratio (the ratio of shortest axis to longest axis). Clay particles are expected to create pores with low aspect ratios, and these crack-like pores will have limited stiffness, whereas sand grains form pores with larger aspect ratios, and these sub-spherical pores will have high stiffness.

The original Kuster–Toksöz theory required the pore concentration to be dilute, such that there is no interaction between pores. In order to avoid this restriction, Xu and White use an effective medium approach, in which the properties of the medium are calculated in stages; at each step a proportion of the porosity is introduced, small enough for the dilute porosity condition to be satisfied, and the output medium properties are used as the input to the next stage of porosity introduction. The outputs from Kuster–Toksöz modelling are the dry rock moduli which can be saturated using Gassmann's equations (Fig. 8.36). The saturated moduli can also be output from the Kuster–Toksöz model but owing to the isolated nature of the pores within the solid this represents a high-frequency solution.

The effect of pore aspect ratio using the Xu-White model is shown in Fig. 8.37. For reference, a curve representing Wyllie's equation is shown, with a good fit for pores with an aspect ratio of about 0.09. This is consistent with the knowledge of consolidated sands in which most of the crack-like pores have been closed during compaction and diagenesis. Whilst direct measurement of pore aspect ratios is possible, for example from thin sections, in practice the pore aspect ratios input to the Xu–White model are effectively fitting parameters, tuned to give results in agreement with observed velocity logs. It should be noted that the mineral modulus of clay is also a variable in the fitting procedure.

Xu and White found typical values of 0.02–0.05 for clay-related pores, 0.12 for sand-related pores, and 0.1 for carbonate-related pores. Typically, single best-fit values for the sand aspect ratio and the clay aspect ratio are determined over a logged section of a few hundred metres. Figure 8.38 shows an example, where there is a good fit to the observed P wave velocity log. In this case there are no logged shear data available, but the Xu–White prediction for the shear log is higher than the Greenberg–Castagna prediction (black curve in shear velocity track) which has been proven to work in this area. After obtaining the required estimates of the aspect ratios from the log fit, the model can be used to predict the effect of varying clay or porosity; it could also be used to replace defective sections of the sonic logs (Section 8.4.3).

As a velocity model, the Xu–White (1995) model is reasonably straightforward and provides insight into the reasons for velocity change. For example it predicts the expected behaviour of dispersed shale in sands. Yan *et al.* (2007) have suggested improvements to the estimation of sand pore aspect ratios by using a relationship with porosity and shale content. Keys and Xu (2002) have shown that it is possible to make an accurate approximation to the Xu–White method which is computationally less intensive, by making certain assumptions about the dry rock properties. The Xu–White (1995) model is, however, purely volumetric, taking no account of where the clay or shale is located within the rock. The reader is referred to Sams and Andrea (2001) for a discussion of the effects of clay distribution in the context of this type of velocity modelling.

Rock physics for seismic modelling

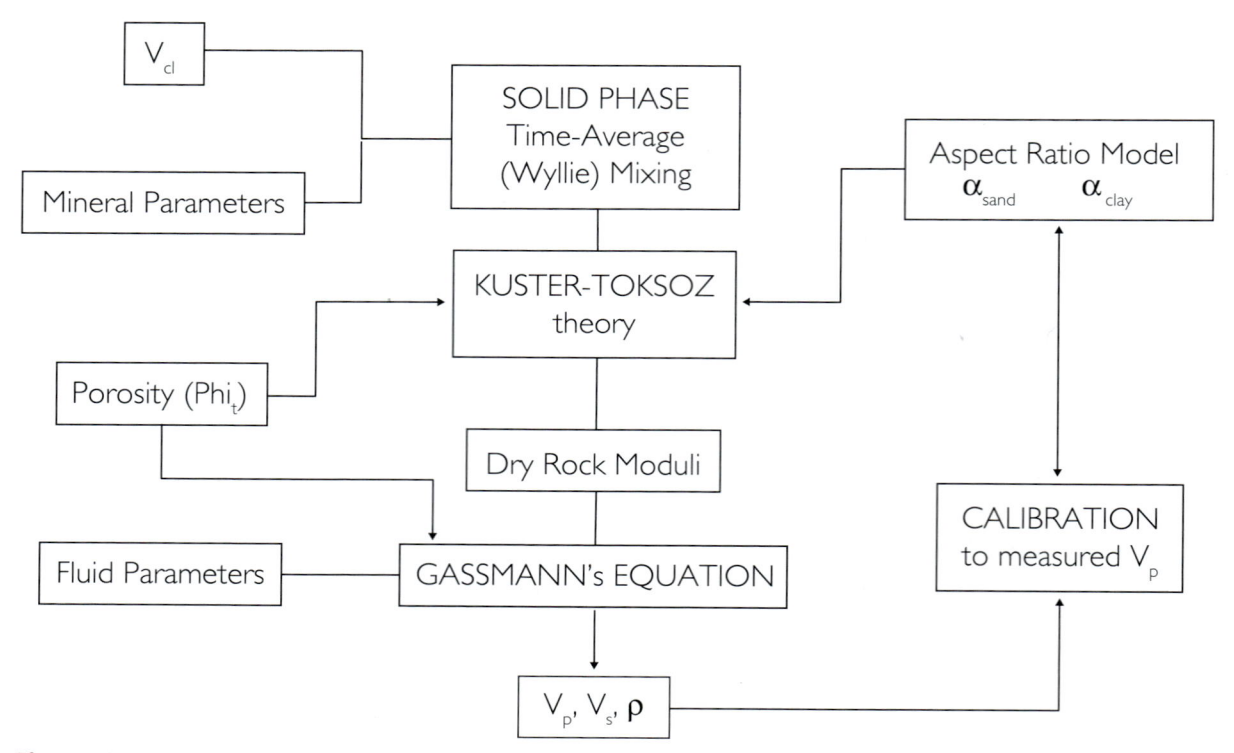

Figure 8.36 Schematic view of the Xu–White (1995) model.

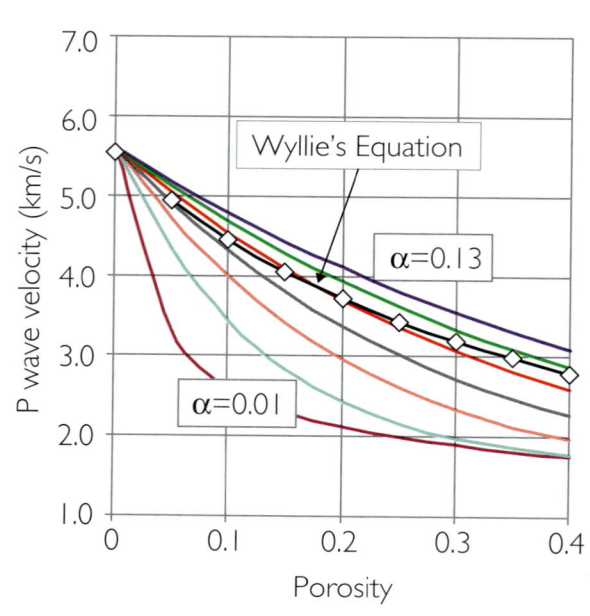

Figure 8.37 Xu–White (1995) model results for a clean sand with varying pore aspect ratio.

Owing to the possibility of parameterising different porosity types, the Xu–White model has found application in carbonates and an extension of the Xu–White model has been presented by Xu and Payne (2009). This proceeds by dividing the total porosity into four types: (1) clay-related pores, as in the Xu–White model, but generally only a small fraction of the porosity as most carbonates are quite clean; (2) interparticle pores, as in the Xu–White model; (3) microcracks, which have low aspect ratios and make the rock weaker; and (4) stiff pores, rounded moldic or vuggy porosity, typically formed by a dissolution process, with high aspect ratio. These pore types are added into an effective medium model in the same way as the Xu–White model. Figure 8.39 shows an example of calculated P wave velocity as a function of porosity and pore type. As with Xu–White, the model predicts rock frame properties and Gassmann substitution is used to calculate the fluid-filled response, with the following refinement. The microcracks have very low local permeability because of their very small size. Pore fluid tends to be trapped in the microcracks and is unable to reach equilibrium as the seismic wave passes through, thus violating the assumptions of the

Requirements for a rock physics study

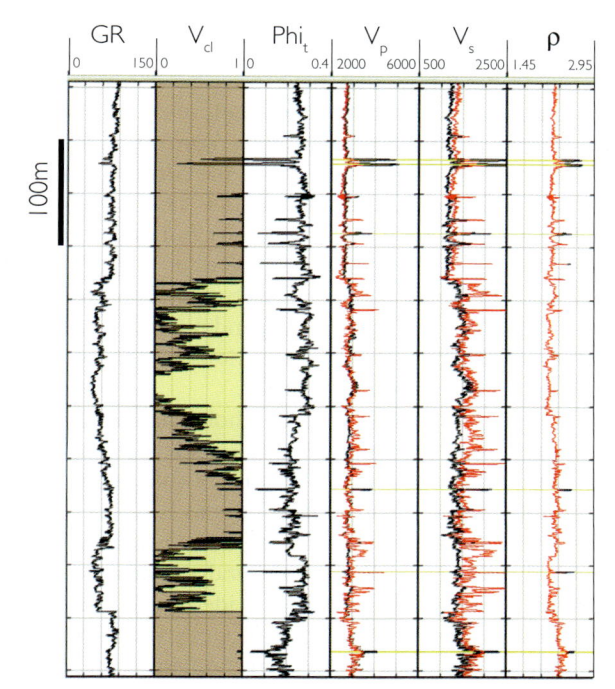

Figure 8.38 Application of the Xu–White model to a 600 m logged sand/shale section: black = log curves, red = Xu–White predictions.

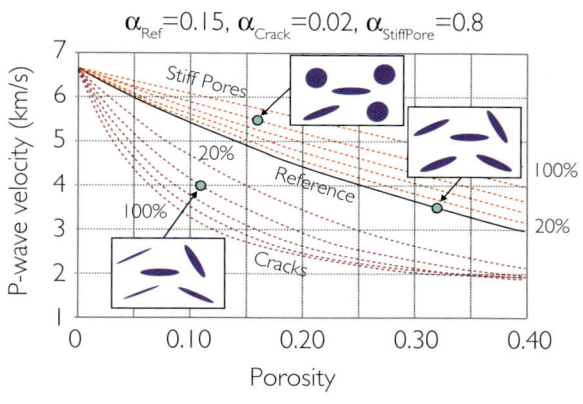

Figure 8.39 Predicted effect of pore type on P wave velocity in carbonates (after Xu and Payne, 2009). The matrix is assumed to be calcite and α is the aspect ratio. The reference curve ($\alpha = 0.15$) represents a system with only interparticle pores, whereas the curves below the reference represent systems with increasing fractions of crack-type pores ($\alpha < 0.15$) and those above it represent increasing fractions of stiff pores ($\alpha > 0.15$).

Gassmann method. Therefore, the microcracks are added into the effective medium model as isolated water-wet pores, and only the interparticle and stiff pores are added by the dry rock Gassmann fluid substitution route. This gives a better fit to observation than applying the dry rock Gassmann calculation to all of the carbonate porosity.

8.3 Requirements for a rock physics study

Although rock physics studies are becoming more routine, the first task for the interpreter is to locate all the necessary information. This usually involves talking with petrophysicists, engineers and geologists.

8.3.1 Data checklist

The following well data are typically required to perform a rock physics study.

- Checkshot data.
- Hole deviation data.
- Stratigraphic tops.
- Composite logs.
- General well parameters – water depth, elevation of log datum.
- Reservoir parameters
  . pore pressure,
  . temperature.
- Drilling mud
  . type – oil based/water based,
  . if water based – salinity of mud.
- Fluid parameters
  . salinity of brine ppm or R_w measurement,
  . gas index of brine 0–1,
  . oil *API*,
  . oil *GOR* (scf/stb or litres/litre),
  . gas gravity.
- Log data in depth (with datum information)
  . sonic, shear sonic, density, other curves (gamma ray, caliper, neutron, resistivity (shallow, medium and deep)).
- Petrophysical information
  . porosity (effective and total), saturation (in the uninvaded reservoir (S_w) and in the invaded zone (S_{xo}), mineral fraction (e.g. shale content),
  . core analysis data if available
    – grain densities, porosities, etc. (with associated report).

In terms of log data, both compressional sonic and density are a minimum requirement. If only a density or sonic log is available then the amount of confident data analysis that can be done will be minimal. This would equally apply if there is significant editing and log prediction required.

8.3.2 Acoustic logs

Given the importance of sonic and density logs to seismic modelling the following presents a discussion of how the tools work as well as practical considerations important in data analysis. Much more detail can be found in petrophysics textbooks (e.g. Ellis and Singer, 2007; Rider and Kennedy, 2011).

8.3.2.1 Density logs

Density is measured from the interaction of gamma rays with the formation. As they pass through the formation, gamma rays in an appropriate energy range are scattered by the electrons in the atoms of the formation material; the resulting attenuation is proportional to the bulk density (after correction for the anomalously high scattering ability of the hydrogen atoms present in water). The logging tool contains a gamma ray source (usually ^{137}Cs) with two or more detectors at short and long spacing from the source. The density tool is pressed against the borehole wall with an arm that extends from the back of the tool, keeping it in position (Fig. 8.40a). In this mode the primary long spaced detector records scattered gamma rays that have travelled through the formation. Usually contact will not be perfect, however, with a thickness of drilling mud or mudcake (clay particles deposited from the drilling mud) between the tool and the formation. This 'stand-off' effect leads to low values of density but is corrected with the use of the short spaced detector that primarily reads the mudcake. The density correction (drho) may be positive or negative depending on the nature of the drilling mud and corrections of up to +/− 0.15g/cc are usually to be trusted.

Display of caliper (i.e. borehole diameter) and density correction logs gives an indication of the quality of a density log. Figure 8.40b shows an example of a log with both good and bad hole. The lowermost sand (marked in yellow) has good hole with the caliper fairly constant and the density correction very low. Poor hole is evident in the uppermost sand, where the caliper indicates a widened hole and the drho log has a large correction; in this case the density log is probably unreliable and needs editing (Section 8.4).

In interpreting density data, it is important to remember that the depth of penetration into the formation is low (typically 5–10 cm), so the tool is

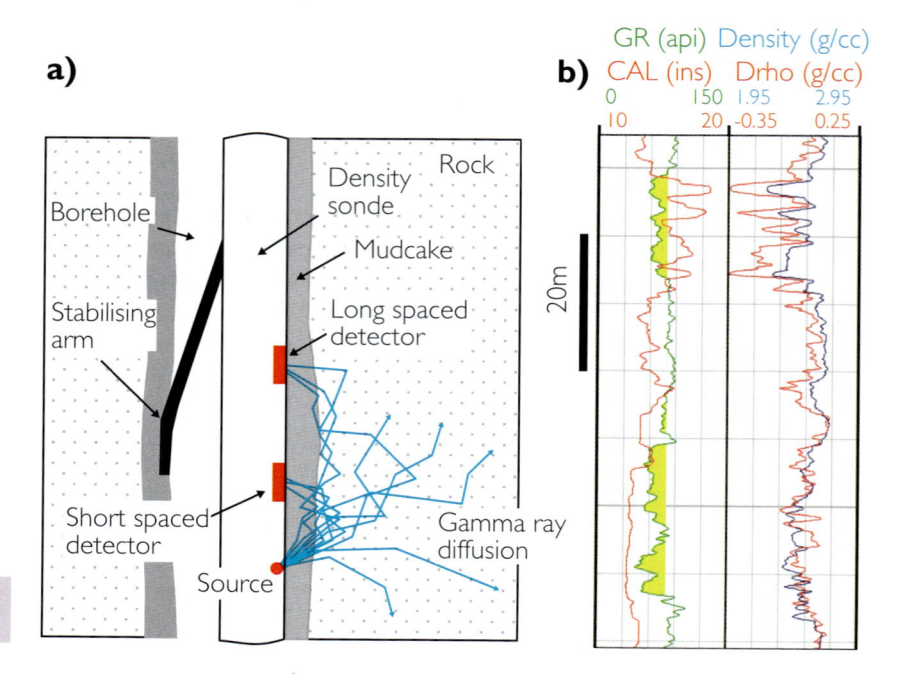

Figure 8.40 Density logging; (a) tool configuration in the borehole (modified after Tittman and Wahl, 1965), (b) an example of compensation in bad hole.

reading in the zone close to the borehole wall where drilling fluids may have invaded permeable formations, displacing the original fluids and perhaps causing significant changes in measured values. For the purposes of seismic modelling these invasion effects should be evaluated and corrections made (Section 8.4.4).

8.3.2.2 Sonic logs

The sonic logging tool is a tool which is run centred in the borehole and stabilised by arms that extend out to the borehole wall. It is constructed with an acoustic source operating at around 3–30 kHz and an array of receivers that detect sonic energy refracted along the borehole wall. There are a range of different types of sonic logging tools, essentially differing in the number and separation of receivers and undergoing significant development over the decades.

The basic elements of sonic logging are shown in Fig. 8.41. Energy from a monopole transmitter (i.e. emitting a pulse of energy isotropically in all directions) generates three distinct arrivals: a low amplitude compressional wave, followed by a larger amplitude shear wave, followed by a Stoneley wave (energy travelling mainly in the borehole) of even larger amplitude. Early sonic tools simply detected compressional wave arrivals (from the first negative excursion of the wave train) and translated the time difference between near and far receiver arrivals into sonic slowness (i.e. μs/ft). Near and far receivers were spaced 3 ft and 5 ft from the transmitter respectively. To make corrections for poor hole conditions two sets of transmitter–receiver arrays were incorporated into the tool and the measurements related to each were averaged. This is what is usually described as the 'conventional' borehole compensated sonic.

In uneven boreholes, for example due to altered shale or washouts, it was found that using a 'long spaced sonic' tool with two transmitters and two receivers each with 2 ft spacing and 8 ft minimum separation between transmitters and receivers, together with a more sophisticated depth referenced compensation technique, gave superior results. The next development was to lengthen the source–receiver separation (up to around 15 ft), introduce a large array of receivers and record the wave train in a time window, giving rise to 'monopole array sonic' tools. Slownesses are derived from the waveforms principally but not exclusively by using semblance/coherency techniques (Fig. 8.42).

A limitation of these tools, however, is that if the shear wave is slower than the mud arrival then no shear arrival is generated. In order to measure shear velocity in 'slow' formations, the dipole source was developed, which generates a 'flexural' wave, essentially an undulation of the borehole wall. This wave shows significant dispersion, i.e. velocity varying with frequency, and in waveform processing a correction is applied to the flexural velocity based on a modelled dispersion curve.

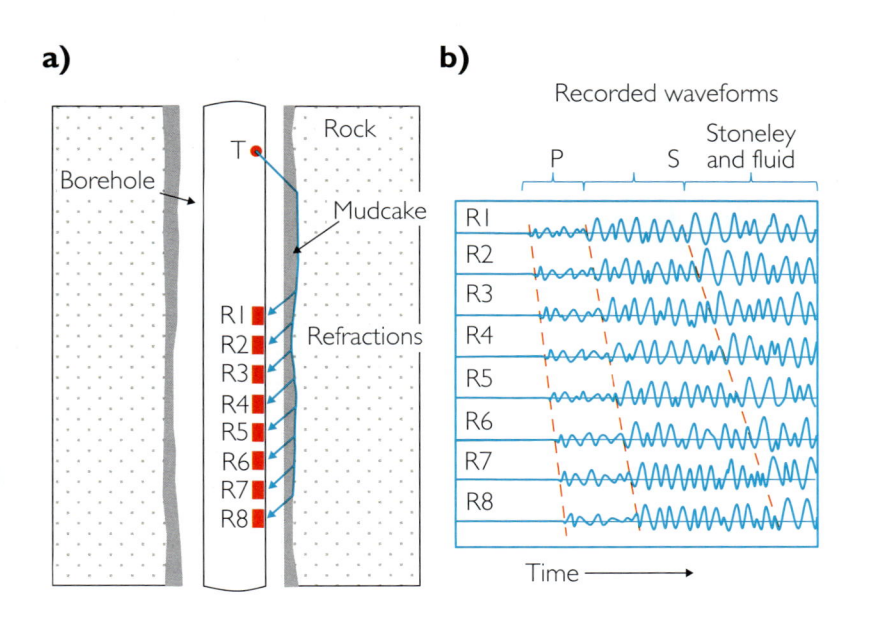

Figure 8.41 The basic elements of sonic waveform logging.

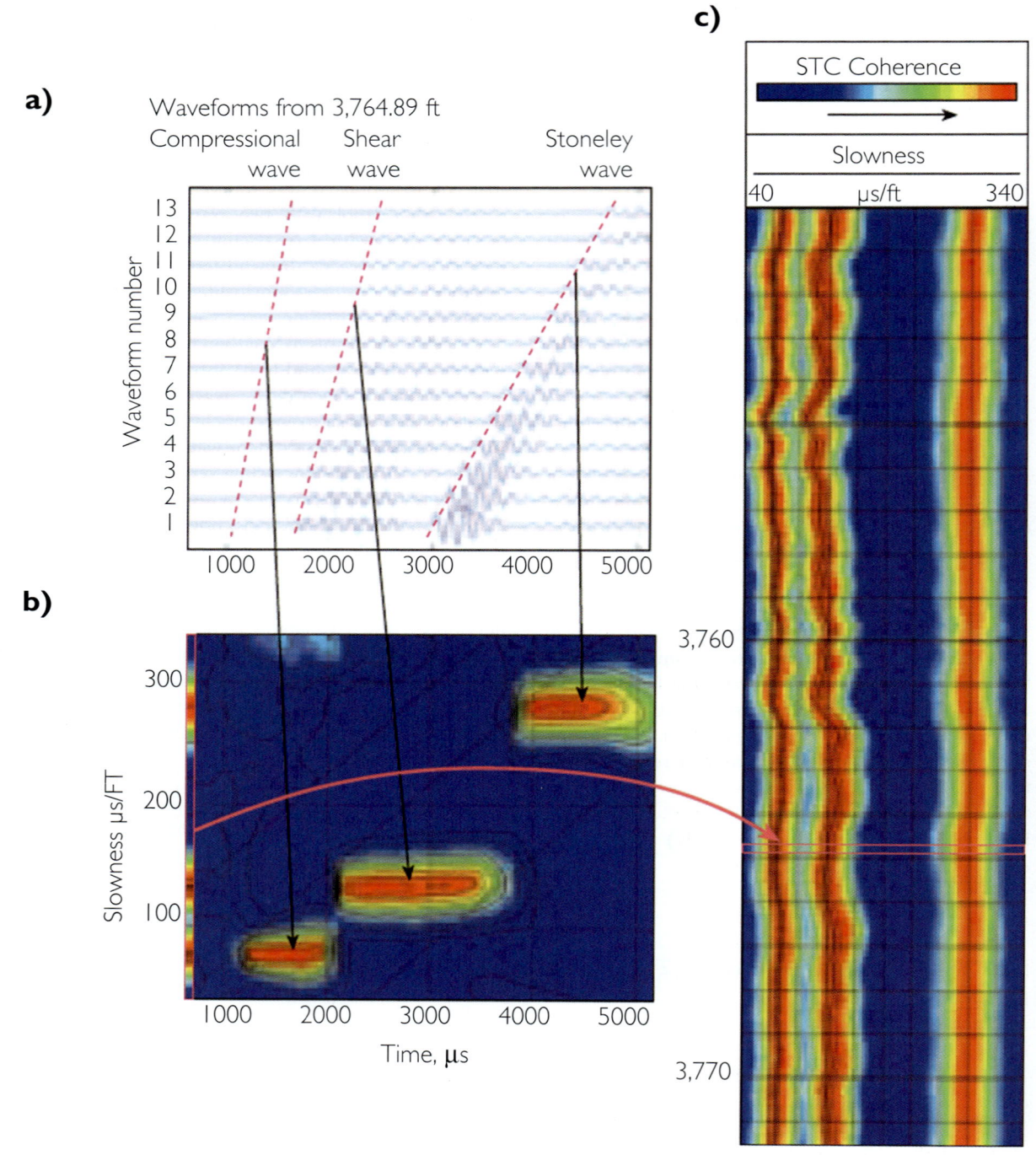

Figure 8.42 Slowness-time-coherence processing of sonic waveforms. Slowness estimates are made by scanning sonic waveforms over time windows and over a range of angles to maximise coherence; (a) waveforms at a particular depth, (b) coherence (red colour = high coherence) shown on time vs slowness (μs/ft) plot, (c) coherence shown on depth scale (after Haldorsen et al., 2006). Figure copyright Schlumberger. Used with permission.

It is not easy to say what the depth of investigation is for the sonic measurement. Although the arrival time of the first signal is along the path with the shortest travel time, the depth of penetration depends on the velocity profile close to the borehole wall as well as the source–receiver separation. For conventional borehole sonic tools the penetration is limited, probably around 5 cm. It may typically be in the range 15–40 cm for longer spaced sonics and dipole sonics.

Modern sonic logging tools are becoming ever more sophisticated and increasingly being used in the understanding of fractures and formation geomechanics. A benefit of the dipole source is that it is directional and can therefore be used to investigate anisotropy. In cross dipole mode the sonic tool can derive information on the anisotropy of the formation based on comparisons of fast and slow shear waves (discussed in Chapters 5 and 7; e.g. Haldorsen *et al.*, 2006). The latest sonic scanner technology is effectively a 3D acoustic imaging tool which can be used not only to measure isotropic velocities but also to characterise intrinsic and stress-induced anisotropy.

8.4 Data QC and log edits

Making sure that the log data are of good quality is a key part of using rock physics to create seismic models. The petrophysicist will typically check the depth registration of various logs and make standard corrections for environmental effects on tools related to factors such as stress, mud weight, pore pressure, temperature and speed of logging. The next step is to identify bad hole effects and other errors followed by replacing bad logs or missing sections.

In some situations there may be a need to scale log responses to maintain consistency across a group of wells. This is most likely to be true of gamma ray logs but caution is advised when considering this process for sonic and density logs. They are compensated tool readings that in good hole have a high degree of accuracy. Applying such scaling to density and sonic logs may remove real differences. A useful summary of the key steps in log quality assessment has been given by Yan *et al.* (2008).

Bad hole effects such as washouts are fairly straightforward to identify from the caliper and density correction logs whilst the effects of filtrate invasion on the density and sonic signature are more subtle. Often the process is iterative, involving comparisons of logs from different wells and pseudo-logs generated from transforms, rock physics models or rock property templates. These comparisons are commonly made in relation to a key reference, such as depth below mudline or stratigraphic zone.

Sometimes it is difficult to identify where log QC stops and log interpretation begins. Indeed these processes are often intertwined. Once a petrophysical evaluation has been generated rock physics models can be fit to the log data, providing a means of evaluating log quality and identifying potential errors. This can be done by comparing predicted and measured curves in the depth domain or/and in the crossplot domain.

A useful tool for both QC and interpretation purposes is the 'rock physics template'(e.g. Ødegaard and Avseth, 2004). Rock physics templates are essentially crossplots such as porosity vs velocity or acoustic impedance vs V_p/V_s with the background rock physics model annotated. Shale and sand end members are indicated as well as hydrocarbon equivalents. An example is shown in Fig. 8.43. The template provides a useful context for understanding the various effects of changing porosity, fluid content, shale content, cement volume and effective pressure described in Chapter 5.

8.4.1 Bad hole effects

Adverse hole conditions (borehole washouts and damaged formations) can lead to bad density and sonic logs, which are a frequent cause of poor well ties. Log editing is therefore an important element of the well tie process. Typical causes of bad hole conditions are the presence of incompetent formations and chemical interaction of mud filtrate and clays. In many instances the caliper log is a key to recognising bad hole. The density tool in particular has low tolerance of poor hole conditions. Figure 8.44 shows an example of zones of caving shales that have led to errors in the density log. In this instance a corrected log has been calculated by using sand and shale densities, mixed in the proportions indicated by the V_{cl} log. Such a scheme is valid of course only where the fluid fill is invariant and there is no depth dependency in the density and sonic measurements.

In contrast to the density log, wireline sonic logs are usually quite tolerant of poor borehole conditions. However, in very bad hole the sonic signal, particularly with conventional borehole sonic logs, can be severely attenuated. This results in the amplitude of the first arrival being too low to trigger the receiver,

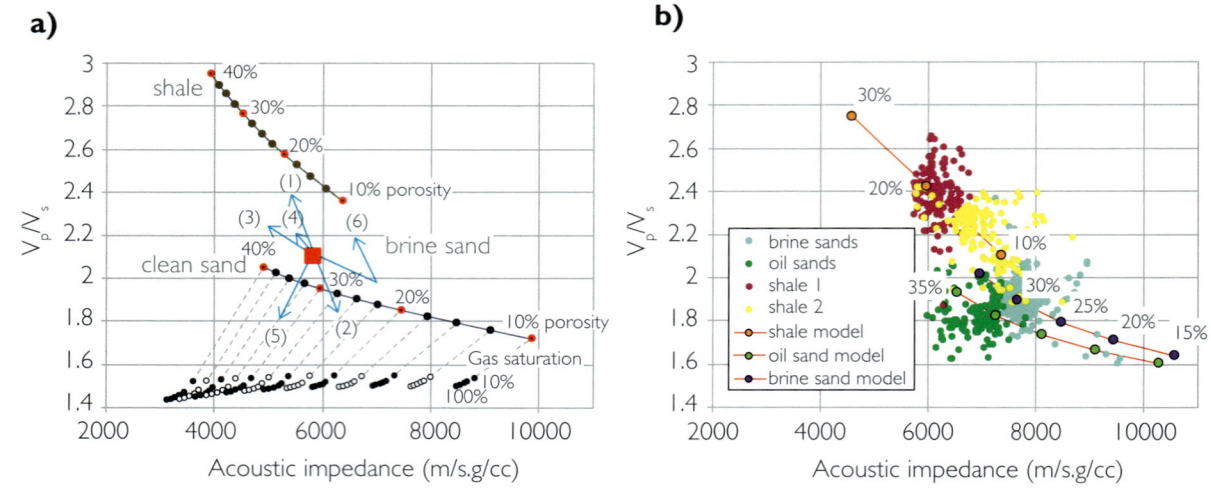

Figure 8.43 Rock physics template examples; a) a template crossplot showing the vectors associated with certain elastic changes relative to a reference brine sand: (1) increasing laminar shale, (2) increasing cementation, (3) increasing porosity, (4) decreasing effective pressure, (5) increasing hydrocarbon saturation, (6) increasing dispersed shale content (modified after Ødegaard and Avseth, 2004), (b) an example of log data plotted from different lithofacies zones, together with rock physics model calibration trends.

which instead records higher amplitude later arrivals. This 'cycle skipping' can lead to sonic transit times that are too high (i.e. velocities too low) (Fig. 8.45).

In Fig. 8.45 the effect of cycle skips in bad hole has been corrected by defining a relationship between deep conductivity (i.e. the reciprocal of the deep resistivity log) and sonic slowness (Burch, 2002). Modern dipole and other advanced sonic tools are more tolerant of bad hole than older designs (Fig. 8.44). Sometimes sonic logs also show noise spikes where the receivers are triggered by, for example, tool movement. Log spikes deserve careful attention to distinguish noise from genuine responses for example related to thin limestone stringers in a shale section. Comparison with other logs such as density is usually helpful. It is also worthwhile checking for potential bad hole effects on the seismic drift curve generated during log calibration (Chapter 4). The bad hole will show up as zones in which sonic velocities are slower than seismic velocities.

8.4.2 V_p and V_s from sonic waveform analysis

Since the early 1990s shear wave logging has become fairly routine. As discussed in Section 8.3.2.2, compressional and shear logs are derived from an analysis of sonic waveforms. In general, rig-based processing products are not the best quality. Final logs are generated from detailed analysis in a dedicated processing centre. In projects where there are a number of wells with variable age shear logs, interpreted by different contractors, it is unlikely that the interpretations will be consistent. In such a situation it is well worth considering re-processing all the wells to obtain a consistent dataset prior to the start of the project.

An initial QC of compressional and shear logs should include checking the Poisson's ratio or V_p/V_s log for anomalously low or high values. Other errors can be picked up on V_p vs V_s crossplots but not all. If the semblance data are available it is worth reviewing these to check the picks. Monopole tools have the limitation that where the shear velocity of the formation is less than the compressional velocity of the drilling mud ('slow' formations), no shear wave is recorded and instead the tool will tend to record the mud velocity. This is easily detected on a V_p vs V_s crossplot, where the shear velocity will be roughly constant and higher than that expected from empirical trends such as Castagna's mudline or sandline (Fig. 8.46).

The mud arrival problem in slow formations was largely solved with the development of the dipole sonic. Figure 8.47a shows an example of dipole data in shales from the mid 1990s, with the shale data mostly located where they might be expected, between Castagna's mudline and sandline. However,

Data QC and log edits

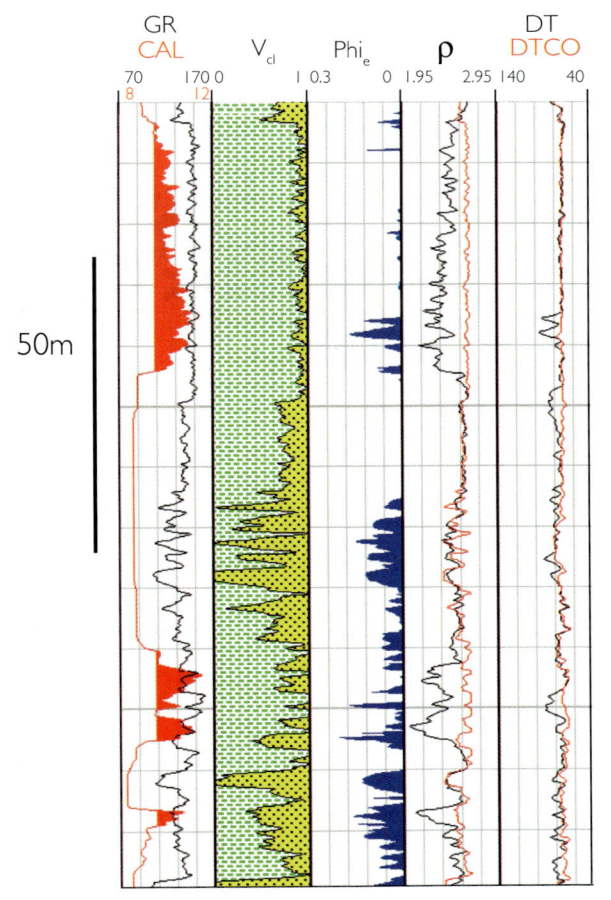

Figure 8.44 Example of shale 'washout' zones on density and sonic logs. Predicted density log is shown in red in column 4. Note how the dipole compressional sonic log (DTCO) is less affected by the washouts than the conventional sonic (DT).

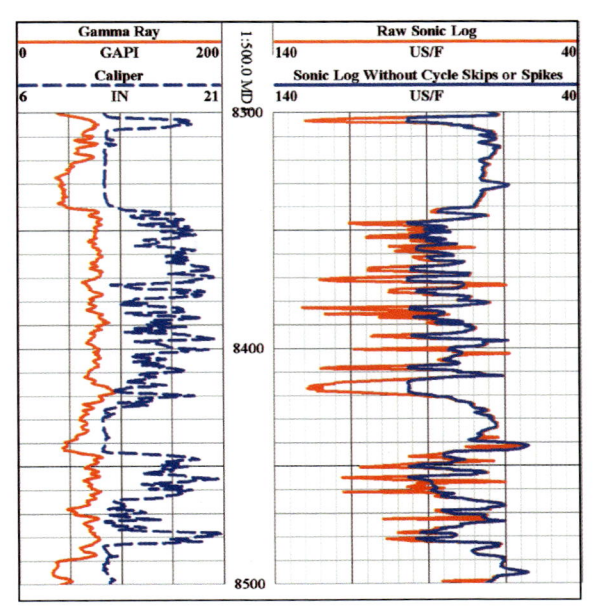

Figure 8.45 The effect of cycle skips on conventional sonic logs (after Burch, 2002; AAPG ©2002, reprinted by permission of the AAPG whose permission is required for further use). The raw sonic log (red curve, right track) shows cycle skips and noise correlating with an enlarged borehole (note caliper log, blue curve left track).

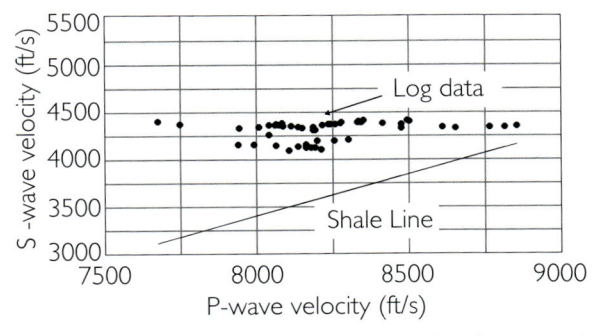

Figure 8.46 Example of monopole array sonic data showing mud arrivals masquerading as shear arrivals in a 'slow' formation.

there is still some interference from mud arrivals at velocities around 4500 ft/s. This would require editing or possibly more appropriately to use the dipole data to determine linear empirical fits for V_s prediction. Figure 8.47b shows an example of poor dipole data in sands, with mud arrival energy and significant scatter in the shear direction. These data are largely unusable but fortunately the fidelity of modern sonic tools is such that this situation does not often arise.

There are some sonic effects that cannot be recognised simply by looking at crossplots. For example, interference from Stoneley waves can bias the V_s log towards lower velocities. Figure 8.48 shows an example. It is also possible for a positive bias to be introduced owing to interference of the shear wave with compressional arrivals. Such effects can only be picked up by detailed phase analysis of the waveform data (e.g. Kozak *et al.*, 2006).

8.4.3 Log prediction

A variety of approaches can be taken to predicting sonic and density logs with a number of log transforms and rock physics models having been described in Section 8.2. Log predictions can provide a useful

Rock physics for seismic modelling

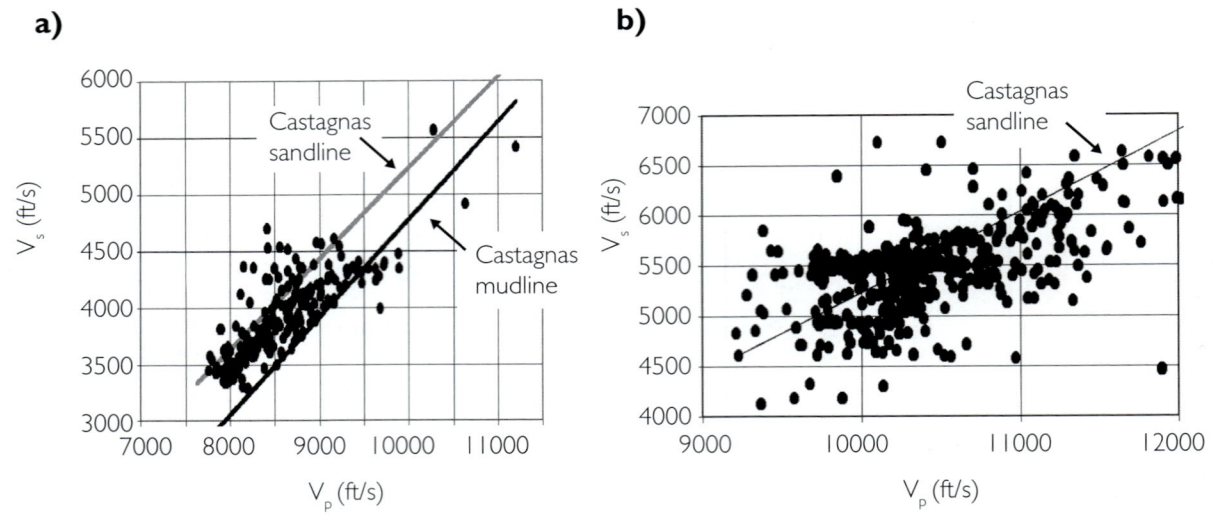

Figure 8.47 Crossplots of velocities from sonic waveforms; (a) an early example of dipole sonic data in shales with shear velocities lower than the compressional velocity of mud. Note the presence of some mud arrival energy at around 4500 ft/s, (b) an example of dipole sonic data from sandstones showing mud arrival energy and significant noise.

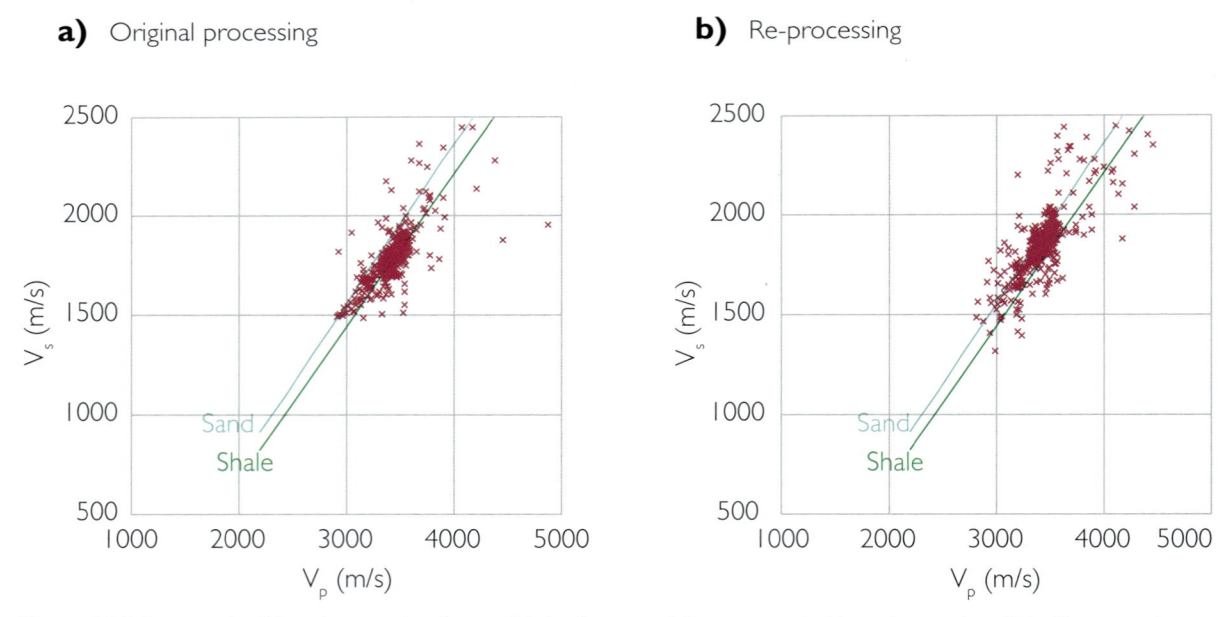

Figure 8.48 An example of Stoneley wave interference, biasing interpreted shear wave velocities to lower values. This effect can only be established with detailed phase analysis.

means of log quality control (i.e. identifying poor sections of log) as well as being a tool for infilling missing sections.

A key issue in the application of both transforms and rock physics models is the need to establish local calibration before applying them. This calibration step clearly requires other well data to be available with the same geology and a similar effective pressure regime. It is important that the geological and rock physics context of transforms and models is correctly understood before application. In particular, rock types and facies need to be discriminated adequately before applying transforms or models. It may be possible, for example, to address sands and shales in the same model (e.g. Xu–White, 1995) but lithologies such as salt, coal and carbonate layers will need to be addressed separately.

A common approach to log prediction is to use multi-linear regression methods (e.g. Hampson et al., 2001). Box et al. (2004) and Box and Lowrey (2003) describe a process in which log editing is approached via a multi-linear regression of data from a large number of wells including all relevant logs and variables (sonic, density, gamma ray, resistivity, depth and pressure). Whilst using several logs rather than one to make a prediction can improve the results the process needs a thorough sense check. There is usually a fairly obvious single transform relationship that forms the basis of the prediction and the question is whether additional logs will help. Figure 8.49 (third column) shows an example where a simple transform of a neutron porosity log has been used to predict a sonic log in a sequence of sands and shales. The addition of the gamma ray (Fig. 8.49, fourth column) makes a slight improvement but the addition of the density log does not (Fig. 8.49, fifth column). As with all prediction methods blind testing and cross validation (e.g. Hampson et al., 2001) are useful in assessing the accuracy of predictions.

Other techniques that are commonly employed for log prediction are neural networks and fuzzy logic. Shear velocity prediction using fuzzy logic and genetic algorithms has been described by Cuddy and Glover (2002). Using neural net or fuzzy logic prediction techniques without adequate QC, however, is likely to be misleading.

An effective test of a log prediction is to perform a well tie. Figure 8.50 illustrates an example where a sonic log was predicted from a density and resistivity log, together with a reasonable well tie based on the predicted sonic.

A good example of log prediction forming a useful quality control on measured data is found in V_s prediction. Comparing the measured Poisson's ratio log with the predicted Poisson's ratio log can identify zones that may benefit from editing. Another important aspect of V_s prediction is the application of empirical trends to data from hydrocarbon bearing wells. The effect of the hydrocarbon on the compressional velocity has to be removed before brine-based empirical transforms can be applied. To do this requires some knowledge of an appropriate stiffness model as described in Section 8.2.3. If shear velocity data are available in nearby wells, dry rock trends can be determined, and then the Gassmann methodology method described in Section 8.2.3 can be employed to derive the 'wet' compressional velocity in the wells. If there are no wells with shear data then a dry rock model needs to be invoked (see Section 8.2.3).

Mavko et al. (1995) describe the application of the critical porosity model to the V_s prediction problem. With elegant simplicity, the critical porosity model effectively scales the difference in the saturated bulk modulus to the difference in the (Reuss mix) fluid modulus at critical porosity (Fig. 8.51):

$$\Delta \kappa_{Gass}(\phi) = \frac{\phi}{\phi_R} \Delta \kappa_R(\phi_R). \tag{8.40}$$

Given that the calculation of K_{sat} requires V_s, a good approximation for the purposes of fluid substitution without V_s is simply to replace K with the M modulus

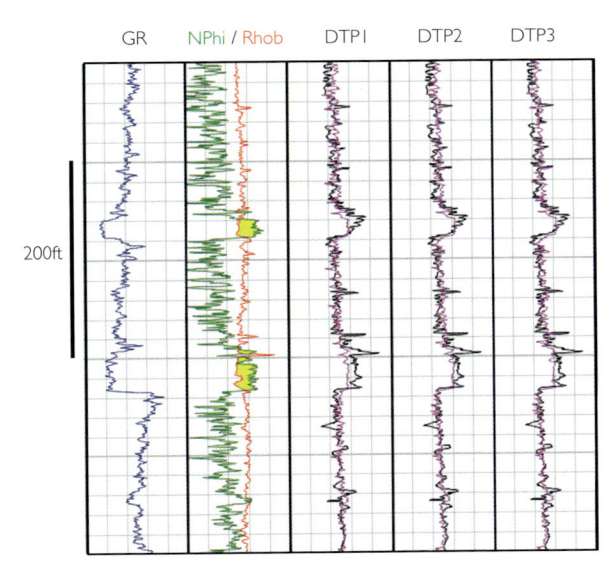

Figure 8.49 Results of multilinear regression to predict compressional slowness (black = original sonic log, purple = prediction), using (1) neutron porosity, (2) neutron porosity and gamma ray, and (3) neutron porosity, gamma ray and density. Note how there is a small improvement in the prediction around the sands with the inclusion of the gamma ray, but no additional improvement with the addition of the density log. Also note how the thin hard carbonate-rich layers (high-density spikes) are not predicted.

Rock physics for seismic modelling

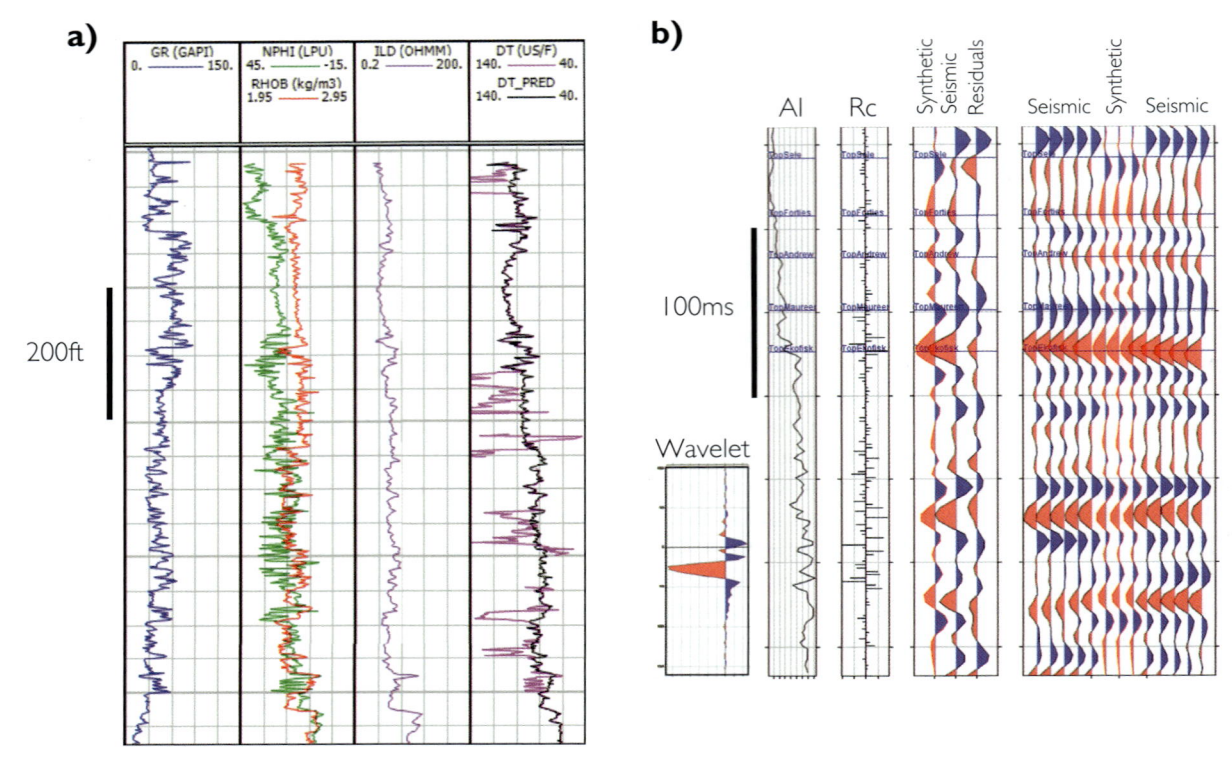

Figure 8.50 Sonic log prediction and QC using well ties; (a) sonic log prediction based on density and resistivity logs (fourth column – purple curve = measured sonic, black = predicted sonic), (b) well tie showing good character match and reasonable wavelet estimation.

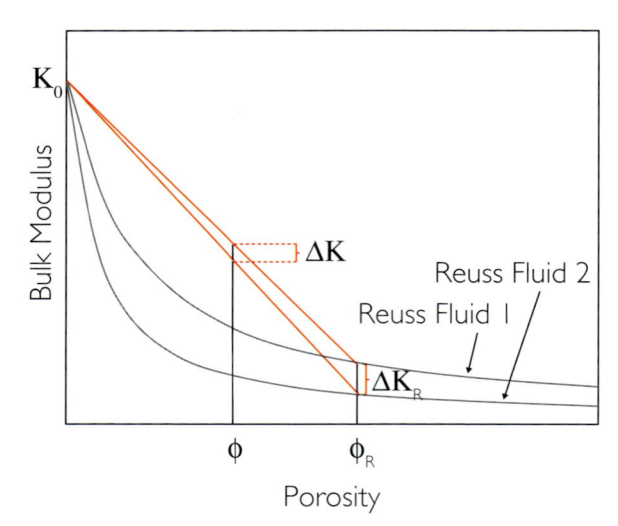

Figure 8.51 Using the critical porosity model to effect fluid substitution without V_s (after Mavko et al., 1995). The change in the bulk modulus (K_{sat}) is proportional to the change in the Reuss average fluid moduli at ϕ_R.

($V_p^2 \rho$). Using the M modulus, the intercept porosity ϕ_R is defined as:

$$\phi_R = \frac{\phi M_0^2 - M_0(1+\phi)M_{fl} + M_{sat}M_{fl}}{(M_0 - M_{sat})(M_0 - M_{fl})}. \qquad (8.41)$$

Another elegant solution for fluid substitution without V_s that is based on the critical porosity model is (M. Kemper, personal communication):

$$V_{p2} = \sqrt{\frac{\rho_1}{\rho_2}V_{p1}^2 - (K_{fl1} - K_{fl2})\frac{\phi}{\phi_R^2}\frac{1}{\rho_2}}, \qquad (8.42)$$

where ρ = bulk density, K_{fl} = fluid bulk modulus, ϕ = porosity, ϕ_R = critical porosity and 1 and 2 refer to initial (hydrocarbon bearing) and substituted (brine bearing) values respectively. In this case the appropriate ϕ_R is determined by optimisation. After fluid substitution to brine, V_s prediction and derivation of Gassmann dry rock parameters, V_p can be predicted and compared to the original log value. The final choice of ϕ_R can be chosen by minimising the error in V_p prediction.

8.4.4 Borehole invasion

The invasion of filtrate into the borehole wall has the potential to alter density and sonic log measurements (e.g. Han and Batzle, 1999; Walls and Carr, 2001). This can be a significant problem in situations where the reflectivity is subtle (e.g. Class IIp type situations). Making corrections for these effects can make a difference to the scaling of extracted wavelets (Vasquez et al., 2004). Whilst correction for invasion is usually necessary for density logs, it is questionable whether sonic logs need to be corrected. The interpreter needs to approach the issue with care owing to the fact that there are a large number of potential variables and the invasion correction solution is often under-constrained.

Owing to the fact that the zone of investigation of the density log is very shallow, the key problem associated with porosity estimation from the density log is the determination of the fluid parameters in the zone in which virgin fluids have been displaced by mud filtrate (i.e. the invaded zone). Typically, invasion is indicated by differences in the shallow, medium and deep resistivities (Fig. 8.52).

Invasion is a complex process resulting from the movement of drilling mud filtrate into the borehole wall under pressure. Invasion proceeds until mudcake builds up and the process stops. The depth of invasion in high-porosity rocks can be limited by rapid mudcake formation, whereas in moderate-porosity rocks, particularly if the hole has been left open for a long time or has been drilled over-balanced (i.e. with a mud weight significantly higher than that required to balance the formation pressure), then invasion can be extensive.

The fluid density of the invaded zone depends on a number of factors including the saturation in the invaded zone, the fluid density of mud filtrate and the density of the virgin formation fluids. In many instances invaded zone fluid parameter estimation may involve a certain degree of trial and error. Core porosities (total porosity calculated in the laboratory) can be invaluable to help constrain effective fluid densities for porosity estimation. Forcing a regression line from the mineral point through density–core porosity data to 100% porosity allows an estimate of the fluid density (Fig. 8.53). Of course much depends on the nature of the dataset as to the perceived accuracy of the extrapolation.

Making corrections for invasion to the density log is straightforward:

$$\rho_{b(corr)} = \rho_b - \left(\left(\phi \cdot \rho_{flx} \right) - \left(\phi \cdot \rho_{flv} \right) \right), \qquad (8.43)$$

where $\rho_{b(corr)}$ = corrected bulk density, ρ_b = original bulk density, ϕ = porosity, ρ_{flx} = invaded zone fluid density, ρ_{flv} = virgin zone fluid density.

Correcting the sonic log for invasion, however, is more complex, requiring Gassmann fluid substitution.

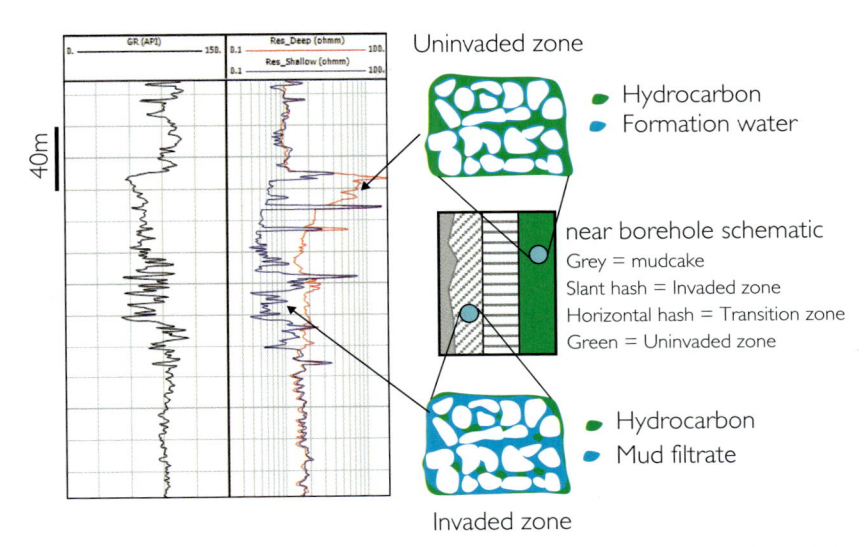

Figure 8.52 Separation of shallow and deep resistivity curves indicating significant borehole invasion in a gas-bearing sand interval.

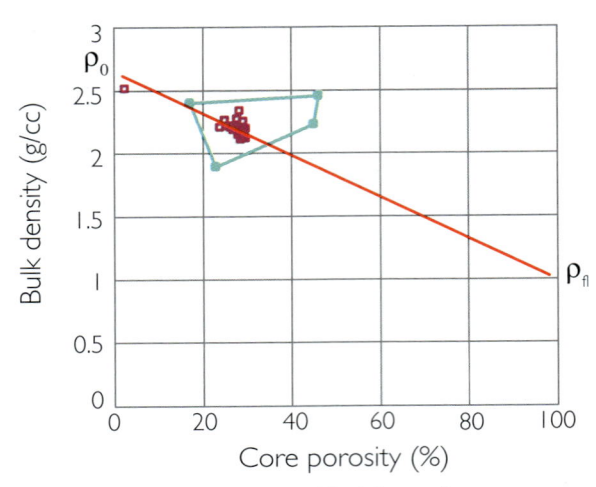

Figure 8.53 Graphic illustration of fluid density determination using density log and core porosity measurements.

Not only is the effective fluid density required but also the effective fluid modulus. The calculation of these properties involves:

- calculation of filtrate elastic parameters,
- estimation of the invaded zone saturation,
- determination of end member fluid parameters,
- choice of a fluid modulus mixing scheme.

Filtrate properties depend on the nature of the drilling mud. Where the saline component of the water-based mud (WBM) filtrate is sodium chloride (NaCl) the equations of Batzle and Wang (1992) may be used to calculate fluid density and modulus. Note that the filtrate density of potassium chloride (KCl) muds is likely to be slightly lower than that of NaCl muds owing to the lower molecular weight of potassium (J. Garnham, personal communication). Typical density values for water-based mud filtrate are around 1 g/cc but can be higher depending on salinity and mud additives. With regard to oil-based muds (OBM), typically up to 50% of OBM filtrate is water with the rest being refined oils and other additives. Filtrate densities are usually about 0.75–0.8 g/cc. To calculate the fluid modulus of OBM filtrate, a mix of diesel (~35 *API*) with no dissolved gas and water usually works well.

Saturation interpretations may come from a resistivity log suite interpretation or a petrophysical 'rule of thumb'. For example, rules of thumb commonly used for invaded zone saturation are as follows.

For water-based mud:

$$S_{xo} = (S_w + 2)/3 \text{ or } S_{xo} = S_w^{0.2} \text{ (Dewan, 1983)}.$$
(8.44)

For oil-based mud:

$S_{xo} = S_w$ if $S_w < 0.5$, otherwise
$S_{xo} = 0.5$ (F. Whitehead, personal communication).
(8.45)

Another approach is to establish an invaded zone saturation based on the estimated fluid density and assumptions about end member fluids. In wells where there are multiple fluid zones and limited variations in reservoir lithology, comparison of results from the different zones may lead to a refinement of the fluid densities that are appropriate for each zone. Relating saturation in the invaded zone to fluid density is straightforward, but the calculation of the fluid modulus is less clear-cut. Besides the definition of the end member fluid properties a key question is what mixing rule should be applied. A starting point is to assume that fluid mixing is homogeneous and that Reuss' equation applies. However, it is possible that in the invaded pore space the saturations are not homogeneous, owing to variable permeability.

The argument for using mixing schemes other than the Reuss average in the invaded zone comes from laboratory results. Studies by Domenico (1977), Knight and Nolen-Hoeksema (1990) and Knight et al. (1998) showed that the velocity vs saturation relationship changes depending on the distribution of gas and brine in the pore space (Fig. 8.54). Where the pore scale saturations were uniform the velocity was fairly constant except for a dramatic drop associated with a small amount of gas. On the other hand, a more gradual decrease in velocity with increasing gas content was observed when the pore scale saturations were heterogenous or 'patchy'. Given the variations in pore space permeability it is conceivable that 'patchy' saturations (on the scale of the logging wavelength) can occur in invaded zones.

Patchy saturation effects can be accommodated in Gassmann's equation by adjusting the fluid stiffness through various modes of fluid mixing. Fluid moduli inversions of various log datasets (using a generalised dry rock model) led Brie et al. (1995) to formulate a relationship (Fig. 8.55) that effectively stiffens the fluid modulus at low gas saturations:

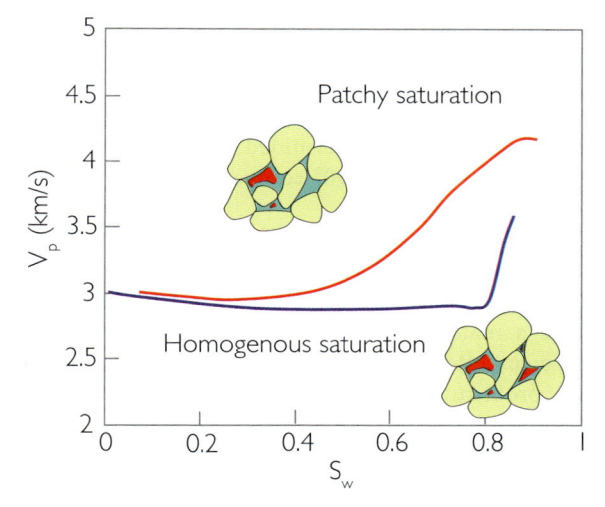

Figure 8.54 Effect of pore scale saturations on ultrasonic velocity (modified after Knight and Nolen-Hoeksema, 1990).

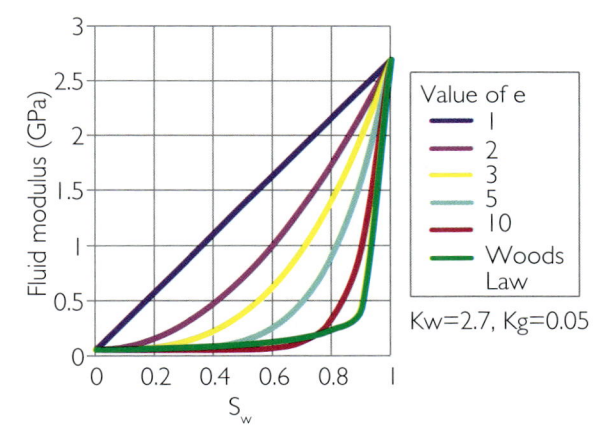

Figure 8.55 Fluid modulus estimation using the method described by Brie *et al.* (1995). See text for discussion. (Copyright 1995, SPE. Reproduced with permission of SPE. Further reproduction prohibited without permission.)

$$K_{fl1} = (K_w - K_g)S_x^e + K_g, \qquad (8.46)$$

where K_{fl1} = fluid modulus, K_w = bulk modulus of brine, K_g = bulk modulus of gas, S_x = invaded zone saturation, e = a number from 1 to 10 (and usually around 2–5).

Other workers favour a Voigt average and the Voigt–Reuss–Hill average to model the patchy saturation effect.

If water-bearing sands have been drilled with a water-based mud, it is unlikely that the difference between mud filtrate and formation water properties will be significant and invasion corrections on the sonic log may not be warranted. However, in flushed gas sands and oil sands drilled with water-based mud or water sands drilled with oil-based mud the possibility of invasion must be considered. In the case of oil-based mud filtrate invading a gas zone, the effective elastic properties of a mixture of water, gas and oil filtrate must carry some uncertainty. One approach might be to mix the oil and water using the Reuss average and then mix this composite with the gas using a stiffer mixing rule (M. Kemper, personal communication).

Once the density, saturation components and end member fluids have been addressed, Gassmann can be implemented. One approach to evaluating whether the sonic needs correcting is to compare dry rock parameter outputs from the following scenarios.

(1) Assuming no invasion effect on sonic: input corrected density, V_p, V_s, S_w, and fluid mixing using the Reuss average.
(2) Assuming invasion effect on sonic: input measured density, V_p, V_s, S_x, and fluid mixing using the Reuss average.

If these approaches do not yield reasonable results then it could be that the Reuss mixing model is inappropriate at logging frequencies and stiffer mixing models are required.

In most situations the workflow will be iterative. For example, where there is a hydrocarbon contact, the dry rock properties above and below the contact should be compared. They should be similar, unless diagenesis has proceeded differently in the hydrocarbon and water legs. If diagenesis differences are significant then they would be manifested as a porosity change at the contact.

Experience has shown that confident invasion corrections can be made in practice only in clean sands with good quality V_p, V_s, and density data and 'hard' evidence of the porosity of the sand (such as core porosity from laboratory measurements). A practical problem with Gassmann's model is that the various parameters are not independent; vary one and it will affect the others. Thus, without independent shear and porosity information the invasion problem is under-constrained. For example, if there is no measured V_s, then it would need to be predicted, but to do this knowledge of the fluid modulus is required.

8.4.5 Sonic correction for anisotropy in deviated wells

Deviated wells can present a range of problems in seismic-to-well ties. One particular problem is that sonic log velocities in shales are generally higher than vertical velocities (Chapter 5) owing to the presence of transverse isotropy (VTI). Misties can result (e.g. Gratwick and Finn, 1995). It is therefore recommended that interpreters consider the potential for an anisotropic influence on sonic velocities in deviated wells and make an appropriate correction to the sonic logs. It should be remembered that it is not really hole angle that is important but relative bed dip (e.g. Vernik, 2008). The reader is recommended the work of Hornby et al. (2003), Rowbotham et al. (2003a), Brevik et al. (2007), Vernik (2008) and Wild (2011).

If cross dipole data are available in a deviated hole then it is possible to obtain the anisotropic constants from an analysis of the sonic slownesses in different directions (e.g. Horne et al., 2012). In the absence of such data the VTI constants may be derived by analysing data from a number of wells, each with different hole deviations through the same geology (Fig. 8.56).

The analysis of log data in shale sections necessarily focusses on averaged sonic data and the detection of a broad trend with increasing hole angle.

Figure 8.57 shows an example from Brevik et al. (2007). It is evident that in these data there is little effect of hole deviation out to an angle of 30° and also that the general level of compressional and shear wave anisotropy (as determined from the fractional increase in velocity at 90° compared to 0°) is of the order of 15%–20%.

The simplest approach to this issue is to determine the anisotropic coefficients in Thomsen's (1986, 2002) anisotropic equations (see Chapter 5, Eq. (5.2)) that make a match to the change of velocity with hole angle (e.g. Rowbotham et al., 2003a; Vernik, 2008). The problem can also be addressed as part of an inversion scheme (e.g. Hornby et al., 2003) or in the application of an anisotropic rock physics model (e.g. Xu et al., 2005).

There is usually significant scatter in the data from a multi-well analysis as shown in Fig. 8.57. This is probably related to lateral variability in shale properties and environmental conditions. The implication is that a variety of fits may be possible but at least it gives a basis for making a correction that can be tested by a well tie. The volume of shale log is used to apply the correction (assuming that clean sands are essentially isotropic, e.g. Wang, 2001a,b). Different authors have used slightly different functions and shale cut-off points but in general no correction is necessary up to 20%–30% V_{sh}, whereas the full

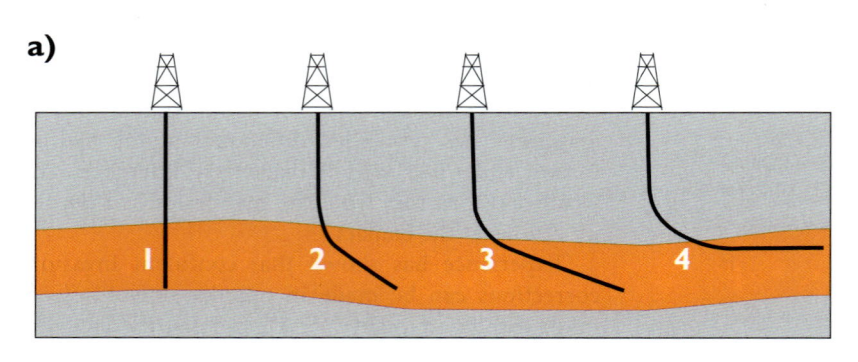

Figure 8.56 Schematic illustration of borehole deviation and sonic slowness in shales; (a) several wells with different hole angles through the same shale formation (orange), (b) decreasing shale slowness (i.e. increasing velocity) with increasing deviation, (c) horizontal vs vertical slowness (after Horne et al., 2012).

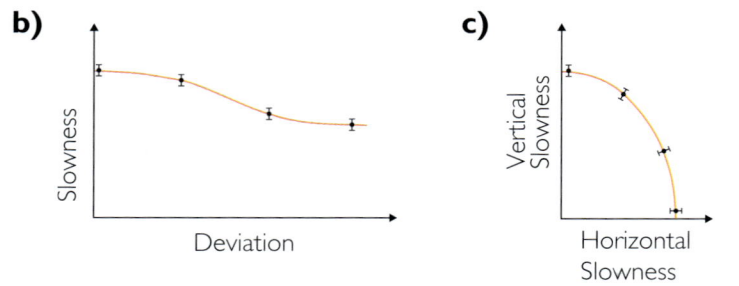

Practical issues in fluid substitution

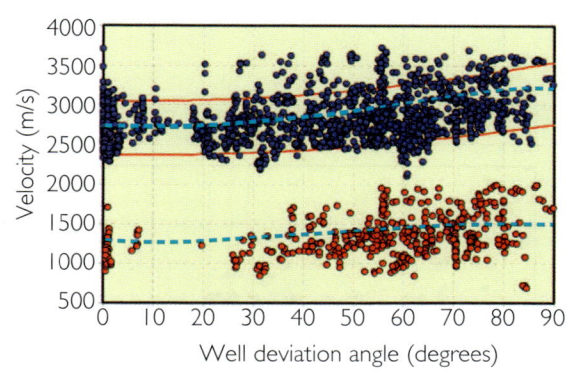

Figure 8.57 Well log velocity data from pure shales in 10 shale formations in two North Sea fields (after Brevik et al., 2007). Red lines indicate +/− 12.5% P wave velocity variations relative to a trend of velocity with deviation angle. Dashed blue lines are velocity predictions based on Thomsen's (1986, 2002) anisotropic equations (Chapter 5) using $\delta = 0.05$, $\varepsilon = 0.18$ and $\gamma = 0.18$.

8.5 Practical issues in fluid substitution

The application of Gassmann's equation is complicated by the fact that real rocks deviate from the simple assumptions inherent in the model. Three commonplace fluid substitution scenarios which require careful thought are shaley sands, laminated sands and tight (gas) sands. There are pitfalls for the unwary. The reader is referred to Sections 8.2.3 and 8.2.5 for background discussions on Gassmann's equation and dry rock trends.

8.5.1 Shaley sands

A common pitfall in the fluid substitution of shaley sands is an exaggerated substitution effect on the compressional velocity and Poisson's ratio logs in low-porosity shale-prone zones (e.g. Skelt, 2004; Simm, 2007). Figure 8.58 shows a typical example where gas has been substituted for water. Note how the Poisson's ratio log in places is close to zero and how some of the shaley zones have a greater magnitude of P wave fluid substitution effect compared to the clean sands. Intuitively this is incorrect. In this particular case the porosity input to Gassmann's equation is effective porosity (derived from the density log using a mix of shale and quartz) and the effective mineral modulus is derived by mixing shale and quartz using the Voigt–Reuss–Hill average (see Section 8.2.1), with shale parameters being estimated from the logs. The normalised modulus plot (see Section 8.2.3) shows that the clean sands have reasonable dry rock values (Fig. 8.58), but as the shale volume increases so does the scatter on the plot. Below about 8% porosity, some of the bulk modulus points are negative, which is not physically possible. Many of the high V_{sh} points are plotting as very soft material, which leads to the large fluid substitution effects on the compressional velocity log.

Such problems are commonly remedied by applying cut-offs, i.e. fluid substitution is done only if the porosity is greater than a particular value and V_{sh} less than a certain value. This may be justified on the basis that at low porosities and high shale volume, the rock is unlikely to be of reservoir quality, but the solution is inelegant, giving rise to saw-tooth effects in the substituted logs where the porosity and V_{sh} logs vary around the cut-off values.

It might be argued that the main issue with the results shown in Fig. 8.58 is that the Gassmann

correction should be applied beyond 70%–80% V_{sh}. In between these cut-offs the correction can be applied linearly.

Any data that might be used to derive anisotropic parameters, such as walk away VSP results, anisotropic velocity measurements on core samples and anisotropic information from time and depth processing, might be useful in constraining the anisotropic model. When there is only a single deviated well with a standard suite of logs, the issue becomes one of trial and error (i.e. make a correction based on a generalised idea of the anisotropy and evaluate the resulting well tie). Empirical approaches, such as those presented by Ryan-Grigor (1997) (see Chapter 5) and by Tsuneyama and Mavko (2005), might be used as the basis for an initial estimation of the anisotropic parameters.

The analysis of log data from deviated wells in which horizontal transverse isotropy (HTI) is present (i.e. vertical fractures) follows a similar workflow in which HTI anisotropic theory (e.g. Hudson, 1981; Schoenberg and Sayers, 1995) is used to justify observed velocities (e.g. Wild, 2011). Whilst anisotropy is often of secondary importance in calculating reflectivities for well ties, the interpreter should consider the possibility that incorporating the anisotropy into the reflectivity calculation may improve the tie, particularly if the seismic has been processed using an anisotropic velocity model. In the case of HTI scenarios, reflectivities need to be calculated for the correct azimuth.

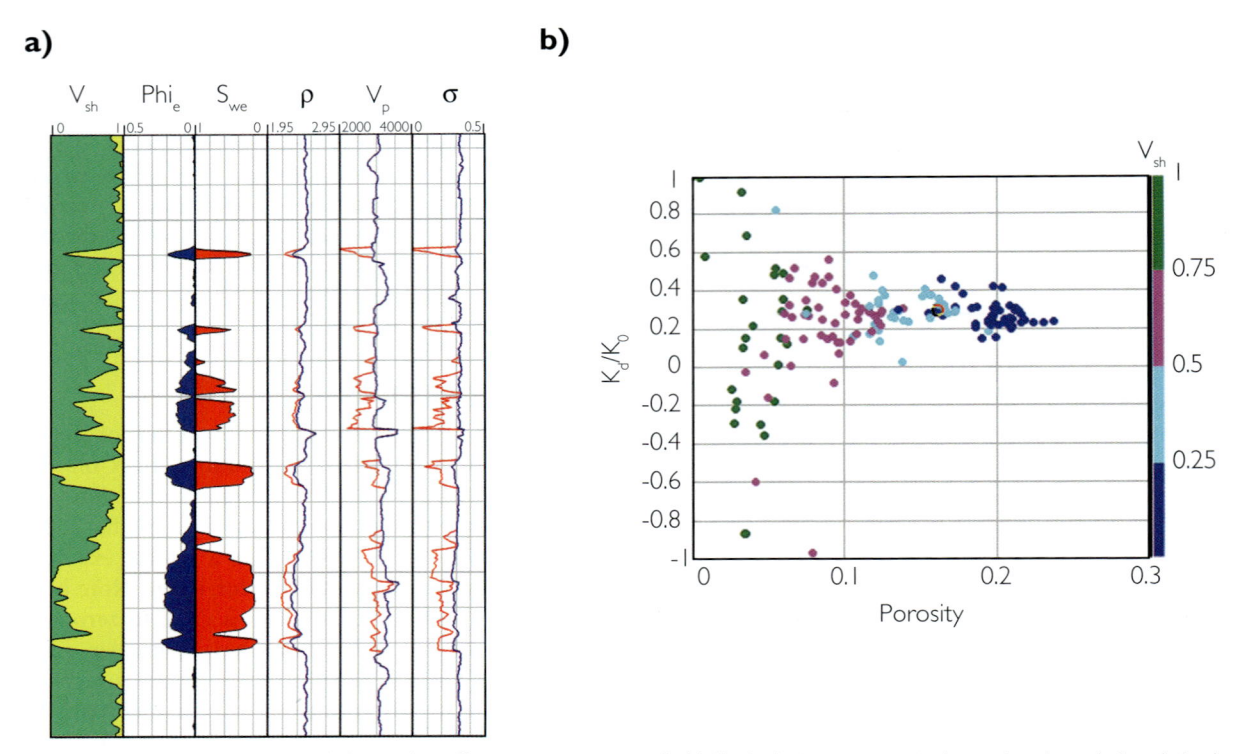

Figure 8.58 Fluid substitution in shaley sands – effective porosity approach. (a) Fluid substitution to gas is done using dry rock data derived from the logs (blue curves = brine, red curves = gas; depth marker spacing is 10 m). Note that there are exaggerated substitution effects on the compressional velocity and Poisson's ratio logs in shale-prone zones. This is the result of inconsistencies in the inputs to Gassmann's equation. (b) Gassmann dry rock properties derived from the brine fill logs in (a). Data points are coloured by shale volume, as indicated by the key on the right. Note that the clean sands give reasonable results but there are erroneous negative dry rock moduli in the shaley zones (after Simm, 2007).

implementation should have been parameterised using total porosity. The problem with a total porosity scheme, however, is the uncertainty over the modulus of dry clay (section 8.2.4). Total porosity was calculated for the dataset shown in Fig 8.58 using densities of 2.6g/cc and 2.65g/cc for clay and quartz respectively and it was assumed that $V_{cl} = 0.7$ V_{sh}. In addition, the water saturation was calculated in terms of the total porosity (Eq. (8.31)). Two separate Gassmann calculations were performed on the data using dry clay moduli of 36 GPa and 10 GPa. The dry rock and fluid substitution results are shown in Figs. 8.59 and 8.60, respectively. It is evident that the stiffness of the shaley sands, and consequently the effect of fluid substitution in shaley sands, depends to a large extent on the moduli of dry clay. The lower value of dry clay modulus appears to give a more reasonable result than the higher dry clay value.

Another approach to avoiding unreasonable results in shaley sands, particularly in the case where effective porosity has been used as an input, is the model-based approach described in Section 8.2.3. Figure 8.61a shows how a trend can be drawn through the mineral point and the sand data of Fig. 8.59. The shape of the curve can be adjusted to reflect the key factors in porosity reduction (i.e. sorting vs cementation). The fluid substitution result is shown in Fig. 8.61b. It is very similar to the total porosity case with low values of dry clay (Fig. 8.60, red curves) and intuitively reasonable given that the clean sands show the largest velocity change and the shaley sands show much smaller effects. There is no single answer, of course, as a range of dry rock trends could have been fit to the data. The model-based approach is a practical solution for deriving reasonable results in exploration settings where log data and petrophysical interpretations may not be optimal, and time is short.

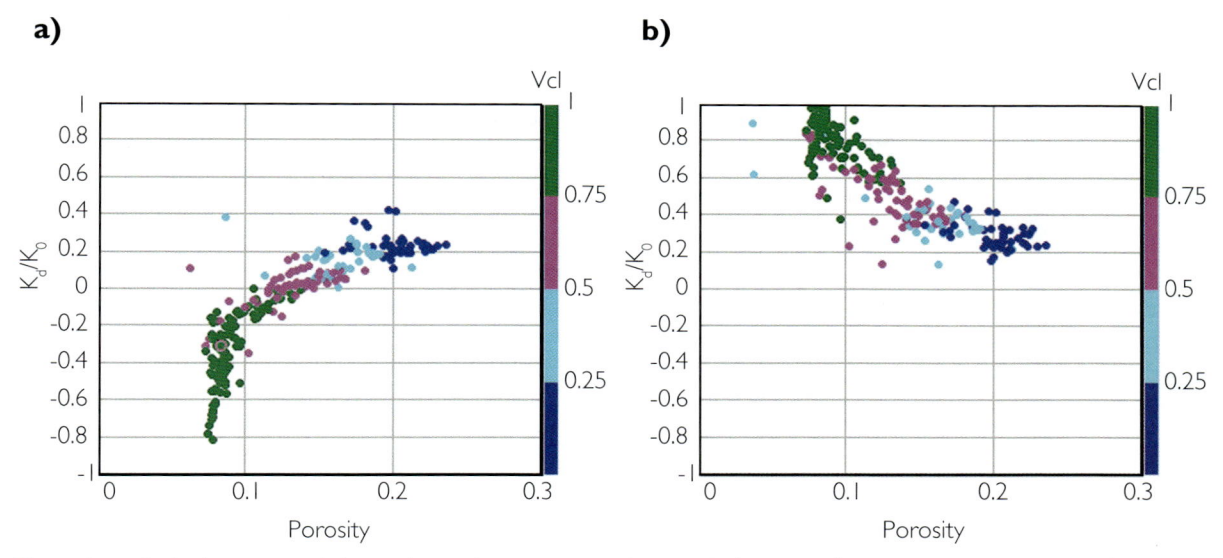

Figure 8.59 Fluid substitution in shaley sands – total porosity approach. Dry rock data inverted using Gassmann's equation with varying clay mineral moduli; (a) K_{cl} = 37.5 GPa, μ_{cl} = 18.7 GPa; (b) K_{cl} = 10 GPa, μ_{cl} = 4.7 GPa. Data points are coloured by shale volume, as indicated by the key on the right (after Simm, 2007).

8.5.2 Laminated sands

It is possible to apply the model-based fluid substitution approach described in the previous section to laminated sand and shale zones and get reasonable results, but conceptually laminated zones require a different approach. Given that the laminae in these reservoirs are below logging and seismic resolution the fluid substitution problem can be simply approached from the viewpoint of mixing between shale, wet sand and hydrocarbon (pay) sand end members (Katahara, 2004). The technique is based on the fact that using the Backus (1962) average formulation for the effective parameters of sequences of thin layers, bulk density and compliance (i.e. inverse of M or μ modulus) vary linearly with shale content (Katahara, 2004). Thus for a laminated sand/shale sequence:

$$\frac{1}{M_{lam}} = \frac{V_{sh}}{M_{sh}} + \frac{1 - V_{sh}}{M_{sand}} \quad (8.47)$$

$$\rho_{lam} = (\rho_{sh} V_{sh}) + (\rho_{sand}(1 - V_{sh})) \quad (8.48)$$

and

$$V_{p\_lam} = \sqrt{\frac{M_{lam}}{\rho_{lam}}}, \quad (8.49)$$

where M = M modulus, ρ = bulk density, 'lam' denotes laminated mix between shale and sand end members, and V_{sh} = volume of shale.

Calculation of V_s follows the same form with M being replaced by μ. End member hydrocarbon sand values are derived using end member wet sand and Gassmann fluid substitution. Practical implementation of the technique on well logs requires a reasonably accurate assessment of volume of shale and also some idea of the parameters of shales and wet sands. Trend curves derived from a collation of data from nearby wells can be used to estimate the end members and the technique is relatively robust in the presence of small errors in defining the end members. Figure 8.62 illustrates the calculation for a laminated interval with 50% V_{sh}.

A more complex alternative to Katahara's laminated model has been presented by Skelt (2004). Skelt's method essentially decomposes the sonic and density data into shale and sand components, then performs fluid substitution in the sand component followed by a re-combination.

8.5.3 Low porosity and permeability sandstones

Of the 'unconventional' reservoirs, fluid substitution may be a technique that could be applied to 'tight' sands (i.e. sandstones that require hydraulic fracture stimulation in order for the reservoir to

Rock physics for seismic modelling

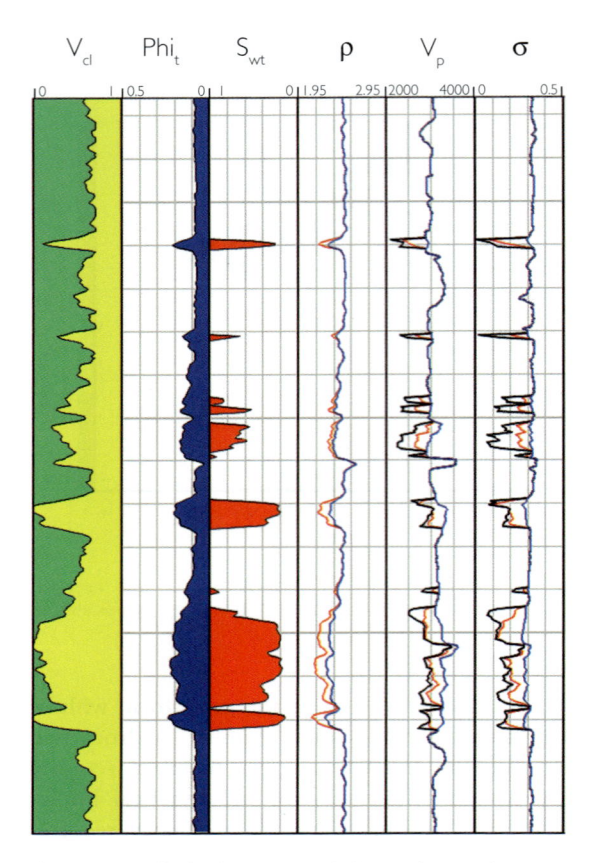

Figure 8.60 Fluid substitution in shaley sands – total porosity approach. Fluid substitution to gas using dry clay moduli of 37.5 GPa (black curves) and 10 GPa (red curves). Depth marker spacing is 10 m (after Simm, 2007).

produce). Tight sands are a heterogenous group of indurated reservoirs comprising shaley, silty unconsolidated sands or clean cemented sandstones with low permeability (less than 0.1 md) and porosity (generally less than 10%). These reservoirs present a significant rock physics challenge. They are characterised by a complex network of primary and secondary porosity with varying levels of connectivity. Smith *et al.* (2009) have described how variations in velocity of tight sands are in large part controlled by variable pore geometries and the presence of microcracks (Fig. 8.63). This effectively means that velocity or seismic amplitude may be relatively poor attributes for predicting porosity. It also means that simple Gassmann fluid substitution is likely to be inappropriate.

Fluid substitution in tight sands needs to be addressed by using rock physics models in which pore shape and the interaction of different pore shapes can be specified. Thus, models such as the Xu–White model (see Section 8.2.7) or the self-consistent model of Berryman (1995) are appropriate. For example, Ruiz and Cheng (2010) show modelling results in which gas within microcracks reduces acoustic impedance significantly, giving a much larger effect than using Gassmann fluid substitution. This result assumes that the gas and brine phases are mixed at the finest scale, throughout the pore volume; a different result would be obtained for a different fluid distribution, such as brine in the soft (crack-like) pores and gas in the stiff (equi-dimensional) pores.

Given the variability of tight sands, rock physics models need to be established on an individual case basis. There are numerous issues in the characterisation of these reservoirs, not least the petrophysical challenges of estimating parameters such as porosity, permeability and water saturation (e.g. Miller and Shanley, 2010). In addition, it is possible that anisotropy due to a preferred alignment of microcracks may be an issue to consider (Smith *et al.*, 2009).

8.6 Rock characterisation and modelling issues

'Rock characterisation' effectively describes the process of bringing together rock physics and petrophysics observations within a geological context. Some workers use the term 'diagnosing' to describe the process of identifying rock types (e.g. friable vs cemented sands) on the basis of elastic behaviour in the context of rock physics theory and physical measurements (e.g. Avseth *et al.*, 1998, 2005). To summarise the various elements, rock characterisation comprises:

- identification of rock types on porosity vs velocity/ moduli crossplots using wet and dry data,
- fitting rock physics models,
- definition of litho-facies,
- fluid substitution,
- derivation of depth (effective pressure) trends,
- generation of depth and depositional environment specific rock physics templates.

As described in this chapter, the construction of a rock physics database requires considerable attention to detail. Below are some key issues that the interpreter should consider in the use of such as database.

Rock characterisation and modelling issues

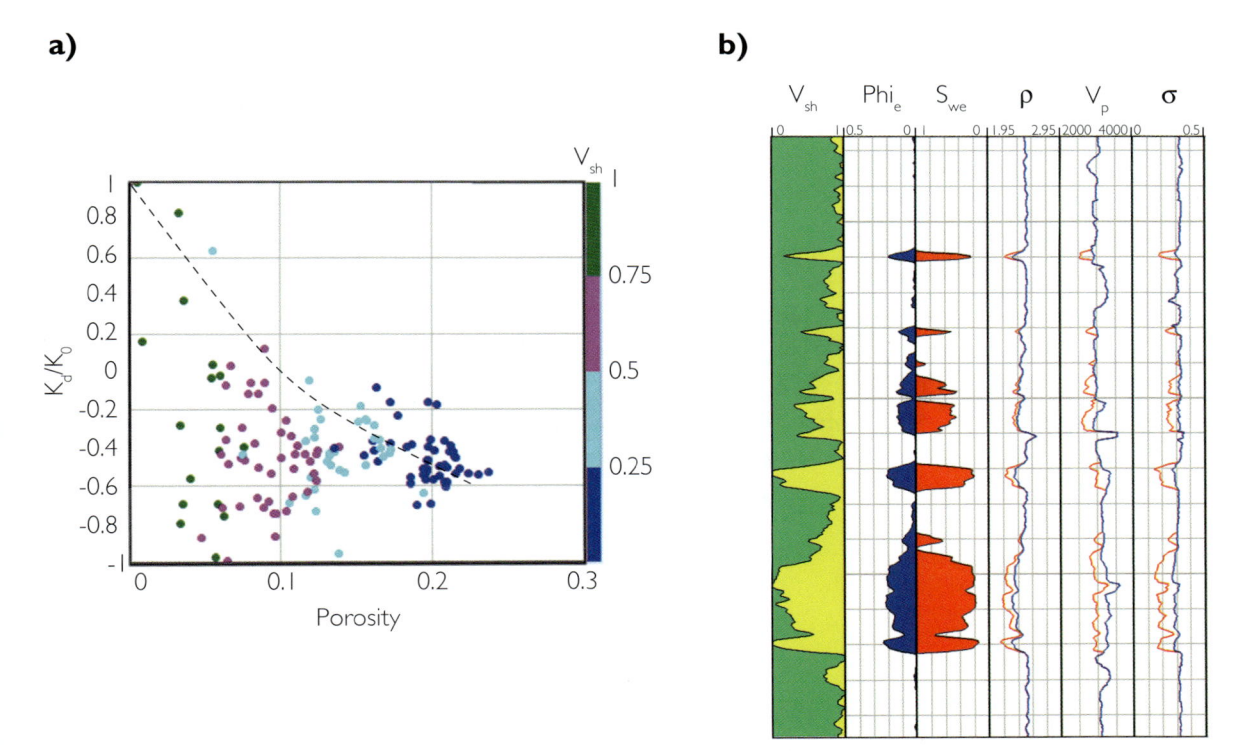

Figure 8.61 Fluid substitution in shaley sands – effective porosity approach with dry rock model. (a) Dry rock data from Fig. 8.58 with a dry rock model (dashed line) superimposed. Data points are coloured by shale volume, as indicated by the key on the right. (b) Fluid substitution to gas using the dry rock model defined in (a): blue curves = brine, red curves = gas, depth marker spacing is 10 m. Note how the effect of fluid substitution on the compressional velocity log varies with porosity and how the substituted Poisson's ratio log is similarly well behaved (after Simm, 2007).

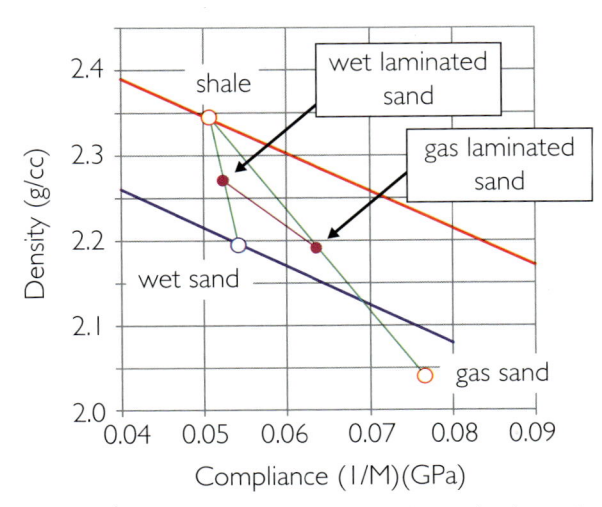

Figure 8.62 Fluid substitution in a laminated interval with 50% V_{sh} using Katahara's (2004) technique. The compliance (inverse of the M or P wave modulus) and density values are linearly scaled between shale and gas sand end members. Red and blue lines are locally determined trends for shale and brine sands respectively.

- Identifying rock types from log data requires that due consideration is given to the key factors that can influence elastic behaviour such as mineral composition, pore geometry, shale content, cement volume and pore pressure (as described in Chapter 5). All geological information needs to be included.
- It is not appropriate to perform fluid substitution on averaged data. Fluid substitution should be undertaken prior to averaging.
- The definition of litho-facies, for example based on discrimination cut-offs, should involve a careful consideration of the geological environment. It is not often that a geological facies scheme, for example based on detailed thin section analysis, can be translated directly into rock physics litho-facies. There is usually some simplification required.
- When assessing variability in elastic parameters between wells and determining statistics for

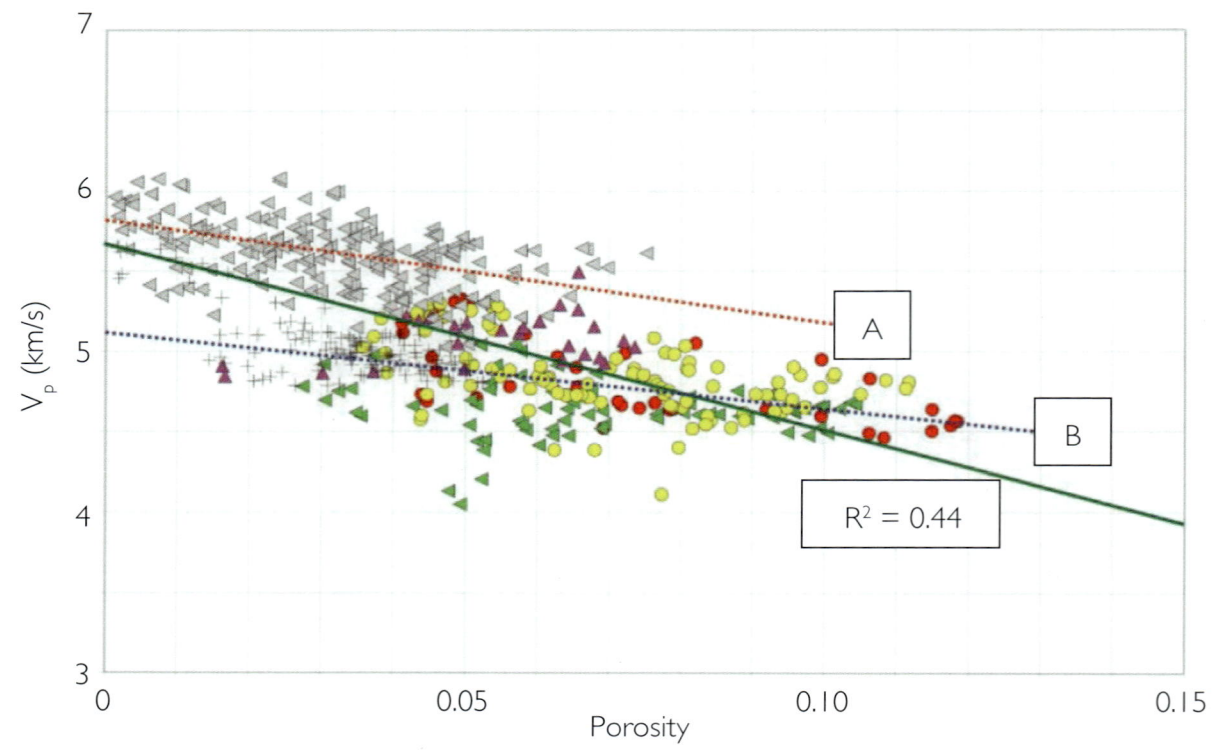

Figure 8.63 Porosity vs velocity for low-porosity clean sandstones from the same formation but different wells (after Smith et al., 2009). Colours represent different wells. It is evident that in general there is a poor correlation between velocity and porosity and a large scatter of P wave velocity. Two groups of data are identified ('A' and 'B') and it is inferred that the principal reason for the difference is that group B have a greater proportion of compliant microcracks than group A. Note how a different effective mineral velocity would be predicted for data groups A and B.

later use in probabilistic applications a distinction needs to be drawn between the natural variability of log data and true geological variability.
- There are a number of important issues in the construction of depth trends from multiple wells. Averaging should be specific to individual lithofacies and it would usually be inappropriate, for example, to combine clean sand and shaley sand data. Equally, understanding the stratigraphic and effective pressure context of the data prior to averaging is of paramount importance.
- Once the rock physics model has been established it is tempting to believe that subsequent fluid or lithology substitution is valid. It is possible that the chosen model is too simplistic. Different models may fit the same data but have different implications. Smith (2011) highlights the need to understand the reasons for porosity change before settling on an appropriate rock physics relationship to describe the data.
- Caution is needed in using rock physics models outside the measured range, for example at depths below which there is no measured log data, or for lithologies that have not been encountered in the wells. Rock physics models can be captivating; one should keep the data in sight.

Chapter 9

Seismic trace inversion

9.1 Introduction

Seismic trace inversion reduces the effect of the wavelet by replacing the seismic with blocks of impedance at a particular time sampling interval (e.g. 4 ms) and their arrangement gives an indication of acoustic layering in the subsurface (Fig. 9.1). Inverted data can be useful to the interpreter in a number of ways, including simplifying stratigraphic relationships and making lithological and fluid related effects more interpretable. An additional benefit is that the results of inversions are more readily appreciated by workers in the non-geophysics disciplines, such as geology and engineering, as the data are presented in recognisable layers.

Inversions to absolute impedance rely on well data for their parameterisation and so they are typically (although not exclusively) carried out in reservoir appraisal and development situations. Insight into the reservoir is gained through combining inversion results with rock physics, geological and petrophysical models, and so inversion results can be directly relevant to the creation of geomodels for reservoir simulation.

In general there are two types of inversion, *deterministic* and *stochastic*. Deterministic inversions are relatively straightforward to generate and are based on the minimisation of the difference between a modelled seismic trace and the actual seismic trace. As such, these types of inversions are smoothed solutions, representing a best estimate within the limits imposed by the bandwidth of the data. This effectively means that in areas where the geology is layered on a scale less than about 1/4 of the seismic wavelength the resulting inversion is likely to be inappropriate for quantitative interpretation. Unfortunately, this scale is important in most reservoirs.

The non-uniqueness of the sub-tuning component in seismic can be addressed through stochastic inversion, which generates multiple impedance realisations at the reservoir model scale. Each realisation matches the seismic trace, honours the statistics of property variation between well datapoints and ties the wells exactly. Realisations can be transformed to reservoir properties (using rock physics relationships and various mapping algorithms) and the results can be analysed in terms of uncertainty as well as connectivity. These types of stochastic inversion products are invaluable for conditioning geomodels in reservoir development but there is usually a significant cost in time and money to produce them.

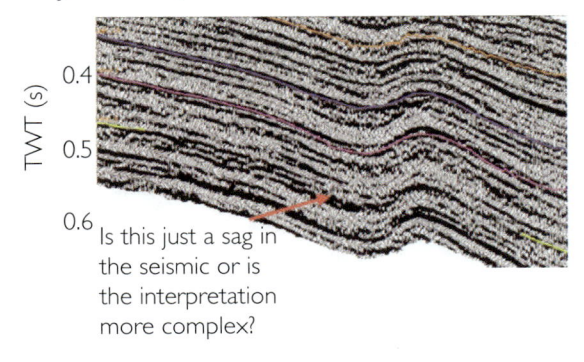

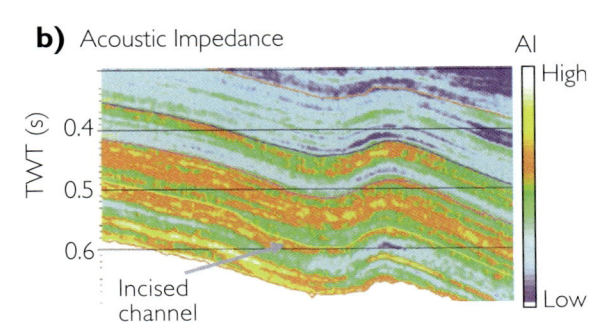

Figure 9.1 A reflectivity section and its acoustic impedance inversion equivalent (after Latimer *et al.*, 2000).

9.2 Deterministic inversion

There are a number of different approaches to deterministic seismic inversion and no attempt will be made here to outline them all. In addition, there is no clear consensus that any particular inversion algorithm is better than the others and it can be argued that careful evaluation of each step along the way is probably more important than choice of algorithm. The main focus will be to outline the general principles and investigate issues of quality control.

9.2.1 Recursive inversion

Early attempts to obtain absolute impedance from seismic involved scaling the seismic section to reflectivity, adding a low-frequency component (derived from an interpolation of well data or stacking velocities scaled to impedance values) and applying the *recursive formula* (i.e. the inverse of the reflection coefficient formula) (e.g. Russell, 1988):

$$AI_{i+1} = AI_i \left[\frac{1+r_i}{1-r_i}\right]. \qquad (9.1)$$

This approach is flawed because the wavelet is not addressed, but it serves as a useful introduction.

9.2.2 Sparse spike inversion

It is better to think of deterministic inversion as the convolutional model in reverse. Thus in Fig. 9.2 the reflectivity is estimated through a deconvolution of the seismic trace with the seismic wavelet. Owing to the fact that the output is a broadband reflectivity sequence, this type of approach is described as a 'broadband' inversion (of which there are a number of different types). To achieve an absolute impedance inversion requires the merging of the reflectivity solution with a low-frequency impedance component. One approach is to input a value for the uppermost impedance layer and apply the recursive formula. In practice, owing to trace-to-trace variability, this requires a model based on well data or a generalised geological concept to constrain the solution.

The observation that the seismic trace can be constructed from a few large reflection coefficients gave rise to an approach called *sparse spike* inversion. The Maximum Likelihood method of Hampson and Russell (1985) and the L1 Norm method of Oldenburg *et al.* (1983) are examples. Figure 9.3 illustrates the principle behind sparse spike inversion, showing the relationship between the well log impedance, reflection coefficient series, the seismic trace and the approximation of the impedance from the sparse spike solution. Sparse spike inversion using constraints to reduce non-uniqueness and find the inverted solution remains a commonly used method.

A strength of sparse spike inversion is that it attempts to produce the simplest model consistent with the seismic data. Thus it pays close attention to

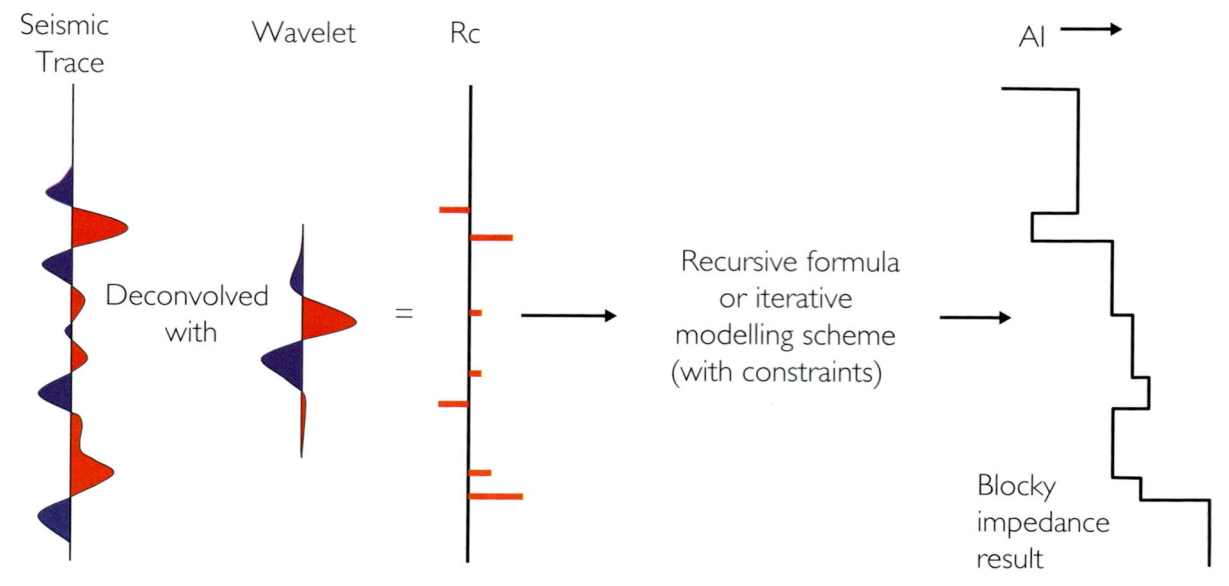

Figure 9.2 The concept of seismic trace inversion to impedance.

Deterministic inversion

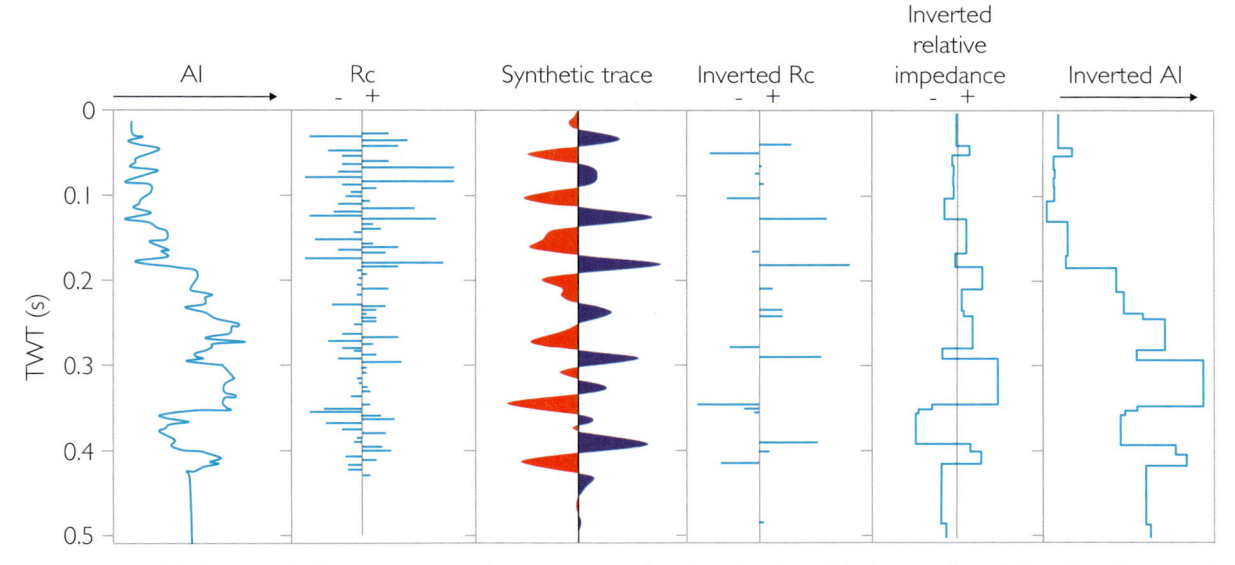

Figure 9.3 Model data example illustrating sparse spike trace inversion (based on data from Oldenburg et al., 1983). Note how the inverted AI trace is a blocky simplification of the well impedance.

the seismic data, and creates impedance changes only when they are required to match the seismic response. Although the seismic is justified in the solution, the result need not be consistent with what is known about the geology from well data or other observations. Thus the inversion result may be an oversimplified version of the subsurface. Other methods of trace inversion provide for the incorporation of geological information with varying levels of sophistication, as will be discussed.

9.2.3 Model-based inversion

Perhaps the most popular broadband inversion technique for which there is readily available commercial software is called *model-based* inversion (e.g. Russell and Hampson, 1991). Model-based inversions use an iterative forward modelling and comparison procedure (see Veeken and Da Silva, 2004, for an overview). The procedure requires a starting model that is subsequently perturbed and checked against the seismic (Fig. 9.4). This starting model may be an interpolation of well data (probably with a low-pass filter applied) or a general trend model based on geological knowledge. Another possible source for a generalised starting model is the seismic stacking velocity cube. This can be converted to interval velocities using the Dix formula and thence to impedance if a velocity–density relation (e.g. Gardner's relation) is assumed.

In practice it might be better to use the seismic interval velocities as a way to extrapolate log-based impedances away from the wells by using a cokriging approach.

The final inversion result is a solution in which the impedance model has been checked against the seismic traces and the errors calculated and minimised. If the wavelet is adequately described then the problem of tuning can be partly addressed. Various algorithms have been employed in the inversion process, e.g. Generalised Linear Inversion (GLI) (Cooke and Schneider, 1983) and the Seismic Lithologic method SLIM (Gelfand and Larner, 1983). Depending on the complexity of the algorithm and the compute power available, the inversion of a complete seismic cube may take hours, or perhaps days, to run.

Model-based inversions typically employ constraints to prevent the impedances in the solution wandering so far from the initial model that they are geologically impossible. These may be either hard boundaries that the solution is not allowed to cross, or soft constraints in whose vicinity a penalty is added to the synthetic error (the measure of the difference between the modelled and real seismic traces) that is being minimised. In the first case limits could be set as a maximum fractional change in the impedance from the starting model values. If the solution found closely adheres to this hard boundary over large TWT ranges, it is an indication that the starting model is

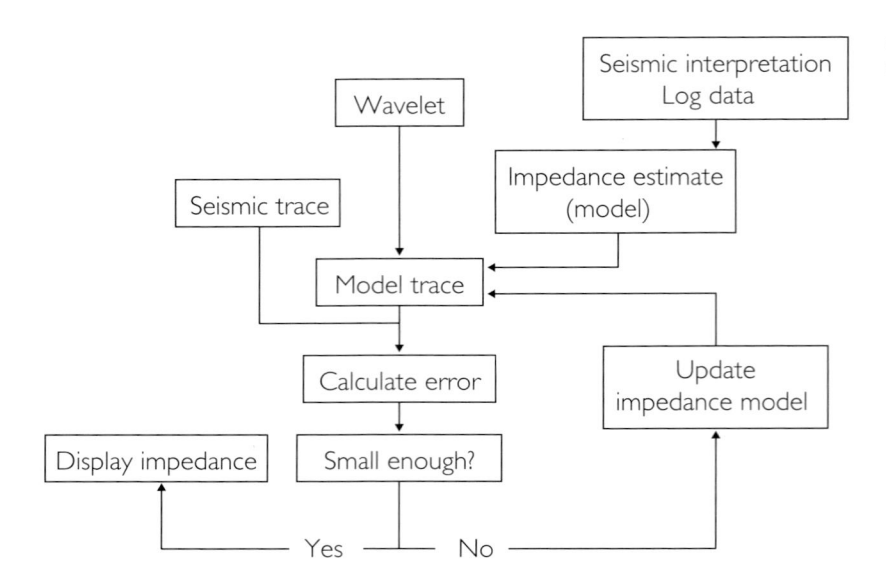

Figure 9.4 Generalised flow-chart for model-based inversion.

not consistent with the seismic data, and examination of the well-to-seismic ties is needed to try to find out whether the well data or the seismic are at fault. In the second case there will be an adjustable weighting parameter that determines the relative weight given to deviations of the impedance model from the starting point on the one hand, and synthetic error on the other. Trials will be needed to establish the best weighting parameter; this is implicitly a judgement about noise levels in the seismic data, and again the well ties are a key to understanding. Usually a range of parameter settings will give equally plausible inversion results. This type of non-uniqueness can easily be explored on a test dataset, giving at least a qualitative appreciation of the range of possible solutions. Some provision is also required in the algorithm to prevent solutions from emerging with high-amplitude oscillations at very high frequencies, such that they would have negligible effect when convolved with the seismic wavelet.

Within these constraints, model-based inversion finds a solution for the impedance trace that minimises the synthetic error. Usually there is a range of impedance trace solutions that provide a fairly good fit to the seismic data. An issue is whether the algorithm is able to find the solution with the lowest possible synthetic error. If the process that updates the model looks only at how small movements away from the starting point affect the synthetic error, then it is likely to find a solution fairly close to the initial model. It will be optimal in the sense that small changes to the final model cause the error to increase.

However, it may be that a more radical change to the model would produce a solution with lower error. Various algorithms are employed to allow the optimisation process to jump out of a local minimum and (it is hoped) eventually find the global optimum with the lowest possible error. An example is *simulated annealing* (e.g. Sen and Stoffa, 1991), in which model changes that increase the synthetic error have a definite probability of being accepted. This probability decreases as the iterations progress, so the process starts out by exploring a range of possible solutions and later homes in on the best of them. The name comes from the analogy with the physical process in which a solid is heated and then slowly cooled until it reaches the global minimum energy state where it forms a crystal. In practice, if the starting model is fairly close to the optimum solution, the global minimum may not be very different from the solution that would have been found by a simpler algorithm that looks only for local improvement.

Figures 9.5–9.9 show an example of a model-based inversion. Horizon picks on the main reflecting interfaces are used to establish the starting model for the inversion (Figs. 9.5 and 9.6). The inversion is run with the appropriate wavelet and constraints. The result is a micro-layer impedance model (Fig. 9.7) which, as a QC step, needs to be closely compared with the actual impedance in the well. A final impedance section is shown in Fig. 9.8. This shows tight (high AI) sands in red and porous (low AI) sands in blue.

Deterministic inversion

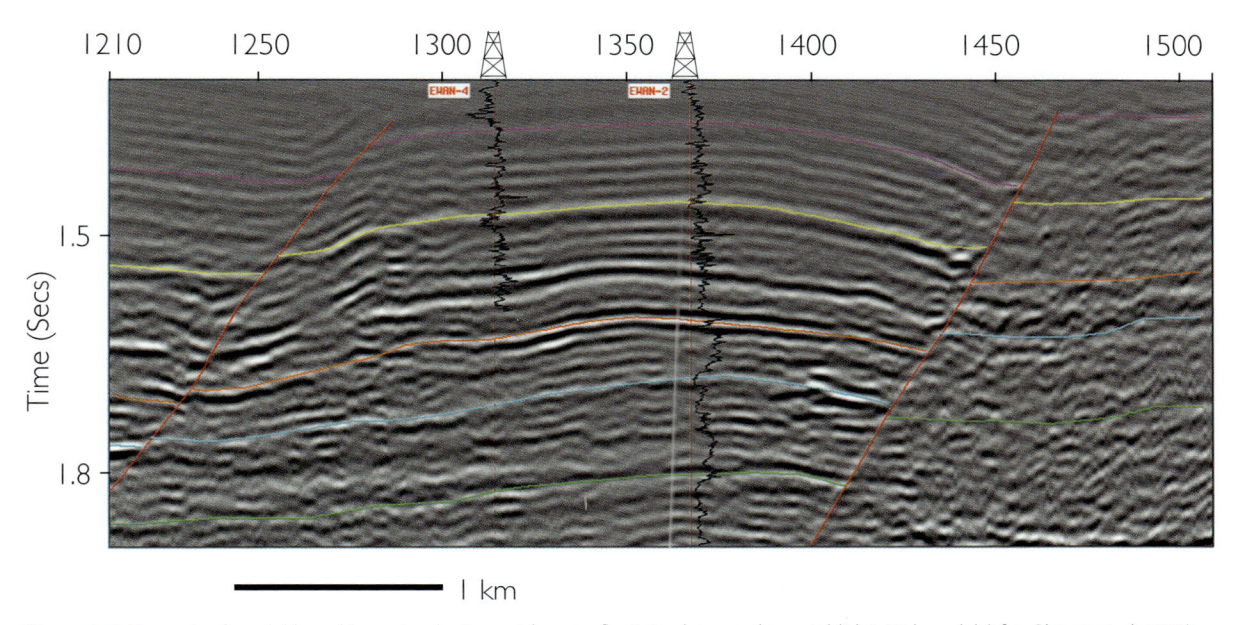

Figure 9.5 Example of model-based inversion; horizon picks on reflectivity data, used to establish initial model (after Pharez et al., 1998).

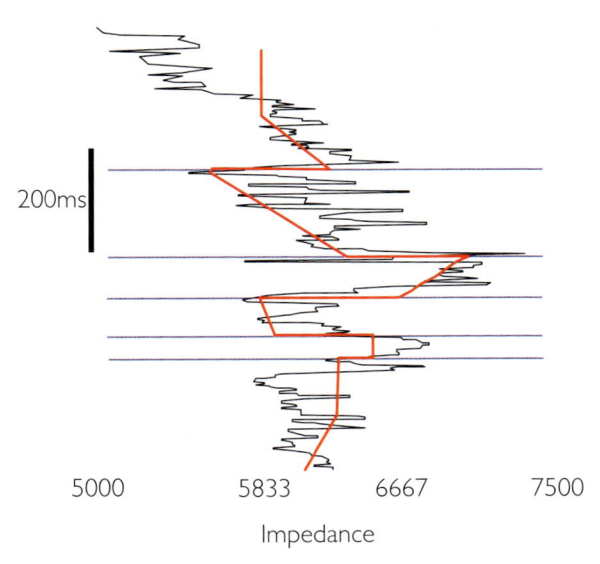

Figure 9.6 Example of model-based inversion; well impedance (black) and macro-layering starting model for the inversion (red).

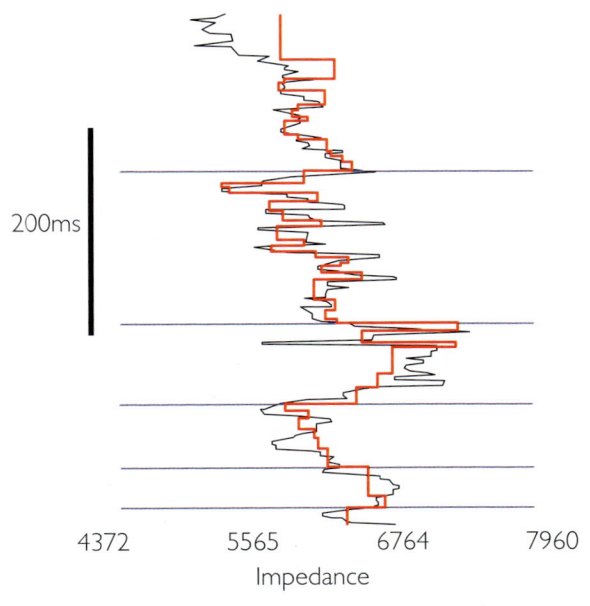

Figure 9.7 Example of model-based inversion; micro-layering output from the inversion (red), well impedance (black).

There are two important approaches to the checking of an inversion. These are impedance prediction at wells and synthetic–seismic error plots. If the initial model was created using the well as input, then the match is always likely to be good. A better test is to use a 'blind' well, not incorporated in the initial model. This is sometimes referred to as 'cross validation'. Blind well tests are only sensible, however, if there are a fairly large number of wells in the project area. If there are only two or three then the omission of one of them greatly loosens the constraint on the model. Figure 9.9 shows an example of a blind well test.

Seismic trace inversion

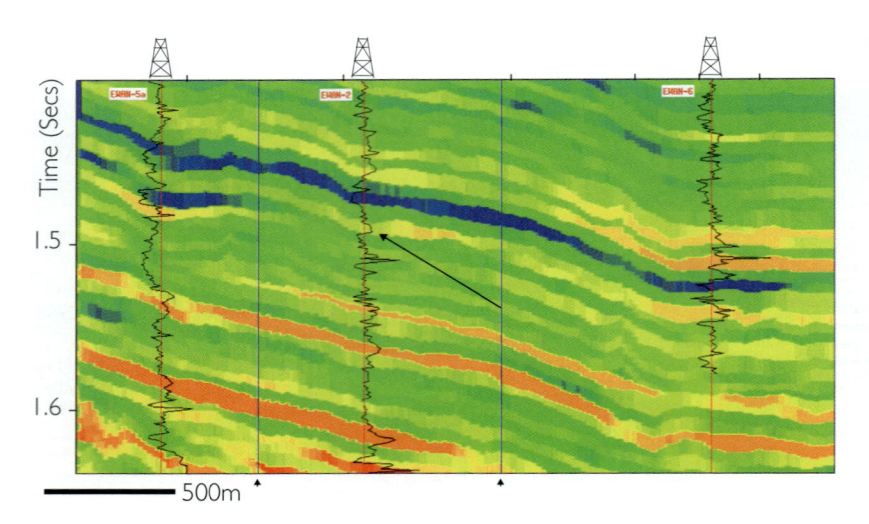

Figure 9.8 Example of model-based inversion; final impedance section, showing tight (high impedance) sands in orange and porous (low impedance) sands in green (after Pharez et al., 1998).

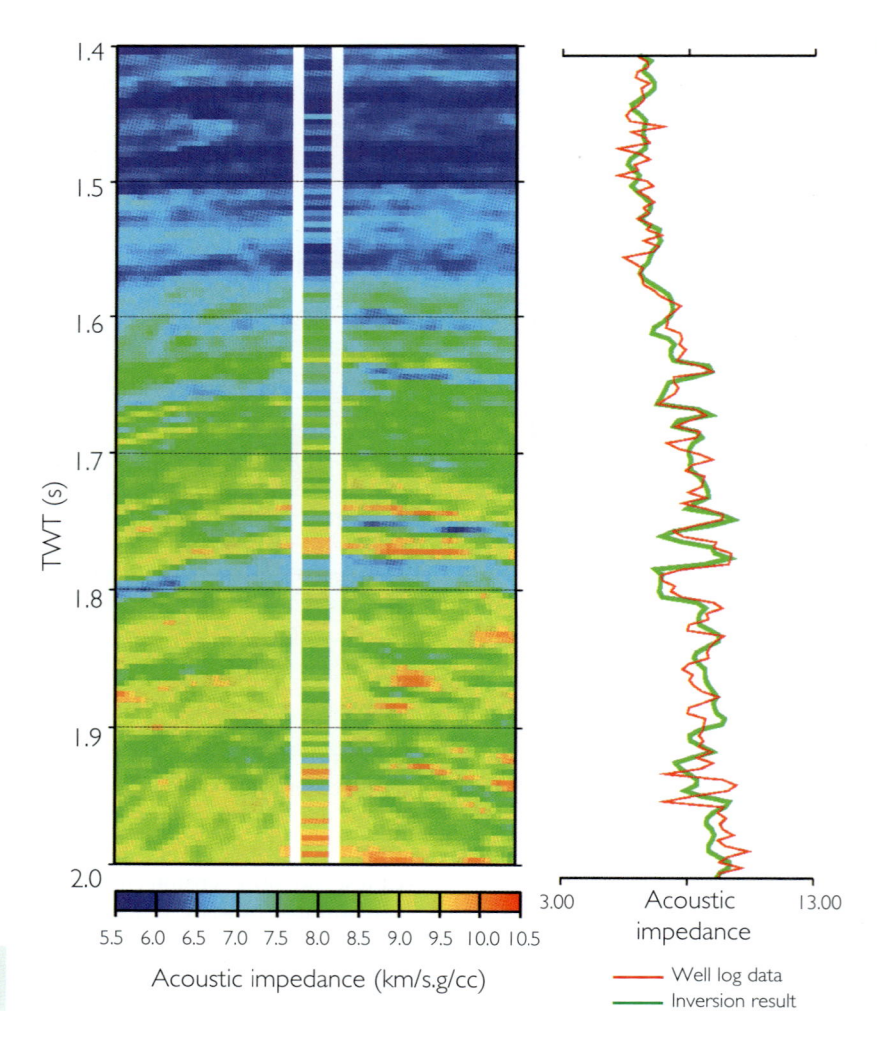

Figure 9.9 Comparison of inverted impedance result with impedance calculated from well logs (after Bach et al., 2000, with kind permission from Springer Science and Business Media).

Note the good fit at the reservoir target at 1.75 s and the poorer fit below, related to the presence of multiple energy in the seismic.

Two separate aspects of the inversion can be checked through comparisons of inverted and logged impedances at wells. Firstly, the scaling of the wavelet can be investigated. If the wavelet scaling is correct then the excursions of impedance around the low-frequency trend should be similar in amplitude for the well data and the impedance result. This scaling is derived from the well ties and might need adjustment if the ties have to be carried out over a different stratigraphic interval or depth range to the inversion target. Secondly, the goodness of fit of the inverted impedance and the logged impedance is to some degree dependent on the frequency cut-off applied to the low-frequency model. As the cut-off frequency is made lower, the inverted impedance is likely to give a worse fit. It should be remembered that a good fit at the wells does not ensure that the inversion is good away from the wells. There may be biases for example introduced by lateral variations in the low-frequency component.

A display of the difference between the input reflectivity section and the synthetic section calculated from the inverted impedance, using the inversion wavelet, is also a useful display for quality control. Ideally the display should have very low amplitudes everywhere with no coherent energy (Fig. 9.10). High amplitudes on the error display may be caused by the inverted impedance hitting a constraint boundary. This display can also be used to evaluate the best frequency cut-off for the background model. As the cut-off is made higher, the synthetic error is likely to increase. Of course, because of non-uniqueness, a small error does not guarantee the right answer. There are an infinite number of impedance volumes that will match the reflectivity input data.

9.2.4 Inversion issues

Often interpreters are not involved in the process of creating an inversion but it is their job to interpret it. Prior to applying geological thought to the inversion products it is necessary to evaluate as best as is possible the usefulness of the inversion. There are a number of distinct issues related to the low-frequency component and the wavelet of which the interpreter should be aware.

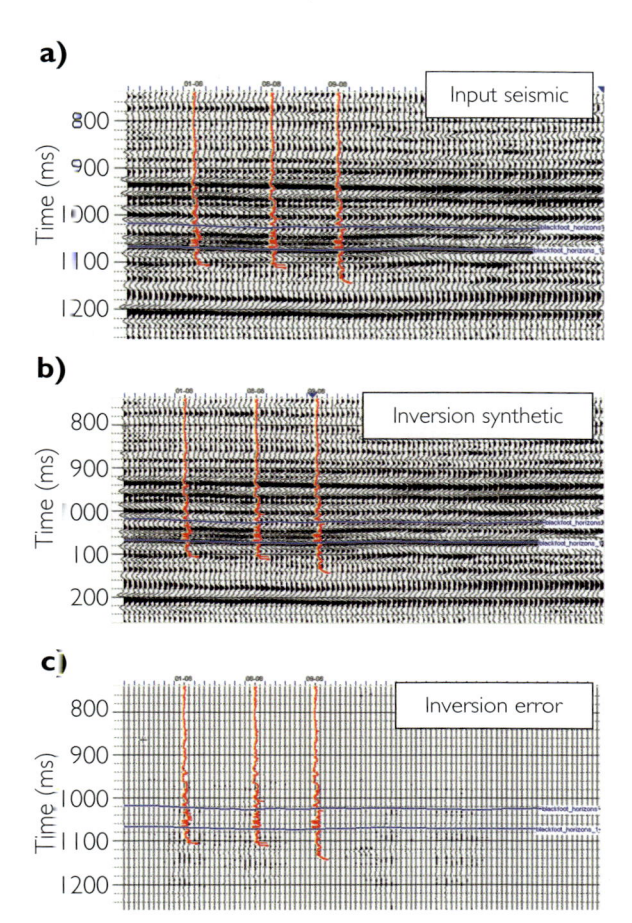

Figure 9.10 Input seismic compared to synthetic generated from inverted impedances; (a) input seismic, (b) synthetic generated from inversion, (c) difference between (a) and (b). Courtesy CGGV Hampson-Russell Software and Services.

9.2.4.1 Background model

The illustration in Fig. 9.11a shows a sparse spike inversion result (Francis and Syed, 2001). Applying a low-pass filter reveals the background model (Fig. 9.11b). It is clear that the pinchout in the final inversion result is quite simply a result of the background model. Applying a bandpass filter to the inversion result and comparing it with bandlimited impedance from seismic (Chapter 5) will reveal how much additional information is provided by the absolute inversion (e.g. Francis and Syed, 2001; Lancaster and Whitcombe, 2000). Often, the two bandlimited impedance products look identical.

9.2.4.2 Low-frequency merge

A particular issue with seismic inversion is the merging of the low-frequency component. Wagner

Seismic trace inversion

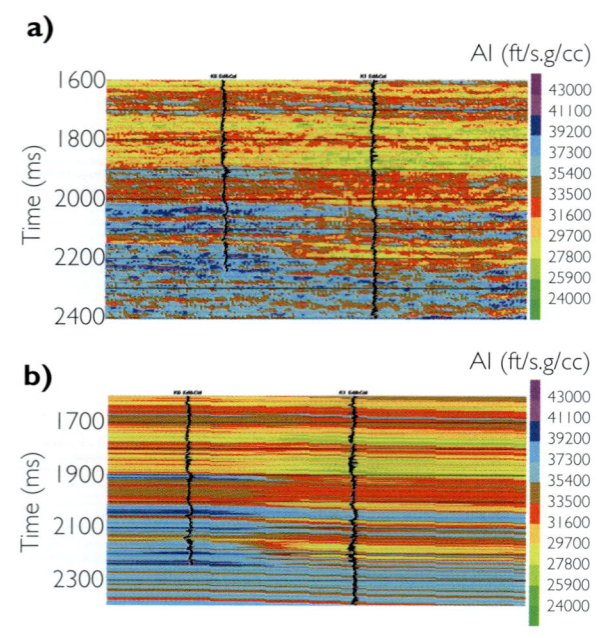

Figure 9.11 An example of a background model adversely influencing a broadband inversion result; (a) model-based inversion, (b) starting model based on interpolation of well data. The pinch-out is located halfway between the wells because of the interpolation methodology used (after Francis and Syed, 2001).

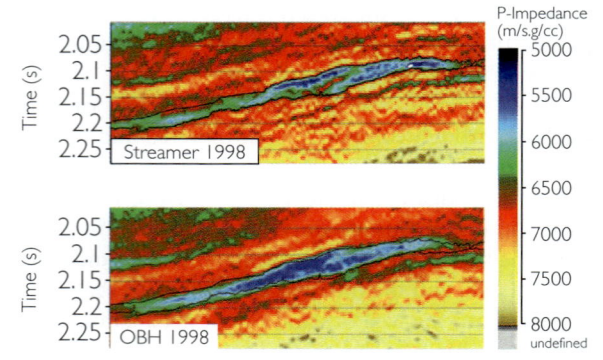

Figure 9.12 Comparison of inverted seismic sections from surface streamer and ocean-bottom hydrophone, after Wagner et al., 2006.

et al. (2006) highlighted the problem of merging the low frequency with a sparse spike inversion result in a comparison of inversions from streamer and ocean-bottom hydrophone (OBH) data. A key issue for these datasets is the difference in frequency content, with the streamer data having a reduced range of frequencies compared to the OBH. A comparison of the inversion results from the two surveys shows significant differences, with the streamer data showing apparently greater resolution with more layers evident in the reservoir section (Fig. 9.12). However, it is probable that the apparent layering in the streamer data is related to residual effects from the bandlimited impedance. The low frequency has not been correctly merged. The wedge model in Fig. 9.13 illustrates the problem. Maps of the impedance results from time-lapse data show dramatic differences between the streamer and OBH data (Fig. 9.14).

9.2.4.3 Wavelet issues

Understanding the wavelet in the data is essential for deriving reliable inversions. Figure 9.15 shows how inversion can be very sensitive to the wavelet used. Although the wavelets shown are very similar there are observable differences in the inversion. In the extreme case where the true phase of the seismic wavelet is 90° different from that used in the inversion, a step change in impedance will become transformed into a thin bed in the inverted result. The accuracy to which wavelet phase needs to be known will vary depending on the data and what the inversion result will be used for, but in general an accuracy of 30° or better is needed. Figure 9.16 shows an example of how an error in phase of 45° can influence the inversion result. Assuming that the impedance of the upper layer is matched, the error in phase changes the general scaling of the impedances although the general form is similar. It is evident that an accurate appreciation of wavelet shape and scaling (Chapter 4) is invaluable in deriving a good inversion.

When several wavelets are available from different well ties, they can be averaged provided their amplitude and phase characteristics are similar (Fig. 9.17). Usually the wavelets would be aligned on the dominant peak or trough before averaging, to allow for small static shifts arising from errors in the time–depth relations. This of course requires the wavelets to have a similar general appearance.

Wavelet averaging would not be the correct thing to do if the wavelet varies systematically either laterally or with depth. As discussed in Chapter 5, variation in both frequency content and phase with depth is to be expected; lateral variation is also possible, for example due to lateral change in absorption. In many cases the variation in depth of the target zone is small across the area of the inversion study, so its effect on the wavelet can generally be ignored. However, lateral variation can be quite significant, for example if there are localised patches of gas-filled sands in the

Deterministic inversion

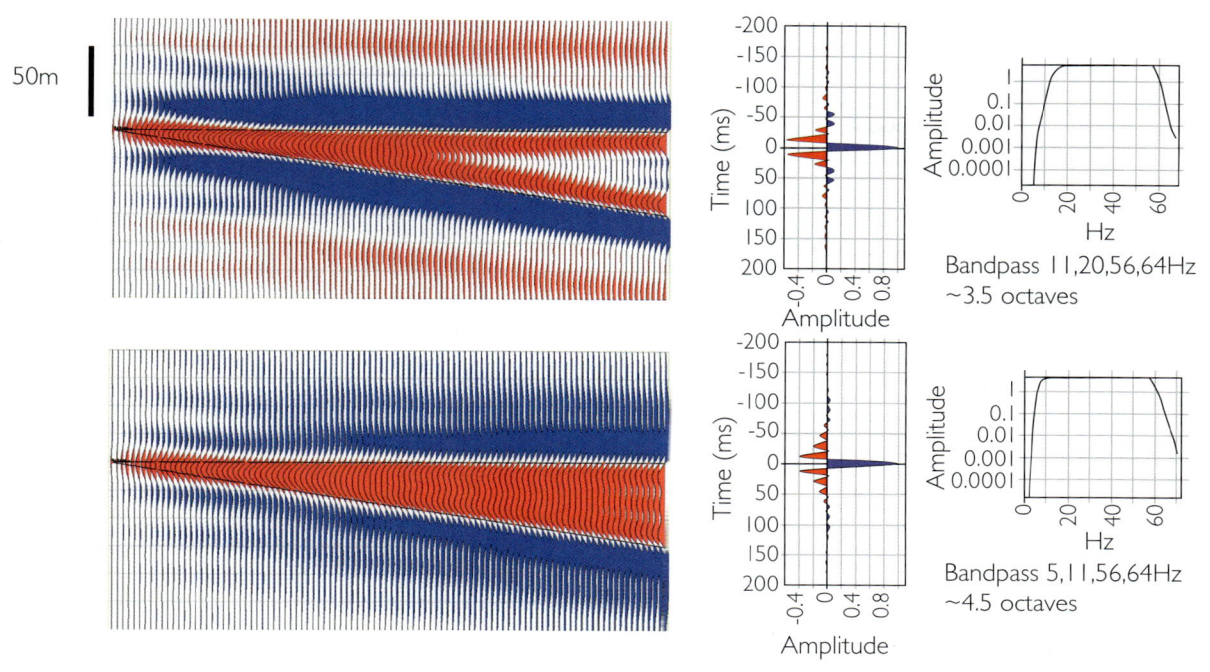

Figure 9.13 Bandlimited impedance wedge models using wavelets of different bandwidth; (top) 11–20–56–64 Hz bandpass filter (simulating surface seismic), (bottom) 5–11–56–64 Hz bandpass filter (simulating ocean-bottom hydrophone).

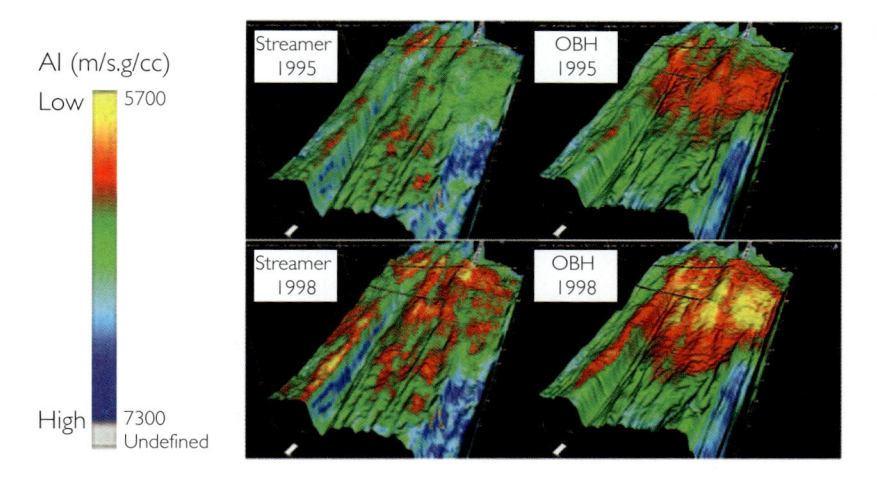

Figure 9.14 Comparison of inversion results from surface streamer and ocean bottom hydrophone; impedance near top reservoir, after Wagner et al., 2006.

overburden. In principle it is possible to use a laterally varying wavelet in the inversion, but in practice it will be hard to define the exact nature of the lateral variation between well control points.

9.2.4.4 Horizon constraints and tuning

In model-based inversions the choice of guiding horizons used in the starting model can have a significant impact on the final result. Figure 9.18 illustrates this effect using a simple wedge model with an artificial well providing impedance control. When only the top surface of the wedge is used (Fig. 9.18a) the tuning effects are not fully removed from the data. Inputting the base of the wedge into the model as well as the top (Fig. 9.18b) gives a good representation of the impedance in the wedge. The interpreter may be wary of inputting top and base horizons to the inversion, however, if the objective is to find a pinch out trap

Seismic trace inversion

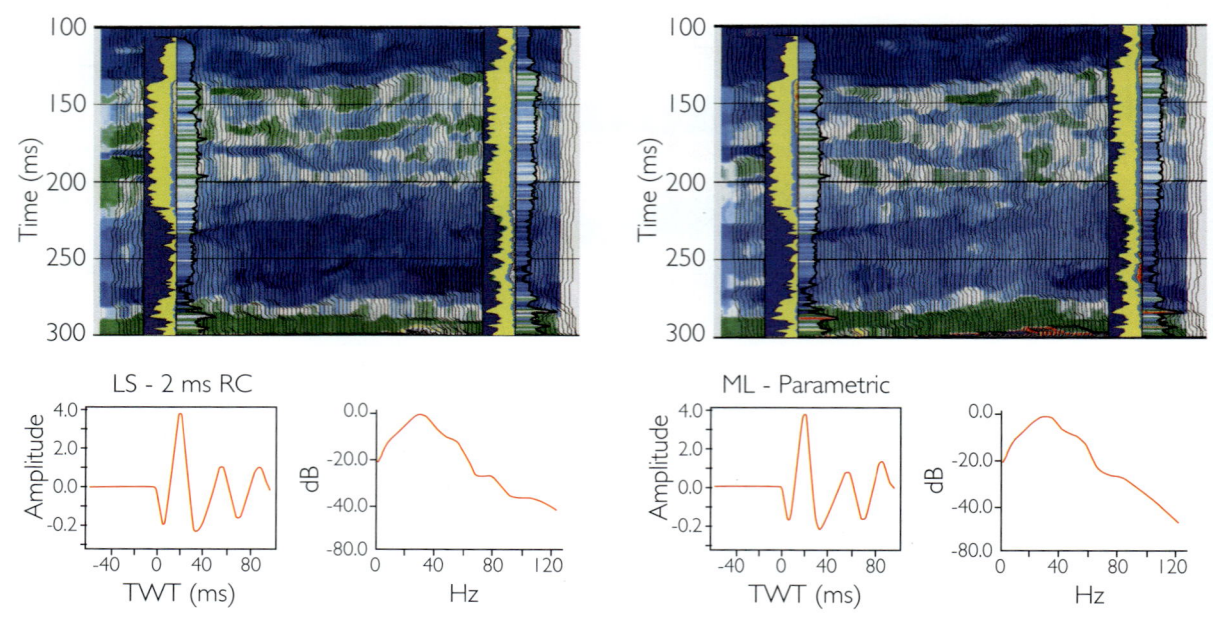

Figure 9.15 Sensitivity of inversion to wavelet shape; the two inversion results (top) were generated using identical parameters except that wavelets were derived using different extraction methods (bottom) (after Ozdemir et al., 1992).

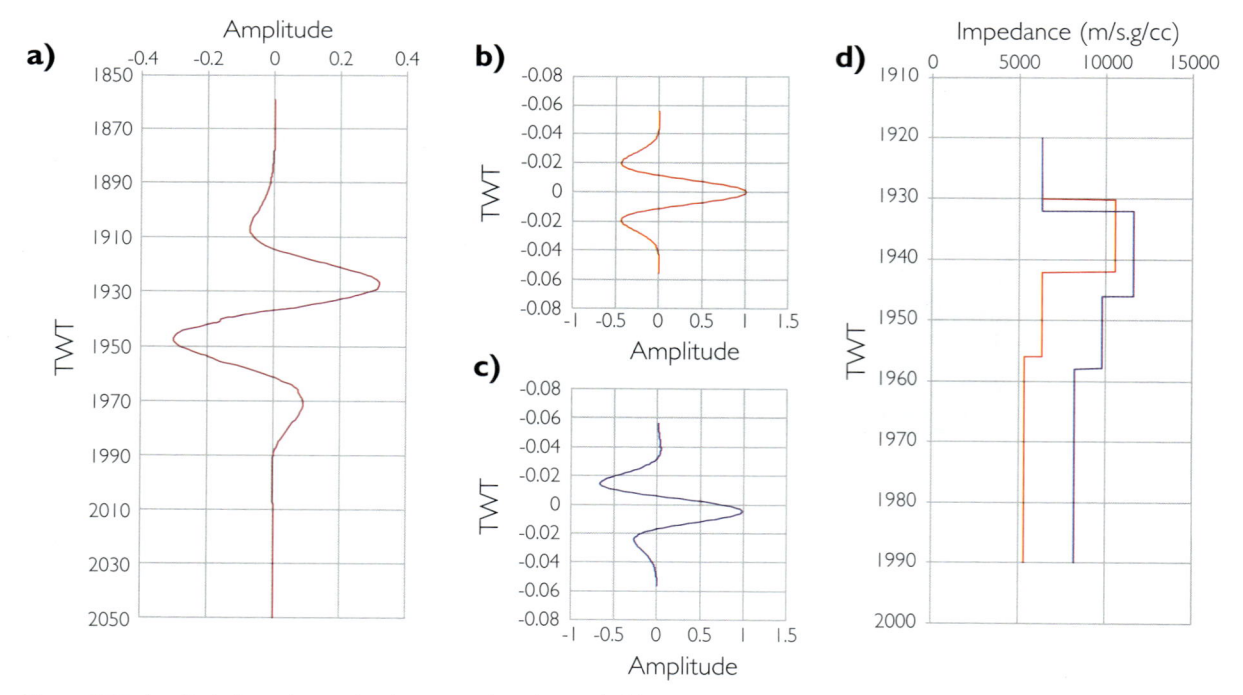

Figure 9.16 An effect of error in wavelet phase on an inversion result; (a) composite seismic response, (b) zero phase wavelet, (c) +45° degree wavelet, (d) red curve assumes the wavelet in (b), blue curve assumes the wavelet in (c). The impedance of the upper layer is kept constant between the two solutions.

Deterministic inversion

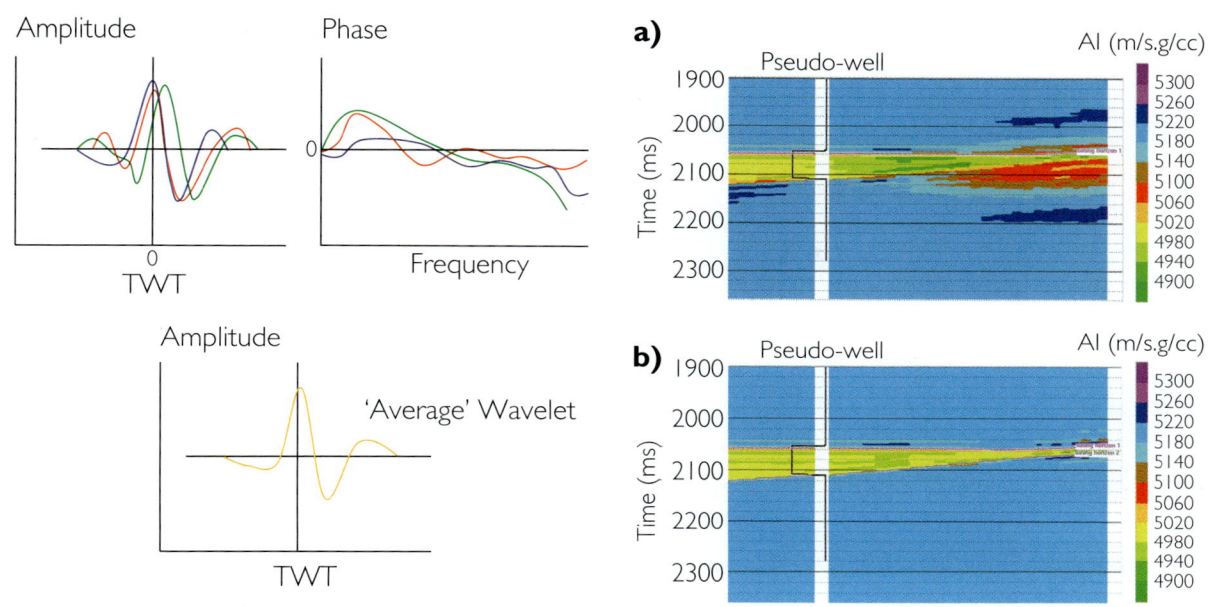

Figure 9.17 Averaging wavelets with similar phase spectra.

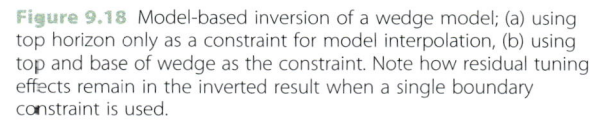

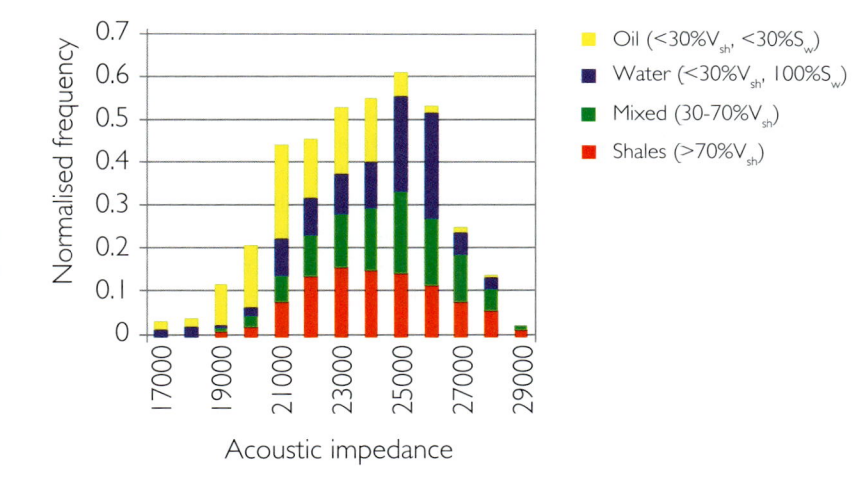

Figure 9.18 Model-based inversion of a wedge model; (a) using top horizon only as a constraint for model interpolation, (b) using top and base of wedge as the constraint. Note how residual tuning effects remain in the inverted result when a single boundary constraint is used.

Figure 9.19 An acoustic impedance histogram based on log data from six wells.

(Fig. 9.11 is a salutary lesson). The conclusion appears to be that inversions which are based on a few widely separated wells are likely to contain residual tuning effects.

9.2.4.5 Vertical scale

Very often, a preliminary study of well impedance values is undertaken prior to deciding whether an inversion is likely to give useful results. Often this will result in a histogram of impedance values for different lithologies and fluid fills, based on log data sampled at a depth spacing of about 15 cm. An example is shown in Fig. 9.19. It would be easy to conclude in this case that there is so much overlap between the various classes that inversion would not be useful in mapping out the oil sands. In fact an inversion clearly shows the oil sand distribution (Fig. 9.20). The reason is that seismic data see

207

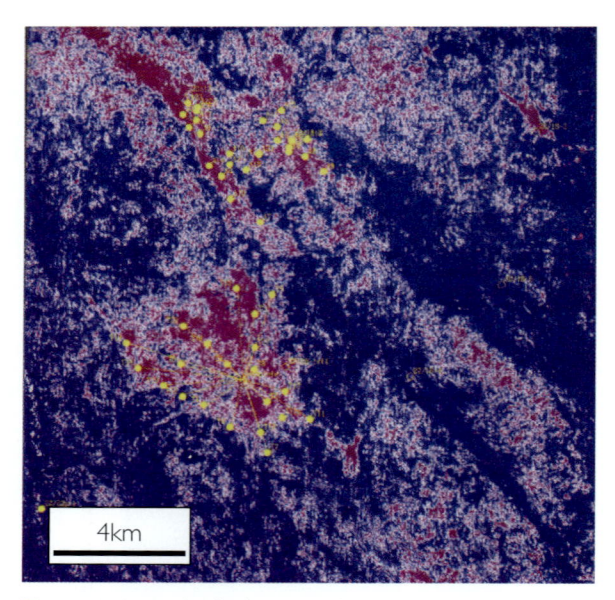

Figure 9.20 Amplitude slice 10 ms below top reservoir on a bandlimited impedance attribute and showing distinct low impedance zones (red) associated with two oil fields.

impedance averaged over tens of metres vertically, and therefore have a different distribution of impedance from the logs (Fig. 9.21). Often the main lithologies will show a greater degree of separation at the seismic scale than at the log scale. It is good practice to make histograms of well data after upscaling the logs to the resolution of the seismic data (for example by Backus averaging, see Chapter 4) when deciding whether inversion is likely to be useful.

9.2.5 Inversion QC checklist

Given the observations made so far, below are some questions that the interpreter might ask before embarking on an inversion interpretation.

- The wavelet:
  - how was it calculated?
  - how well is the wavelet understood, particularly its phase?
- Well ties:
  - how good is the match between well synthetics and the seismic reflectivity traces particularly over the target (e.g. reservoir) interval?
- Background model:
  - is the low frequency predictable or is it a possible source of bias in impedance values?
  - are there recognisable artefacts related to the background model (e.g. lateral variations such as pinchouts that are only contained in the background model)?
  - what does the comparison of a bandlimited impedance (e.g. coloured inversion) dataset and the inversion with bandpass applied indicate?
- Impedance prediction at wells:
  - do the high-frequency variations in impedance fit with the well data?
  - are there artefacts associated with the presence of residual tuning effects remaining in the data?
- Low-frequency merge:
  - are there potential bandlimited effects in 'thick' units that may be the result of errors in merging the low-frequency component?
- Scale and resolution
  - is the choice of inversion appropriate for the purpose of the study?

9.2.6 Bandlimited vs broadband

For deterministic inversions a fundamental choice lies between bandlimited and broadband methods. The former are more robust in the sense that they are staying close to the original seismic input, and may be entirely satisfactory if the geology is fairly blocky (e.g. sands and shales in a turbidite sequence). If there are gradational changes in thicker beds, then some attempt to supply a background model is probably needed in order to be able to interpret the impedance volume in geological terms. This may be much less robust given that the background model depends on a geological construct rather than directly observed data.

9.2.7 Inversion and AVO

Prior to the 1990s it was commonly held amongst geophysicists that a full stack could be simply inverted to acoustic impedance, with the AVO effects essentially ignored. With increased knowledge about reflectivity changes with offset it became clear, however, that AVO needs to be taken into account. Early attempts to incorporate AVO into seismic inversion were termed 'elastic inversion' and there were essentially two approaches.

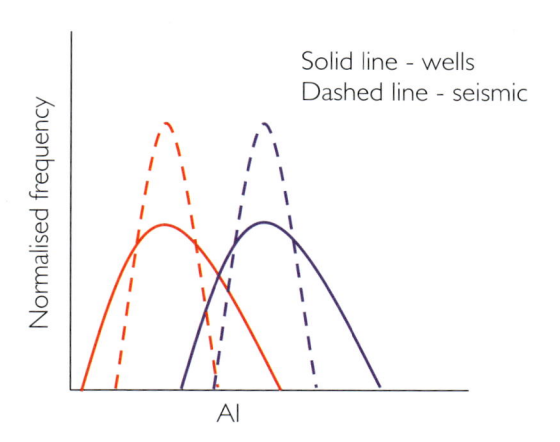

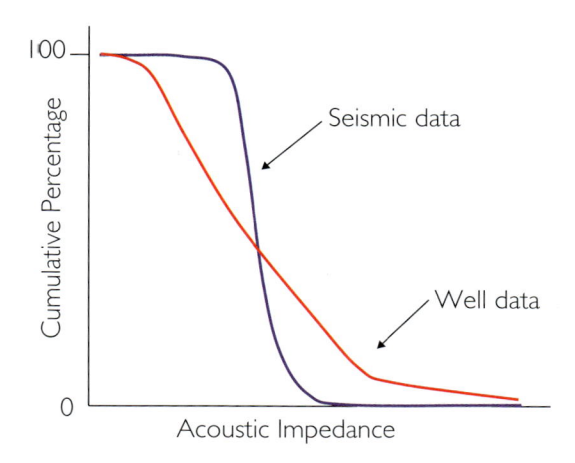

Figure 9.21 Impedance distributions, well and seismic.

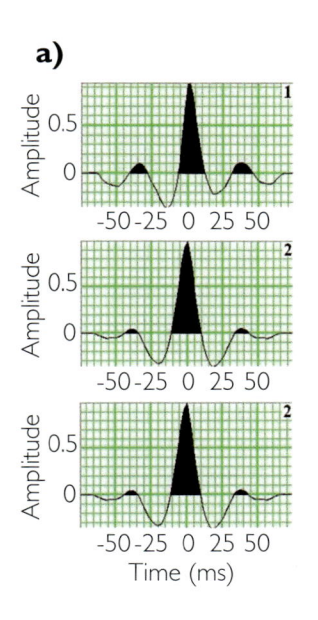

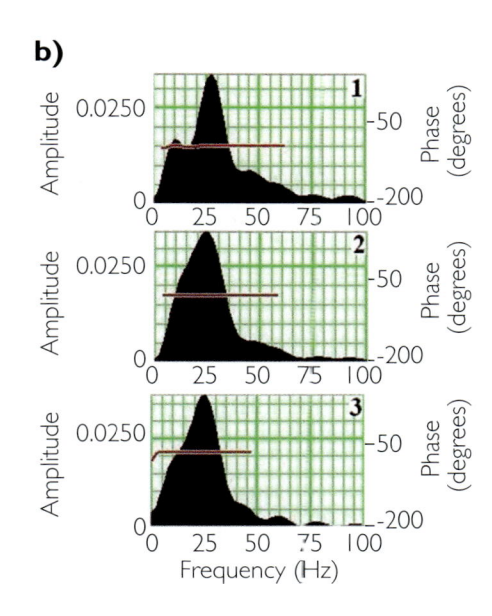

Figure 9.22 Wavelets and their spectra for near, mid and far stacks (top to bottom); (a) wavelets and (b) amplitude and phase spectra (after Sirotenko, 2009).

9.2.7.1 Elastic inversion – the Connolly approach

The 'Connolly' method uses elastic impedance (Chapter 5) as the basis of the starting model for inversion. Single stacks can be inverted using a background model generated with an impedance calculated at the appropriate angle. Wavelets are extracted from each stack so that separate inversions are effectively scaled to the model generated at wells (Fig. 9.22). This removes the need for any offset balancing correction (Chapter 6) and effectively accounts for any phase and frequency differences between stacks.

Initially, separate inversions were performed on near and far stack datasets and their results combined (using projections, Chapter 5) to extract fluid and lithology information (e.g. Simm et al., 2002). The development of the extended elastic impedance concept (Chapter 5) means that the fluid and lithology projections can now be performed prior to inversion. So, for example, pseudo-gamma ray inversions might be generated from a reflectivity 'stack' projected to an angle that correlates with gamma ray (Chapter 5) (e.g. Neves et al., 2004). A subsequent development in elastic inversion was joint or simultaneous inversion of near and far angle stacks (e.g. Rasmussen et al., 2004) in which additional constraints were applied, such as the co-dependency of P and S, in order to refine the inversion.

9.2.7.2 Elastic inversion – the Fatti approach

The 'Fatti approach' (after Fatti et al., 1994) is based on the extraction of fitting coefficients from pre-stack data using two- or three-term Aki–Richards approximations. A re-write of the Aki–Richards three-term equations is shown below. By introducing constraints based on a P wave velocity cube (i.e. offset to angle calculations and invoking a V_p/V_s transform (such as Castagna's mudrock line (Chapter 8)) the Zoeppritz approximation can be sufficiently constrained to obtain P and S reflectvities (two-term) or P, S and density reflectivities (three-term) from least-squares fitting:

$$R_{PP}(\theta) = c_1 R_{P0} - c_2 R_{S0} - c_3 R_D, \quad (9.2)$$

where $c_1 = 1 + \tan^2\theta$, $c_2 = 8\left(\frac{V_s}{V_p}\right)^2 \sin^2\theta$,
$c_3 = \frac{1}{2}\tan^2\theta - 2\left(\frac{V_s}{V_p}\right)^2 \sin^2\theta$, $R_{P0} = \frac{1}{2}\left[\frac{\Delta V_P}{V_P} + \frac{\Delta \rho}{\rho}\right]$,
$R_{S0} = \frac{1}{2}\left[\frac{\Delta V_S}{V_S} + \frac{\Delta \rho}{\rho}\right]$ and $R_D = \frac{\Delta \rho}{\rho}$.

Reflectivities extracted from the gathers are subsequently inverted to acoustic impedance and shear impedance, for example using the model-based inversion technique. The subsequent development was for simultaneous inversion of Z_p (acoustic impedance) and Z_s (shear impedance) from R_p and R_s (e.g. Pendrel et al., 2000).

9.2.7.3 Elastic inversion – intercept and gradient approach

Elastic inversion can also be approached from the perspective of the AVO intercept and gradient (White, 2000). The AVO intercept, which is the normal incidence P wave reflection coefficient R_p, and gradient G can be related to the normal incidence S wave reflection coefficient R_s as follows:

$$G = R_P - 8\gamma^2 R_S + \varepsilon \frac{\Delta \rho}{\rho}, \quad (9.3)$$

where V_p, V_s and ρ are the average values of the P wave velocity, S wave velocity and density at the interface, $\gamma = \frac{V_s}{V_p}$, $\Delta \rho$ is the density contrast across the interface, and $\varepsilon = 2\gamma^2 - \frac{1}{2}$. In most cases ε is small and the density term can be adequately approximated by a Gardner relation of the type $\rho = kV_P^n$. In this case R_s is linearly related to R_p and G:

$$R_S = aR_p + bG, \quad (9.4)$$

where $a = \frac{\left(1+\varepsilon\frac{2n}{1+n}\right)}{8\gamma^2}$ and $b = \frac{1}{-8\gamma^2}$.

It is thus possible to create trace volumes of the P and S wave reflection coefficients from AVO intercept and gradient and invert them individually to P and S wave impedance using exactly the same methods as discussed previously for acoustic impedance inversion. As with the Fatti approach a smoothed background model of V_s/V_p is required as an additional input to the calculation of R_s. The equations above are a salient reminder that the shear component in elastic inversion is directly related to the AVO gradient. Noise in the gradient (Chapters 5 and 6) will translate into errors in the impedance estimation. This is why data conditioning is considered to be so important for AVO inversion.

9.2.7.4 Pre-stack simultaneous inversion

The two-step process of reflectivity estimation followed by model-based inversion is now commonly replaced by one-step pre-stack simultaneous inversion algorithms deriving Z_p and Z_s directly (e.g. Ma, 2002; Hampson et al., 2005; Russell et al., 2006; Fig. 9.23). Clearly a good quality control is the match of the gathers input to the inversion with the synthetic gathers generated from the inversion result (Fig. 9.24). Once Z_p and Z_s volumes have been created, they can be easily manipulated to create other useful volumes such as $V_p/V_s (= Z_p/Z_s)$. It is also possible to create impedance volumes $\lambda \rho$ and $\mu \rho$ which are the product of density and the Lamé elastic constants λ and μ: $\lambda \rho = Z_P^2 - 2Z_S^2$ and $\mu \rho = Z_S^2$. These might be regarded as fluid and lithology volumes respectively (Goodway et al., 1999). As with AVO projections (Chapter 5) various adaptive combinations of acoustic and shear impedance can be created to highlight lithology and fluid (e.g. Espersen et al., 2000; Russell et al., 2006).

To stabilise the inversion process, it is usual to supply background models of the relation between Z_p and Z_s and between Z_p and density; the inversion calculates differences from this background trend (Fig. 9.25). This may be problematic if several different lithologies are present in the zone of interest, with very different V_p/V_s ratios. Pre-stack simultaneous inversion is commercially available in a number of software packages and is widely used. A typical result is shown in Fig. 9.26.

Deterministic inversion

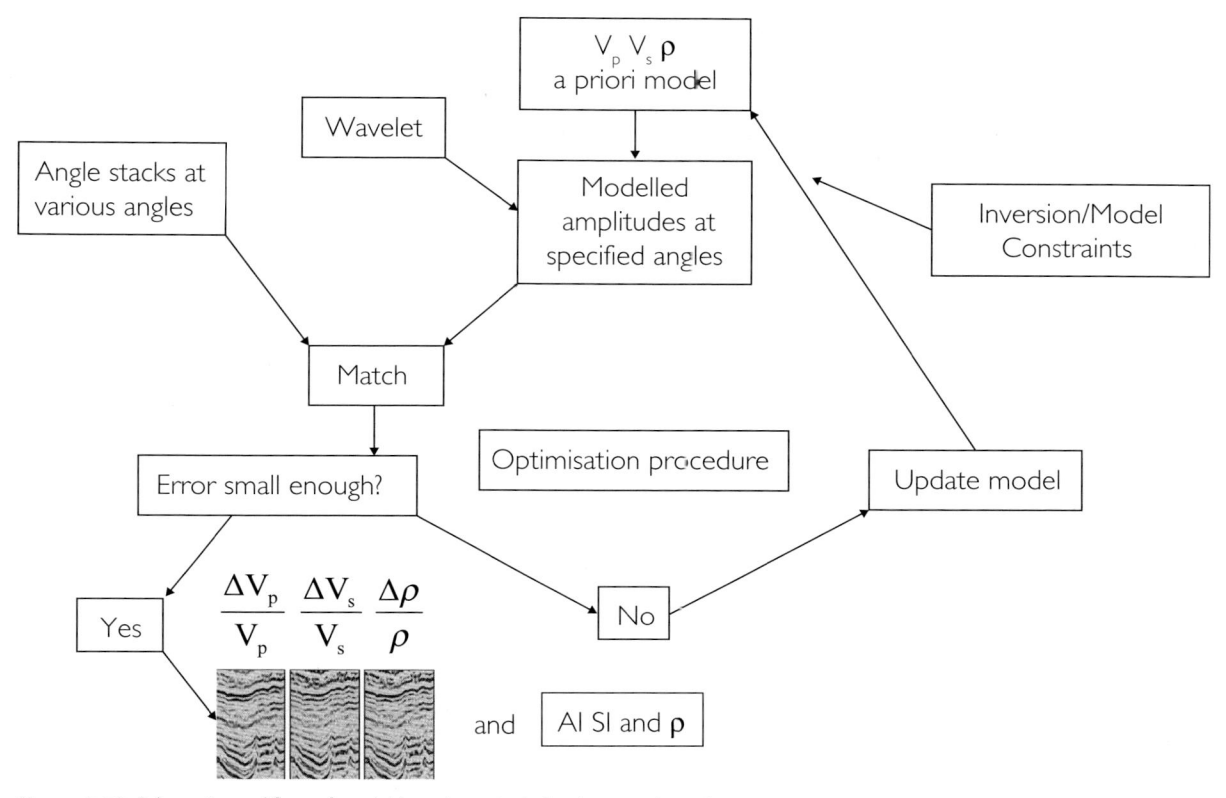

Figure 9.23 Schematic workflow of model-based pre-stack simultaneous inversion.

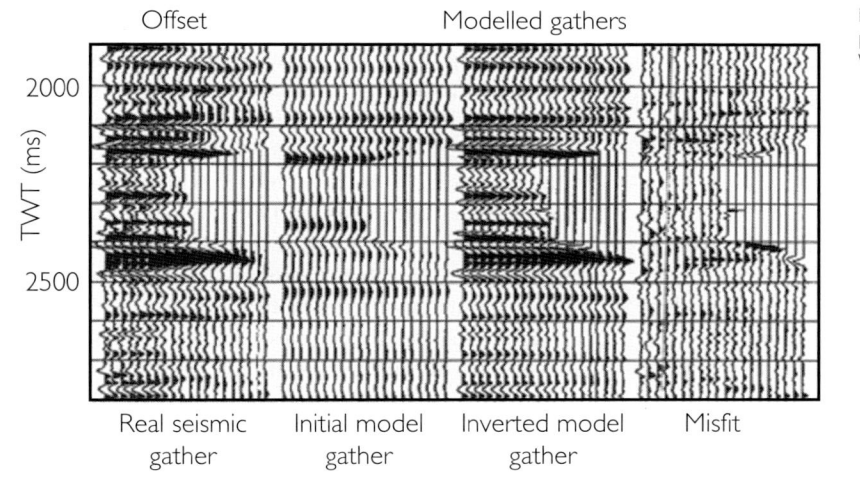

Figure 9.24 Gather match from pre-stack simultaneous inversion (after Veeken and Da Silva, 2004).

9.2.7.5 Density inversion

Inversion for the density term (Eq. (9.2)) offers the possibility of being able to recognise low gas saturations (e.g. Kelly et al., 2001; Skidmore et al., 2001; Roberts et al., 2005) (Fig. 9.27). These low saturations are of no commercial interest so it is important if possible to distinguish them from high saturations during prospect generation. Density varies linearly with saturation and so there will be no density anomaly associated with low gas saturations. However, in

Seismic trace inversion

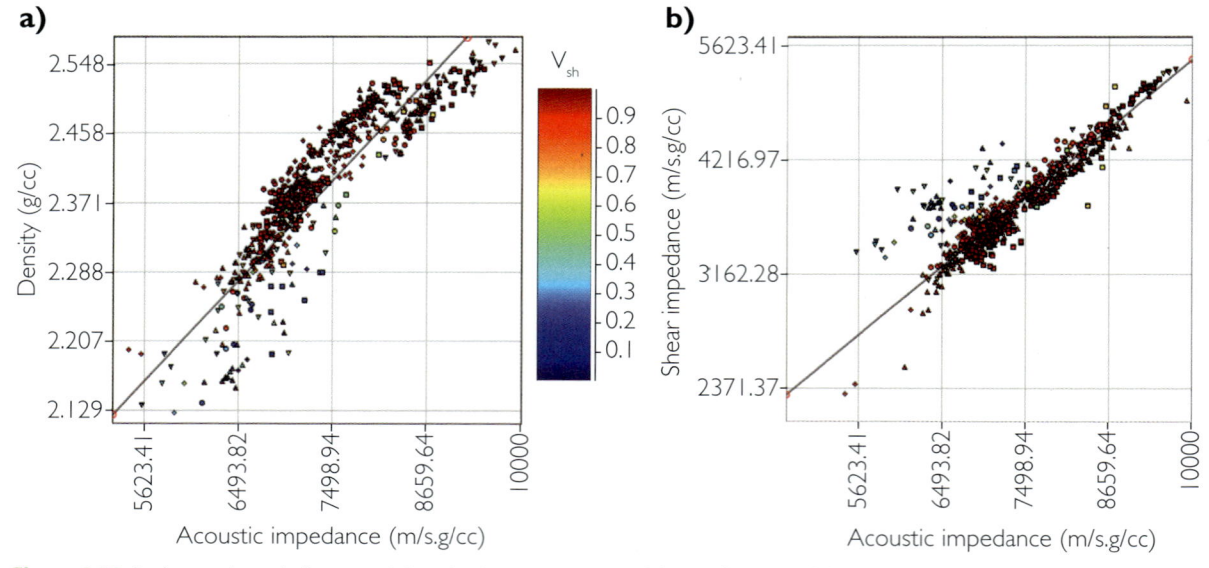

Figure 9.25 Background trends for constraining simultaneous inversion; (a) AI vs density and (b) AI vs SI plotted on log – log axes (courtesy Ikon Science).

terms of the AVO equations, density reflectivity is a relatively minor effect and only becomes important at far angles where there tends to be significant noise and uncertainty in the interpretation of the amplitude signature. If it is at all possible to extract density information from seismic, very good data are required with incidence angles beyond 40°.

9.2.8 Issues with quantitative interpretation of deterministic inversions

Because the seismic trace is bandlimited, there are a large number of impedance models which when converted to reflectivity and convolved with the wavelet would give a satisfactory match to the seismic traces. The inversion methods discussed so far can produce only a blocky average impedance solution. They will work best where the subsurface contains fairly 'thick' layers (so that the lack of high frequencies in the seismic data is not important), and where the layering is conformal with limited lateral variability (so that the low-frequency model can be established from limited well control).

As illustrated in the previous discussion there is considerable potential for bias to be incorporated into the inversion results, particularly through the low-frequency model (i.e. the DC component of the impedance). In addition to this, however, the smoothing inherent in deterministic inversions is also an issue that can cause both bias in volume calculation and errors in mapped connectivity in reservoirs with 'thin' beds (Francis, 2006a; Sancervero et al., 2005).

Figure 9.28 illustrates the issue with a logged section of a thin-bedded reservoir in which the original log (Fig. 9.28a) is upscaled to the seismic scale using a Backus average calculation (note that the sand is shown in blue). It is evident that the upscaled log (Fig. 9.28b) cannot be used to reliably predict the pay in the reservoir. If the threshold determined from the logs is applied to the upscaled log the net sand will be under predicted. In practice the interpreter faces a number of issues that can make it difficult to be precise about impedance calibration. For example, it is unlikely that an absolute impedance inversion will recover the Backus averaged log exactly owing to problems in removing tuning effects, and this can potentially offset the negative effects of smoothing. The more wells that are available, the better the calibration can be understood. Certainly the interpreter should be aware that predicting net sand from thin zones with deterministic inversions is likely to be prone to an unpredictable bias.

Connectivity is another issue that is affected by smoothing. Figure 9.29 from Francis (2006a) illustrates how smoothing applied to a structure map has

Stochastic inversion

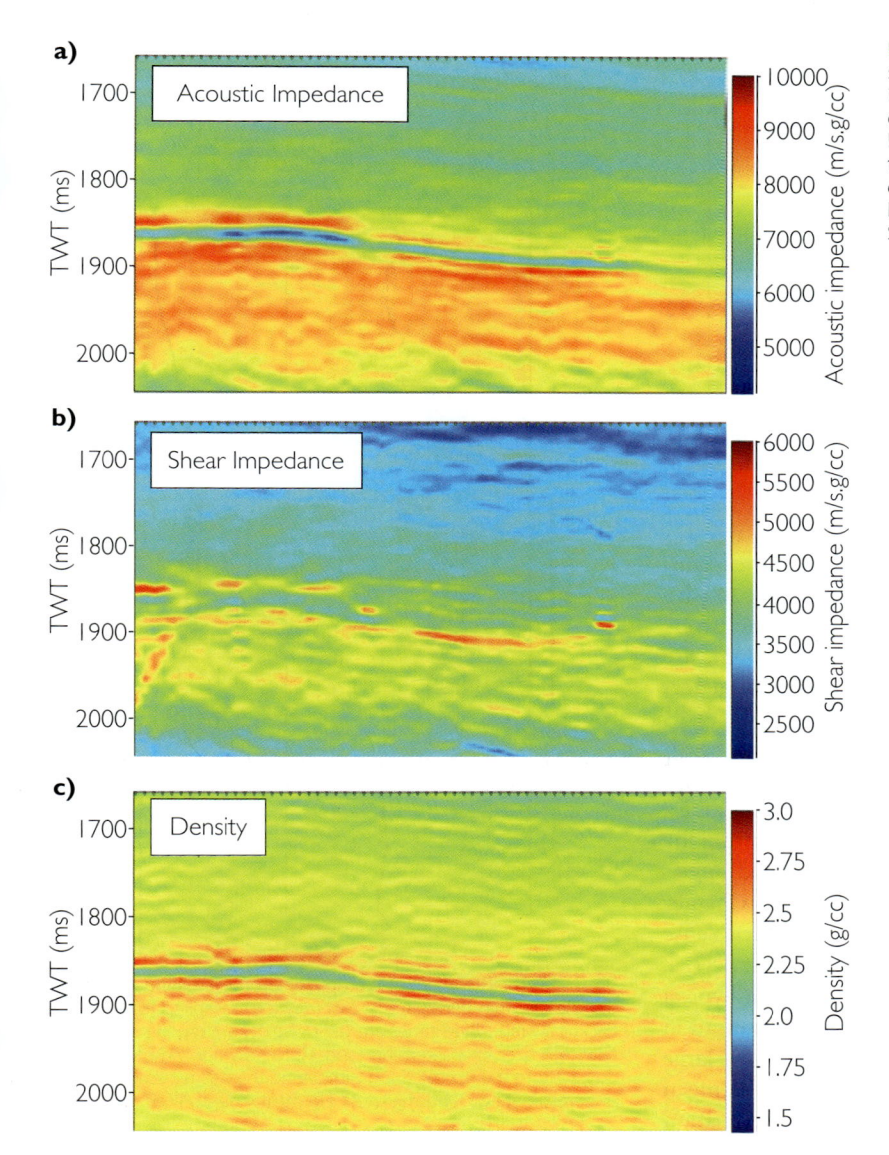

Figure 9.26 Typical results from simultaneous inversion; (a) acoustic impedance, (b) shear impedance and (c) density. In this case, density is determined largely from the correlation with AI, rather than independently from the seismic data. Note that the low AI and density layer is a hydrocarbon sand (courtesy Ikon Science).

the effect of lowering the gross rock volume in the closure as well as enhancing the connectivity.

9.3 Stochastic inversion

Whilst deterministic or best-estimate inversion obtains a minimised solution of the inverse problem, stochastic inversion techniques attempt to describe the potential variability of inverse solutions. Unlike deterministic inversion, therefore, a stochastic inversion does not provide a single 'optimal' solution. Multiple realisations of the subsurface impedance are generated, the synthetics from which all tie the seismic as well as honouring both the well data, the statistical properties of the impedances as well as any spatial model constraints. Given a sufficient number of realisations, the mean is close to the deterministic or best estimate inversion.

Stochastic inversions that utilise geostatistics (i.e. geostatistical inversions) typically use a small sampling increment (e.g. 1 ms) allowing the inversion results to be integrated with reservoir models. Indeed, experience has shown that integrating the seismic into geological models using geostatistical inversion

Seismic trace inversion

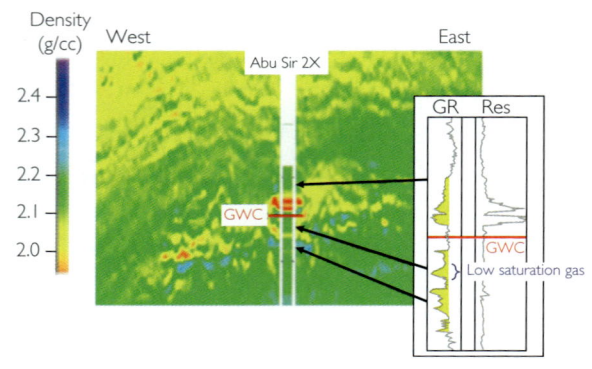

Figure 9.27 Density inversion at a well, showing commercial gas saturations in red and yellow colours and low gas saturation/non-gas responses in green (after Roberts *et al.*, 2005).

techniques enables greater control on the uncertainties in reservoir models (e.g. Rowbotham *et al.*, 2003b). Other stochastic methods, such as the Bayesian inversion technique of Gunning and Glinsky (2003), are more closely constrained to layer thicknesses determined by the bandwidth of the data.

Figure 9.30 illustrates an example of several impedance realisations, all fitting at the wells and honouring a geostatistical model determined from well data and a lateral spatial variability model. Note that following the discussion of angle-dependent impedance and angle-independent elastic parameters presented in Chapter 5, the term 'impedance' is used here in a generic sense to represent any quantity inverted from seismic (i.e. including V_p/V_s ratio). Figure 9.31 shows a typical example of a single stochastic impedance realisation and also the mean and standard deviation of a large number of realisations. The smoothness of the mean solution (Fig. 9.31b) compared to the single realisation (Fig. 9.31a) is a general feature of all stochastic inversions as is the increase in the standard deviation (Fig. 9.31c) away from the well control, consistent with honouring the data at the well.

An analysis of stochastic realisations can typically provide:

(1) probability of a particular facies occurring at a given location, for example oil sands having an impedance below a particular threshold,
(2) statistical distributions of volumes and areas,
(3) indications of likely connectivity, for example, whether a particular area of low-impedance gas sand samples is likely to be connected to a well penetration.

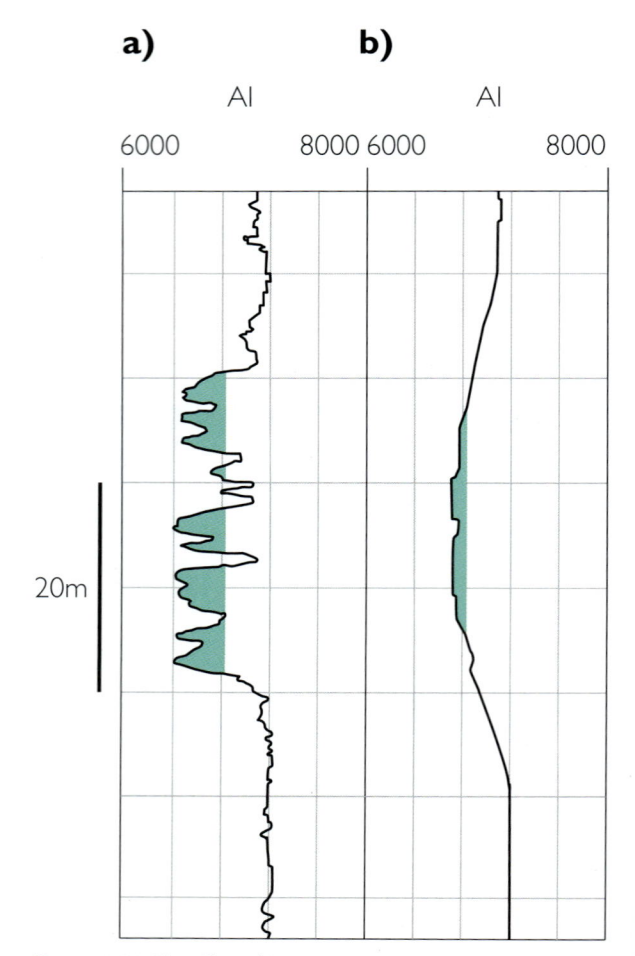

Figure 9.28 The effect of log upscaling on estimating sand thickness in thin beds; (a) 0.152 m (raw log) sampling (b) 20 m Backus average. Blue zones are values below an impedance value of 6865 m/s.g/cc, characterising sand. Note how applying the threshold to the upscaled log (simulating the seismic scale) gives an underestimate of sand thickness. Horizontal depth lines are 10 m.

Numerous workflows exist in stochastic inversion, from inversions that utilise geostatistics to generate multiple realisations of impedance in the time domain to simultaneous realisations of facies and reservoir properties conditioned by a geo-seismic model and presented within a depth referenced geo-cellular grid. Stochastic inversion is a rapidly developing area of geophysics and the discussion presented below attempts to outline the key elements and approaches. The reader is referred to the work of Dubrule (2003) and Bosch *et al.*, (2010) for useful reviews.

The first successful application of geostatistics in seismic inversion was presented by Haas and Dubrule

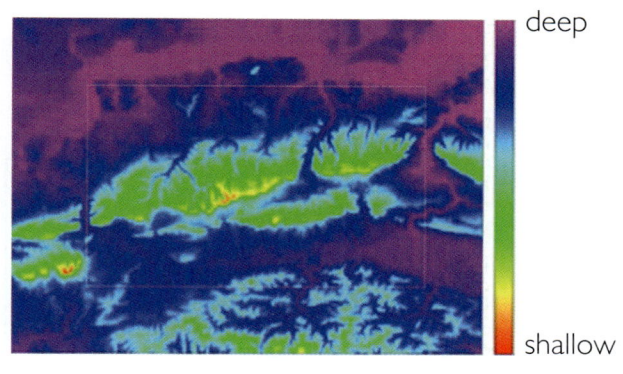

Topographic surface with false datum

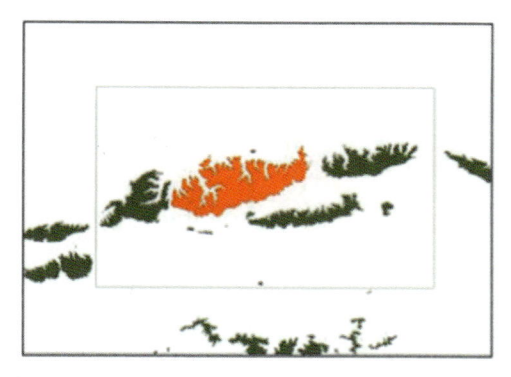

No smoothing GRV=300, A=984

3x3 moving average GRV=278, A=970

5x5 moving average GRV=261, A=957

Figure 9.29 Truncation of a topographic surface with a pseudo-hydrocarbon contact, followed by calculation of area A and Gross Rock Volume GRV above the contact, initially with no smoothing and then with two levels of smoothing. The smoothing causes the area and GRV of the central area (red) to decrease, and the highest smoothing creates erroneous connectivity to the west (from Francis, 2006a).

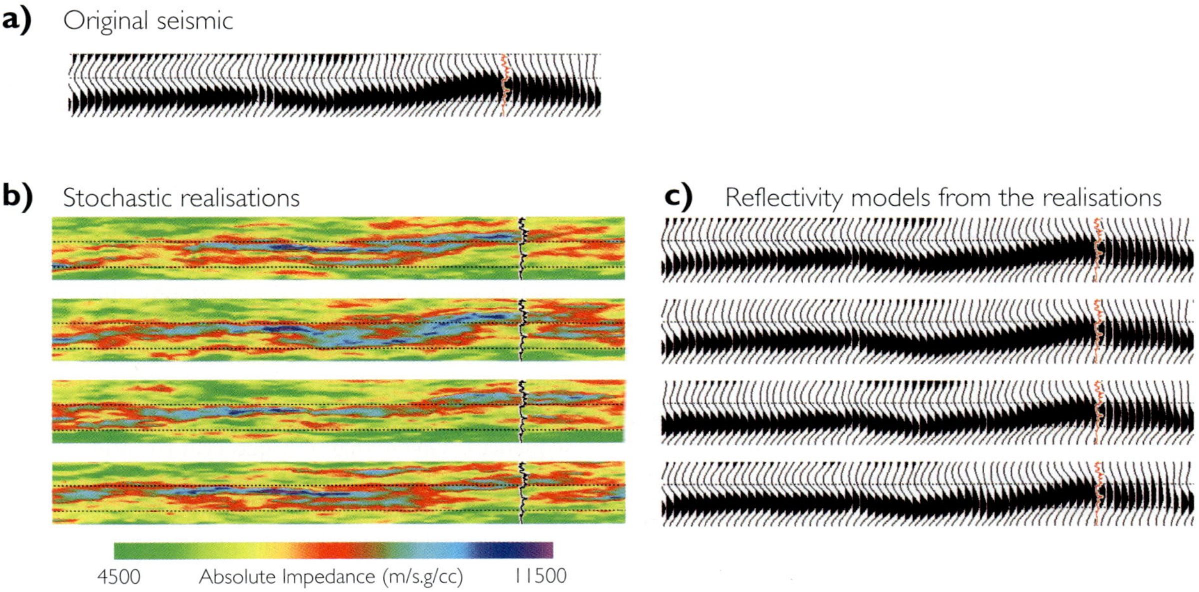

a) Original seismic

b) Stochastic realisations

c) Reflectivity models from the realisations

4500   Absolute Impedance (m/s.g/cc)   11500

Figure 9.30 Separate realisations from a geostatistical inversion, all of which fit the data at the wells and honour the geostatistical model (after Francis, 2006b).

Seismic trace inversion

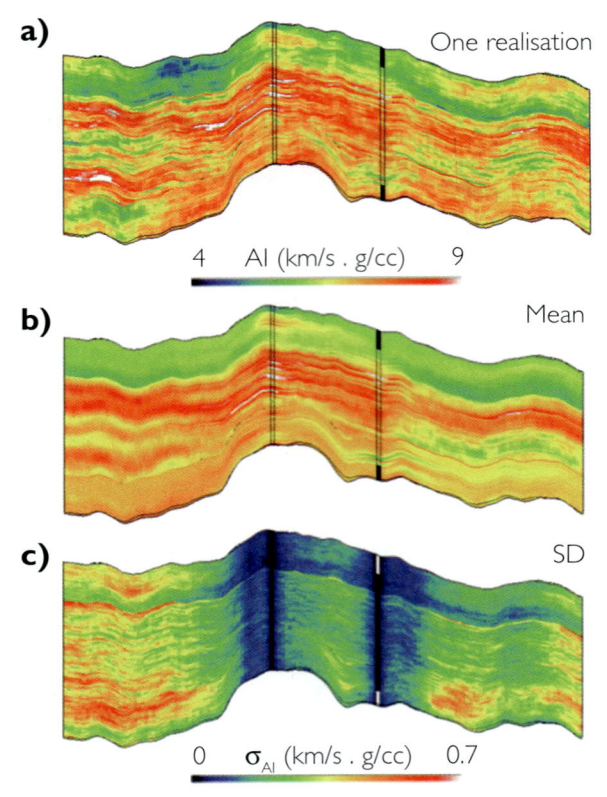

Figure 9.31 An example of geostatistical inversion products; (a) single realisation, (b) mean of the realisations and (c) standard deviation of the realisations. Note how the mean solution (b) is smoothed compared to the single realisation (a) and how the standard deviation increases away from the well control, consistent with honouring the data at the well (after Lamy et al., 1999).

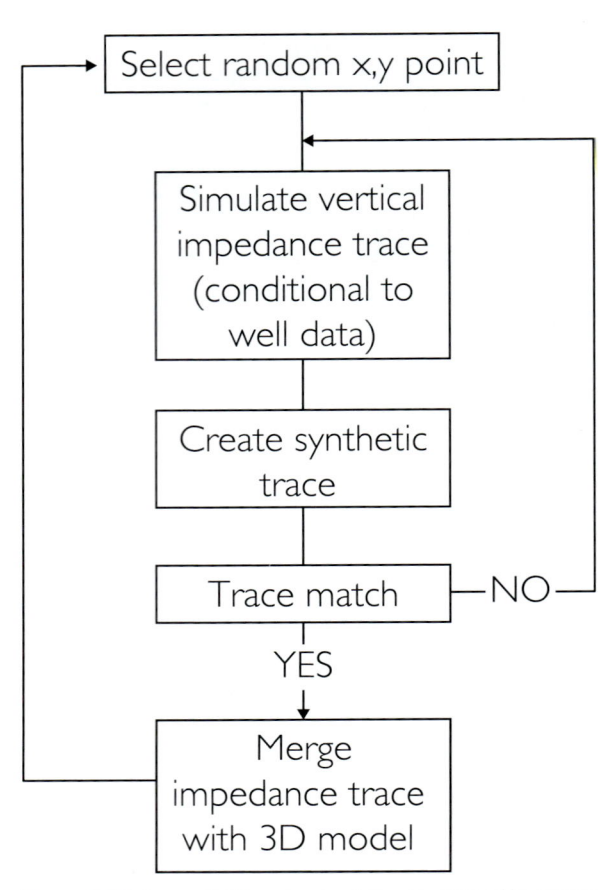

Figure 9.32 Sequential Gaussian simulation constrained by seismic data (based on Haas and Dubrule, 1994).

(1994). The methodology, referred to as *Sequential Gaussian Simulation*, is based on a random search method and works as follows. A particular trace is chosen at random from the set of seismic traces in the volume to be inverted and a large number of impedance trace realisations are simulated. The simulation estimates an impedance value at each pixel by kriging of well data to determine a value with its Gaussian distribution, and subsequent sampling using Monte Carlo. For each simulated impedance trace, reflectivity coefficients are calculated and convolved with the seismic wavelet. When a satisfactory trace match is obtained the impedance trace is incorporated as a new control point. Another seismic trace is chosen at random and the process repeated. A single global realisation is completed when all the traces have been inverted (Fig. 9.32). Another realisation can then be computed in the same way, using a different set of random choices.

Geostatistical inversion is commonly done within a 3D structural/stratigraphic geo-cellular grid (e.g. Marion et al., 2000; Rowbotham et al., 2003b). This allows appropriate control of the histogram model and spatial relationships in sedimentary layers and forms an effective basis for the integration of numerous types of data. Sequence stratigraphy and biostratigraphy provide the framework for correlations between wells and extrapolation away from wells, whilst depositional environments control the location, trends and proportions of facies within the sequences. Given that the geo-cellular grid is constructed in depth, it is important that there is an accurate 3D time–depth relationship.

An important aspect of any geostatistical approach is the assumption that the impedances from wells are statistically representative and the spatial descriptors are adequately defined. It is clearly important to have

Stochastic inversion

adequate well control. Critically, the variograms control the vertical variance and lateral continuity of the impedance realisations. High-frequency information in the realisations is constrained principally by the vertical variogram, with the low frequencies determined by the mean and standard deviation within each layer. Typically it is straightforward to derive vertical variograms from well logs, but there are usually not enough wells to determine the lateral variogram accurately. In many cases use is made of amplitude or impedance maps (e.g. a slice parallel to top reservoir on reflectivity or coloured impedance data) from which the lateral dimensions of stratigraphic features are inferred. Geological analogue information can also be used to define the lateral variogram. Careful consideration and sensitivity analysis is required. It should be remembered that the lateral variogram has a significant impact on any connectivity interpretation based on the simulation results. Usually the variations of impedance are not identical in every direction so anisotropies must be taken into account. Figure 9.33 illustrates examples of vertical and lateral variograms.

Impedance traces are selected on the basis of satisfying an objective function. Various goodness of fit

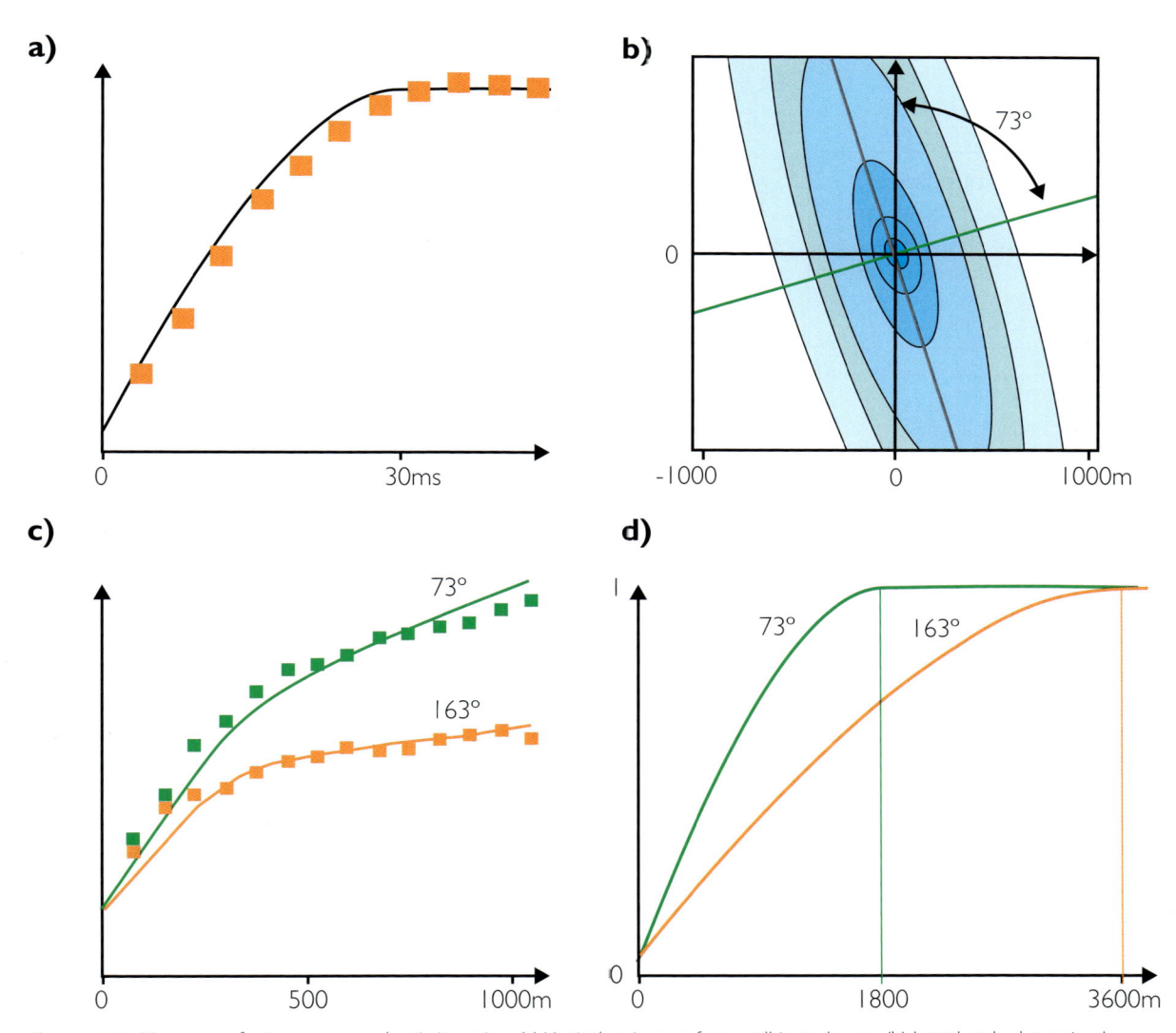

Figure 9.33 Variograms for input to a stochastic inversion; (a) Vertical variogram from well impedances, (b) lateral scale determined from seismic amplitude map with anisotropy, (c) horizontal variogram for the main axes of anisotropy identified in (b), (d) final lateral variogram (from Haas and Dubrule, 1994).

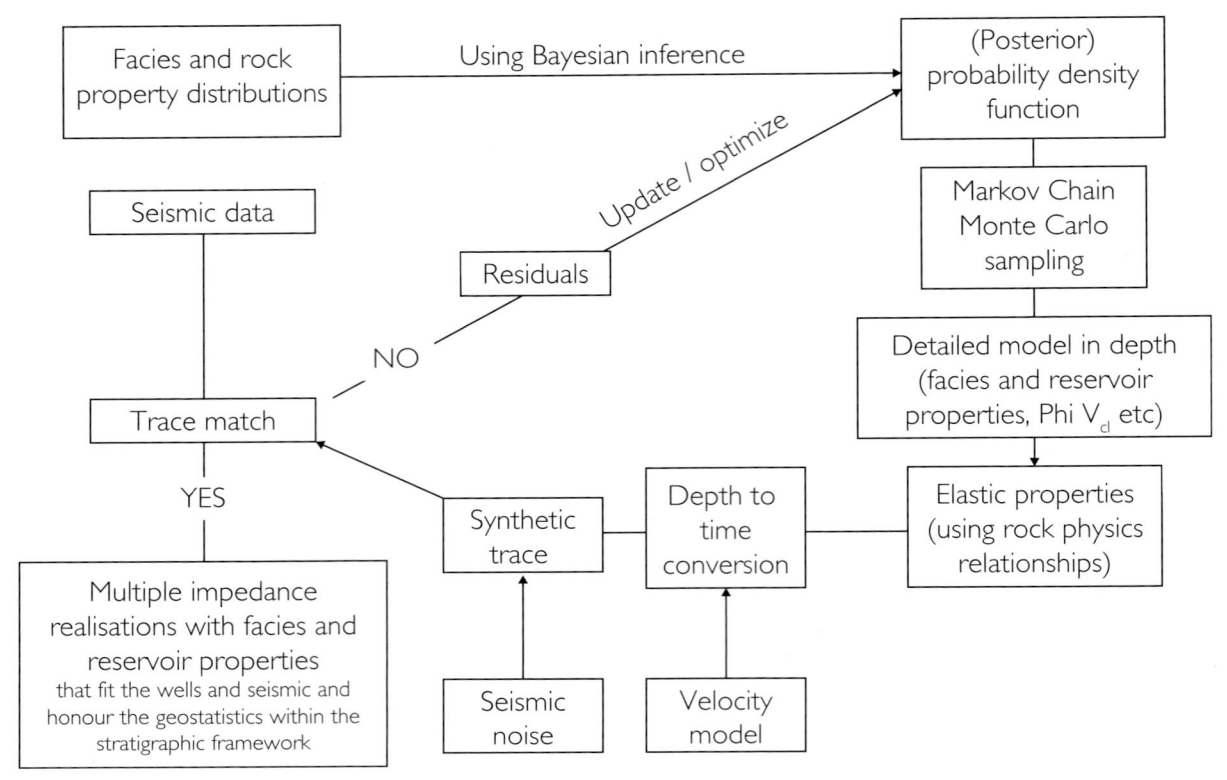

Figure 9.34 A stochastic inversion workflow incorporating Bayesian inference (derived from the work of Saussus and Sams (2012) and Sams et al. (2011)).

measures might be used, such as absolute error, mean squared error or correlation coefficient or combinations thereof. Typically a cross-correlation coefficient similar to that used in well ties is calculated. As in deterministic inversion an analysis of the residuals is a useful QC as well as the use of 'blind' well testing. Data quality is an important issue determining methodology and parameter choice. With deterioration in seismic quality the seismic has less of a constraining effect and the realisations increasingly resemble the results of kriging with well data.

It is not immediately obvious how many realisations are sufficient for geostatistical inversion results to be statistically meaningful. Effectively, there need to be enough realisations to generate smooth cumulative distributions of parameters derived from the realisations. The minimum number of realisations is usually of the order of 100 or greater. Whilst Sequential Gaussian Simulation is a useful way of understanding how geostatistical inversion works, as a practical implementation it has the drawback of being computationally slow. Some authors have implemented faster algorithms such as Markov Chain Monte Carlo, a random walk method in which the algorithm is guided (e.g. Contreras et al., 2005), or simulated annealing (e.g. Debeye et al., 1996; Torres-Verdin et al., 1999), but creating enough simulations in a time efficient manner remains a significant challenge. Francis (2002) adopts a hybrid approach, referred to as Multi-Point Stochastic Inversion (MPSI), in which traditional inversion methods are used to condition model realisations to the seismic in the time domain and within the seismic bandwidth. Although a variety of geostatistical methods might be used to generate these realisations, a Fast Fourier Transform based spectral frequency method is preferred.

Since the mid 1990s there have been significant developments in the way that geostatistical inversion is approached particularly from the view point of data integration and the conditioning of reservoir models with seismic and prior rock physics information. In a sequential workflow populating reservoir models with seismic conditioned reservoir properties is done by transforming the stochastic impedance realisations with various statistical approaches such as linear regression functions, collocated cokriging (Doyen *et al.*, 1996) and Bayesian classification (e.g. Coulon *et al.*, 2006). However, a relatively early development was the co-simulation of lithofacies with impedance (e.g. Torres-Verdin *et al.*, 1999; Sams *et al.*, 1999)). The logic of using the seismic directly to condition reservoir properties in the 3D model via the facies concept is compelling (e.g. Saussus and Sams, 2012). For example, ensuring that elastic properties are consistent say with a saturation height function is difficult unless saturation is part of the inversion model (Sams *et al.*, 2011). Notionally, applying all constraints simultaneously will lead to tighter integration and a more robust and consistent reservoir model.

Since the work of Buland and Omre (2003) and Buland *et al.* (2008), Bayesian inference has increasingly been used to frame the model space from which stochastic realisations are drawn (e.g. Eidsvick *et al.*, 2004; Moyen and Doyen, 2009; Escobar *et al.*, 2006; Merletti and Torres-Verdín, 2010). In Bayesian reasoning, the model space (or *posterior* distribution) is the product of the *prior* information (i.e. the distributions and spatial characteristics of elastic parameters from well data, plus other constraining geological information) and the *likelihood* of the prior given the seismic observation (i.e. as determined by the synthetic match to seismic). Added levels of sophistication have been incorporated into the prior model by some authors, for example with the use of Markov random fields or Bayesian networks to define dependencies and spatial couplings (e.g. Eidsvick *et al.*, 2004; Rimstad and Omre, 2010). Figure 9.34 shows a workflow based on the work of Saussus and Sams (2012) and Sams *et al.* (2011), and Fig. 9.35 illustrates typical outputs.

Given the level of data conditioning required and the use of sophisticated algorithms, stochastic inversion is usually performed by specialists on fields in the appraisal or development stages. Although Bayesian inversion software code has been made publicly available (e.g. Gunning and Glinsky, 2003) there are as yet

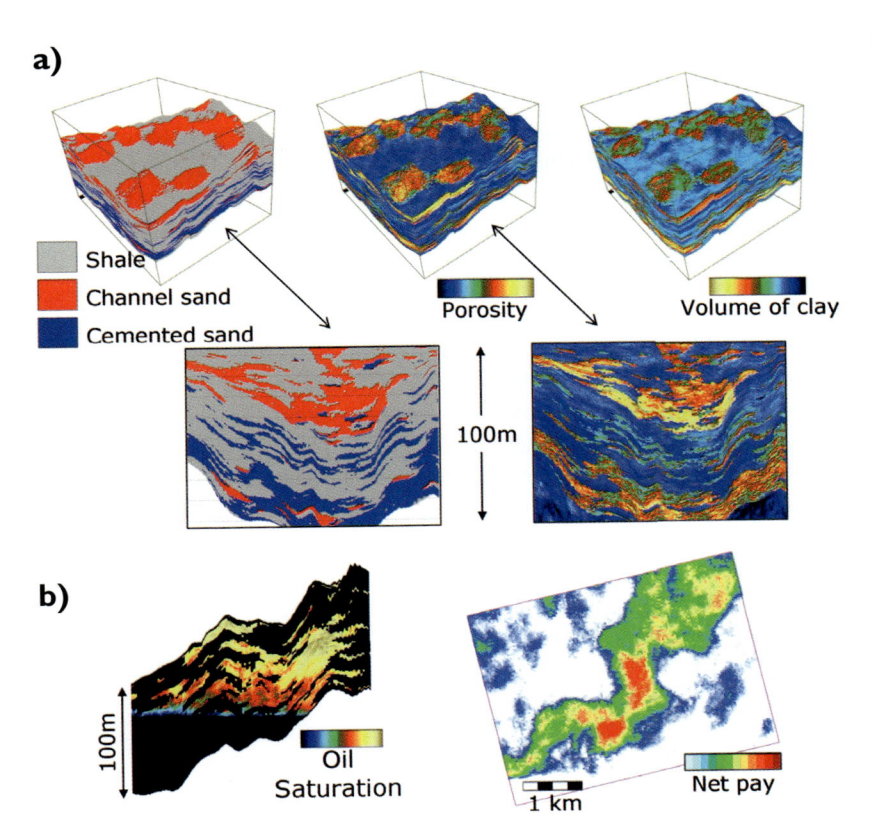

Figure 9.35 Volumes output from a single geostatistical inversion realisation include facies, petrophysical parameters (i.e. porosity, volume of shale and hydrocarbon saturation). The realistic detail, heterogeneity and geological plausibility arise from the inversion being conditioned to geological as well as elastic parameters (after Saussus and Sams, 2012).

Seismic trace inversion

```
┌─────────────────────────┐
│ Statistical rock physics │
└───────────┬─────────────┘
            ↓
┌──────────────────┐    ┌──────────────────┐    ┌──────────────────┐    ┌──────────────────────┐
│  Pseudo-wells    │    │ Trace matching   │    │ Posterior for    │    │ Spatial statistics   │
│ (likelihood      │───▶│ (likelihood      │───▶│ reservoir        │    │ for reservoir        │
│  relating        │    │  relating elastic│    │ properties from  │    │ properties           │
│  reservoir       │    │  properties to   │    │ seismic          │    │                      │
│  properties to   │    │  seismic data)   │    │ waveforms        │    │                      │
│  elastic prop.)  │    │                  │    │                  │    │                      │
└──────────────────┘    └──────────────────┘    └──────────────────┘    └──────────────────────┘
         ↑                                                ↓                       ↓
┌──────────────────┐                             ┌──────────────────┐    ┌──────────────────────┐
│ Vertical stats   │                             │ Conjunction of   │───▶│ Multiple realisations│
│ for reservoir    │                             │ reservoir        │    │ of reservoir         │
│ properties       │                             │ properties       │    │ properties           │
└──────────────────┘                             └──────────────────┘    └──────────────────────┘
                                                          ↑
┌──────────────────┐    ┌──────────────────┐    ┌──────────────────┐    ┌──────────────────────┐
│ Reservoir prop.  │    │ Seismic facies   │    │ Posterior for    │    │ Kriged well data     │
│ from analogue    │───▶│ interpretation   │───▶│ reservoir        │◀───│ (optional)           │
│ data             │    │ (likelihood      │    │ properties from  │    │                      │
│ (likelihood)     │    │  relating macro- │    │ seismic facies   │    │                      │
│                  │    │  facies to       │    │ and well data    │    │                      │
│                  │    │  seismic data)   │    │                  │    │                      │
└──────────────────┘    └──────────────────┘    └──────────────────┘    └──────────────────────┘
```

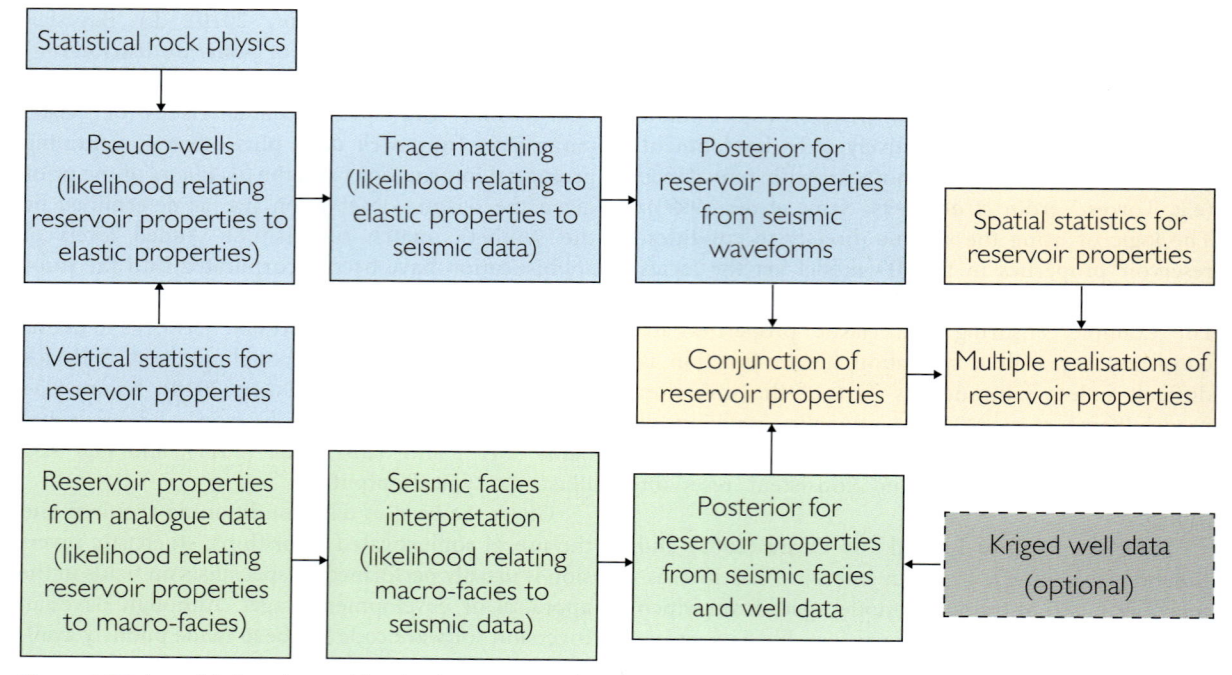

Figure 9.36 A possible inversion workflow involving trace matching with reservoir property statistics being combined with separately mapped facies data (after Connolly, 2012).

no commercially available software packages. To a large extent Bayesian applications are still very much in a research phase.

Adoption of new geostatistical techniques is often hampered by the fact that geophysicists view them sceptically as the application of sophisticated random number generators in a black box (Rowbotham et al., 2003b). Some authors seek more transparency in the process, particularly so that there is a broad understanding amongst the different disciplines that make up asset teams of how different data sets are used. For example Connolly (2012) proposes an approach in which uncertainties for each data type are evaluated separately and then combined towards the end of the process. The methodology (Fig. 9.36) involves generating a large number of synthetic traces (with data from a Bayesian posterior distribution) and matching these to the seismic (i.e. a 1D fitting procedure which takes no account of spatial dependencies). Generally speaking this is similar to the *incremental net pay thickness* technique of Neff (1990a, 1990b, 1993) and Burge and Neff (1998), the HIT cube technique of Ayeni *et al.* (2008) and the physics-driven inversion approach of Spikes *et al.* (2008). Given the intimate relationship to the background rock physics model, each synthetic trace defines an arrangement of reservoir properties, with the model fits providing statistical properties of each reservoir parameter at the trace location. Litho-facies are separately mapped from the seismic using a combination of seismic attributes, geostatistical interpolation and analogue information, with mean and standard deviation parameters being assigned to rock properties such as net-to-gross. Lateral variograms can then be utilised in the combination of facies and reservoir property maps, with the final map set providing a basis for constraining further geostatistical simulation.

Chapter 10

Seismic amplitude applications

10.1 Introduction

This chapter focuses on the specific role of amplitude interpretation for the purpose of reservoir evaluation. A number of techniques based on seismic amplitude have been described in previous chapters, including AVO analysis and various inversion techniques, and the interpreter's choice will depend to a large extent on the quality and type of data available as well as the problem at hand. In general, relative techniques, such as AVO analysis for defining fluid related anomalies, are appropriate in the exploration phase whereas in development projects more sophisticated techniques requiring significant well control, such as deterministic and stochastic inversion, are warranted. At each stage the interpreter can attempt to use seismic amplitude to describe the critical aspects of the reservoir as well as defining limits of uncertainty. It should be noted that to be consistent with the aim of the book in terms of the physical interpretation of seismic amplitude, purely statistical or 'data mining' type techniques have not been included in the discussion.

In field development, seismic interpretation is an integral part of the field geological model; key reservoir parameters that can be characterised include geological facies, reservoir properties (including porosity, net-to-gross and water saturation) and reservoir geometry and connectivity. In addition to the static element of the reservoir, time-lapse seismic offers the chance to evaluate dynamic changes. Seismic amplitudes also have a central role in the risk evaluation of prospects characterised by DHIs. In addition, modern amplitude technologies can produce results with reasonable certainty such that they can be incorporated into the process of reserves determination.

To a large extent the following discussion uses published examples for illustration of the range of applications of seismic amplitude. This has the disadvantage that the authors are not in a position to make critical comment, as not all relevant information is published. Inevitably, pitfalls or drawbacks in the various techniques are not always laid bare. The reader is urged toward scepticism but with an open and enquiring mind. There are no silver bullets with seismic amplitude technologies and there is no substitute for personal experience.

10.2 Litho/fluid-facies from seismic

A key tool for the interpreter, at least in the exploration stage, is AVO analysis (Chapter 7). Whether the approach is simply to look for anomalous signatures on partial stack data, identify 'down to the left' data clusters on the AVO crossplot, or generation of AVO projections (Chapter 5), the primary idea behind AVO analysis is to identify anomalous reflectivity that may be related to the presence of hydrocarbons. These signatures need to be put into a clear geological context. In some areas, lithological projections can assist in establishing the sedimentary framework, though this is not always the case. AVO projection techniques effectively provide a means of scanning the data in search of anomalies and their geological context.

In areas where there is well control it is possible to evaluate the detailed facies context of AVO signatures. Well log analysis that incorporates the variability of elastic parameters within each facies inevitably shows that AVO signatures (defined by intercept and gradient) are to some extent non-unique. If the well log statistics are considered representative then it is possible to model the probabilities for each facies for given AVO signatures (Fig. 10.1). Following a non-trivial calibration step (Chapter 6) the statistical model can then be used in conjunction with techniques such as fuzzy logic (Cuddy, 1998), neural networks (e.g. Trappe and Hellmich, 2000) or Bayesian classification (e.g. Mukerji *et al.*, 2001) to predict litho-facies from

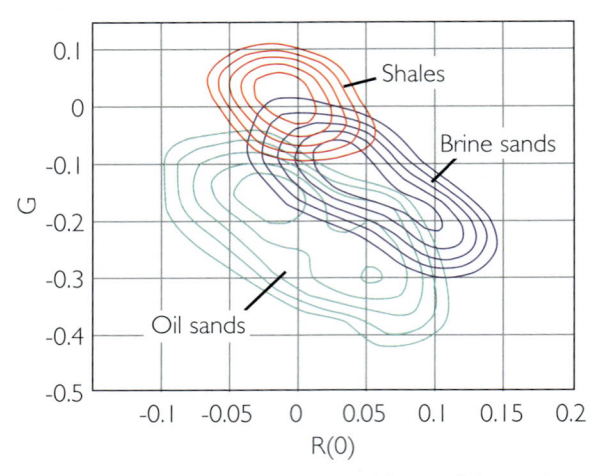

Figure 10.1 An AVO crossplot generated from well data and based on a single interface AVO model, showing different facies with iso-probability lines defining the probability density function for each (from Avseth et al., 2001).

seismic. The application of these techniques should give superior discrimination compared to AVO projections. Avseth et al. (2001, 2003) describe a statistical AVO approach which uses a single interface model for generating intercept and gradient pairs for given boundary types (see Fig. 5.51). A drawback to this type of modelling of course is that the effects of reflector interference or of seismic noise have not been accounted for. However, this could be addressed by generating the interpretation model from pseudo-wells in which seismic noise has been included (e.g. Sams and Saussus, 2007).

Given that deterministic inversion reduces the effects of the wavelet and attempts to back out absolute values of impedance, a popular approach to the facies identification problem is the use of crossplot templates of 'impedance type' properties such as AI vs EI (e.g. Mukerji et al., 2001), or angle-independent properties such as AI and SI (Vernik et al., 2002), AI and V_p/V_s (e.g. Ødegaard and Avseth, 2004; Lamont et al., 2008) or $\lambda\rho$ vs $\mu\rho$ (e.g. Goodway et al., 1997). These templates can provide the basis for probabilistic interpretation in a similar way to that shown in Fig. 10.1. Figure 10.2 illustrates an example of the use of a Bayesian classification scheme to estimate the probability of gas sands from pre-stack simultaneous inversion results. Key issues to address in this type of approach are the upscaling of the log data to create the appropriate rock model for a given target zone and the quality of the pre-stack seismic data and well ties. Such approaches are relatively straightforward to implement with modern software, but given that the inversion results are constrained to the seismic bandwidth the results may not be reliable when the reservoir comprises thin beds.

A similar approach has been used by Michelena et al. (2009) to estimate the probability of channels in a fluvial system characterised by sands with low porosity and permeability and a large degree of vertical and lateral compartmentalisation. Following characterisation of the well logs in terms of dominant facies (e.g. channels (multi- and single-story) and floodplains (sandy and shaley)), a conditional probability model was generated from upscaled P and S impedance logs and applied to the inverted data. The probability model is estimated from crossplots as shown in Fig. 10.3a. Depth slices through the 3D channel probability volumes show credible channel patterns (Fig. 10.3b). The results also illustrate the sensitivity to the choice of well data used in the modelling, with a more optimistic result being obtained when the wells chosen for the probability model have a relatively high N:G (Fig. 10.3c).

In areas with enough well control, the thin-bed limitations of deterministic inversions might be addressed with geostatistical inversion techniques (Chapter 9). Figure 10.4 shows an example of sand probability predicted from geostatistical impedance realisations. In this case the probability at each sample point is simply the proportion of the realisations that have an impedance below a particular cut-off. These realisations can also be used to investigate the potential connectivity of reservoirs. For example, Fig. 10.5 illustrates three stochastic realisations with impedance values below a threshold value and connected to well penetrations.

Another approach to generating facies realisations from seismic is the use of indicator simulation techniques (e.g. Doyen et al., 1994; Perez et al., 1997). Figure 10.6 shows an example where channel and non-channel facies have been predicted on the basis of:

(1) seismic amplitude distributions for the two facies,
(2) estimated proportions of channel sands and non-channel shales,
(3) a spatial covariance model.

It is clear from the illustration that the connectivity of channel sands is highly dependent on the covariance range.

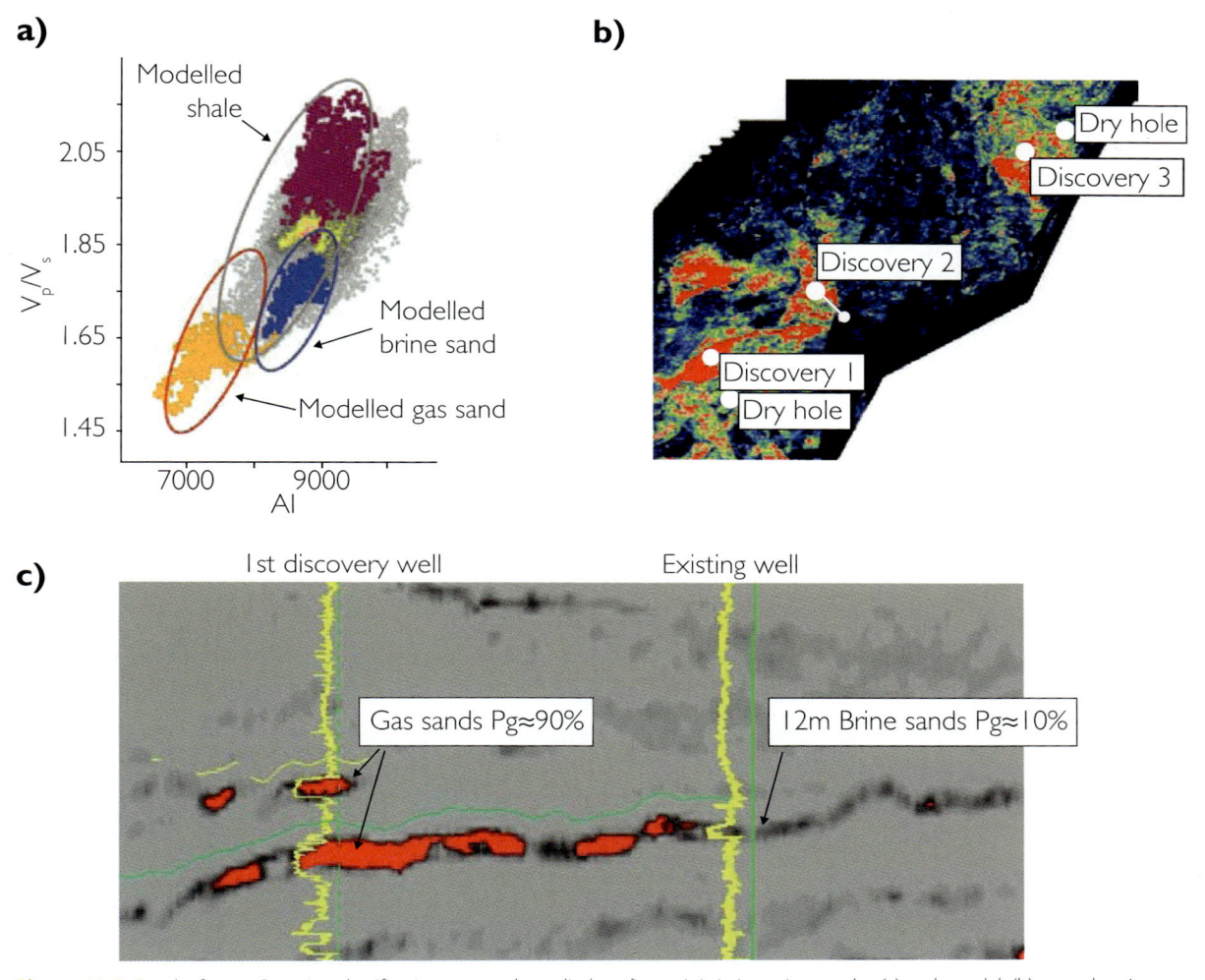

Figure 10.2 Results from a Bayesian classification approach applied to deterministic inversion results; (a) rock model, (b) map showing probability of gas sands occurring (high probability = red) and (c) seismic section showing the location of a discovery well downdip from a well with water sands and based on the probability prediction (after Lamont et al., 2008).

10.3 Reservoir properties from seismic

It is commonplace for seismic amplitudes and inverted impedances to be transformed into reservoir properties through the use of linear and nonlinear regression techniques. Specific reservoir properties such as porosity, net-to-gross and water saturation may be estimated, although permeability prediction is usually based on a relationship with porosity or facies. Mapping of reservoir properties can be enhanced by the use of geostatistical techniques. Whilst linear regression techniques are straightforward to implement the interpreter needs to be aware that there may be a potential for bias in the use of well log averages or simply as a result of inadequacies in well sampling. In situations where the reservoir is a single discrete body with thicknesses above and below tuning, net pay thickness estimation may be possible.

10.3.1 Reservoir properties from deterministic inversion

When there is adequate well control, deterministic inversion can offer advantages over AVO analysis, particularly in regard to the calibrated nature of the inversion products and improved resolution (e.g. Ross, 2010). Techniques such as pre-stack simultaneous inversion (Chapter 9) offer the possibility of

Seismic amplitude applications

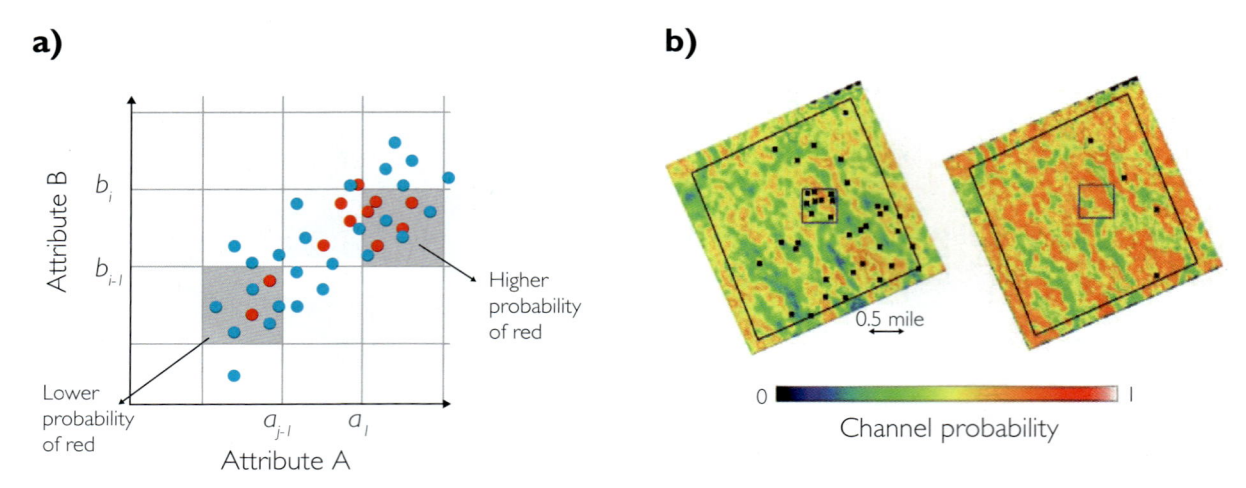

Figure 10.3 Estimating probabilities of channel sands in the fluvial Lance Formation of the Green River basin, onshore USA; (a) principle behind the probability modelling (attributes A and B are P and S impedance respectively), (b) depth slices through 3D channel probability cubes (left – using 39 wells to generate the conditional probability model and right – using four wells), after Michelena et al. (2009).

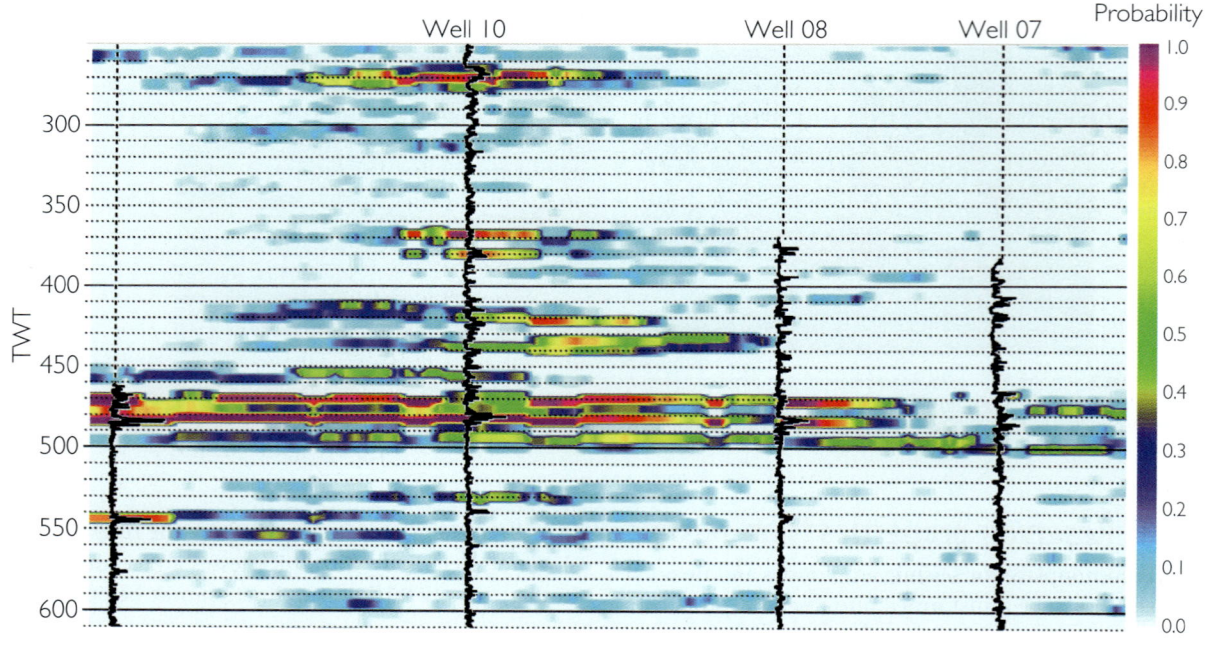

Figure 10.4 Probability of sand occurrence based on 50 stochastic realisations (after Francis 2002).

addressing the ambiguities that arise in AVO analysis. For example in situations where wet sands have low impedance, pre-stack simultaneous inversion may be able to help in differentiating the relative contributions of porosity and saturation as well as accounting for thickness effects on amplitude.

A good example of the potential of pre-stack simultaneous inversion for reservoir property mapping has been described from the Marlin Field in the Gulf of Mexico by Russell et al. (2006). Both full stack amplitudes and the fluid factor AVO attribute (effectively a projection at $\chi=26.2°$) show relatively poor

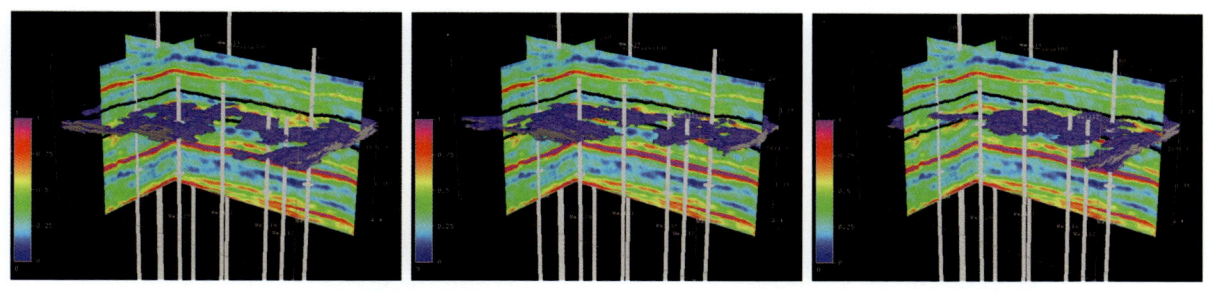

Figure 10.5 Three stochastic realisations with low impedance 'geobodies' connected to wells (after Francis and Hicks, 2010).

a)

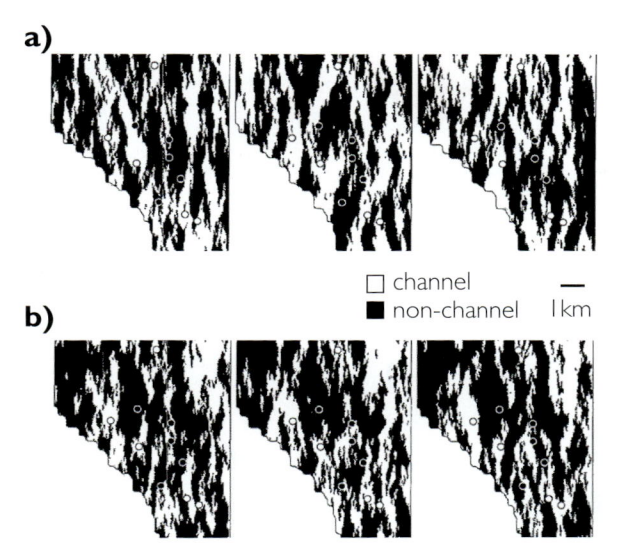

b)

☐ channel
■ non-channel   ⊢ 1 km

Figure 10.6 Channel sand models generated for the Ness formation in Oseberg Field, Offshore Norway using wells and seismic amplitudes. N–S and E–W correlation ranges are (a) 1000 m and 600 m and (b) 2000 m and 600 m respectively. After Doyen et al., 1994. Copyright 1994, SPE. Reproduced with permission of SPE. Further reproduction prohibited without permission.

discrimination of fluids (Fig. 10.7). P-impedance, S-impedance and density were inverted from pre-stack gather data using a model-based inversion approach (Figs. 10.8a–c). Empirical transforms based on the available well data were used to predict rock properties from the inverted products. N:G, porosity and water saturation maps are shown in Figs. 10.8d–f.

10.3.2 Simple regression, calibration and uncertainty

In many cases the simplest approach in mapping reservoir properties from seismic is to apply a transform to the seismic attribute map based on cross-plotting of seismic and well data (Fig. 10.9). This approach assumes that there is no spatial dependence of the relationship. In order to make the final map fit the wells a residual correction surface would need to be applied. More sophisticated approaches will be discussed below but there are a number of issues that the interpreter should be aware of prior to deriving a functional dependence of seismic attributes and reservoir properties. These include the role of synthetic models, the nature of calibration and the nature of uncertainty.

Simply crossplotting attribute values at well locations and data from the wells may not be an optimum approach. At least in the first instance it is useful to demonstrate the relationship between the seismic attribute and the well data through synthetic modelling. At this stage careful consideration needs to be given to the effect of facies as well as spatial controls on any possible attribute/reservoir property relationship. For example, Stanulonis and Tran (1992) describe how porosity can be linearly related to seismic amplitude in the Lisburne pool on the North slope of Alaska, but find that the relationship varies in separate geographical regions due to factors such as sub-cropping geology, the presence of fractures and variation in fluid phase (i.e. oil vs gas). Another issue to investigate in determining a seismic/rock property function is the uncertainty that arises from the limited resolution of seismic. For example, beds below a certain thickness might not be evident on seismic but these beds may be effective reservoir and may have been used to calculate rock property values from well logs.

Following the determination of an attribute/rock property relationship using synthetics the next step is calibration of well and seismic values. When crossplotting mapped seismic attribute and well property data there is inevitably a question over which value to take from a seismic map: is it the closest value, an

Seismic amplitude applications

a) Full stack amplitude

b) Fluid factor

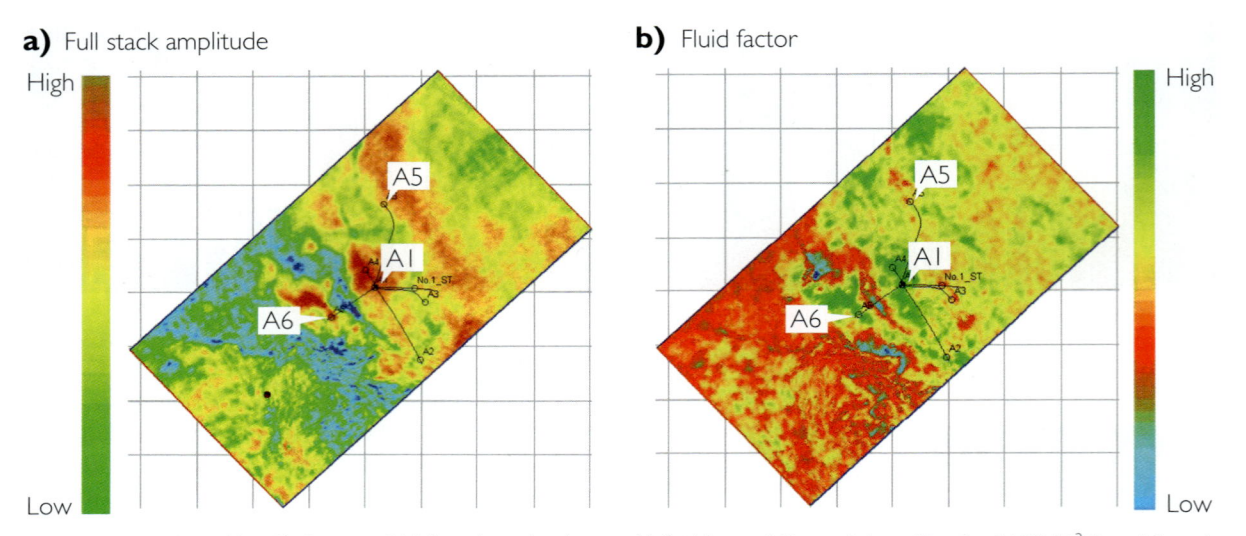

Figure 10.7 Marlin Field, Gulf of Mexico; (a) full stack amplitude map, (b) fluid factor AVO map (where $FF = R_p - 1.16(V_s/V_p^2)R_s$, and R_p and R_s are derived by using the Fatti et al. (1994) method). Note that well A6 has 25 m gas column in a 45 m gross section, well A1 has 40 m gas and well A5 has 33 m of wet sand (after Russell et al., 2006, with permission from the CSEG Recorder).

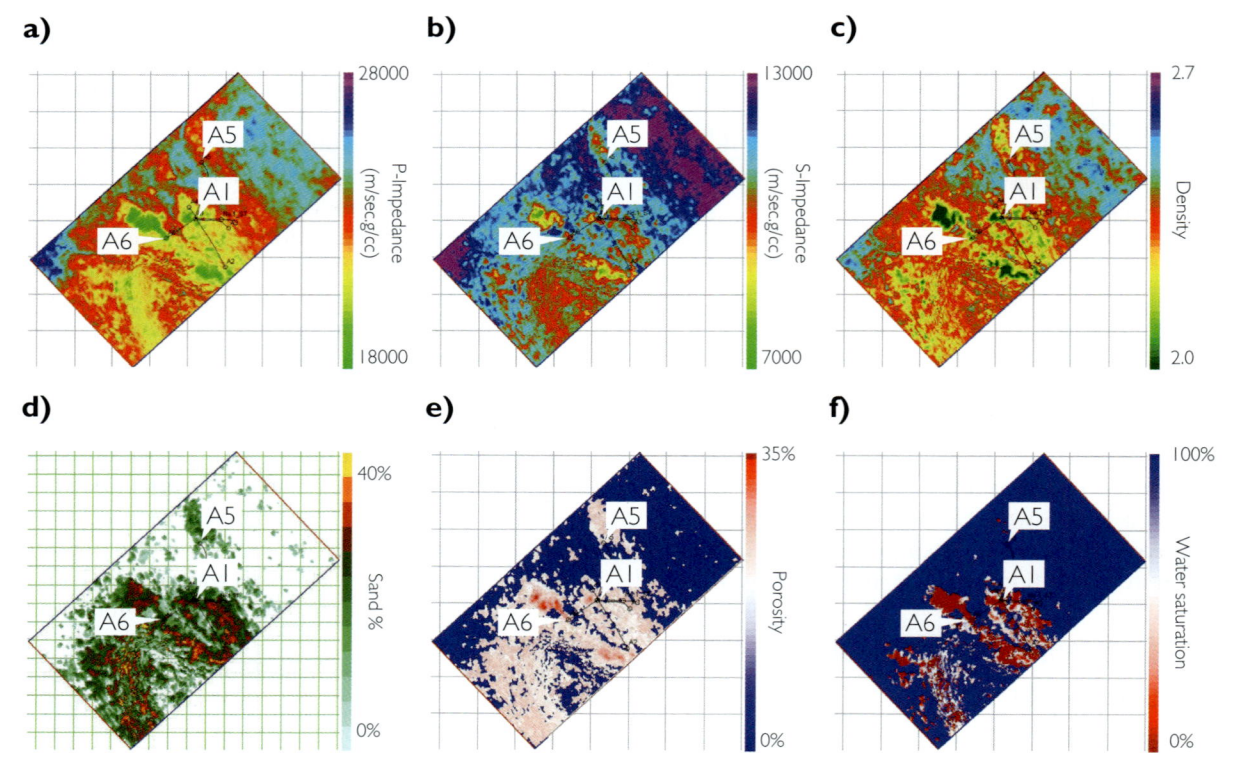

Figure 10.8 Simultaneous inversion products from Marlin Field, Gulf of Mexico; (a) P impedance, (b) S impedance, (c) density, (d) N:G based on inverted density, (e) porosity based on inverted P impedance, (f) saturation based on inverted density (after Russell et al., 2006, with permission from the CSEG Recorder).

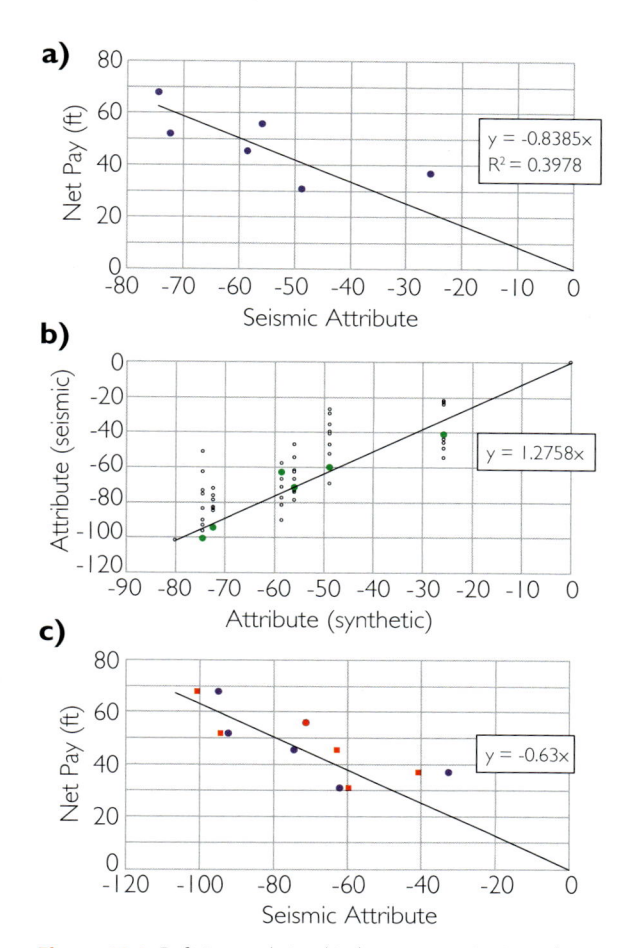

Figure 10.9 Defining a relationship between a seismic attribute and a reservoir property; (a) function estimated from well synthetics, (b) the calibration step, estimating the scaler for matching the seismic and synthetic data and taking into account the potential for mis-positioning of seismic and wells (the data plotted are the nine nearest grid points for each well, with the nearest grid point to the well coloured blue), (c) final function with both synthetic and seismic data displayed.

average over an area or the value at the best match location? The answer will be case-dependent, with the interpreter evaluating well tie issues, such as wavelet shape and scaling uncertainties, as part of the decision-making process. The potential uncertainty should be appreciated and documented.

Figure 10.9 illustrates the issue of obtaining an attribute/reservoir property relationship. A simple linear regression fit of a seismic attribute and reservoir property has been defined from well synthetics (Fig. 10.9a). The calibration step (Fig. 10.9b) shows the synthetic attribute plotted against a range of seismic values within ~40 m of the well location (based on nine grid point values). This example comes from a fairly shallow reservoir, with minimal overburden or seismic imaging issues, so the nearest points to the wells (coloured blue) define a reasonable scaler relating the well points to the seismic. Following calibration, the final function is shown in Fig. 10.9c incorporating both synthetic and seismic data. Uncertainties in the prediction owing to the variability of rock properties might be incorporated into the crossplot model by the generation of pseudo-wells.

Mukerji and Mavko (2008) discuss a potential pitfall in the use of averages for determining an attribute/reservoir property relationship. Figure 10.10 illustrates the problem. In Fig. 10.10a various N:G models for a thin bed scenario (i.e. below tuning) are generated using average values of acoustic impedance for sands and shales. These scenarios are averaged, using the Backus average, and normal incidence synthetics are generated. The model suggests that there is a linear relationship between amplitude and N:G (Fig. 10.10b). However if the impedance distributions of sands and shales are used (Fig. 10.10c), together with Monte Carlo simulation to generate the N:G scenarios prior to upscaling then the result suggests a different (non-linear) relationship between amplitude and N:G. Effectively, for this distribution of impedances, N:G predicted from amplitude would be less than the 'average of averages' model. There is also the possibility that some of the scenarios would give amplitudes of the opposite polarity. In addition, considering distributions gives an inherent description of probability.

A number of practical insights can be drawn from this example. Certainly one conclusion is that an 'average of averages' model is likely to be inappropriate both in terms of potential bias in the prediction of N:G but also in the nature of the attribute/rock property relationship (i.e. linear vs nonlinear). In practice, the linear model assumption could result in either positive or negative bias. For example, consider a practical situation in which wells have been drilled on the brightest amplitudes represented in Fig. 10.10d and it is assumed that amplitude and N:G are linearly related, possibly as a result of the type of modelling shown in Figs. 10.10a,b. If the green, blue and red curves in Fig. 10.10d represent the real geological situation then a linear assumption based on the 'high-amplitude' well data would give N:G predictions for the lower amplitudes which are too low.

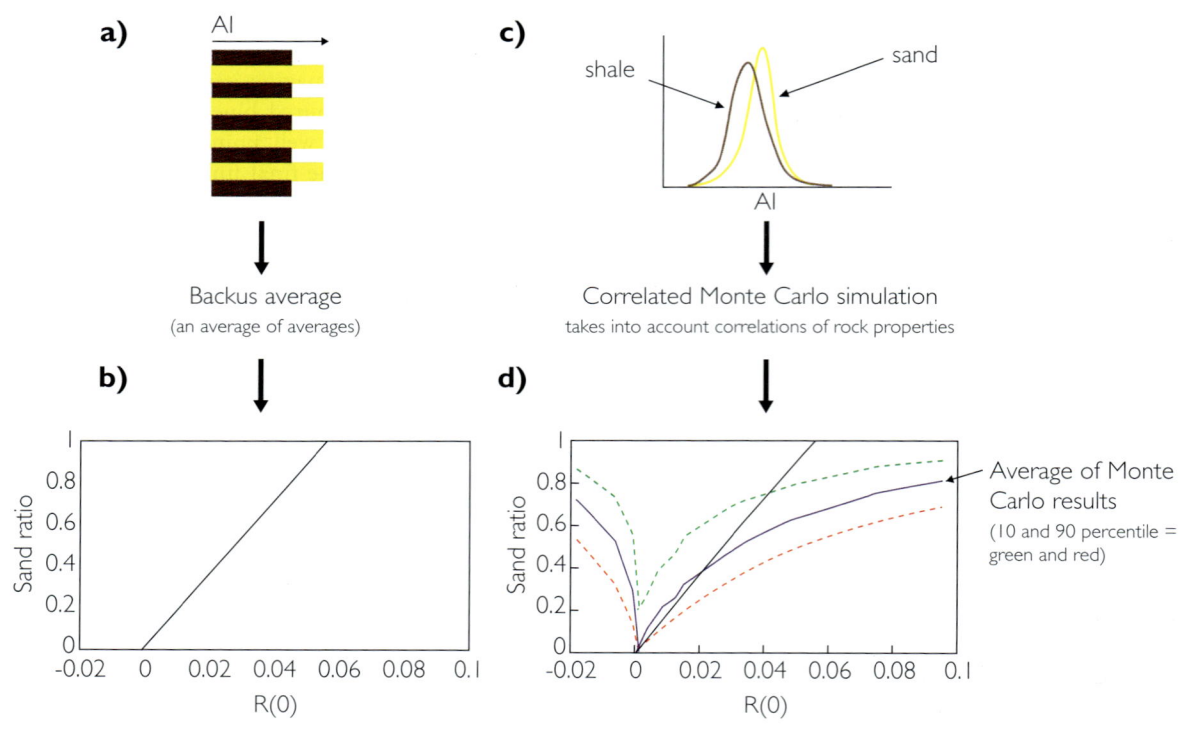

Figure 10.10 The flaw of averages; (a) model using sand and shale acoustic impedance averages from the logs and thickness below tuning, (b) relationship of N:G varying with normal incidence reflectivity for the model in (a), (c) distributions of AI for sand and shales, (d) N:G vs normal incidence reflectivity results from a set of models generated using the distributions in (c) and Monte Carlo sampling (after Mukerji et al., 2008).

The Mukerji and Mavko (2008) example also highlights the role of well data in the nature of uncertainty. All too often the perception of uncertainty is gauged from the scatter of points around the regression line on the crossplot. However, given the potential for bias in functions that are derived simply by line fitting through a few points it is necessary to try and place the well results in the context of geological variance. To quote Connolly (2010), 'Uncertainty is too complex to be adequately captured by analysing the results from a small number of wells'. Given that well data are often limited, a key aspect of statistical rock physics is the extension of the model dataset through the application of techniques such as Monte Carlo simulation (e.g. Mukerji et al., 2001) and the generation of pseudo-wells (e.g. Connolly and Kemper, 2007). This provides the context for understanding uncertainty, at least in terms of the available data. Ultimately, of course, other uncertainties also need to be considered, including the effect of seismic noise.

10.3.3 Reservoir property mapping using geostatistical techniques

Application of a single regression function to a seismic attribute map assumes spatial independence. Various forms of kriging, the traditional geostatistical approach to data interpolation, can be used to address the issues of fit to wells and spatial variability. Kriging is an interpolation technique which uses a spatial variogram to determine the mapped value (e.g. Dubrule, 2003). As illustrated in Section 9.3, the variogram is a way of describing the spatial continuity of the data, capturing the rapidity with which the variable changes as a function of lateral or vertical distance. A large number of wells is required to make this an effective method for interpolating well data; with only a few wells it is not possible to estimate the horizontal variogram. When building reservoir models, variograms estimated for geological analogues may be used, perhaps from cases that can be studied at outcrop where measurements can be made

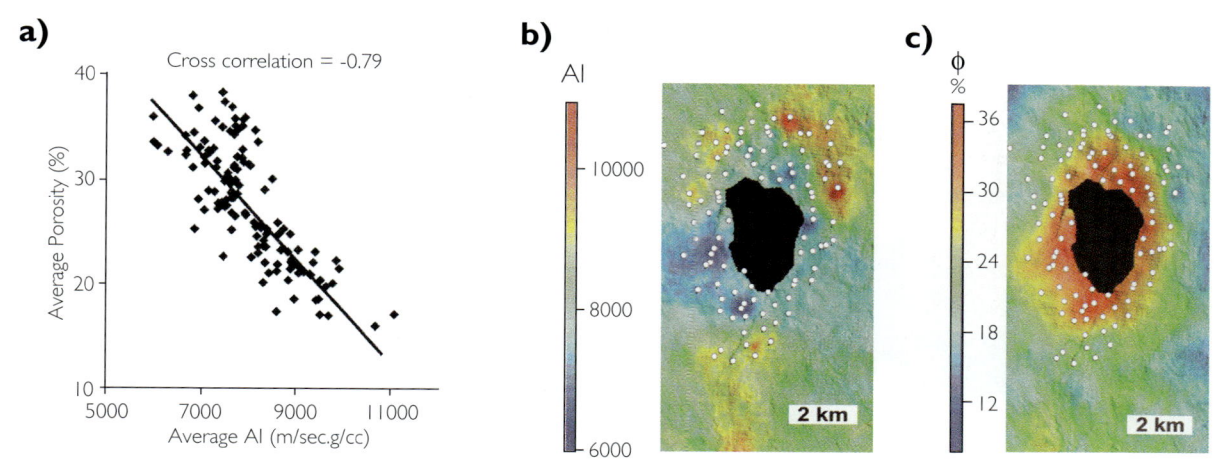

Figure 10.11 Ekofisk Field porosity mapping (after Doyen et al., 1996; copyright 1996, SPE. Reproduced with permission of SPE. Further reproduction prohibited without permission). (a) Crossplot of average porosity from wells vs average seismic impedance for the upper part of the Tor chalk formation, (b) average seismic impedance map, (c) porosity map generated from collocated cokriging. Note that the centre of the map is data-free owing to the presence of a gas cloud (colour illustrations from Dubrule, 2003).

with fine sampling over a large lateral extent. Alternatively, information derived from seismic attributes can be used to define the lateral variability. Techniques such as collocated cokriging and kriging with external drift are examples of kriging that combine seismic attribute data with well data and provide a single deterministic solution that fits the wells. Kriging with external drift is usually applied in depth conversion as there is an implicit assumption that the secondary variable (e.g. stacking velocities) provides low-frequency information on the primary variable (e.g. average velocity). Collocated cokriging effectively provides a weighted average of the kriged estimate (in which a spatial co-variance or variogram model is specified) and the estimate derived from the linear regression of the rock property and seismic attribute. The weighting factor is related to the correlation coefficient between the seismic attribute and the reservoir property (Doyen et al., 1996; Fig. 10.11).

10.3.4 Net pay estimation from seismic

The interpreter has a number of choices when deciding on a technique for interpreting hydrocarbon in place. Traditional approaches are based on gross rock volume calculations from depth structure maps. Stock Tank Oil Initially In Place (STOIIP) or Gas Initially In Place (GIIP) can then be calculated by making assumptions about porosity, thickness, net to gross ratio (N:G), and formation volume factor. It is possible, however, to use seismic directly for estimating hydrocarbon in place. For example, one approach might be to use data from a deterministic inversion, with hydrocarbon sand thickness estimated from a relationship between impedance and hydrocarbon pay. As described in Chapter 9, this approach requires a large number of wells to understand the various effects of smoothing and residual tuning effects that may bias the estimation. Stochastic inversion techniques may address these pitfalls but these also require good well control and may not be practical in exploration scenarios. Fortunately, there are relatively simple approaches that may be applicable, dependent on the geology, for inferring hydrocarbon pay sand thickness from seismic and these can be utilised readily with commonly available software.

10.3.4.1 Amplitude scaling techniques

Simple techniques for hydrocarbon pay estimation can be applied in situations where the reservoir is an isolated low impedance unit of limited thickness (generally less than about 60 ms; Brown et al., 1984, 1986; Connolly, 2007). These techniques effectively assume a bi-polar arrangement of sand and shales, each with a single value of impedance, with net sand being predicted from mapped seismic amplitude by using the following assumptions:

(1) there is a first order linear relationship between N:G and amplitude for a given apparent thickness (Chapter 5);

Seismic amplitude applications

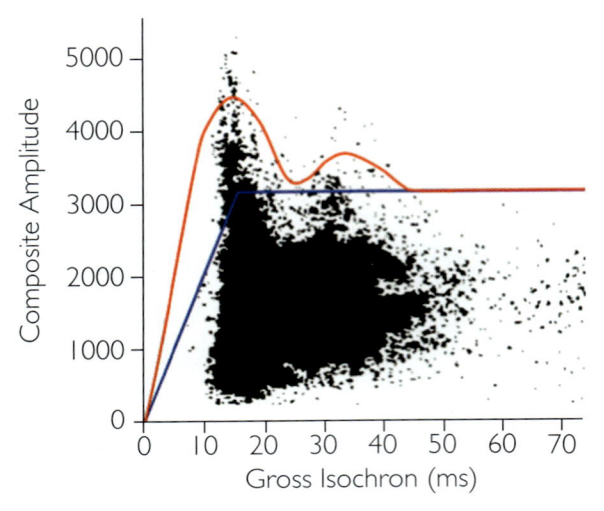

Figure 10.12 Seismic crossplot example of thickness versus composite amplitude with simple tuning curve (red) and baseline (blue). Redrawn from Brown et al. (1986).

(2) the amplitude scaler varies with apparent thickness to account for interference effects (Chapter 3).

Sometimes this approach is described as 'amplitude de-tuning' and effectively provides an approximation of a 'net pay' component.

Brown et al. (1984, 1986) introduced a technique applicable to seismic reflectivity data in which a simple tuning curve is used as a reference to 'de-tune' the amplitudes (Fig. 10.12). The curve may be based on the wavelet in the data or it may be approximated by eye from the shape of the distribution. The tuning curve is positioned on the plot either to encompass the majority of points or to 'calibrate' to wells. A 'no-tuning' baseline, with constant amplitude above tuning thickness and tending to zero below tuning, is also defined. Tuning effects are then removed by multiplying each point with the baseline/tuning curve ratio appropriate for the given apparent thickness, effectively normalizing the data to the baseline amplitude. Net sand or net pay can then be calculated by multiplying the 'corrected' amplitude map by the apparent thickness. In practice, this technique tends to overestimate clean sand thickness below tuning and becomes inappropriate when two sands close to tuning thickness are present (Simm, 2009).

Notwithstanding the limitations of amplitude scaling this type of technique provides the interpreter with an important initial sense check, and helps in avoiding the error of predicting actual thickness from apparent thickness in 'thin' beds. An example is shown in Fig. 10.13 from a gas field with a low-impedance clean sand reservoir. Two wells have been drilled. Well K was the discovery well and drilled a thin (4 m) gas column. Well L subsequently encountered a 13 m column. The tuning plot (blue points) was generated from top and base seismic picks and a tuning curve (generated using the wavelet in the data) has been scaled to the highest amplitudes on the plot. With respect to the well K location, a pre-drill error might be to interpret the seismic thickness as the actual thickness (i.e. 10 m, whereas the well found 4 m). Scaling the data with a tuning curve, however, gives a net pay thickness prediction of around 2 m, assuming that the sand is clean (Fig. 10.13b). If the reservoir is composed of sand and shale layers it is possible, following the observations of Meckel and Nath (1977) (see Fig. 5.30), that the amplitude/net pay relationship is more linear; in which case the predicted net pay would have been around 3 m. In either case the net pay would have been slightly underestimated as opposed to being severely overestimated. At the location of well L the net pay thickness would have been calculated pre-drill by taking the ratio of the trough amplitude to the scaled tuning curve amplitude and multiplying by the apparent thickness (i.e. (1500/2250) × 20 ms = 13.4 ms (13.4 ms/2000) × 1690 m/s = 11.3 m), giving a thickness close to that encountered in the well.

An improvement on the amplitude scaling technique has been published by Connolly (2007). Connolly's technique utilises bandlimited impedance in the form of coloured inversion, although other approaches to bandlimited impedance can be used (Chapter 5). The net pay prediction problem is framed in terms of a net pay equation, modified for seismic application:

Net pay = seismic N : G × apparent thickness,

where seismic N:G = average bandlimited impedance × scaler.

The method can be understood in relation to a simple wedge model, calculated using a bandpass wavelet consistent with the bandlimited impedance (Fig. 10.14). Average bandlimited impedance ($ABLI$) and apparent thickness are calculated relative to the zero crossings associated with the reflections from top and base of the wedge. Given that the net sand

Reservoir properties from seismic

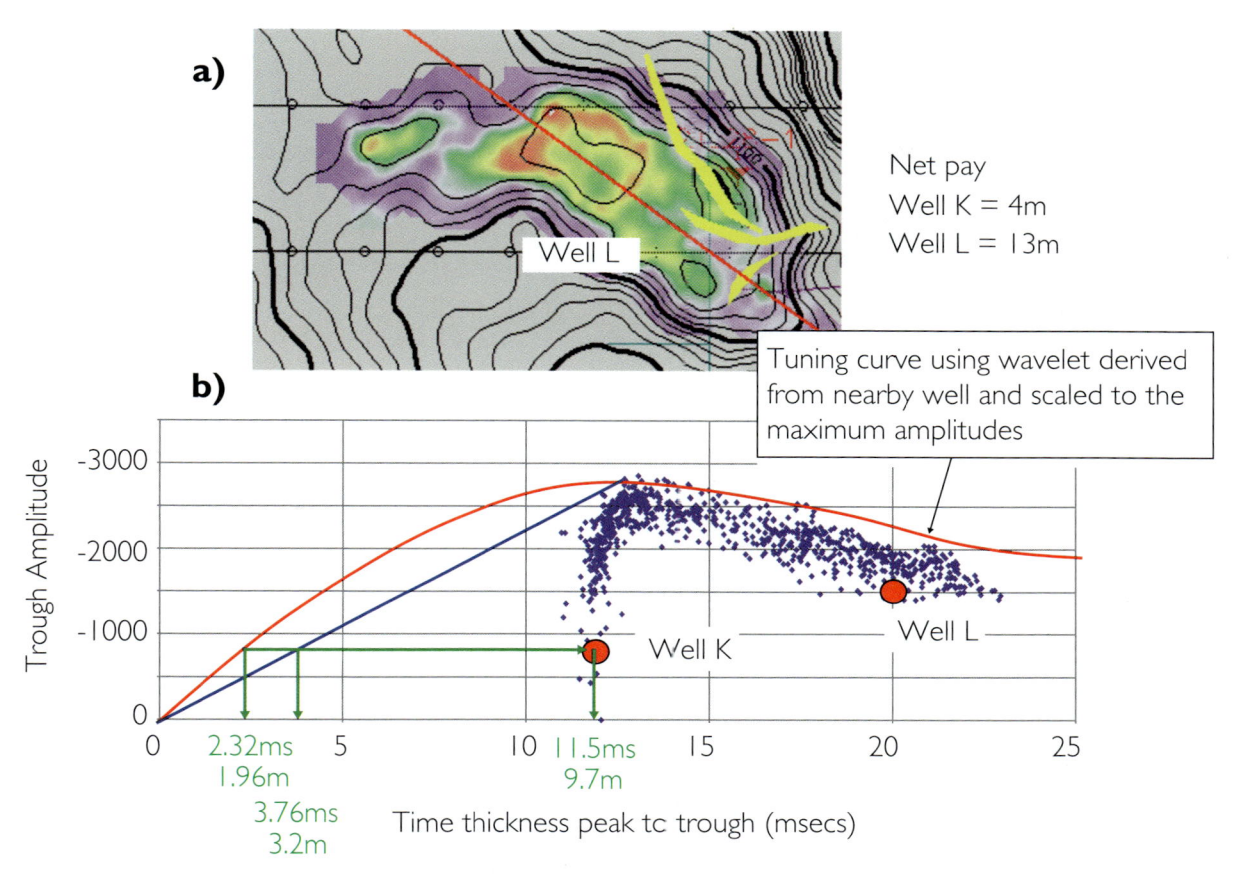

Figure 10.13 Net pay prediction using simple amplitude scaling; (a) composite amplitude map (reds are high amplitudes) with two-way time structure and showing locations of wells L and K; (b) tuning plot from seismic together with tuning curve scaled to the maximum amplitudes. Courtesy RPA Ltd.

thickness is known in the model the variation of the scaler with apparent thickness is straightforward to derive. Plotting the inverse of the scaler on the seismic tuning plot gives a 'calibration curve' which provides a useful graphical appreciation of the calculation (i.e. net pay = (ABLI/calibration curve value)*apparent thickness).

To illustrate how the net pay prediction technique works, a model was generated by using data based on a real field example (Fig. 10.15), including both thin and thick layering situations. The model comprises gas sands and shale, in which the acoustic properties are held constant between the wells, and there are no hydrocarbon contacts. For each of the model wells, apparent thickness and average bandlimited impedance were calculated. The scaler function was then calibrated as described above. Figure 10.15d demonstrates the viability of the technique. In practice, if there are wells available to generate the coloured inversion operator but there are no pay values to guide the fitting, the scaler would be calibrated such that the calibration curve would be positioned above the data points on the average bandlimited impedance versus seismic thickness crossplot (Connolly et al., 2002). The positioning of the curve would reflect the general understanding of N:G.

A useful simplification of Connolly's (2007) technique is to assume that the scaler varies linearly with apparent thickness (Simm, 2009). This may be appropriate for example when there are no wells and the effective bandpass wavelet is unknown. A linear scaler assumption leads to the following:

Net pay = (ABLI × apparent thickness^2) ÷ constant.

Fig 10.16 illustrates that the simplification works quite well with the model dataset.

Seismic amplitude applications

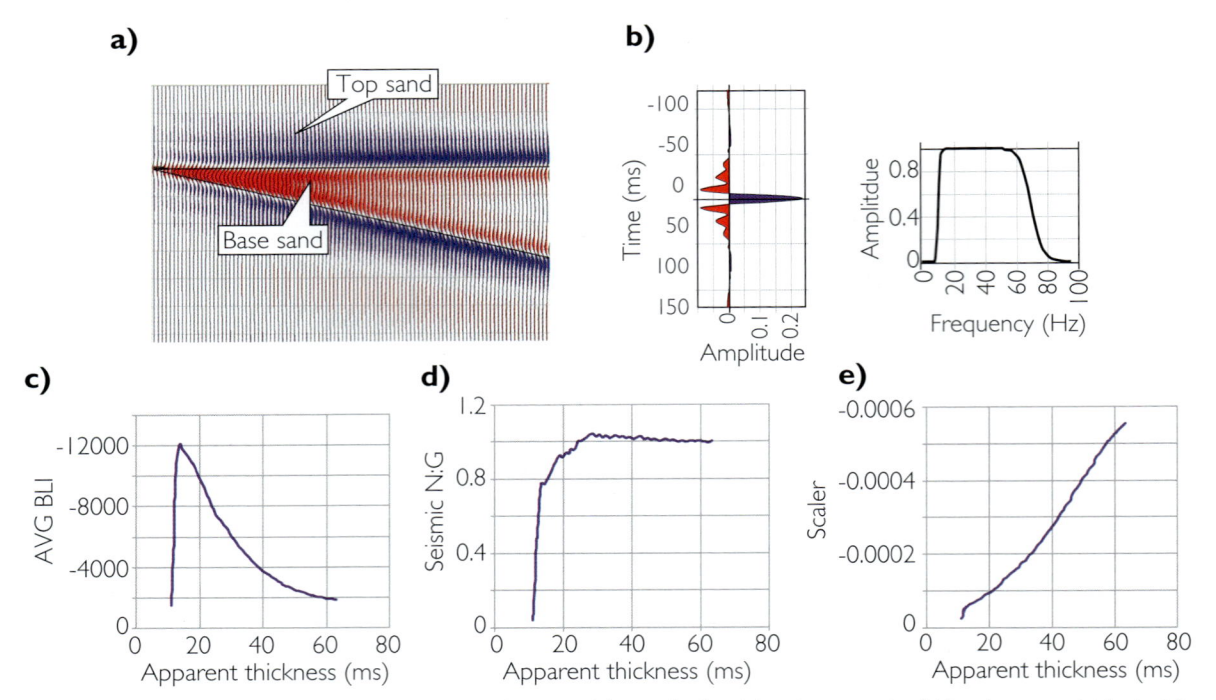

Figure 10.14 Low impedance wedge model; (a) impedance model convolved with bandpass wavelet, (b) bandpass wavelet (output from the coloured inversion process), (c) tuning curve, (d) seismic N:G, (e) scaler (after Simm, 2009).

The application of Connolly's (2007) method is particularly appropriate for deep sea turbidite sequences. Figure 10.17 illustrates an example from Offshore West Africa. A bandlimited impedance far-stack seismic cube is the basis for the net pay interpretation of the lower reservoir unit shown in Fig. 10.17a. Apparent thickness and average bandlimited impedance are calculated between the zero crossings picked on the seismic (Figs. 10.17b,c). Seismic N:G is determined from the bandlimited impedance map multiplied by the scaler obtained from a simple wedge model using the bandpass wavelet implicit in the coloured inversion. Net pay is simply calculated from the product of the seismic N:G and apparent thickness maps. Note how the sweet spots on the final net pay map are located to the north of the main area of bright impedance.

The successful implementation of Connolly's (2007) technique will in practice depend on a number of geological and data factors, the prime requirements being:

(1) good lateral and vertical fidelity of seismic amplitude;
(2) bi-polar geology in which sands have one impedance and shales have a higher impedance;
(3) there is no interference between the reservoir and events above or below;
(4) the zero-crossing picks are clear;
(5) there is a single hydrocarbon phase (variable errors will be expected around contacts: Simm, 2009);
(6) the thickness of the reservoir is no greater than a half cycle of the lowest frequency component, e.g. 60 ms at 8 Hz low cut frequency (Connolly, 2007).

Estimates of net pay based on Connolly's (2007) technique are likely to be an underestimate in areas where there are thin sands below seismic resolution. This may be significant in reservoirs with subtle dip where hydrocarbons extend away from the visible edge of the amplitude anomaly. A rough estimate of the distance can be made from the smallest thickness calculated from the visible amplitude anomaly and the dip of the top of the reservoir. For low-relief structures, there may be a significant contribution to the volume of hydrocarbons in place hidden in this invisible pay zone.

The squared term inherent in Connolly's approach means that the accuracy of the thickness measurement is critical. In situations where there is difficulty in

Reservoir properties from seismic

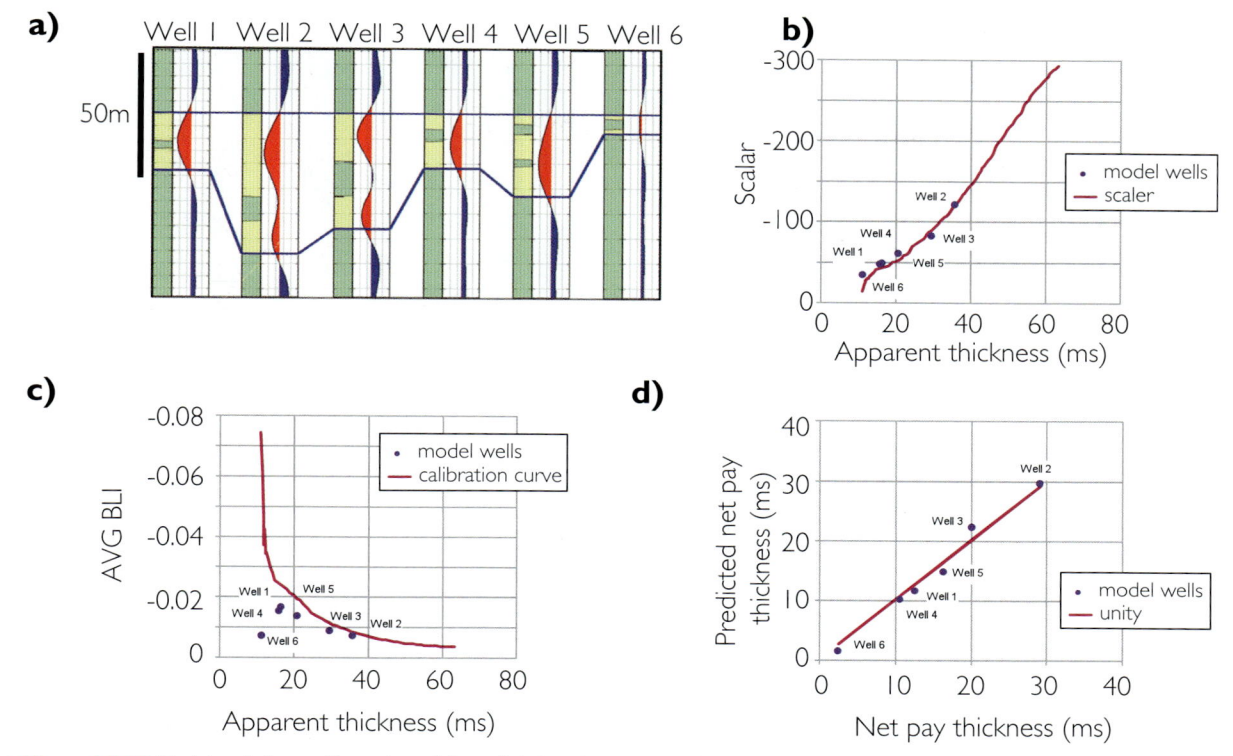

Figure 10.15 Model well dataset illustration of Connolly's (2007) net pay prediction technique; (a) model dataset based on a real field example (green = shale, yellow = sand), (b) scaler curve scaled to minimise error at wells, (c) tuning plot with calibration curve (i.e. inverse of scaler), (d) net pay prediction using scaler compared with actual net pay in model wells (from Simm 2009).

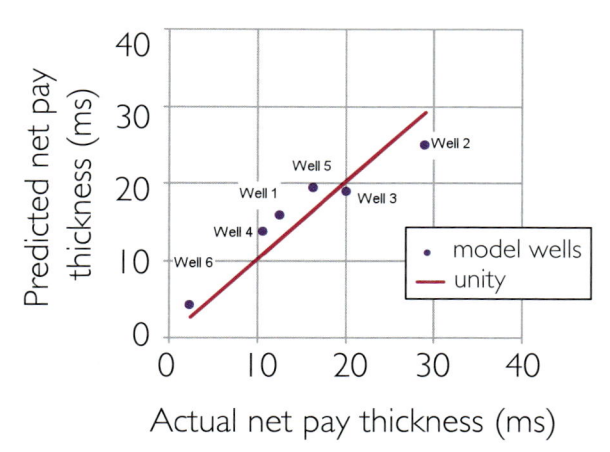

Figure 10.16 Model well dataset: net pay thickness prediction using Connolly's (2007) technique with a linear scaler (after Simm, 2009).

horizon picking, it may be more appropriate to use an attribute that is less sensitive to layer thicknesses. One such attribute is sum of negative amplitudes (*SNA*). Figure 10.18 illustrates a bandlimited wedge model contrasting average bandlimited impedance with sum of negative amplitudes. It is evident that whilst *SNA* shows a tuning effect, the curve has a different form and tuning occurs at around 130 ft in contrast to 50–60 ft with average bandlimited impedance.

10.3.4.2 Pseudo-wells and uncertainty

Neff (1990a, 1990b, 1993) considered the amplitude scaling methodology of Brown *et al.* (1984, 1986) to be too simplified and determined that the problem of analysing the complex relationships between trough-to-peak amplitudes, isochrons and reservoir parameters requires a more rigorous approach. Clearly the nature of the relationship between amplitude and thickness is specific to particular geological scenarios. Thus it is useful to generate pseudo-wells that are consistent with known geological relationships. Initially this was done by varying the amount of pay in well-based synthetic models but this can be extended using statistical rock physics and geological modelling of layers and facies. Figure 10.19 shows a case in which thickening of the pay interval is associated with

Seismic amplitude applications

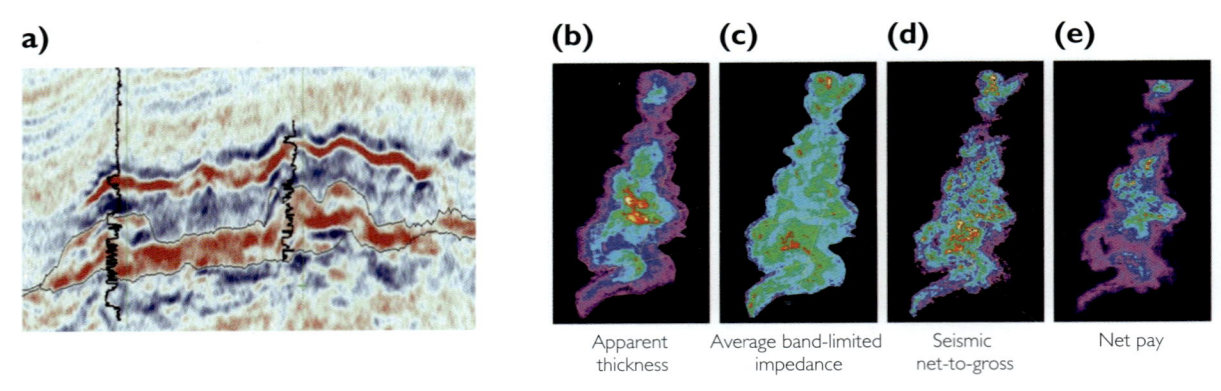

Figure 10.17 A West Africa example of net pay estimation; (a) far-stack coloured inversion, (b) apparent thickness, (c) average bandlimited impedance, (d) seismic N:G and (e) net pay thickness (after Connolly, 2007).

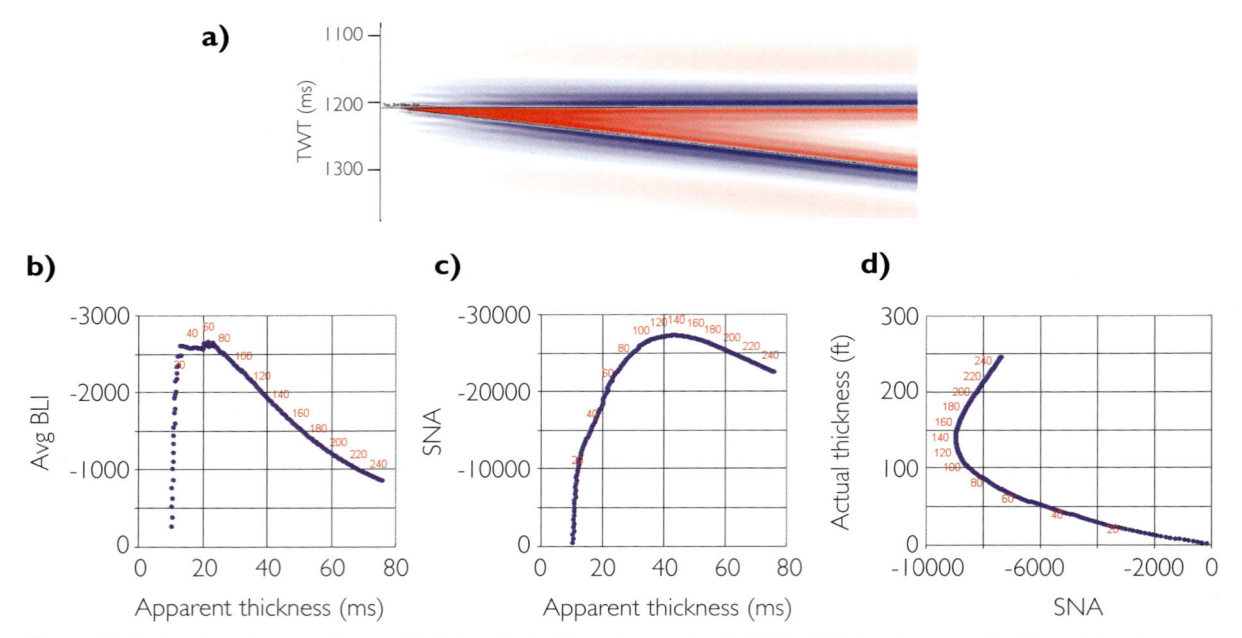

Figure 10.18 Low-impedance wedge model; (a) bandlimited impedance using 4–14–75–95 Hz bandpass wavelet, (b) typical tuning curve generated from average bandlimited impedance, (c) tuning curve for sum of negative amplitudes (SNA), (d) relationship between SNA and actual sand thickness. Red numbers = net sand in feet.

increase in amplitude but a decrease in apparent thickness, contrary to a simple tuning model.

Many pseudo-wells can be used to populate apparent thickness vs seismic amplitude space providing a template for interpretation. Fervari and Luoni (2006) have used this technique to generate a net pay template for a turbidite reservoir (Fig. 10.20). It should be noted that the net pay lines shown in Fig. 10.20 are effectively averages (i.e. there is significant overlap of net pay solutions). Thus, it would be relatively straightforward to split the plot up into various thickness/amplitude classes and use the model data for quantitative estimation of uncertainty.

With respect to Connolly's net pay estimation technique, Connolly and Kemper (2007) and Connolly (2010) address the problem of uncertainty in terms of pseudo-wells. The workflow (shown in Fig. 10.21) comprises the following steps.

- Determine layer configuration of sands and shales for a given reservoir interval (Connolly and Kemper (2007) use a fractal model conditioned by the well data).

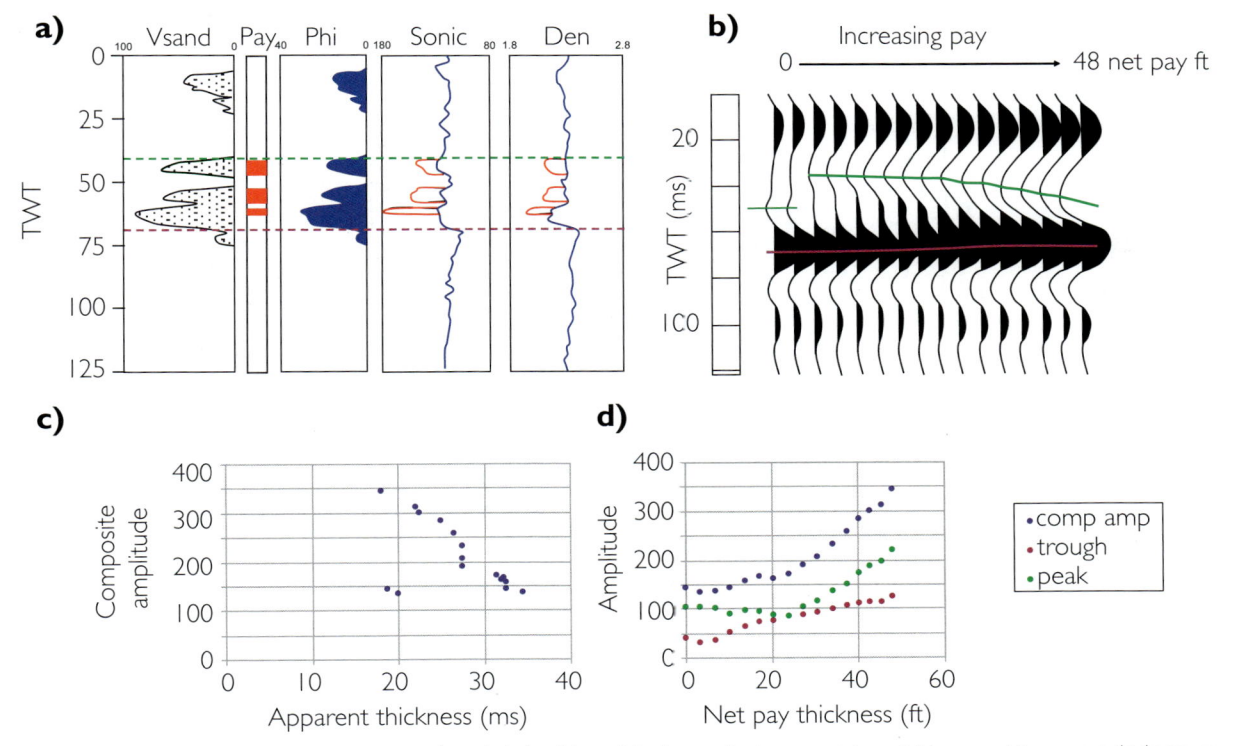

Figure 10.19 (a) Log plot showing porous sands and shales, (b) modelled traces for incremental pay thicknesses, (c) apparent thickness vs composite amplitude plot, (d) net pay thickness vs amplitude (re-drawn from Neff, 1990).

- Populate sands and shales with velocity and density data from probability density functions, and constraining cross correlations, derived from rock physics analysis.
- Calculate impedance traces at the appropriate AVO angle.
- Convolve impedance traces with the correctly scaled bandpass wavelet relevant to the bandlimited impedance dataset.
- Pick upper and lower zero crossings and calculate average bandlimited impedance and apparent thickness.

Uncertainty is then estimated from the pseudo-well data in the form of standard deviations of seismic N:G (Fig. 10.22).

The logical application of pseudo-wells is to perform seismic inversion through trace matching (e.g. Burge and Neff, 1998; Ayeni et al., 2008; Spikes et al., 2008). An attractive aspect of trace matching is that when pseudo-wells provide simultaneous matches to multiple angle stacks then the result provides all petrophysical and geological parameters (including net pay thickness) associated with the match. Unlike traditional approaches to trace inversion (Chapter 9) no further inversion steps (for example from impedance to reservoir property) are required. In addition, multiple matches can be used to assess uncertainty.

10.4 Time-lapse seismic

'Time-lapse' or '4D' seismic refers to the use of seismic data to detect changes in the reservoir over time. Typically the technique is used in hydrocarbon production but it can also be used for other purposes such as monitoring gas or carbon dioxide storage. The time-lapse premise is straightforward, with a 'baseline' survey typically acquired before production and successive 'monitor' surveys acquired at periods (months or years apart) during the life of the field. Individual surveys may be interpreted as separate 'snapshots' but most modern time-lapse projects aim to create differenced seismic volumes. Provided that the seismic processing has optimised the time-lapse signal, the differences between the

Seismic amplitude applications

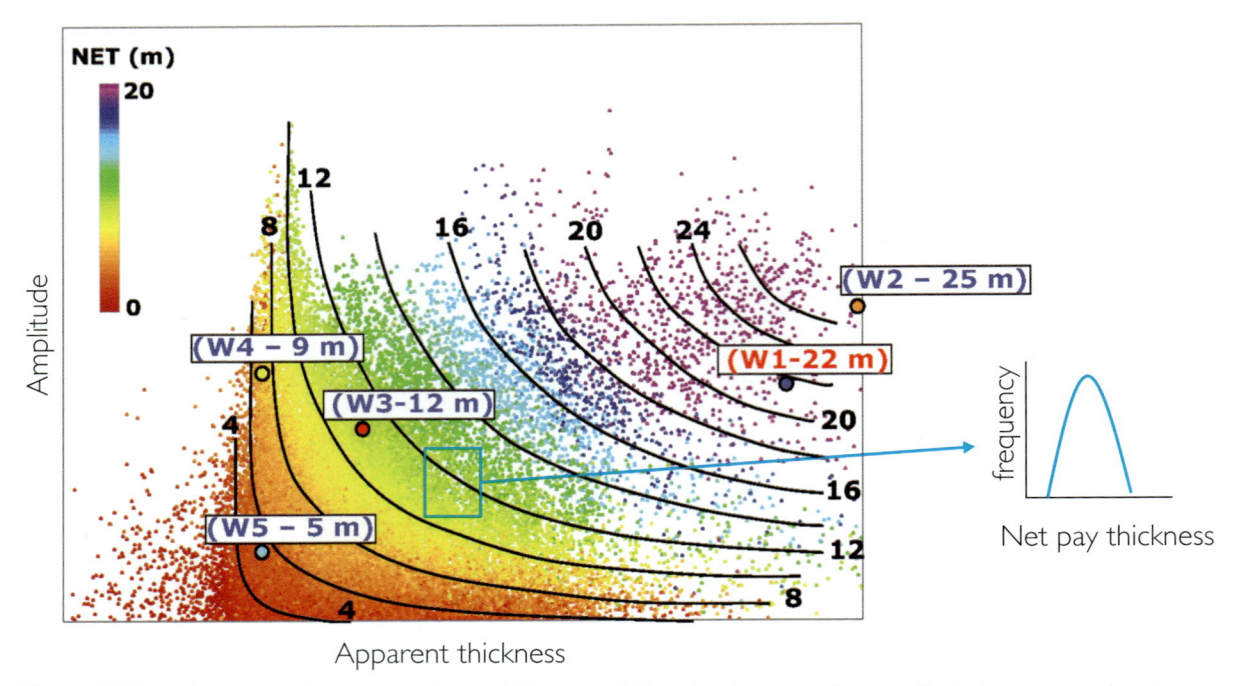

Figure 10.20 Model data template generated for a turbidite reservoir. Note that the net pay lines are effectively averages as there is considerable overlap. Blue wells were blind tests whilst the red well was the basis for the pseudo-well data in the model. After Fervari and Luoni (2006).

surveys effectively highlight where production effects have occurred (Fig. 10.23). Amplitude maps can give a dramatic indication of large scale changes over the field area (Fig. 10.24). Detailed interpretation of the time-lapse signal in terms of fluid flow (e.g. identifying barriers and estimating sweep efficiency in reservoir compartments) and the identification of remaining reserves requires careful reference to the baseline reservoir description and geological model. The results can be used to optimise infill drilling and manage injection/production strategies. The reader is referred to excellent summaries of time-lapse technology given by Jack (1997), Calvert (2005) and Johnston (2013).

Figure 10.25 shows a view of the time-lapse process in which a seismic floodmap based on time-lapse seismic interpretation is used to update the field simulation. In some instances, the level of integration is even greater than that implied in the figure, with the reservoir simulation being fully integrated with the seismic observations via a 'shared earth model' (Riddiford and Goupillot, 1993). A good example of this level of discipline integration has been described from Draugen field (Guderian *et al.*, 2003), where

dynamic and static reservoir models have a high degree of consistency. Many modern time-lapse projects seek to 'close-the-loop' (Gutteridge *et al.*, 1994) by comparing seismic models generated from the simulation with the acquired data. In this way history matching of the simulation can be validated or adjusted. rock physics modelling (Chapters 5 and 8) provides the basis for interdisciplinary consistency (e.g. Gawith and Gutteridge, 2007).

Time-lapse seismic is a relatively new technology founded largely through advances in the understanding of reservoir rock physics in the late 1980s and early 1990s (e.g. Wang, 1997a). Prior to this time, time-lapse effects had generally been described from fields on land with fairly large changes in compressibility, a classic example being the fire flood project described by Greaves and Fulp (1987). It was uncertain whether or not there was sufficient signal or detection capability to make it a key technology for offshore developments particularly in conventional oil reservoirs. From tentative beginnings, various studies, largely on North Sea fields, essentially proved technical viability. Subsequently, experience shows that the net value gain associated with the use of

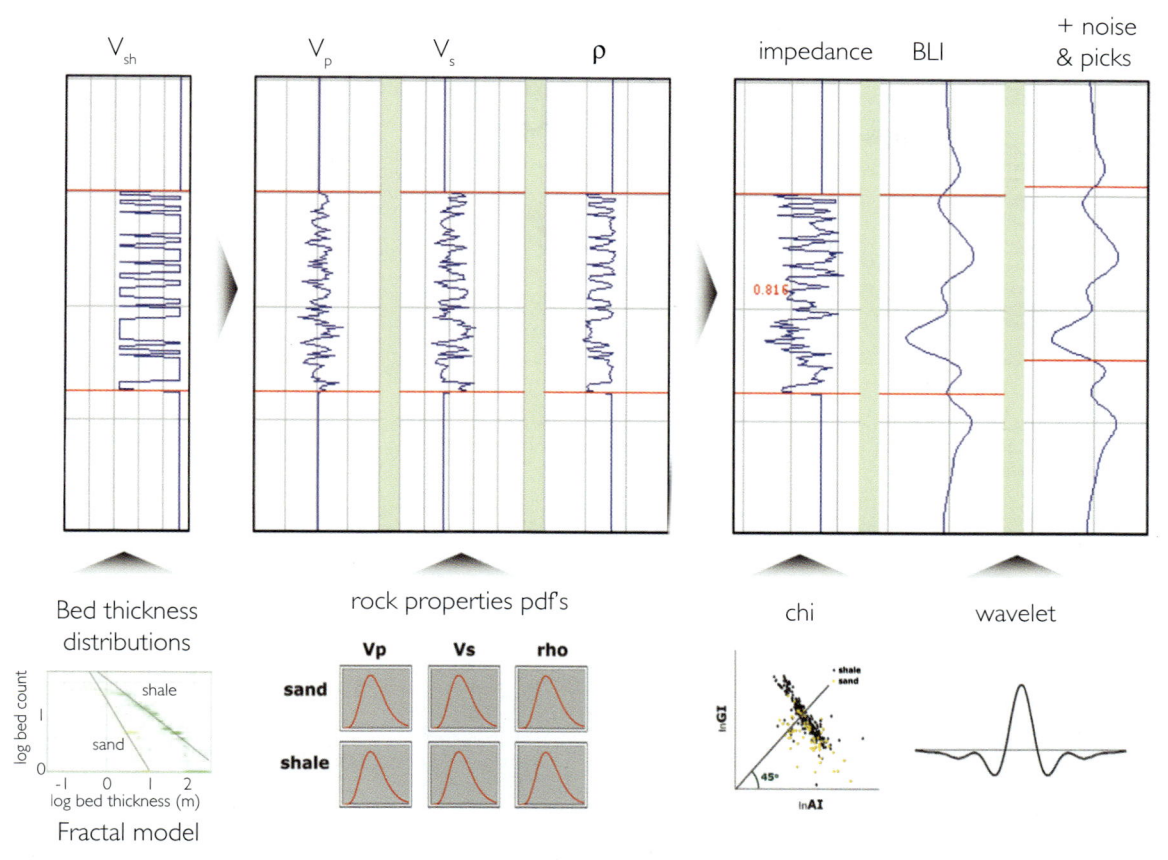

Figure 10.21 Generation of pseudo-wells for assessing uncertainty in Connolly's net pay estimation technique (after Connolly, 2010).

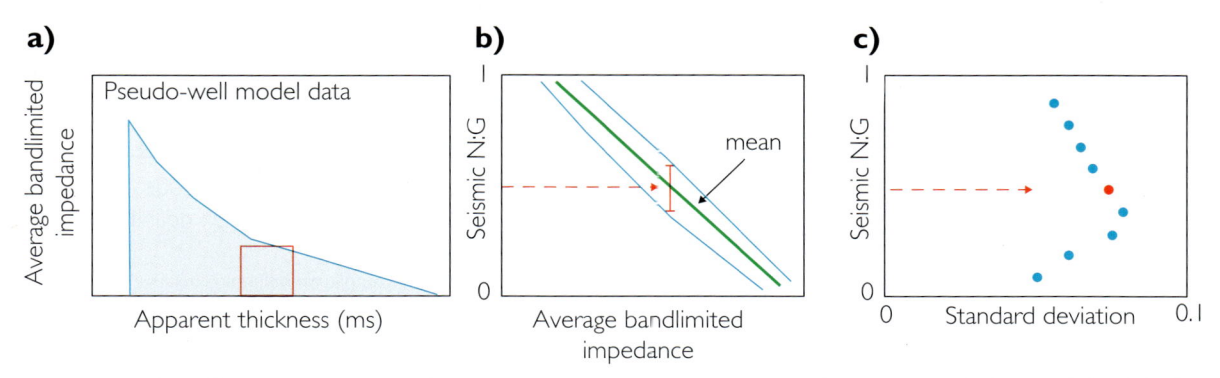

Figure 10.22 Connolly net pay estimation technique: estimating uncertainty in seismic N:G. (a) An apparent thickness is defined, (b) data in the thickness class are crossplot with average bandlimited impedance vs seismic N:G – for a given value of Seismic N:G the standard deviation around the mean is determined, (c) the standard deviation is calculated across the range of seismic N:G. The mapped values of apparent thickness and seismic N:G are then used to derive a seismic N:G standard deviation map.

time-lapse can be significant (e.g. Verbeek et al., 1999; Koster et al., 2000) and the industry now perceives the use of seismic in reservoir monitoring as an integral part of the field development plan, at least in offshore production where the costs are high. Justifying the expenditure is required on an individual project basis, of course, and each company has their own way of determining the value of 4D information (e.g. Pinto

Seismic amplitude applications

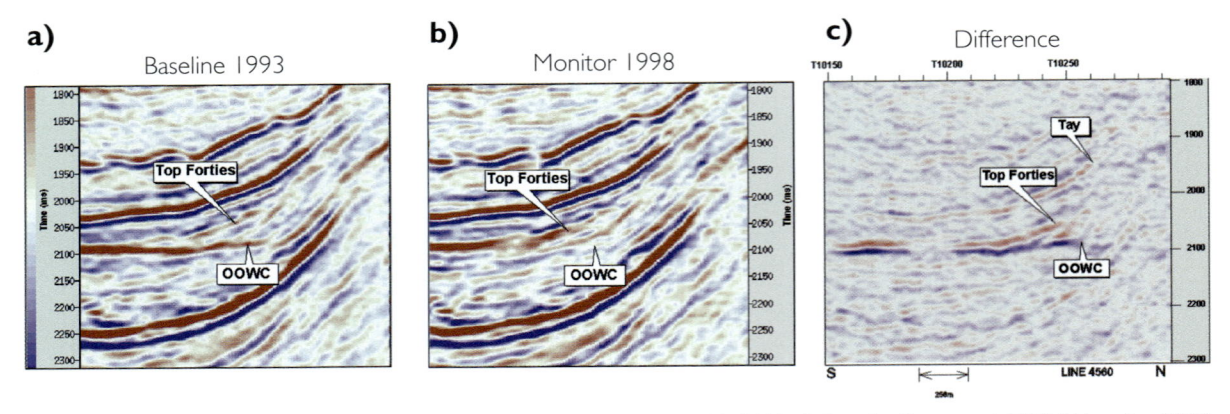

Figure 10.23 An example of time-lapse seismic in a producing oil field (Gannet C Field, North Sea, after Koster *et al.*, 2000); (a) baseline (1993) survey showing clear oil water contact, (b) monitor (1998) survey with no obvious flat reflection at original contact and brightening at top Forties reservoir; (c) difference section.

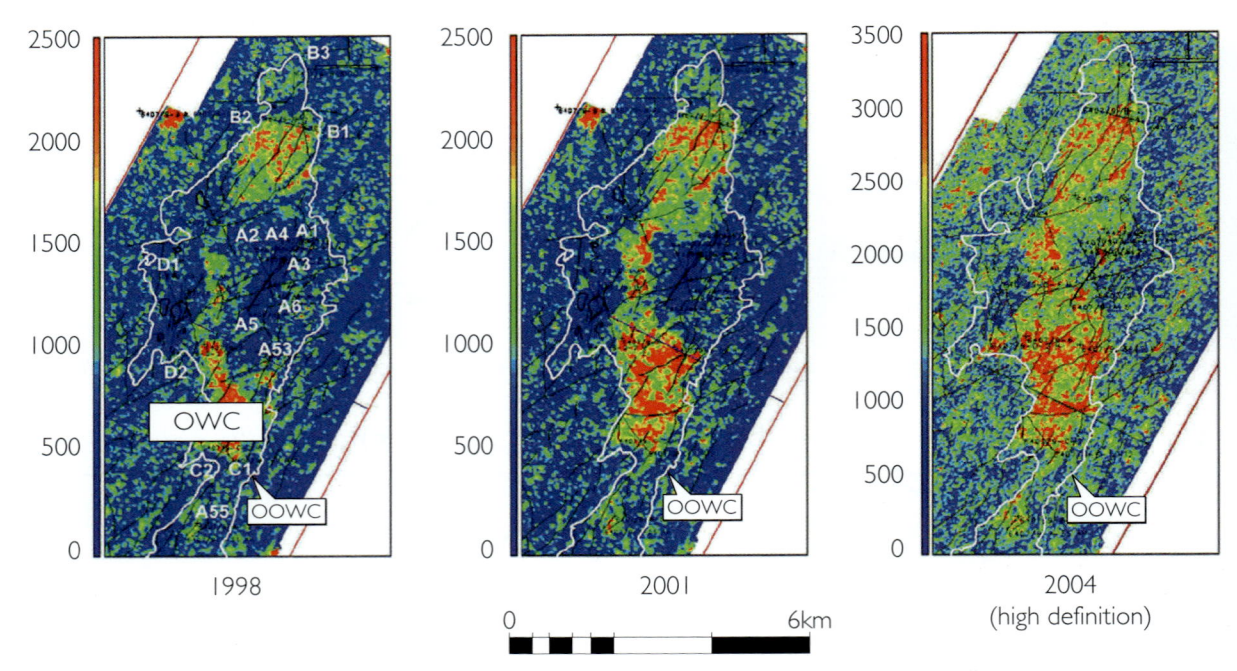

Figure 10.24 Time-lapse maps from Draugen Field, offshore Norway (1990–2004): RMS seismic amplitude difference (relative to 1990) showing 1998, 2001 and 2004 water encroachment. There is a higher noise level for the 2004 map because of differences in acquisition geometry. OOWC = original oil/water contact (after Mikkelsen *et al.*, 2008. Copyright 2008, SPE; reproduced with permission of SPE, further reproduction prohibited without permission).

et al., 2011; Fehintola and Olatunbosun, 2011). However, whilst there are a number of different types of feasibility studies that can be carried out, it is not always straightforward to determine the precise cost-benefit. Time-lapse can confirm an existing perception of field performance or it might yield surprises which can be used to adapt and optimise reservoir development strategy.

Any change in the reservoir will have an effect on acoustic behaviour; the main question for assessing the feasibility of time-lapse is whether these changes are large enough to be detected on seismic given the perceived levels of seismic repeatability. Key reservoir factors include:

- state of consolidation,
- porosity,

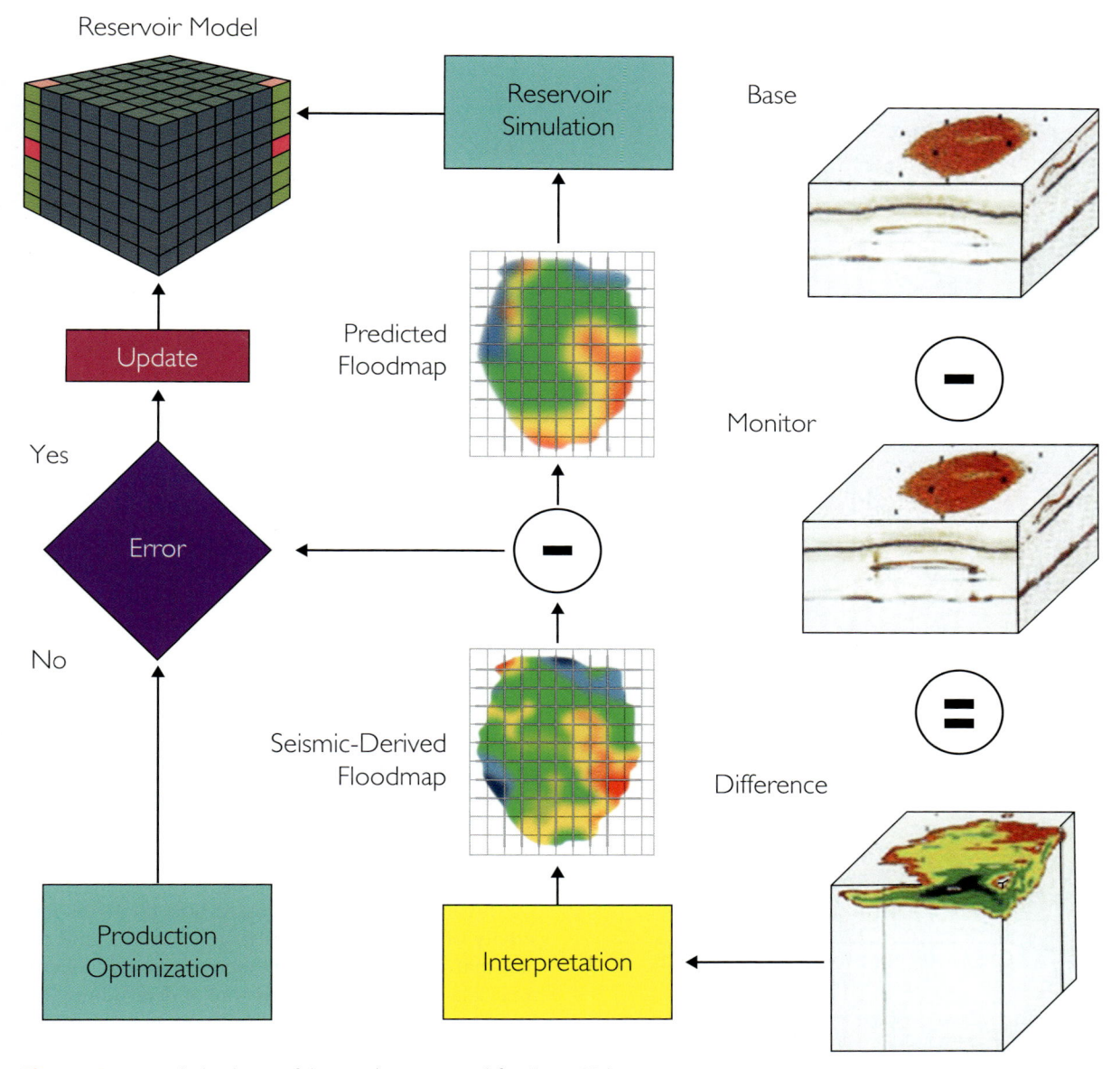

Figure 10.25 An idealised view of the time-lapse process (after King, 1996).

- stress sensitivity of the dry frame modulus,
- pore fluid phase and compressibility,
- reactivity of minerals and fluids.

Large time-lapse signals are likely with

- saturation and pressure changes in high-porosity unconsolidated reservoirs,
- saturation changes from gas to water or generation of solution gas as pressure drawdown takes an oil reservoir below the bubble point,
- saturation changes associated with oils with high API and gas-oil-ratio,
- large temperature changes for example in a steam or fireflood (e.g. Jenkins et al., 1997),
- significant porosity reduction associated with compacting reservoirs (such as is seen in North Sea Chalk reservoirs, e.g. Menghini, 1988; Haugvalstad et al., 2011),
- rocks with extensive sets of open fractures.

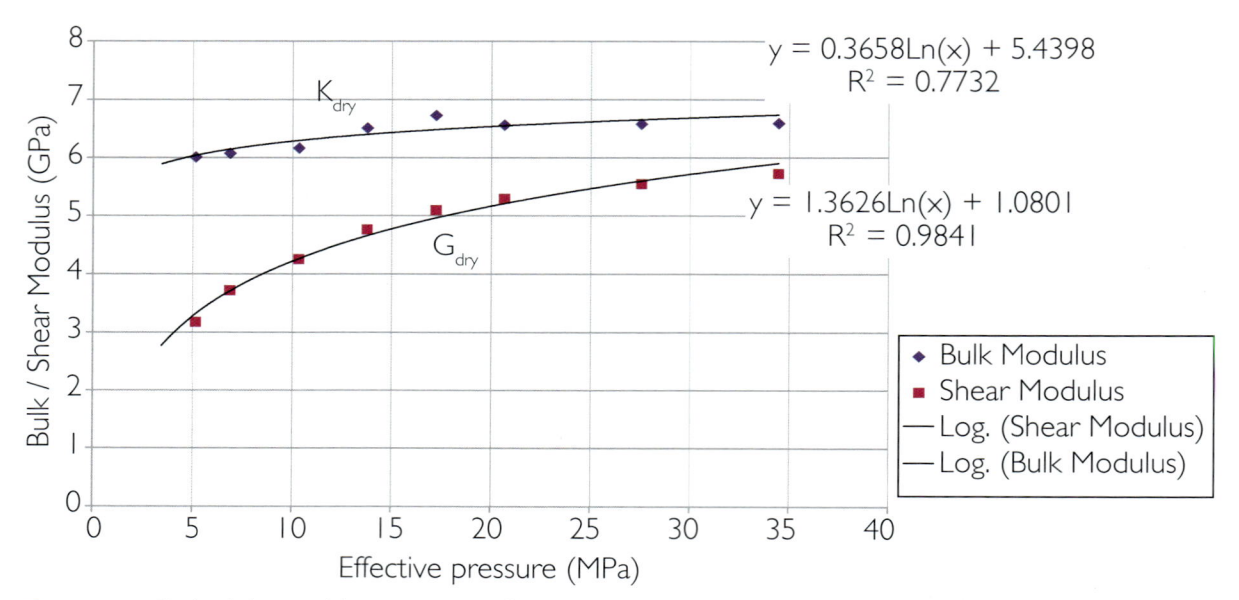

Figure 10.26 Fluid and shear modulus variation with effective pressure from laboratory measurements: Schiehallion Field, West of Britain (after Meadows *et al.*, 2005).

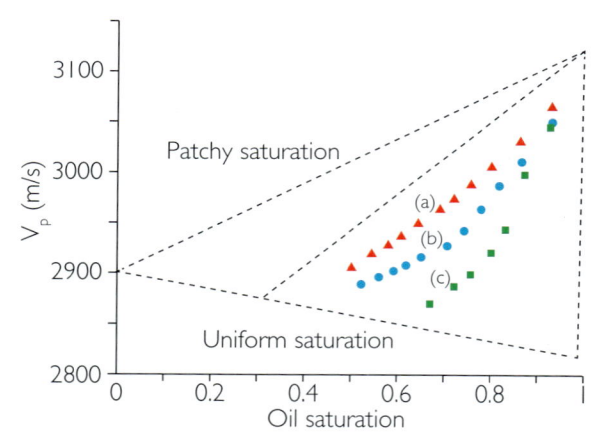

Figure 10.27 Experimental results showing patchy saturation effects related to gas injection into an oil reservoir. Cases (a),(b) and (c) are different permeability scenarios with different layering characteristics but essentially permeability improves from (a) to (c). After Sengupta and Mavko, 2003.

More subtle signals are likely to result from
- pressure changes in consolidated oil reservoirs,
- saturation changes in low *API* oil reservoirs under waterflood,
- saturation changes in carbonate oil reservoirs (e.g. Wang *et al.*, 1991).

Early feasibility modelling of time-lapse effects in conventional oil reservoirs treated the problem simply in terms of parameter changes in Gassmann's equation (Chapter 8), for example changing the fluid parameters associated with saturation change from virgin to residual saturation and accounting for the pressure change on the fluid modulus via the Batzle and Wang (1992) equations. It was realised, however, that two important factors were overlooked; namely the sensitivity of the dry rock frame to changes in stress during production and the potential for heterogeneous saturation distributions.

In Chapter 8 it was discussed how the stress sensitivity of sandstones shows large variations and the fact that in the absence of reservoir-specific laboratory data there is no obvious way of determining the likely response for a given sandstone (e.g. MacBeth, 2004; see also Chapter 8). Thus, acquiring laboratory measurements from core data (e.g. Fig. 10.26) is necessary to reduce the uncertainty. Dry rock modulus changes tend to have greater importance in the depletion of overpressured reservoirs compared to consolidated reservoirs.

Producing a reservoir can lead to inhomogeneous saturations throughout the pore space. This patchy saturation effect (Chapter 8) has been described from well log analysis (e.g. Dvorkin *et al.*, 1999; Caspari *et al.*, 2011) and it is envisioned that this might occur on a seismic scale (e.g. Sengupta and Mavko, 1998). The effect is most important in the case of gas, for

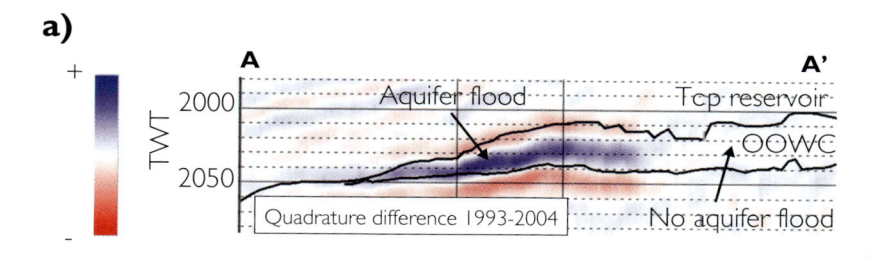

Figure 10.28 Time-lapse monitoring of a water flood front (after Staples et al., 2007); (a) quadrature difference section (i.e. reflectivity difference phase rotated by 90°) with blue loop effectively representing impedance increase within the reservoir (note that top reservoir and original oil water contact picks have been made on the baseline seismic survey), (b) map of maximum quadrature difference or impedance increase showing the form of the flood front.

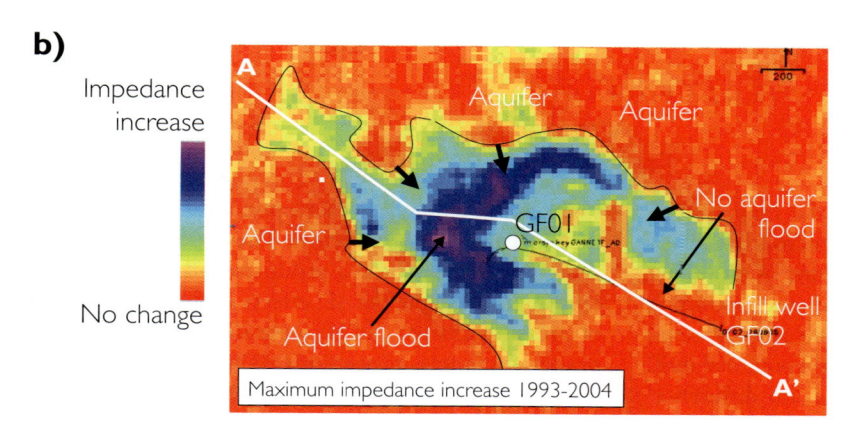

example where a gas reservoir is flushed with water or where gas is injected into an oil reservoir (Fig. 10.27). Effectively, these patchy saturation effects might be included in seismic modelling through the saturation vs fluid modulus relationship (see Chapter 8). Factors which appear to influence 'patchiness' of the fluid distribution (Sengupta and Mavko, 2003), are:

- permeability distribution,
- relative permeability of the fluids,
- irreducible saturations,
- density contrasts of the fluid phases,
- wettability of the rock,
- fluid properties.

A usual assumption in time-lapse modelling is that the mineral component of the rock is considered inert (i.e. the fluid substitution process does not change the nature of the dry rock frame). However, certain fluids and minerals can be reactive. For example, Vanorio et al. (2008, 2011) describe changes in the dry rock frame related to carbon dioxide injection. In brine-saturated sandstones the dry rock frame can be stiffened due to the precipitation of salts at grain contacts and in small pore throats, whereas in carbonates the dry rock frame might be softened by the dissolution of micro-crystalline matrix by carbon dioxide rich water. It has also been observed that the Gassmann assumption of the fluid independence of the shear modulus may not be correct for carbonates (e.g. Japsen et al., 2002; Baechle et al., 2003). Both shear modulus weakening and strengthening have been described from laboratory experiments in carbonates.

Determining the appropriate seismic display for highlighting these various changes in the reservoir inevitably requires a consideration of AVO. The magnitude of fluid-related impedance change typically increases with increasing incidence angle and far stacks may provide a greater signal than full stacks (e.g. Boyd-Gorst et al., 2001). Time-lapse changes will affect both the amplitudes of reflections, owing to changes in impedance, and the timing of seismic reflections, owing to the change in velocity. In normally pressured and consolidated reservoirs the dominant time-lapse effect is on seismic amplitudes. Figure 10.28 shows an example of a relatively simple time-lapse signature from the Gannet F field in the North Sea, related to a water flood. The zone of water flooding is clearly seen on the time-lapse difference section (Fig. 10.28a) generated from two quadrature (i.e. 90° rotated) seismic volumes. Note that the top of the reservoir has been

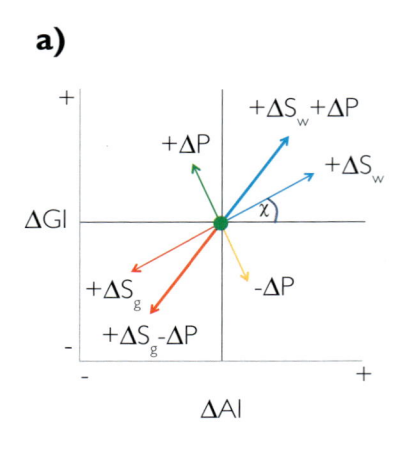

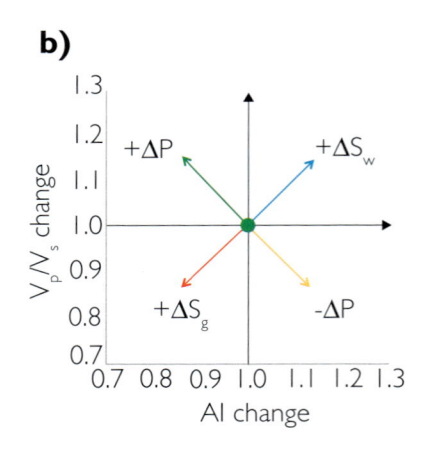

Figure 10.29 Pressure and saturation effects on an initially oil filled sandstone reservoir; (a) AI vs GI crossplot showing changes in AI and GI relative to an initial oil filled sand (green point) (modified after Ricketts et al., 2008); (b) AI vs V_p/V_s crossplot showing change as the ratio of monitor to baseline values (modified after Andersen et al., 2006. Copyright 2006, SPE, reproduced with permission of SPE, further reproduction prohibited without permission). Note that P = pore pressure, S_w = water saturation and S_g = gas saturation.

picked on the baseline survey and the map display (Fig. 10.28b) effectively shows the encroachment of the water aquifer into the field. The zone of no difference to the right in Fig. 10.28a and corresponding (red) area in the bottom of Fig. 10.28b is interpreted as a zone of remaining oil, representing a potential infill target. Clearly the interpretation of a 'no difference' zone depends to a large extent on the understanding of the reservoir based on the baseline interpretation. Relatively low N:G zones which are poorly connected to the main reservoir units might give a similar signature.

The whole range of reflectivity and inversion techniques described so far in this book can be brought to bear on time-lapse interpretation. For example, Connolly's (2007) net pay technique might be used to estimate the change in net pay from the types of responses shown in Fig. 10.28. Saturation and pressure changes are often approached from the perspective of AVO, either in the reflectivity or inverted impedance domain (e.g. Tura and Lumley, 1999; Landrø, 2001). The AVO component can help in addressing potential interpretation ambiguities, for example determining the difference between a rise in impedance due to pressure drawdown and an increase in water saturation due to oil production.

Figure 10.29 shows a schematic diagram illustrating the relative elastic effects of pressure and saturation on a consolidated sandstone reservoir. If an oil sand positioned at the origin of the $\Delta AI/\Delta GI$ plot (Fig. 10.29a) is replaced by water, the point will move to the upper right (thin blue vector), whereas replacing oil by gas moves the point along the thin red vector. Pressure changes are likely to be orthogonal to the fluid change vectors. The combined effect of increasing water saturation and pore pressure (i.e. during water injection) will give a response that falls between the fluid and pressure vectors. The opposite will occur if the pressure drops at a producer and oil is replaced by gas. In principle the 4D effects of fluid change and of pressure change can be separated by using appropriate chi angle projections (Chapter 5). Relative changes in saturation will be optimised at relatively small projection angles whereas pressure effects will require large projection angles (e.g. Nunes et al., 2009). Inevitably this means that the pressure signals are likely to be more prone to noise (Chapter 5). Using absolute impedance inversions, changes in saturation and pressure can be deduced by the relative changes in acoustic impedance and Poisson's ratio (Fig. 10.29b).

Clearly, the fact that time-lapse signatures are often tuned responses gives rise to potential ambiguities in interpretation. For example, in a layered reservoir where the sweep is uneven it may not be possible to determine in which layer the time-lapse effects have been generated. Stochastic inversion methods discussed in Chapter 9 could be used to address this problem (e.g. Gawith and Gutteridge, 2001; Veire et al., 2007).

Rather than estimate reservoir changes directly from AVO signatures or inverted impedances, some authors approach the problem by using nonlinear multi-attribute methods in which correlations between production data (i.e. changes in fluid saturation and effective pressure) and a variety of seismic attributes are established (e.g. Sønneland et al., 1997; MacBeth et al., 2004, Ribeiro and Macbeth, 2006). Figure 10.30 shows an example of saturation and pressure mapping based on such an approach. Ribeiro and Macbeth (2006) describe time-lapse inversion directly for the fluid modulus and pore pressure. Uncertainty estimation can be incorporated into the

Time-lapse seismic

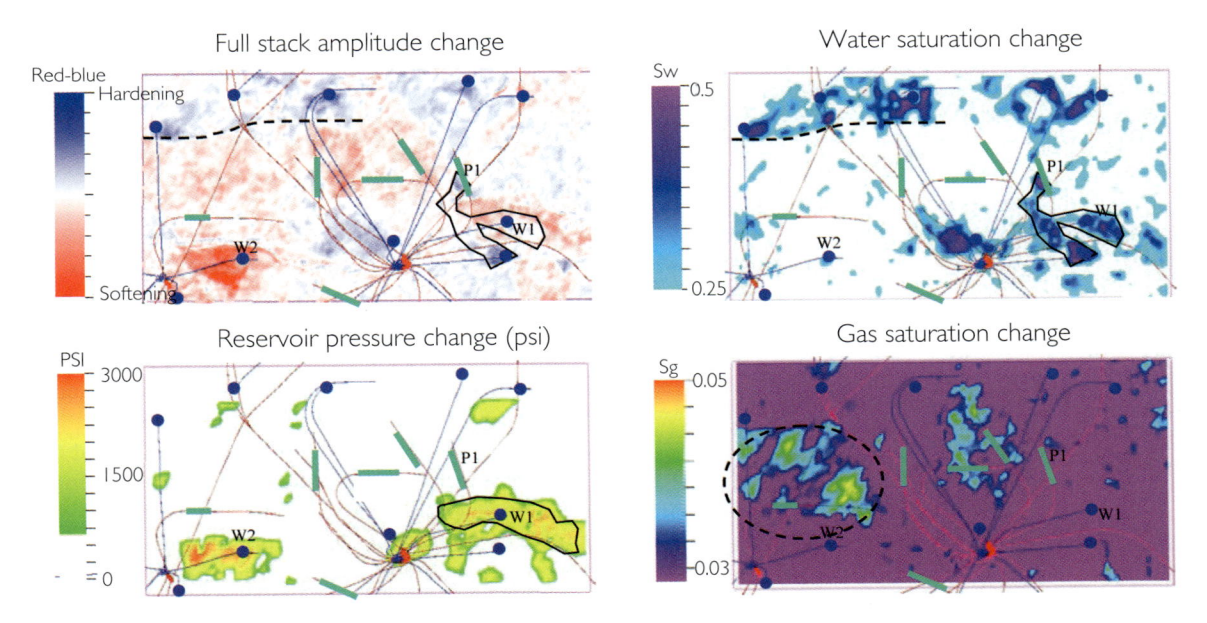

Figure 10.30 Schiehallion Field time-lapse mapping using a multi-attribute calibration of production data (after Floricich et al., 2006). The attributes used in this example were far amplitude, intra-reservoir time-stretch (i.e. time-shifts computed in a window below the reservoir minus time-shifts computed in a window above the reservoir), near amplitude, full amplitude and instantaneous frequency.

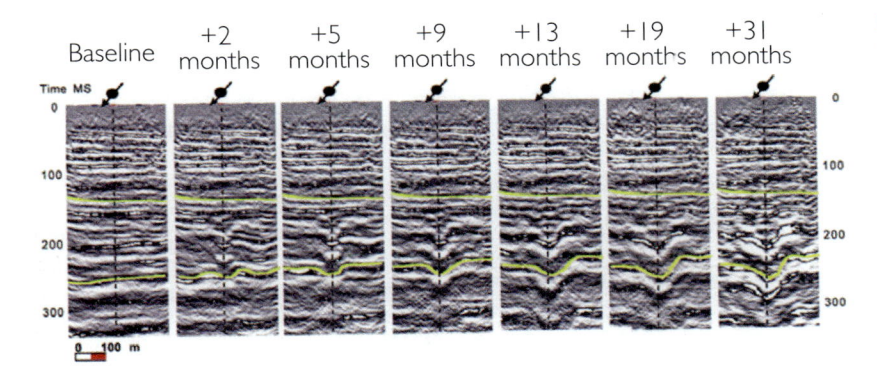

Figure 10.31 Vertical seismic sections from baseline and monitor surveys in the Duri Field, Indonesia. The yellow lines show the top and base of the steam injection interval. Note how the reflection time of the base increases as the steamflood progresses in time, whilst the data do not change above the steam zone (after Jenkins et al., 1997).

workflow for example by the introduction of Bayesian concepts (e.g. Floricich et al., 2006).

Travel-time differences have typically been used to analyse time-lapse situations with large changes in fluid compressibility. For example, Fig. 10.31 illustrates travel-time changes on base reservoir reflections with progression of a steamflood. However, timing effects can also be associated with geomechanical changes in the reservoir (i.e. porosity reduction) and in the overburden as a stress response to pressure depletion in the reservoir. These phenomena have been described from a range of producing fields including North Sea Chalk fields which compact during production (e.g. Nickel et al., 2003; Herwanger et al., 2010) and high-pressure/high-temperature fields (e.g. Hatchell and Bourne, 2005; Hawkins et al., 2007). Figure 10.32 shows the time-lapse time-shift effect related to subsidence on Ekofisk field (after Nickel et al., 2003). Geomechanical time-lapse effects can be modelled using rock physics relationships (e.g. Hatchell and Bourne, 2005) and these can be compared with the time-lapse seismic result (Fig. 10.33). Time-shifts are now being specifically incorporated into time-lapse inversion methodologies (e.g. Bhakta and Landrø, 2013).

Confident identification of time-lapse effects generally requires high levels of seismic repeatability.

Seismic amplitude applications

A commonly used generalised measure of repeatability is the normalised root mean square (*NRMS*) (Kragh and Christie, 2002). This is calculated in zones where no production changes have taken place and is defined as:

$$NRMS = \frac{2RMS(a-b)}{RMS(a)+RMS(b)},$$

where a and b are the two surveys being compared. Typical values for *NRMS* are around 10%–30%. However, what is acceptable will depend on numerous factors including acquisition and processing as well as the magnitude of the time-lapse signal.

A key issue in achieving acceptable levels of repeatability is that the acquisition should reproduce the source and receiver positioning as accurately as possible. This is quite difficult in the case of marine streamer data, because currents in the seawater cause the receiver streamer to be pushed to one side rather than being towed in a straight line directly behind the ship, an effect usually called *streamer feathering*. These currents often vary significantly with time due to tidal or larger-scale circulation effects. Matching feathering angles and overlapping streamer coverage as well as shooting infill lines on the basis of repeatability rather than simple fold (i.e. coverage) criteria are typical strategies used to address this situation. Better receiver-position repeatability is possible where receiver cables are deployed on the ocean floor (e.g. Beasley *et al.*, 1997). For this type of acquisition the accuracy of cable deployment as well as receiver/sea

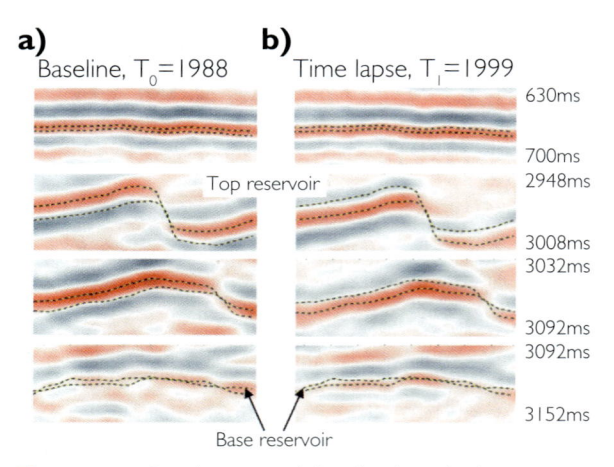

Figure 10.32 Time-lapse time shifts related to subsidence on Ekofisk Field North Sea (after Nickel *et al.*, 2003). The figure shows baseline sections (a) and monitor sections (b) for various two-way-time intervals. The dashed lines are seismic picks made on the baseline survey; comparisons of baseline and monitor sections illustrate the timing effect related to reservoir compaction and subsidence.

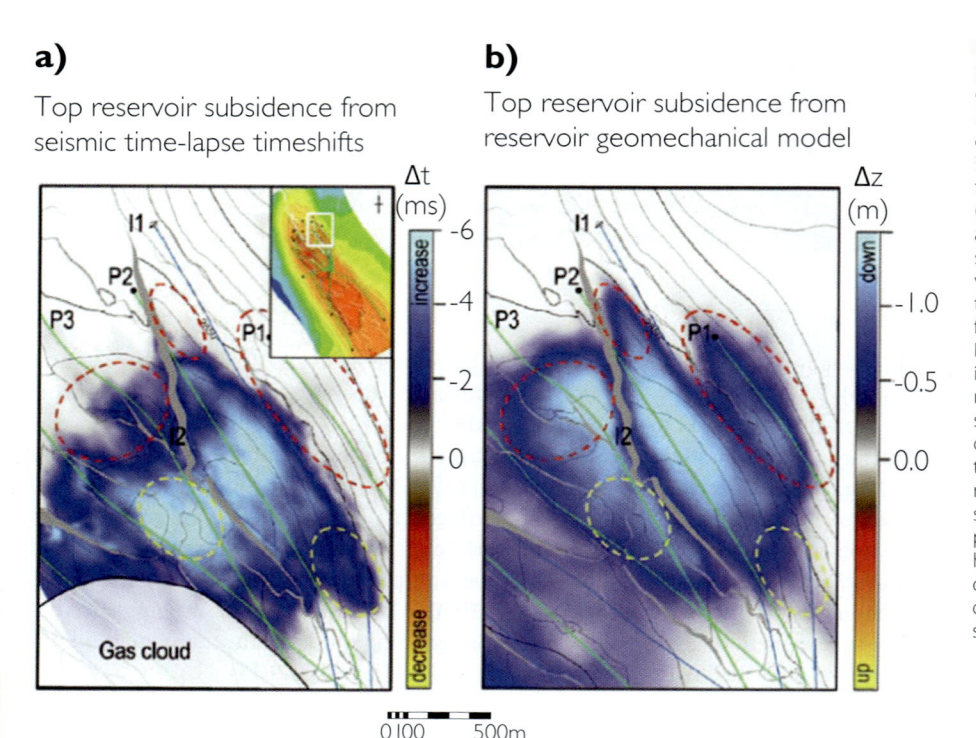

Figure 10.33 Observed and predicted reservoir compaction on South Arne Field, offshore Denmark (after Herwanger *et al.*, 2010); (a) compaction-induced travel time changes to the top-reservoir reflector (time-lapse time-shifts Δt) as a measure for top-reservoir subsidence, (b) predicted vertical displacement Δz of the top-reservoir surface from a geomechanical model. Note that a 6 ms increase in travel time corresponds roughly to 1.5 m top-reservoir subsidence. Yellow areas = observed subsidence larger than predicted subsidence, red areas = observed subsidence is less than predicted subsidence. Note how faults exert a large degree of control on both observed and predicted subsidence.

floor coupling are important success factors. The same is true for ocean-bottom nodes (e.g. Beaudoin and Ross, 2007). Further improvements in repeatability might be gained from permanent seabed receiver installations entrenched in the seabed (e.g. Jack *et al.*, 2010). On land, repeatability of acquisition geometry is not a problem; instead, the main difficulty lies in changes to source and receiver coupling and scattering in the weathered zone, which may vary through the year with changes in soil saturation and position of the water table. Pevzner *et al.* (2011) describe an *NRMS* value of about 20% for a time-lapse survey on the Otway CO_2 sequestration project in Australia.

Seismic processing for time-lapse aims to minimise differences in the amplitude, phase and timing for non-reservoir reflections and thereby enhance signals related to production differences (e.g. Ross *et al.*, 1996; Rickett and Lumley, 2001). Experience has shown that it is more beneficial to combine the monitor survey processing with re-processing of the baseline survey than attempt to match different surveys with different acquisition and processing (e.g. Johnston, 2013). Cross-equalisation tools such as space and time-variant amplitude envelope balancing, matched filter corrections for amplitude, bandwidth, phase and static shifts, residual migration and residual time alignment (including static shifts and warping) can then be more effectively used to enhance the repeatability.

A good example of increasing repeatability of marine surface seismic surveys over time has been presented by Haugvaldstad *et al.* (2011) (Figs. 10.34 and 10.35). This concerns the Ekofisk Field in the North Sea, over which five marine seismic surveys have been acquired. The first survey was in 1989 (dual source, dual streamer), the second in 1999 (two sources, four streamers), and the third in 2003 (single source, eight streamers). In 2006 a survey was shot with the same configuration as in 2003, though with no attempt

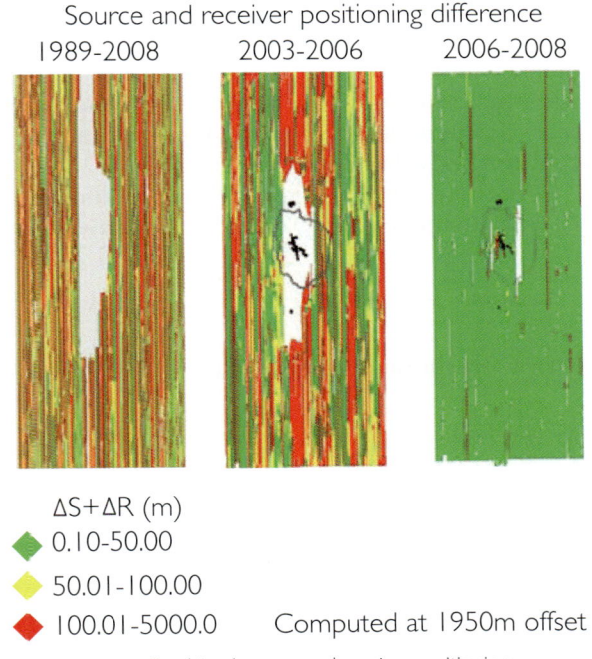

Figure 10.34 Combined source and receiver positioning difference maps for different pairs of surveys over Ekofisk, from Haugvaldstad *et al.* (2011).

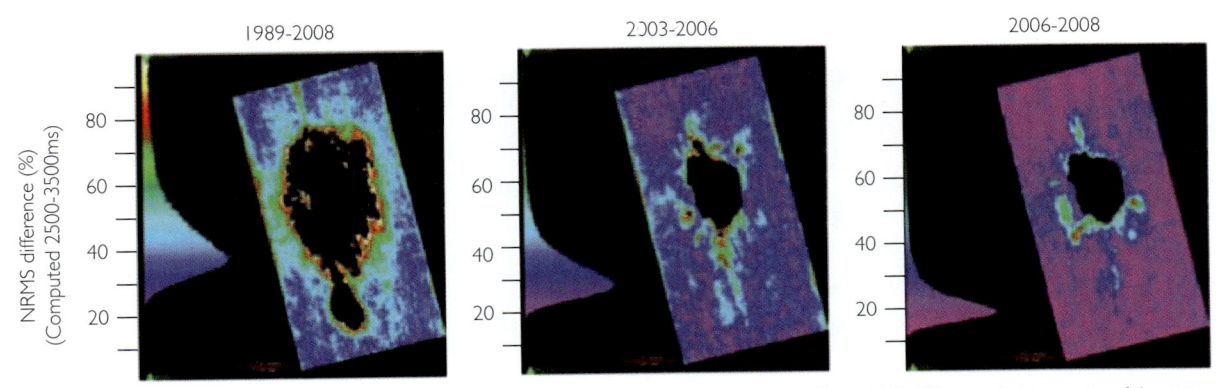

Figure 10.35 NRMS difference maps over Ekofisk, from Haugvaldstad *et al.* (2011). The areas of large RMS difference in the centre of the maps are genuine production-related effects (Haugvaldstad *et al.*, 2011).

to replicate the 2003 source and receiver locations. Finally, a survey was shot in 2008 with steering of the source and streamer employed in order to replicate the 2006 source and receiver locations as closely as possible, resulting in much greater consistency (Fig. 10.34). *NRMS* difference between surveys in the area not affected by production has evolved from 30%–40% for 1989–2008, through 20%–30% for 2003–2006 and finally to 10%–20% for 2006–2008 with duplication of source and receiver positions (Fig. 10.34). In this particular case, it was decided that six-monthly monitor surveys would be appropriate for field management, and in 2010 a large array of permanent seabed receivers was deployed, ensuring lower future acquisition costs and enhanced receiver repeatability.

Prior to a time-lapse project being sanctioned it is usual for a feasibility study to be carried out.

A simple approach which can be used to rank as well as to identify specific issues for individual fields is the scorecard approach of Lumley et al. (1997). Two scorecards are completed, a reservoir scorecard which objectively identifies the relative magnitudes of rock physics changes during production and a more subjective seismic scorecard (Tables 10.1 and 10.2). Lumley et al. (1997) suggest that the reservoir scorecard requires a >60% score for time-lapse to be considered as a viable option. Given the advances in seismic repeatability it is likely that this threshold is too conservative. Marsh et al. (2003) present a simpler ranking tool based on practical experience of time-lapse projects, in which the normal incidence reflectivity of a notional contact is referenced against different acquisition and processing strategies.

Following the initial screening phase, more detailed modelling is also carried out to evaluate whether the production effects can be seen against the noise background with a level of detail that will be of any use to the reservoir engineer. Models for various production scenarios would typically include modelled gathers based on well logs as well as 2D models. However, in order to address issues such as lateral sweep efficiency it is common for the reservoir simulation to be used to generate a 3D model. This type of approach is time consuming, requiring a detailed rock physics model for both reservoir and non-reservoir units and a consideration of how to address differences in scale between geological and reservoir models.

10.5 Amplitudes in prospect evaluation

Traditionally, the risk (or chance of success) on a hydrocarbon prospect is determined by identifying

Table 10.1. Reservoir and seismic scorecard for assessment of time lapse seismic feasibility (after Lumley et al., 1997)

Time-lapse feasibility scorecard	Ideal
RESERVOIR	
Dry bulk modulus	5
Fluid compressibility contrast	5
Fluid saturation change	5
Porosity	5
Predicted impedance change	5
Reservoir total	25
SEISMIC	
Image quality	5
Resolution	5
Fluid contacts	5
Repeatability	5
Seismic total	5
TOTAL SCORE	45

Table 10.2. Detailed reservoir scorecard for assessment of the feasibility of time lapse seismic (after Lumley et al., 1997)

Reservoir scorecard							
	Score	5	4	3	2	1	0
Dry bulk modulus	Gpa	<3	3–5	5–10	10–20	20–30	30+
Fluid compressibility contrast	% change	250+	150–250	100–150	50–100	25–50	0–15
Fluid saturation change	% change	50+	40–50	30–40	20–30	10–20	0–10
Porosity	%	35+	25–35	15–25	10–15	5–10	0–5
Predicted impedance change	% change	12+	8–12	4–8	2–4	1–1	0
Travel-time change	#samples	10+	6–10	4–6	2–4	1–2	0

the probability of there being an effective trap, with a reservoir which has been charged by hydrocarbons migrating from a mature source rock. Thus there are numerous geological elements which need to be evaluated both separately and together as part of the petroleum system. The presence of reliable seismic indications of the presence of hydrocarbon in the seismic (i.e. a DHI), can suggest, however, that all the individual components are present and the risking process becomes focussed on the validity of the DHI. It is well documented that drilling success rates are generally higher in basins where DHI recognition is possible (e.g. Rudolph, 2001).

In practice, both geological and DHI risking are commonly undertaken. The DHI evidence can be used to modify individual risking elements, such as the risk on hydrocarbon migration, or to condition the overall risk (e.g. Roden et al., 2005; Forrest et al., 2010) or indeed it may be used to override the geological risk altogether. The process, however carefully prescribed, is often subjective; two different interpreters may not assess the significance of the observations in exactly the same way. It is also not free of an emotive element; using the label 'DHI' may imply more certainty of hydrocarbon presence and a much narrower range of possible outcomes than is warranted (Citron and Rose, 2001).

Clearly, amplitude-related observations need to be subjected to a rigorous evaluation before assigning significance in terms of risk. This will include an assessment of the validity of a DHI interpretation as well as evaluating the play and risking context. It is often overlooked that DHI evaluations are to some degree dependent on the knowledge of the play. The interpreter who has been working with seismic amplitudes in a basin for several years is perhaps more likely to recognise what is and what is not a DHI more accurately than an interpreter who is new to the basin.

10.5.1 An interpreter's DHI checklist

A checklist questionnaire is a useful tool in checking the validity of a DHI and commonly forms the basis of a 'DHI Index' (Roden et al., 2012). An example of such a questionnaire is shown in Table 10.3. These types of questions can be arranged into a scorecard for the purpose of creating a DHI risk.

A simple approach to guide a DHI risking exercise is to use a play specific success matrix (e.g. Citron and Rose, 2001). In this case (Fig. 10.36), the axes are defined as the knowledge of the play and the confidence in recognition of a DHI. The checklist questionnaire would help in arriving at the appropriate position in the matrix and the assigned probabilities would be specific to a particular play, data types and analytical techniques. Populating the matrix of course requires access to data for a sample of prospects. This is relatively straightforward for large operators drilling a large number of wells but much more difficult for smaller companies.

10.5.2 A Bayesian approach to prospect risking

The DHI risk is often used to lower the prospect risk where a valid DHI is present but it can also be used to increase the risk in situations where a DHI is absent, but where one would be expected if significant quantities of hydrocarbons were present. Combining DHI information with general geological risking can effectively be handled using Bayes' theorem, using the following form:

$$P(hc|dhi) = \frac{P(dhi|hc)P(hc)}{P(dhi|hc)P(hc)+P(dhi|nohc)P(nohc)},$$

where

$P(hc|dhi)$ = the probability of hydrocarbons given the observed DHI,

$P(dhi|hc)$ = the probability of occurrence of the observed DHI if hydrocarbons are present,

$P(dhi|nohc)$ = the probability of occurrence of the observed DHI if no hydrocarbons are present,

$P(hc)$ = the *a priori* probability of hydrocarbons ignoring the DHI (i.e. the 'geological' chance of success),

$P(nohc) = (1 - P(hc))$ is the *a priori* probability of no hydrocarbons.

To illustrate how this works, suppose that there is a drillable prospect in a well understood play where the chance of finding hydrocarbons using geological risking in a particular trap is 0.3 (i.e. a typical risk for a drillable prospect in a relatively mature basin). A convincing DHI is present, such that it is very unlikely it would be seen if no hydrocarbons are present; thus $P(dhi|hc)$ might be assigned a value of 0.7 and $P(dhi|nohc)$ might be 0.2 (note that the two probabilities do not need to add up to 1 given that they are essentially unrelated). The chance of success would be calculated as $(0.7 \times 0.3)/(0.7 \times 0.3 + 0.2 \times 0.7) = 0.6$. This high chance of success indicates that the DHI has transformed the perception of the prospect. If there

Table 10.3. A checklist for assessing the validity of a DHI

Data quality
- Is the seismic data quality good enough to interpret relative amplitudes (high signal-to-noise ratio, good event continuity)?
- Have the processing parameters been checked for processes that would significantly distort amplitude behaviour?
- Are migration errors and other imaging artefacts (such as multiples) a significant issue?
- Is there a high degree of confidence in the phase and polarity of the data (i.e. based on well ties)?
- If AVO is a critical element of the play:
    - can reflections be traced from near to far traces on seismic gathers?
    - Is residual moveout a problem?
- Is the approach to AVO analysis appropriate for the given data quality?

Trap definition
- Is the trap adequately defined and consistent with the expected style for this play?

Well control/analogues
- Are there enough well data of good quality to calibrate the seismic data?
- Have the logs been appropriately conditioned?
- Have appropriate reflectivity/impedance models been generated for both pay and non-pay scenarios?
- Is there good reason to believe that the assumptions in the modelling (including the low-frequency component in seismic inversions) hold over the prospect area?

Amplitude relationships
- Are there clear relative/absolute differences in amplitude across the prospect area that fit with the hydrocarbon rock physics model, particularly from downdip to updip across the region of the contact and justifying the interpretation of dim spots, bright spots and phase reversals, as well as other features such as low-frequency shadow zones and velocity sags? Generally four-way dip traps show more DHI characteristics than stratigraphic traps.

Conformance of amplitude to structure
- Do the amplitudes show a high degree of conformance with a particular depth contour?

Flat spot
- If a flat spot exists, does it show discordance to stratigraphy?
- Are apparent terminations present?
- Does the flat spot have amplitude, polarity and AVO behaviour in accordance with the modelled hydrocarbon response?

Confidence in the play
- Confidence in the stratigraphic identification of seismic reflections is high.
- Has the target been drilled before in this type of situation (i.e. is there a good (local) analogy)?
- Are there reliable statistics from well results available for the target of interest?
- Is the effect of variation in lithology, fluid fill and bed thickness as a function of reflection angle well understood? Note that a single well may provide only limited calibration; the chance of the lithology being significantly different from that at the well may be quite high.

was similar geological risking but a much more equivocal DHI, for example a weak flat spot, giving a value of 0.6 for $P(dhi|hc)$ and a value of 0.4 for $P(dhi|nohc)$, then the chance of success would be $(0.6 \times 0.3)/(0.6 \times 0.3 + 0.4 \times 0.7) = 0.39$. Thus, the equivocal DHI has not changed the perception of the prospect very much. A similar calculation can deal with the risk of low (non-commercial) gas saturations; in this case $nohc$ would include sub-commercial saturations. Bright soft amplitudes on a flank, consistent with the presence of gas but not showing conformance to structure or evidence of a gas column from a flat spot, might then have $P(dhi|hc)$ equal to $P(dhi|nohc)$ at say 0.6, and the chance of success is calculated as $(0.6 \times 0.3)/(0.6 \times 0.3 + 0.6 \times 0.7) = 0.3$. In this case the amplitudes are making no modification to the geological risk.

	GOOD		
N/A	20%	50%	
N/A	10%	25%	Knowledge of play
N/A	N/A	12%	
		POOR	
LOW		HIGH	
	Confidence in DHI		

Figure 10.36 An example of a simple DHI risking matrix. Note that the numbers in the cells are play specific; in this case they are modelled on the Yegua example presented by Allen and Peddy (1993).

To see what happens if a DHI is predicted from modelling but it is absent in the seismic data, then the same equation would be used but now with *nodhi* instead of *dhi*. So P(*nodhi*|*hc*) might be 0.2, and the chance of no DHI given no hydrocarbons might be 0.9; then the chance of hydrocarbons given no DHI would be (0.2 × 0.3)/(0.2 × 0.3 + 0.9 × 0.7) = 0.09. The absence of the DHI has substantially reduced the perception of the chance of success.

The *a priori* risk, based on traditional geological reasoning, has its influence on the outcome. Suppose the geological chance of success is only 0.1 and a convincing DHI is present as in the first example, so that P(*dhi*|*hc*) is 0.7 and P(*dhi*|*nohc*) is 0.2. The chance of hydrocarbons would then be calculated as (0.7 × 0.1)/(0.7 × 0.1 + 0.2 × 0.9) = 0.28. Even a convincing DHI will therefore be allocated a modest chance of success if the geological model is sceptical of hydrocarbon presence.

The Bayesian approach outlined above is a useful sense check, particularly for comparing with probabilities derived from subjective compromises between geological and DHI risking. Stratigraphic prospects which are entirely driven by seismic amplitudes present a difficulty for this approach, however, as it would be difficult to assign a trap risk independent of the amplitude information.

10.5.3 Risking, statistics and other sense checks

Given the business importance of prospect risking most oil companies do not tend to publish results. Thus, the literature is fairly impoverished when it comes to detailed case studies. Some general papers have been published and a few observations are worth highlighting.

An early example of the role of DHI information in drilling results from the Yegua trend, onshore Texas has been published by Allen and Peddy (1993) and Allen *et al.* (1993). This is a mature gas play comprising shallow high-porosity sands that when gas-filled give bright spots on stacked sections and increasing amplitude with offset on pre-stack gathers. Shale on brine sand reflections generally have the opposite polarity to shale on gas sand, and show decreasing amplitude with offset (Fig. 10.37). Eighty-four wells were drilled on AVO anomalies described from 2D data. Commercial success rates improved dramatically, from 5%–10% to 50% when AVO techniques were employed (Fig. 10.38). This is encouraging, but of course it does not follow that use of AVO will bring similar improvements elsewhere. Interestingly, around 10% of the AVO anomalies were considered to have had low validity. In the majority of failure cases sand was encountered (Fig. 10.39). Reasons for failure included:

- problems with 2D data quality
    - imaging effects, including out of plane effects, migration/Fresnel zone issues and seismic noise,
- polarity and phase issues,
- low gas saturation,
- tuning,
- lithology effects
    - wet sands giving Class III AVO
    - sand quality (e.g. tight sands)
    - misidentification of coal signature
    - shale/shale AVO, possibly related to anisotropy?

A more global approach to amplitude statistics has been undertaken by the Rose & Associates DHI Consortium (e.g. Roden *et al.*, 2005, 2012; Forrest *et al.*, 2010). Partner companies contribute to a database which in 2012 comprised 217 wells, with roughly

Seismic amplitude applications

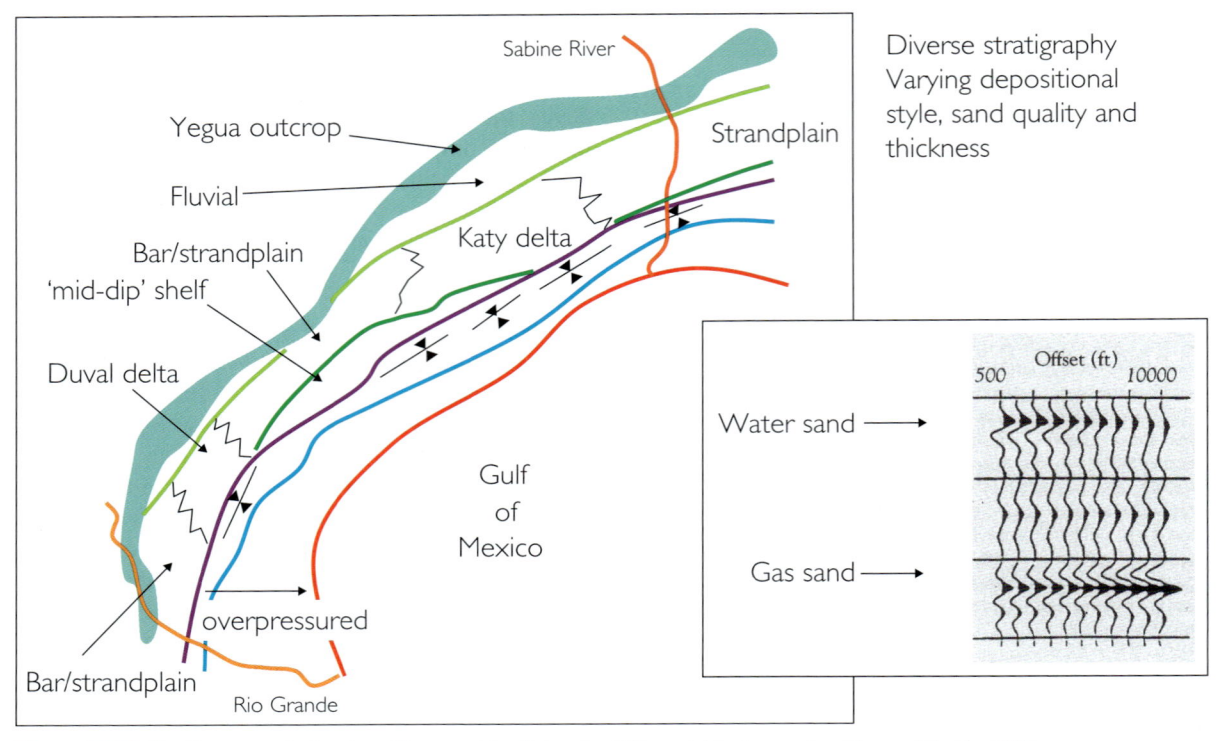

Figure 10.37 The Yegua trend of onshore Texas and the first-order AVO model (redrawn after Allen and Peddy, 1993).

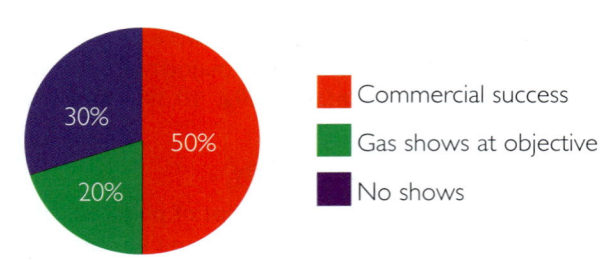

Figure 10.38 AVO results from the Yegua trend (after Allen et al., 1993).

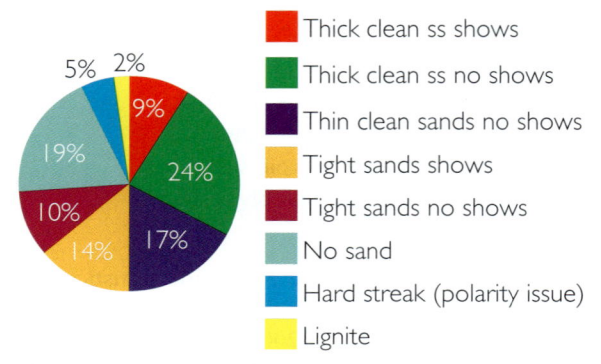

Figure 10.39 Yegua trend: geological outcome in cases of failure of AVO pay prediction (after Allen and Peddy, 1993).

equal numbers of successes and failures. Key parameters have been documented including geologic setting, seismic and rock physics data quality, DHI characteristics, pre drill risk estimates and drilling results. The reservoirs range in age from Triassic to Pleistocene. Most of the wells come from AVO Class III (75%) or Class II (22%) settings. For the Class III examples the most important DHI characteristics for assessing the impact on chance of success were (in order of importance):

- conformance of amplitude to structure on stacked or far-offset seismic data,
- phase or character change at downdip edge of the anomaly,
- amplitude consistency (uniformity) within the mapped target area on stacked data,
- flat spots,
- AVO response anomalous compared to events above and below.

For the Class II AVO prospects, the most important characteristics were:

- conformance of amplitude to structure on far-offset seismic data,

- amplitude consistency within the mapped target area, on gathers, far offset or far-angle stacks,
- AVO observations on gathers or offset/angle stacks: checking that the AVO response is indeed Class II and not affected by noise, inadequate NMO correction, multiples and processing artefacts,
- AVO behaviour is anomalous compared to the same seismic event outside closure,
- compatibility of observed AVO response with prediction from well or modelling based on rock physics.

Interestingly, the success rate within Class II prospects is higher than with Class III prospects, probably because of the discrimination provided by the far stack in the case of phase reversal signatures (Chapter 7). The analysis of failures in the DHI consortium dataset shows many similarities with the Allen and Peddy (1993) study; wet sands (49%), low saturation gas (23%), tight reservoirs (11%) and no reservoir (17%).

Roden et al. (2010) and Forrest et al. (2010) echo a conclusion made by Alexander and Lohr (1998) that prospects with very good DHI signatures tend to be over-risked whereas poor prospects tend to be under-risked. High risk (<20%) prospects tend to work only 5% of the time, whereas mid range prospects (25%–60%) tend to be successful 35%–75% of the time.

In contrast to the Yegua example exploration using seismic amplitudes in the area West of Shetlands appears to be more challenging (e.g. Loizou, 2005; Lamers and Carmichael, 1999). Of the 40 prospects in which amplitude was a primary pre-drill factor, around 25% proved to be discoveries. Most of the failures are due to lack of a valid trap, with the amplitude signatures principally being related to lithological effects. Determining amplitude conformance proves to be difficult in channelised sandstone environments (Lamers and Carmichael, 1999). Comparable success rates to the Yegua example have only been attained on valid structural closures. It is evident that successful exploration in this area requires a good understanding of stratigraphy as well as defining a valid trap in a good position to receive charge.

The presence of a DHI has implications for volume calculations as well as risk; for example limiting the area of the trap to the AVO anomaly and/or to the thickness derived from net pay analysis. As a consequence, hydrocarbon in-place assessments based on DHI observations tend to have narrower ranges than those based on geological considerations of trap integrity and spill points. Roden et al. (2012) suggest using a DHI index to weight the volumetric outputs from the two approaches.

10.6 Seismic amplitude technology in reserves estimation

The reporting of hydrocarbon reserves, for the purpose of assigning value to companies or projects, follows strict guidelines laid out by stock market regulators, primarily the Securities and Exchange Commission (SEC) of the United States. The SEC rules specify, for example, that the area of reservoir considered as proved includes undrilled portions of the reservoir that can with 'reasonable certainty' be judged to be continuous with the area identified by drilling and to contain economically producible oil or gas on the basis of available geoscience and engineering data. Also, in the absence of data on fluid contacts, proved reserves are limited by the lowest known hydrocarbons as seen in a well penetration, unless geoscience, engineering or performance data and reliable technology establishes a lower contact with reasonable certainty. Reliable technology is defined as a grouping of one or more technologies that have been field tested and have been demonstrated to provide reasonably certain results with consistency and repeatability in the formation being evaluated or in an analogous formation. These disclosure rules are closely aligned to the Petroleum Resources Management System (PRMS) that has been published under the sponsorship of the Society of Petroleum Engineers (SPE) and others (Lorenzen et al., 2012).

Given the developments in seismic fidelity and amplitude interpretation over the past ten years or so it is clear that seismic amplitude technologies may have more of a role to play in establishing proven reserves. Some examples of the use of seismic in practice have been given by Kloosterman and Pichon (2012). For example, a common problem is the need to assess an accumulation that consists of several fault blocks, only some of which have been drilled. Then the PRMS guidelines (which should be consulted for additional detail) indicate that seismic amplitude anomalies may be used to support reservoir and fluid continuity across the faults provided that:

Seismic amplitude applications

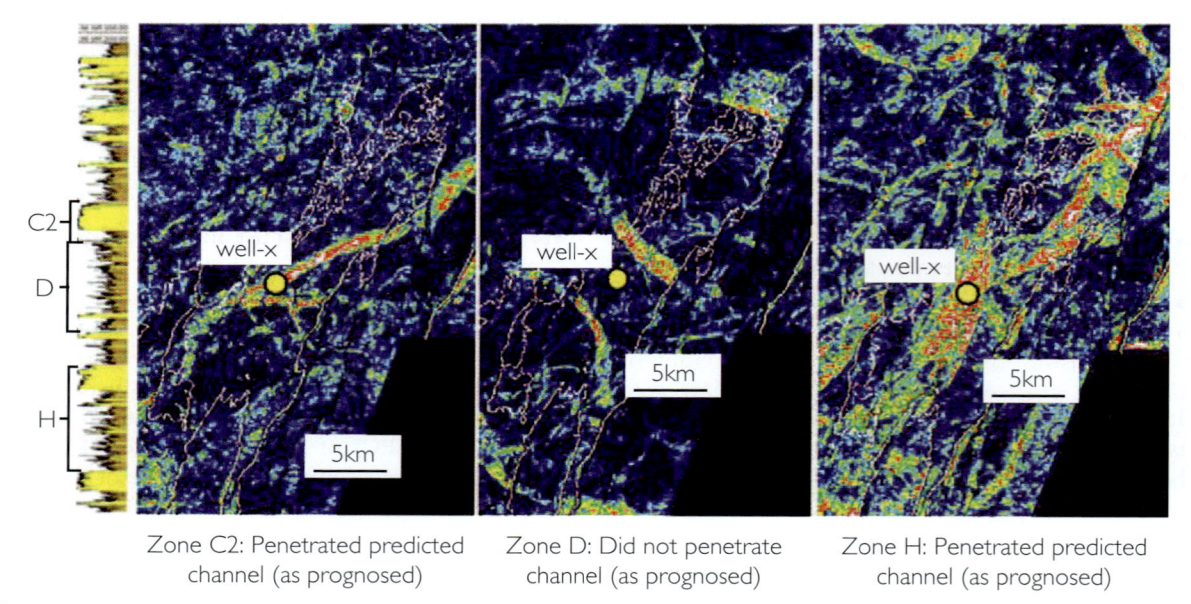

Figure 10.40 Seismic amplitude predicting reservoir and fluid properties at well-x, from Kloosterman and Pichon (2012).

Zone C2: Penetrated predicted channel (as prognosed)

Zone D: Did not penetrate channel (as prognosed)

Zone H: Penetrated predicted channel (as prognosed)

- within the drilled fault block, well data demonstrate a strong tie between the hydrocarbon-bearing reservoir and the seismic anomaly,
- fault throw is less than reservoir thickness over part of the hydrocarbon-bearing section across the fault,
- the seismic anomaly is spatially continuous across the fault.

The PRMS gives criteria for the use of seismic amplitudes anomalies and flat spots in increasing confidence in fluid contacts which can be summarised as:

- the flat spot or seismic amplitude anomaly is clearly visible in 3D seismic and not related to imaging issues,
- well data (logs, pressures, production test data) demonstrate a strong tie between the calculated hydrocarbon-water contact (not necessarily drilled) and the seismic flat spot or downdip edge of the seismic anomaly,
- the flat spot or downdip edge of the amplitude anomaly fits a structural contour within the reservoir fairway.

Evidence on the reliability of the seismic data can come from assessment of data quality, modelling of

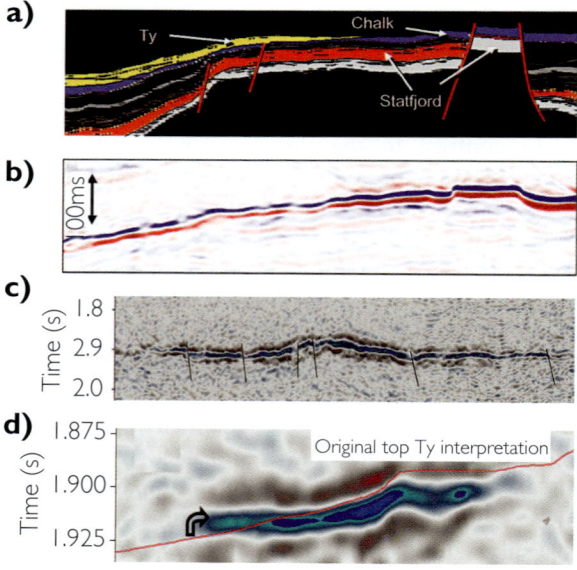

Figure 10.41 Ringhorne Field, Norwegian North Sea. Top: (a) geological section, (b) seismic section. The Ty reservoir sand is almost invisible on the seismic section, where the response is dominated by the high-impedance Chalk. (c) Cross-section from the 4D difference volume, phase rotated by 90° to pseudo-impedance, showing a strong response from the Ty sand body where oil has been replaced by water in the course of production, (d) detail of difference section showing the original top reservoir pick (after Johnston and Laugier, 2012).

response in different fluid and reservoir scenarios, and the track record of prediction accuracy. Figure 10.40 shows an example from Kloosterman and Pichon (2012) where high seismic amplitudes could be directly linked to stacked channel belts on the basis of 16 wells penetrating 40 sands. In this case, the prognosis of hydrocarbon sand in the C2 and H zones but not in the D zone was in accordance with the actual well results, demonstrating a reasonable level of certainty in the interpretation of seismic amplitudes.

The PRMS also recognises the uncertainty in predicting reservoir properties from seismic data, noting in particular that if seismic inversion is used then there will be a low-frequency component that is not uniquely constrained by the data and that this should be taken into account in the uncertainty analysis.

Time-lapse (4D) seismic can also be a useful tool in reserves estimation. Figure 10.41 shows an example where the reservoir is almost invisible on conventional seismic (Figs. 10.41a,b) but is clearly apparent on a 4D difference section (Figs. 10.41c,d). In this case the 4D data led to revision of the extent of the reservoir sand and hence an increase in the reserves volumes.

References

Adcock, S. (1993). In search of the well tie: what if I don't have a sonic log? *The Leading Edge*, **12**, 1161–1164.

Ahmadi, Z., Sawyers M., Kenyon-Roberts S. et al. (2002). Paleocene. In *The Millenium Atlas: Petroleum Geology of the Central and Northern North Sea*, ed. D. Evans, C. Graham, A. Armour and P. Bathurst. Geological Society, London.

Aki, K. and Richards, P. G. (1980). *Quantitative Seismology*. San Francisco: Freeman.

Al-Chalabi, M. (1974). An analysis of stacking, rms, average and interval velocities over a horizontally layered ground. *Geophysical Prospecting*, **22**, 458–475.

Alexander, J. A. and Lohr, J. R. (1998). Risk analysis: lessons learned. *SPE 49030*.

Alkhalifa, T. (1997). Velocity analysis using nonhyperbolic moveout in transversely isotropic media. *Geophysics*, **62**, 1839–1854.

Allen, J. L. & Peddy, C. P. (1993). *Amplitude variation with offset: Gulf Coast case studies*. SEG Geophysical Developments Series **4**.

Allen, J. L., Peddy, C. P. and Fasnacht, T. L. (1993). Some AVO failures and what (we think) we have learned. *The Leading Edge*, **12**, 162–167.

Andersen, T., Zachariassen, H., Øye T. et al. (2006). Method for conditioning the reservoir model on 3D and 4D elastic inversion data applied to a fluvial reservoir in the North Sea. *SPE 100190*.

Anderson, J. K. (1999). The capabilities and challenges of the seismic method in chalk exploration. In *Petroleum Geology of Northwest Europe: Proceedings of the 5th Conference*, ed. A. J. Fleet and S. A. R. Boldy. London: Geological Society.

Angeleri, G. P. and Loinger, E. (1984). Phase distortion due to absorption in seismograms and VSP. *Geophysical Prospecting*, **32**, 406–424.

Anstey, N. A. (1980). *Seismic Exploration for Sandstone Reservoirs*. Boston: IHRDC.

(1982). *Simple Seismics*. Houston: Baird Petrophysical.

Anstey, N. A. and O'Doherty, R. F. (2002). Cycles, layers, and reflections. *The Leading Edge*, **21**, 44–51, 152–158.

Aram, R. B. (1999). West Greenland versus Voring Basin: comparison of two deepwater frontier exploration plays. In *Petroleum Geology of Northwest Europe: Proceedings of the 5th Conference*, ed. A. J. Fleet and S. A. R. Boldy. London: Geological Society.

Archer, S. H., King, G. A., Seymour, R. H. and Uden, R. C. (1993). Seismic reservoir monitoring – the potential. *First Break*, **11**, 391–397.

Archie, G. E. (1942). The electrical resistivity log as an aid in determining some reservoir characteristics. *Petroleum Transactions of AIME*, **146**, 54–62.

Athy, L. F. (1930). Density, porosity, and compaction of sedimentary rocks. *AAPG Bulletin*, **14**, 1–24.

Avseth, P. (2000). *Combining rock physics and sedimentology for seismic reservoir characterisation of North Sea turbidite systems*. PhD thesis, Stanford University.

Avseth, P., Dræge, A., Van Wijngaarden, A.-J., Johansen, T. A. and Jørstad, A. (2008). Shale rock physics and implications for AVO analysis: a North Sea demonstration. *The Leading Edge*, **27**, 788–797.

Avseth, P., Dvorkin, J., Mavko, G. and Rykke, J. (1998). Diagnosing high porosity sands for reservoir characterisation using sonic and seismic. *SEG Annual Meeting Abstract*.

Avseth, P., Flesche, H. and Van Wijngaarden, A.-J. (2003). AVO classification of lithology and pore fluids constrained by rockphysics depth trend. *The Leading Edge*, **22**, 1004–1011.

Avseth, P., Mukerji, T., Jørstad, A., Mavko, G. and Veggeland, T. (2001). Seismic reservoir mapping from 3-D AVO in a North Sea turbidite system. *Geophysics*, **66**, 1157–1176.

Avseth, P., Mukerji, T. and Mavko, G. (2005). *Quantitative Seismic Interpretation*. Cambridge University Press.

Ayeni, G., Huck, A. and de Groot, P. (2008). Extending reservoir property prediction with pseudo-wells. *First Break*, **27**(11), 57–62.

Bach, T., Esperen, T. B., Hinkley, R., Pillet, W. R. and Rasmussen, K. B. (2000). Inversion of seismic AVO data. *Methods and Applications of Inversion, Lecture Notes in Earh Sciences* **92**, 31–41, Springer.

Backus, G. E. (1962). Long wave elastic anisotopy produced by horizontal layering. *J. Geophys. Res.*, **67**, 3327–4440.

Bacon, M., Simm, R. and Redshaw, T. (2003). *3D Seismic Interpretation*. Cambridge University Press.

Baechle, G., Weger, R., Eberli, G. and Boyd, A. (2007). Pore type and fluid substitution effects on acoustic properties in carbonates. *EAGE/SEG summer workshop 'Challenges in Seismic Rock Physics'*, Beijing.

Baechle, G. T., Weger, R. J., Eberli, G. P., Massaferro, J. L. and Sun, Y.-F. (2003). Changes in shear moduli in carbonate rocks: Implications for Gassmann applicability. *SEG Annual Meeting Abstract*.

Bailey, T. and Dutton, D. (2012). An empirical Vp/Vs shale trend for the Kimmeridge Clay of the Central North Sea. *EAGE Annual Meeting Abstract*.

Banik, N. C. (1987). An effective anisotropy parameter in transversely isotropic media. *Geophysics*, **52**, 1654–1664.

Barley, B. (1985). Well data in seismic processing. *EAGE Annual Meeting Abstract*.

Barton, J. and Gullette, K. (1996). Reconnaissance amplitude versus offset techniques in the Niger Delta. *AAPG Bulletin*, **80**, 1272.

Batzle, M. L. and Wang, Z. (1992). Seismic properties of pore fluids. *Geophysics*, **57**, 1396–1408.

Batzle, M., Zadler, B., Hofmann, R. and Han, D. (2004). Heavy oils – seismic properties. *SEG Annual Meeting Abstract*.

Beasley, C. J., Chambers, R. E., Workman, R. L, Craft K. L. and Meister, L. J. (1997). Repeatability of 3-D ocean-bottom cable seismic surveys. *The Leading Edge*, **16**, 1281–1285.

Beaudoin G. and Ross, A. A. (2007). Field design and operation of a novel deepwater, wide azimuth node seismic survey. *The Leading Edge*, **26**, 494–503.

Berryman, J. G. (1995). Mixture theories for rock properties. In *American Geophysical Union handbook of physical constants*, ed. T. J. Ahrens. New York: AGU.

Bhakta, T. and Landrø, M. (2013). Discrimination between pressure-saturation changes in compacting reservoirs using time-lapse amplitudes and travel time. *EAGE Annual Meeting Abstract*.

Biot, M. A. (1956). Theory of propagation of elastic waves in a fluid saturated porous solid. I. Low frequency range and II. Higher frequency range. *Journal of the Acoustic Society of America*, **28**, 168–191.

(1962). Mechanics of deformation and acoustic propagation in porous media. *Journal of Applied Physics*, **33**, 1482–1498.

Blache-Fraser, G. and Neep, J. (2004). Increasing seismic resolution using spectral blueing and coloured inversion: Cannonball Field, Trinidad. *SEG Annual Meeting Abstract*.

Blom, F. and Bacon, M. (2009). Application of direct hydrocarbon indicators for exploration in a Permian-Triassic play, offshore the Netherlands. *First Break*, **27**, 37–44.

Blow, R. A. and Hardman, M. (1997). Calder Field appraisal well 110/7a-8, East Irish Sea Basin. In *Petroleum Geology of the Irish Sea and Adjacent Areas*, ed. N. S. Meadows, S. P. Trueblood, M. Hardman and G. Cowan. Geological Society of London Special Publication 124.

Bosch, M., Mukerji, T. and Gonzalez, E. F. (2010). Seismic inversion for reservoir properties combining statistical rock physics and geostatistics: a review. *Geophysics*, **75**, A165–A176.

Box, R. and Lowrey, P. (2003). Reconciling sonic logs with checkshot surveys: stretching synthetic seismograms. *The Leading Edge*, **22**, 510–517.

Box, R., Maxwell, L. and Loren, D. (2004). Excellent synthetic seismograms through the use of edited logs: Lake Borgne Area, Louisiana, U.S. *The Leading Edge*, **23**, 218–224.

Boyd-Gorst, J., Fail, P. and Pointing, L. (2001). 4-D time lapse reservoir monitoring of Nelson Field, Central North Sea: successful use of an integrated rock physics model to predict and track reservoir production. *The Leading Edge*, **20**, 1336–1350.

Brevik, I., Ahmadi, G. R., Hatteland, T. and Rojas, M. A. (2007). Documentation and quantification of velocity anisotropy in shales using wireline log measurements. *The Leading Edge*, **26**, 272–277.

Brie, A., Pampuri, F., Marsala, A. F. and Meazza, O. (1995). Shear sonic interpretation in gas bearing sands. *SPE 30595*.

Brown, A. R. (2001). Data polarity for the interpreter. *The Leading Edge*, **20**, 549.

(2004). *Interpretation of three-dimensional seismic data*. AAPG Memoir **42**, SEG Investigations in Geophysics **9**, sixth edition.

Brown, A. R., Wright, R. M., Burkart, K. D. and Abriel, W. L. (1984). Interactive seismic mapping of net producible gas sand in the Gulf of Mexico. *Geophysics*, **49**, 686–714.

Brown, A. R., Wright, R. M., Burkart, K. D., Abriel, W. L. and McBeath, R. G. (1986). Tuning effects, lithological effects and depositional effects in the seismic response of gas reservoirs. *Geophysical Prospecting*, **32**, 623–647.

Buland, A. and Omre, H. (2003). Bayesian linearized AVO inversion. *Geophysics*, **68**, 185–198.

Buland, A., Kolbjørnsen, O., Hauge, R., Skjæveland, Ø. and Duffaut, K. (2008). Bayesian lithology and fluid prediction from seismic pre-stack data. *Geophysics*, **73**, C13–C21.

References

Burch, D. (2002) Seismic to well ties with problematic sonic logs. *AAPG Explorer*, Part 1 February, Part 2 March.

Burge, D. W. and Neff, D. B. (1998). Well based seismic lithology inversion for porosity and pay thickness mapping. *The Leading Edge*, **17**, 166–171.

Burnett, M. D., Castagna, J. P., Méndez-Hernández, E. et al. (2003). Application of spectral decomposition to gas basins in Mexico. *The Leading Edge*, **22**, 1130–1134.

Bussian, A. E. (1983). Electrical conductance in a porous medium. *Geophysics*, **48**, 1258–1268.

Calvert, R. (2005). Insights and methods for 4D reservoir monitoring and characterisation. SEG distinguished Instructor Series no. 8.

Cambois, G. (1998). AVO attributes and noise: pitfalls of crossplotting. *SEG Annual Meeting Abstract*.

(2001). AVO processing: myths and reality. *SEG Annual Meeting Abstract*.

Campbell, A., Fryer, A. and Wakeman, S. (2005). Vertical seismic profiles – more than just a corridor stack. *The Leading Edge*, **24**, 694–697.

Campbell, S. J. and Gravdal, N. (1995). The prediction of high porosity chalks in the East Hod field. *Petroleum Geoscience*, **1**, 57–70.

Caspari, E., Müller, T. M. and Gurevich, B. (2011). Time-lapse sonic logs reveal patchy CO_2 saturation in-situ. *Geophysical Research Letters*, **8**, L13301.

Castagna, J. P. (1993). AVO analysis – tutorial and review. In *Offset dependent reflectivity – theory and practice of AVO analysis*, ed. J. P. Castagna and M. M. Backus. Tulsa: SEG.

Castagna, J. P., Batzle, M. L. and Eastwood, R. L. (1985). Relationships between compressional-wave and shear-wave velocities in clastic silicate rocks. *Geophysics*, **50**, 571–581.

Castagna, J. P., Batzle, M. L. and Kan, T. K. (1993). Rock physics – the link between rock properties and AVO response. In *Offset Dependent Reflectivity – Theory and Practice of AVO Analysis*, ed. J. P. Castagna and M. M. Backus. Tulsa: SEG.

Castagna, J. P. and Smith, S. W. (1994). Comparison of AVO indicators: A modelling study. *Geophysics*, **59**, 1849–1855.

Castagna, J. P. and Swan, H. W. (1997). Principles of AVO crossplotting. *The Leading Edge*, **16**, 337–342.

Castagna, J. P., Swan, H. W. and Foster, D. J. (1998). Framework for AVO gradient and intercept interpretation. *Geophysics*, **63**, 948–956.

Chacko, S. (1989). Porosity identification using amplitude variations with offset: examples from South Sumatra. *Geophysics*, **54**, 942–951.

Chakraborty, A. and Okaya, D. (1995). Frequency-time decomposition of seismic data using wavelet-based methods. *Geophysics*, **60**, 1906–1916.

Chapman, M. (2008). Nonlinear seismic response of rock saturated with multiple fluids. *SEG Annual Meeting Abstract*.

Chapman, M., Liu, E. and Li, X.-Y. (2006). The influence of fluid-sensitive dispersion and attenuation on AVO analysis. *Geophys. J. Int.* **167**, 89–105.

Chiburis, E. F. (1984). Analysis of amplitude versus offset to detect gas/oil contacts in the Arabian Gulf. *SEG Annual Meeting Abstract*.

(1987). Studies of amplitude versus offset in Saudia Arabia. *SEG Annual Meeting Abstract*.

(1993). AVO applications in Saudi Arabia. In *Offset Dependent Reflectivity – Theory and Practice of AVO Analysis*, ed. J. P. Castagna and M. M. Backus. Tulsa: SEG.

Chopra, S., Alexeev, V. and Lanteigne, J. (2004). New VSP wavefield separation methods. *CSEG National Convention Abstract*.

Chopra S., Chemingui, N. and Miller, R. D. (2005). An introduction to this special section – carbonates. *The Leading Edge*, **24**, 488–489.

Christiansen, F. G., Bojesen-Koefoed, J. A., Chalmers, J. A. et al. (2001). Petroleum geological activities in West Greenland in 2000. *Geology of Greenland Survey Bulletin*, **189**, 24–33.

Citron, G. P. and Rose, P. R. (2001). Challenges associated with amplitude-bearing, multiple-zone prospects. *The Leading Edge*, **20**, 830–838.

Clark, V. A. (1992). The effect of oil under in-situ conditions on the seismic properties of rocks. *Geophysics*, **57**, 894–901.

Cliff, D. C. B., Tye, S. C. and Taylor, R. (2004). The Thylacine and Geographe Gas Discoveries, offshore Eastern Otway basin. *Appea Journal*, **44**, 441–461.

Connolly, P. (1999). Elastic impedance. *The Leading Edge*, **18**, 438–452.

(2007). A simple, robust algorithm for seismic net pay estimation. *The Leading Edge*, **26**, 1278–1282.

(2010). *Robust Workflows for Seismic Reservoir Characterisation*. SEG distinguished lecture.

(2012). *Seismic inversion in a geological Bayesian framework*. EAGE Conference 'Integrated Reservoir Modelling: are we doing it right?', Dubai, UAE.

Connolly, P. and Kemper, M. (2007). Statistical uncertainty of seismic net pay estimations. *The Leading Edge*, **26**, 1284–1289.

Connolly, P., Schurter, G., Davenport, M. and Smith, S. (2002). Estimating net pay for deep-water turbidite channels offshore Angola. *EAGE Annual Meeting Abstract*.

Contreras, A., Torres-Verdín, C. and Fasnacht, T. (2007). Sensitivity analysis of data-related factors controlling AVA simultaneous inversion of partially stacked seismic amplitude data: Application to deepwater hydrocarbon reservoirs in the central Gulf of Mexico. *Geophysics*, 72(1), C19–C29.

Contreras, A., Torres-Verdin, C., Kvein, K., Fasnacht, T. and Chesters, W. (2005). AVA stochastic inversion of pre-stack seismic data and well logs for 3D reservoir modelling. *EAGE Annual Meeting Abstract*.

Cooke, D. A. and Schneider, W. A. (1983). Generalized linear inversion of reflection seismic data. *Geophysics*, 48, 665–676.

Coulon, J.-P., Lafet, Y., Deschizeaux, B., Doyen, P. M. and Duboz, P. (2006). Stratigraphic elastic inversion for seismic lithology discrimination in a turbiditic reservoir. *SEG Annual Meeting Abstract*.

Cowan, G., Swallow, J., Charalambides, P., *et al.* (1998). *The Rosetta P1 Field: a fast track development case study, Nile Delta, Egypt*. 14th Egyptian General Petroleum Company Conference, Cairo, Egypt.

Crain, E. R. (2013). *Crain's Petrophysical Handbook*. www.spec2000.net.

Crampin, S. (1990). The potential of shear wave VSPs for monitoring recovery: a letter to management. *The Leading Edge*, 9, 50–52.

Crampin, S., Bush, I., Naville, C. and Taylor, D. B. (1986). Estimating the internal structure of reservoirs with shear-wave VSPs. *The Leading Edge*, 5, 35–39.

Cuddy, S. (1998). The application of the mathematics of fuzzy logic to the geosciences. *SPE 49470*.

Cuddy S. J. and Glover P. W. J. (2002). The application of fuzzy logic and genetic algorithms to reservoir characterization and modeling. In *Soft Computing for Reservoir Characterization and Modeling*, ed. P. M. Wong, F. Aminzadeh and M. Nikravesh. Heidelberg: Springer.

D'Angelo, R. M., Brandal, M. K. and Rørvik, K. O. (1997). Porosity detection and mapping in a basinal carbonate setting, offshore Norway. In *Carbonate Seismology*, ed. I. Palaz and K. J. Marfurt, SEG Geophysical Developments 6.

Dasgupta, R. and Clark, R. A. (1998). Estimation of Q from surface seismic reflection data. *Geophysics*, 63, 2120–2128.

Debeye, H. W. J., Sabbah, E. and van der Made, P. M. (1996). Stochastic Inversion. Presented at 65th Annual SEG meeting.

Dennis, H., Bergmo, P. and Holt, T. (2005). Tilted oil-water contacts: modelling the effects of aquifer heterogeneity. In *Proceedings of 6th Petroleum Geology of Northwest Europe Conference*, ed. A. G. Doré and B. A. Vining. London: Geological Society.

Dewan, J. T. (1983). *Essentials of Modern Open-hole Log Interpretation*. Tulsa: Pennwell.

Dix, C. H. (1955). Seismic velocities from surface measurements. *Geophysics*, 20, 68–86.

Domenico, S. N. (1977). Elastic properties of unconsolidated porous sand reservoirs. *Geophysics*, 42, 1339–1368.

Dong, W. (1996). A sensitive combination of AVO slope and intercept for hydrocarbon indication. *EAGE Annual Meeting Abstract*.

Dong, W., Tura, A. and Sparkman, G. (2003). An introduction – carbonate geophysics. *The Leading Edge*, 22, 637–638.

Dos Santos, W. L. B., Ulrych, T. J. and De Lima, O. A. L. (1988). A new approach for deriving pseudovelocity logs from resistivity logs. *Geophysical Prospecting*, 36, 83–91.

Downton, J. and Gunderson, J. (2005). Fluid substitution without S-wave velocity information in hydrocarbon saturated reservoirs. *SEG Annual Meeting Abstract*.

Doyen, P. M., den Boer, L. D. and Pillet, W. R. (1996). Seismic porosity mapping in the Ekofisk field using a new form of collocated cokriging. *SPE 36498*.

Doyen, P. M., Psaila, D. E. and Strandenes, S. (1994). Bayesian sequential indicator simulation of channel sands from 3D seismic data in the Oseberg field, Norwegian North Sea. *SPE 28382*.

Dragoset, W., Verschuur, E., Moore, I. and Bisley, R. (2010). A perspective on 3D surface-related multiple elimination. *Geophysics*, 75, A245 –A261.

Dubrule, O. (2003). *Geostatistics for Seismic Data Integration in Earth Models*. Houten: EAGE.

Dutta, T., Mavko, G., Mukerji, T. and Lane, T. (2009). Compaction trends for shale and clean sandstone in shallow sediments, Gulf of Mexico. *The Leading Edge*, 28, 590–596.

Dvorkin, J., Gutierrez, M. A. and Nur, A. (2002). On the universality of diagenetic trends. *The Leading Edge*, 21, 40–43.

Dvorkin, J., Mavko, G. and Nur, A. (1999). Overpressure detection from compressional- and shear-wave data. *Geophysical Research Letters*, 26, 3417–3420.

Dvorkin, J. P. and Mavko, G. (2006). Modeling attenuation in reservoir and nonreservoir rock. *The Leading Edge*, 25, 194–197.

Dvorkin, J., Moos, D., Packwood, J. L. and Nur, A. M. (1999). Identifying patchy saturation from well logs. *Geophysics*, 64, 1756–1759.

Dvorkin, J. A. and Nur, A. (1996). Elasticity of high porosity sandstones: Theory for two North Sea datasets. *Geophysics*, 61, 1363–1370.

References

Dvorkin, J., Nur, A. and Yin, H. (1994). Effective properties of cemented granular material. *Mechanics Material*, **18**, 351–366.

Dvorkin, J., Wall, J., Uden, R. et al. (2004). Lithology substitution in fluvial sand. *The Leading Edge*, **23**, 108–112.

Eberli, G. P., Baechle, G. P., Anselmetti, F. S. and Incze, M. L. (2003). Factors controlling elastic properties in carbonate sediments and rocks. *The Leading Edge*, **22**, 654–660.

Edgar, J. A. and van der Baan, M. (2011). How reliable is statistical wavelet estimation? *Geophysics*, **76**, V59–V68.

Eidsvik, J., Avseth, P., Omre, H., Mukerji, T. and Mavko, G. (2004). Stochastic reservoir characterization using prestack seismic data. *Geophysics*, **69**, 978–993.

Eissa, M. A., Castagna J. P. and Leaver, A. (2003). AVO detection of gas-producing dolomite trends in nonproducing limestone. *The Leading Edge*, **22**, 462–468.

Ellis, D. V. and Singer, J. M. (2007). *Well Logging for Earth Scientists*. Dordrecht: Springer.

Escobar, I., Williamson, P., Cherrett, A. et al. (2006). Fast geostatistical stochastic inversion in a stratigraphic grid. *SEG Annual Meeting Abstract*.

Espersen, T. B., Veggeland, T., Pedersen, J. M. and Bolding, K. (2000). The Lithology cube. *SEG Annual Meeting Abstract*.

Fatti, J. L., Smith, G. C., Vail, P. J., Strauss, P. J. and Levitt, P. R. (1994). Detection of gas in sandstone reservoirs using AVO analysis: a 3-D seismic case history using the Geostack technique. *Geophysics*, **59**, 1362–1376.

Faust, L. Y. (1953). A velocity function including lithologic variation. *Geophysics*, **18**, 271–288.

Fehintola, T. O. and Olatunbosun, G. (2011). Business impact of time-lapse data in deepwater Niger Delta. *The Leading Edge*, **30**, 662–665.

Fervari, M. and Luoni, F. (2006). Quantitative characterization of thin beds: a methodological contribution using conventional amplitude and seismic inversion. *First Break*, **24**(9), 53–62.

Fischer, J. S. and Good, W. E. (1985). An economic approach to sonic error correction: the EASElog process. *SEG Annual Meeting Abstract*.

Floricich, M., Macbeth, C., Stammeijer, J. et al. (2006). A new technique for pressure-saturation separation from time-lapse seismic – Schiehallion case study. *EAGE Annual Meeting Abstract*.

Forrest, M., Roden, R. and Holeywell, R. (2010). Risking seismic amplitude anomaly prospects based on database trends. *The Leading Edge*, **29**, 570–574.

Foster, D. J., Smith, S. W., Dey-Sarkar, S. and Swan, H. W. (1993). A closer look at hydrocarbon indicators. *SEG Annual Meeting Abstract*.

Francis, A. M. (2002). Deterministic Inversion: Overdue for Retirement? Presented at PETEX 2002 Conference and Exhibition, London, UK.

Francis, A. (2006a). Understanding stochastic inversion: part 1. *First Break*, **24**(11), 69–77.

(2006b). Understanding stochastic inversion: part 2. *First Break*, **24**(12), 79–84.

Francis, A. M. and Hicks, G. J. (2010). *Reservoir Connectivity and Fluid Uncertainty Analysis using Fast Geostatistical Seismic Inversion*. GEO2010 9th Middle East Geoscience Conference and Exhibition, Manama, Bahrain.

Francis, A., Millwood Hargrave, M., Mulholland, P. and Williams, D. (1997). Real and Relict Direct Hydrocarbon Indicators in the East Irish Sea Basin. In *Petroleum Geology of the Irish Sea and Adjacent Areas*, ed. N. S. Meadows, S. P. Trueblood, M. Hardman and G. Cowan. Geological Society of London Special Publication 124.

Francis, A. M. and Syed, F. H. (2001). Application of relative acoustic impedance inversion to constrain extent of E sand reservoir on Kadanwari Field. SPE/PAPG Annual Technical Conference, Islamabad.

Froner, B., Purves, S. J., Lowell, J. and Henderson, J. (2013). Perception of visual information: the role of colour is seismic interpretation. *First Break*, **31**, 29–34.

Gardner, G. H. F, Gardner, L. W. and Gregory, A. R. (1974). Formation velocity and density: the diagnostic for stratigraphic traps. *Geophysics*, **39**, 770–780.

Gassmann, F. (1951). Elastic waves through a packing of spheres. *Geophysics*, **16**, 673–685.

Gawith, D. and Gutteridge, P. (2001). Model-based interpretation of time-lapse seismic, using stochastic matching. *EAGE Annual Meeting Abstract*.

(2007). Redefining what we mean by shared earth model. *First Break*, **25**, 73–75.

Geertsma, J. and Smit, D. C. (1961). Some aspects of elastic wave propagation in fluid saturated porous solids. *Geophysics*, **26**, 169–181.

Gelfand, V. and Larner, K. (1983). Seismic lithology modelling. *The Leading Edge*, **3**(11), 30–35.

Gherasim, M., Crider, R., Davis, S. et al. (2010). Q-compensation study at the Thunder Horse Field, Gulf of Mexico. *The Leading Edge*, **29**, 408–413.

Gluyas, J. G., Robinson, A. G., Emery, D., Grant, S. M. and Oxtoby, N. H. (1993). The link between petroleum emplacement and sandstone cementation. In *Petroleum Geology of Northwest Europe: Proceedings of the 4th Conference*, ed. J. R. Parker. London: Geological Society.

Goodway, W., Chen, T. and Downton, J. (1997). Improved AVO fluid detection and lithology discrimination using Lamé petrophysical parameters; $\lambda\rho$, $\mu\rho$ and λ/μ fluid stack from P and S inversions. *SEG Annual Meeting Abstract*.

(1999). Rock parameterisation and AVO fluid detection using Lamé petrophysical factors – λ, μ and λρ, μρ. *EAGE Annual Meeting Abstract*.

Goodway, W., Perez, M., Varsek, J. and Abaco, C. (2010). Seismic petrophysics and isotropic–anisotropic AVO methods for unconventional gas exploration. *The Leading Edge*, **29**, 1500–1508.

Gratwick, D. and Finn, C. (1995). What's important in making far-stack well-to-seismic ties in West Africa. *The Leading Edge*, **14**, 739–745.

Gray, D., Anderson, P., Logel, J., Delbecq, F., Schmidt, D. and Schmid, R. (2012). Estimation of stress and geomechanical properties using 3D seismic data. *First Break*, **30**(3), 59–68.

Gray, D., Roberts, G. and Head, K. (2002). Recent advances in determination of fracture strike and crack density from P-wave seismic data. *The Leading Edge*, **21**, 280–285.

Greaves, R. J. and Fulp, T. J. (1987). Three dimensional seismic monitoring of an enhanced oil recovery process. *Geophysics*, **52**, 1175–1187.

Grechka, V. (2009). *Applications of Seismic Anisotropy in the Oil and Gas Industry*. Houten: EAGE.

Greenberg M. L. and Castagna J. P. (1992). Shear wave velocity estimation in porous rocks: theoretical formulation, preliminary verification and applications. *Geophysical Prospecting*, **40**, 195–209.

Gregory, A. R. (1977). Aspects of rock physics from laboratory and log data are important to seismic interpretation. In *Seismic Stratigraphy – Applications to Hydrocarbon Exploration*, ed. C. E. Payton, AAPG memoir 26.

Guderian, K., Kleemeyer, M., Kjelstad, A., Pettersson, S. E. and Rehling, J. (2003). Draugen field – successful reservoir management using 4D seismic. *EAGE Annual Meeting Abstract*.

Guitton, A. and Verschuur, D. J. (2004). Adaptive subtraction using the L1-norm. *Geophysical Prospecting*, **52**, 27–38.

Gulunay, N., Gamar, F., Hoeber, H. et al. (2007). Robust residual gather flattening. *SEG Annual Meeting Abstract*.

Gunning, J. and Glinsky, M. (2003). Bayesian seismic inversion delivers integrated sub-surface models. *EAGE Annual Meeting Abstract*.

Gutteridge, P. A., Gawith, D. and Tang, Z. (1994). *Closing the Loop – seismic modelling from reservoir simulation results*. EAPG Conference, Vienna 1994.

Haas, A. and Dubrule, O. (1994). Geostatistical inversion – a sequential method of stochastic reservoir modelling constrained by seismic data. *First Break*, **12**, 561–569.

Hacikoylu, P., Dvorkin J. and Mavko, G. (2006). Resistivity–velocity transforms revisited. *The Leading Edge*, **25**, 1006–1009.

Haldorsen, J. B. U., Johnson, D. L., Plona, T. et al. (2006). Borehole acoustic waves. In *Oilfield Review*. Houston: Schlumberger.

Hall, S. and Kendall, J. M. (2003). Fracture characterization at Valhall: Application of P-wave AVOA analysis to a 3-D ocean-bottom data set. *Geophysics*, **68**, 1150–1160.

Hamilton E. L. (1956). Low sound velocities in high porosity sediments. *Journal of the Acoustic Society of America*, **28**, 16–19.

Hampson, A. J. and Russell, B. (1985). Maximum-likelihood seismic inversion (abstract no. SP-16). *National Canadian CSEG meeting, Calgary, Alberta*.

Hampson, D., Russell, B. and Bankhead, B. (2005). Simultaneous inversion of pre-stack seismic data. *SEG Annual Meeting Abstract*.

Hampson, D. P., Schuelke, J. S. and Quirein, J. A. (2001). Use of multiattribute transforms to predict log properties from seismic data. *Geophysics*, **66**, 220–236.

Han, D. (1986). *Effects of porosity and clay content on acoustic properties of sandstones and unconsolidated sediments*. PhD dissertation, Stanford University.

Han, D.-H. and Batzle, M. (1999). Fluid invasion effects on sonic interpretation. *SEG Annual Meeting Abstract*.

(2000a). Velocity, density and modulus of hydrocarbon fluids – data measurement. *SEG Annual Meeting Abstract*.

(2000b). Velocity, density and modulus of hydrocarbon fluids – empirical modelling. *SEG Annual Meeting Abstract*.

(2002). Fizz water and low gas saturated reservoirs. *The Leading Edge*, **21**, 395–398.

Han, D.-H., Nur, A. and Morgan, D. (1986). Effect of porosity and clay content on wave velocity in sandstones. *Geophysics*, **51**, 2093–2107.

Harris, P. E., Kerner, C. and White, R. E. (1997). Multichannel estimation of frequency-dependent Q from VSP data. *Geophysical Prospecting*, **45**, 87–109

Harvey, P. J. (1993). Porosity identification using amplitude variations with offset in Jurassic carbonate, offshore Nova Scotia. *The Leading Edge*, **12**, 180–184.

Hashin, Z. and Shtrikman, S. (1963). A variational approach to the elastic behaviour of multiphase minerals. *J. Mech. Phys. Solids*, **11**, 127–140.

Hatchell, P. and Bourne, S. (2005). Rocks under strain: strain-induced time-lapse time shifts are observed for depleting reservoirs. *The Leading Edge*, **24**, 1222–1225.

References

Haugvaldstad, H., Lyngnes, B., Smith, P. and Thompson, A. (2011). Ekofisk time-lapse seismic – a continuous process of improvement. *First Break*, **29**(9), 113–120.

Hawkins, K., Howe, S, Hollingworth, S. *et al.* (2007). Production-induced stresses from time-lapse timeshifts: a geomechanics case study from the Franklin & Elgin fields. *The Leading Edge*, **26**, 655–662.

Helmore, S, Merry, A. P. and Humberstone, I. (2007). Application of frequency split structurally oriented filtering to seismic whitening and seismic inversion workflows. *EAGE Annual Meeting Abstract*.

Hendrickson, J. (1999). Stacked. *Geophysical Prospecting*, **47**, 663–706.

Herwanger J. V., Sciøtt, C. R., Frederiksen, R. *et al.* (2010). Applying time-lapse seismic methods to reservoir management and field development planning at South Arne, Danish North Sea. In *Petroleum Geology: From mature basins to New Frontiers – Proceedings of the 7th Petroleum Geology Conference*, ed. B. Vining and S. Pickering. Geological Society, London.

Hicks, W. G. and Berry, J. E. (1956). Application of continuous velocity logs to determination of fluid saturation in reservoir rocks. *Geophysics*, **21**, 739–754.

Hill, R. (1952). The elastic behaviour of a crystalline aggregate. *Proc. Phys. Soc. London*, **A65**, 349–354.

Hilterman, F. (1990). Is AVO the seismic signature of lithology? A case history of Ship Shoal South Addition. *The Leading Edge*, **9**(6), 15–22.

(2001). Seismic amplitude interpretation. *SEG/EAGE distinguished instructor short course no 4*.

Hilterman, F., Sherwood, J. W. C., Schellhorn, R., Bankhead, B. and Devault, B. (1998). Identification of lithology in the Gulf of Mexico. *The Leading Edge*, **17**, 215–222.

Ho, S. M., Lee, S. S. and Purnell, W. (1992). Comparison of p-wave AVO techniques for locating zones of fractured dolomite within non-reservoir limestone. *SEG Annual Meeting Abstract*.

Höcker, C. and Fehmers, G. (2002). Fast structural interpretation with structure-oriented filtering. *The Leading Edge*, **21**, 238–243.

Hornby, B. E., Howie, J. M. and Ince, D. W. (2003). Anisotropy correction for deviated-well sonic logs: application to seismic well tie. *Geophysics*, **68**, 464–471.

Horne, S., Walsh, J. and Miller, D. (2012). Elastic anisotropy in the Haynesville Shale from dipole sonic data. *First Break*, **30**(2), 37–41.

Hosken, J. W. J. (1988). Ricker wavelets in their various guises. *First Break*, **6**(1), 24–33.

Hossain, Z., Mukerji, T. and Fabricius, I. L. (2012). Vp-Vs relationship and amplitude variation with offset modelling of glauconitic greensand. *Geophysical Prospecting*, **60**, 117–137.

Hottmann, C. E. and Johnson, R. K. (1965). Estimation of formation pressures from log-derived shale properties. *Journal of Petroleum Technology*, **17**, 717–722.

Hudson, J. A. (1981). Wave speed and attenuation of elastic waves in material containing cracks. *Geophysical Journal of the Royal Astronomical Society*, **64**, 133–150.

Hunt, L., Reynolds, S., Brown, T. *et al.* (2010). Quantitative estimate of fracture density variations in the Nordegg with azimuthal AVO and curvature: a case study. *The Leading Edge*, **29**, 1122–1137.

Isaacson, E. S. and Neff, D. B. (1999). A, B AVO cross plotting and its application in Greenland and the Barents Sea. In *Petroleum Geology of Northwest Europe: Proceedings of the 5th Conference*, ed. A. J. Fleet and S. A. R. Boldy. London: Geological Society.

Ishiyama, T., Ikawa, H. and Belaid, K. (2010). AVO applications for porosity and fluid estimation of carbonate reservoirs offshore Abu Dhabi. *First Break*, **29**, 93–100.

Jack, I. (1997). Time-lapse seismic in reservoir management. Distinguished lnstructor Short Course, Society of Exploration Geophysicists.

Jack, I., Barkved, O. I. and Kommedal, J. H. (2010). The life of field seismic system at Valhall, Norwegian North Sea, in *Methods and applications in reservoir geophysics*, ed. D. H. Johnston. SEG Investigations in Geophysics no. 15, 581–625.

Japsen, P., Høier C., Rasmussen, K. B. *et al.* (2002). Effect of fluid substitution on ultrasonic velocities on chalk plugs: South Arne field, North Sea. *SEG Annual Meeting Abstract*.

Jenkins, S. D., Waite, M. W. and Bee, M. F. (1997). Time lapse monitoring of the Duri steamflood: a pilot and case study. *The Leading Edge*, **16**, 1267–1274.

Jenner, E. (2002). Azimuthal AVO: methodology and data examples. *The Leading Edge*, **21**, 782–786.

Johns, M. K., Wang D. Y., Lu, C. P. *et al.* (2008). P wave azimuthal AVO in a carbonate reservoir: an integrated seismic anisotropy study. *SPE Reservoir Evaluation and Engineering*, **11**, 719–725.

Johnston, D. (2013). Practical applications of time-lapse seismic data. SEG distinguished Instructor Series no. 16.

Johnston, D. H. and Laugier, B. P. (2012). Resource assessment based on 4D seismic and inversion at Ringhorne Field, Norwegian North Sea. *The Leading Edge*, **31**, 1042–1048.

Jones, I. F. (2010). *An Introduction to Velocity Model Building*. Houten: EAGE Publications.

Jorstad, A., Mukerji, T. and Mavko, G. (1999). Model-based shear-wave velocity estimations versus empirical regressions. *Geophysical Prospecting*, 47, 785–797.

Kabir, M. M. N. and Marfurt, K. J. (1999). Toward true amplitude multiple removal. *The Leading Edge*, 18, 66–73.

Kaderali, A., Jones, M. and Howlett, J. (2007). White Rose seismic with well data constraints: a case history. *The Leading Edge*, 26, 742–754.

Kallweit, R. S. and Wood L. C. (1982). The limits of resolution of zero phase wavelets. *Geophysics* 47, 1035–1046.

Katahara, K. (2004). Fluid substitution in laminated shaly sands. *SEG Annual Meeting Abstract*.

(2008). What is shale to a petrophysicist? *The Leading Edge*, 27, 738–741.

Keho, T., Lemanski, S. and Raja, B. (2001). The AVO hodogram: Using polarization to identify anomalies. *The Leading Edge*, 20, 1214–1224.

Kelly, M., Skidmore, C. and Ford, D. (2001). AVO inversion, Part 1: isolating rock property contrasts. *The Leading Edge*, 20, 320–323.

Kennett, B. L. N. (1983). *Seismic Wave Propagation in Stratified Media*. Cambridge University Press.

Keys, R. G. and Xu, S. (2002). An approximation for the Xu-White velocity model. *Geophysics*, 67, 1406–1414.

Kim, D. Y. (1964), *Synthetic velocity log. Paper presented at 33rd Annual International SEG meeting*, New Orleans.

King, G. (1996). 4D seismic improves reservoir management decisions. *World Oil*, 217(3), 75–80.

Klarner, S. and Klarner, O. (2012). Identification of Paleo-Volcanic Rocks on Seismic Data. In *Updates in Volcanology – A Comprehensive Approach to Volcanological Problems*, ed. F. Stoppa. Rijeka: InTech.

Kleyn, A. H. (1983). *Reflection Seismic Interpretation*. Dordrecht: Springer.

Kloosterman, H. J. and Pichon, P.-L. (2012). The role of geophysics in petroleum resources estimation and classification – new industry guidance and best practices. *The Leading Edge*, 31, 1034–1040.

Knight, R., Dvorkin, J. and Nur, A. (1998). Acoustic signatures of partial saturation. *Geophysics*, 63, 132–138.

Knight, R. and Nolen-Hoeksema, R. (1990). A laboratory study of the dependence of elastic wave velocities on pore scale fluid distribution. *Geophysics Research Letters* 17, 1529–1532.

Koefoed, O. (1981). Aspects of vertical seismic resolution. *Geophysical Prospecting*, 29, 21–30.

Koster, K., Gabriels, P., Hartung, M. et al. (2000). Time-lapse seismic surveys in the North Sea and their business impact. *The Leading Edge*, 19, 286–293.

Kozak, M., Kozak, M. and Williams, J. (2006). Identification of mixed acoustic modes in the dipole full waveform data using instantaneous frequency-slowness method. *SPWLA 47th Annual Logging Symposium*.

Kragh, E. and Christie, P. (2002). Seismic repeatability, normalised rms and predictability. *The Leading Edge*, 21, 640–647.

Krief, M., Garat, J., Stelingwerf, J. and Ventre, J. (1990). A petrophysical interpretation using the velocities of P- and S-waves (full waveform sonic). *The Log Analyst*, 31, 355–369.

Kuster, G. T. and Toksöz, M. N. (1974). Velocity and attenuation of seismic waves in two-phase media: Part 1. Theoretical formulation. *Geophysics*, 39, 587–606.

Lamers, E. and Carmichael, S. M. M. (1999). The Palaeocene deepwater sandstone play, West of Shetland, in: *Petroleum Geology of Northwest Europe, Proc. of 5th Conference*, ed. A. J. Fleet and S. A. R. Boldy. Geological Society, London.

Lamont, M. G., Thompson, T. A. and Bevilacqua, C. (2008). Drilling success as a result of probabilistic lithology and fluid prediction – a case study in the Carnarvon Basin, WA. *APPEA Journal*, 48, 31–42.

Lamy, P., Swaby, P. A., Rowbotham, P. S., Dubrule, O. and Haas, A. (1999). From Seismic to Reservoir Properties with Geostatistical Inversion. *SPE Reservoir Eval. & Eng.*, 2(4), 334.

Lancaster, S. J. and Connolly, P. A. (2007). Fractal layering as a model for coloured inversion and blueing. *EAGE Annual Meeting Abstract*.

Lancaster, S. and Whitcombe, D. (2000). Fast-track 'coloured' inversion. *SEG Annual Meeting Abstract*.

Landrø, M. (2001). Discrimination between pressure and fluid saturation changes from time-lapse seismic data. *Geophysics*, 66, 836–844.

Latimer, R. B., Davison, R. and Van Riel, P. (2000). An interpreter's guide to understanding and working with seismic-derived acoustic impedance data. *The Leading Edge*, 19, 242–256.

Lazaratos, S. and Finn, C. (2004). Deterministic spectral balancing for high-fidelity AVO. *SEG Annual Meeting Abstract*.

Leaney, W. S. (2008). Polar anisotropy from walkway VSPs. *The Leading Edge*, 27, 1242–1250.

Leaney, W. S., Miller, D. E. and Sayers, C. (1995). Fracture induced anisotropy and multiazimuthal walkaways. 3rd SEGJ/SEG Intnl Symp.

Levy, S. and Oldenburg, D. W. (1987). Automatic phase correction of common-midpoint stacked data. *Geophysics*, 52, 51–59.

References

Li, Y., Downton, J. and Goodway, W. (2003). Recent applications of AVO to carbonate reservoirs in the Western Canadian sedimentary basin. *The Leading Edge*, **22**, 670–675.

Lindsey, J. P. (1989). The Fresnel zone and its interpretational significance. *The Leading Edge*, **8**(10), 33–39.

Liner, C. and Fei, T. (2007). The Backus number. *The Leading Edge*, **26**, 420–426.

Loizou, N. (2005). West of Shetland exploration unravelled – an indication of what the future may hold. *First Break*, **23**(10), 53–59.

Lorenzen, R., Purewal, S. and Etherington, J. (2012). Introduction to the Petroleum Resources Management System and the implications for the geophysical community. *The Leading Edge*, **31**, 1028–1032.

Lumley, D. E., Behrens, R. A. and Wang, Z. (1997). Assessing the technical risk of a 4-D seismic project. *The Leading Edge*, **16**, 1287–1291.

Lynn, H. B. (2004). The winds of change: Anisotropic rocks – their preferred direction of fluid flow and their associated seismic signatures. *The Leading Edge*, **23**, 1156–1162, 1258–1268.

Lynn, H. and Michelena, R. J. (2011). Introducton to this special edition: Practical applications of anisotropy. *The Leading Edge*, **30**, 726–730.

Lynn., H. B, Simon, K. M, Bates, C. R. *et al.* (1995). Use of anisotropy in P-wave and S-wave data for fracture characterization in a naturally fractured reservoir. *The Leading Edge*, **14**, 887–893.

Ma, X.-Q. (2002). Simultaneous inversion of prestack seismic data for rock properties using simulated annealing. *Geophysics*, **67**, 1877–1885.

Macbeth, C. (2002). *Multi-component VSP analysis for applied seismic anisotropy. Handbook of Geophysical Exploration* **26**. Pergamon.

MacBeth, C. (2004). A classification for the pressure–sensitivity properties of a sandstone rock frame. *Geophysics*, **69**, 497–510.

Macbeth, C., Jakubowitz, H., Kirk, W., Li X-Y. and Ohlsen, F. (1999). Fracture related amplitude variations with offset and azimuth in marine seismic data. *First Break*, **17**, 13–26.

MacBeth, C., Soldo, J. and Floricich, M. (2004). Going quantitative with 4D seismic analysis. *EAGE Annual Meeting Abstract*.

Marion, D., Insalaco, E., Rowbotham, P., Lamy, P. and Michel, B. (2000). Constraining 3D static models to seismic and sedimentological data: a further step towards reduction of uncertainties. *Europec SPE 65132*.

Marion, D., Mukerji, T. and Mavko, G. (1994). Scale effects on velocity dispersion: from ray to effective medium theories in stratified media. *Geophysics*, **59**, 1613–1619.

Marion, D., Nur, A., Yin, H. and Han, D. (1992). Compressional velocity and porosity of sand–clay mixtures. *Geophysics*, **57**, 554–563.

Marsh, M. J., Whitcombe, D. N., Raikes, S. A., Parr, R. S. and Nash. T. (2003). BP's increasing systematic use of time-lapse seismic strategy. *Petroleum Geoscience*, **9**, 7–13.

Martin., M. A. and Davis., T. L. (1987). Shear wave birefringence: a new tool for evaluating fractured reservoirs. *The Leading Edge*, **6**, 22–28.

Mavko, G., Chan, C. and Mukerji, T. (1995). Fluid substitution: estimating changes in V_p without knowing V_s. *Geophysics*, **60**, 1750–1755.

Mavko, G. and Mukerji, T. (1995). Seismic pore space compressibility and Gassmann's relation. *Geophysics*, **60**, 1743–1749.

Mavko, G., Mukerji, T. and Dvorkin, J. (1998). *The Rock Physics Handbook*. Cambridge University Press.

McCann, C. and Southcott, J. (1992). Laboratory measurements of the seismic properties of sedimentary rocks. In *Geological Applications of Wire Line Logs 2,* ed. A. Hurst, P. F. Worthington and C. Griffiths. Special publications of the Geological Society of London, 65.

Meadows, M., Adams, D., Wright, R. *et al.* (2005). Rock physics analysis for time-lapse seismic at Schiehallion Field, North Sea. *Geophysical Prospecting*, **53**, 205–213.

Meckel, L. D. and Nath, A. K. (1977). Geologic considerations for stratigraphic modelling and interpretation. In *Seismic Stratigraphy – Applications to Hydrocarbon Exploration*, ed. C. E. Payton. AAPG Memoir, 26, 417–438.

Menghini, M. L. (1988). Compaction monitoring in the Ekofisk area Chalk fields. *OTC 5620*.

Merletti, G. D. and Torres-Verdín, C. (2010). Detection and spatial delineation of thin-sand sedimentary sequences via joint stochastic inversion of well logs and 3D pre-stack seismic amplitude data. *SPE 102444*.

Michelena, R. J., Godbey, K. S. and Angola, O. (2009). Constraining 3D facies modeling by seismic-derived facies probabilities: example from the tight-gas Jonah Field. *The Leading Edge*, **28**, 1470–1477.

Mikada, H., Becker, K., Moore, J. C. *et al.* (2002). Proceedings of the Ocean Drilling Program, *Initial Reports* Volume **196**.

Mikkelsen, P. L., Guderian, K. and du Plessis, G. (2008). Improved reservoir management through integration of 4D-seismic interpretation, Draugen Field, Norway. *SPE Reservoir Evaluation and Engineering*, **11**(1), 9–17.

Miller M. and Shanley, K. (2010). Petrophysics in tight gas reservoirs – key challenges still remain. *The Leading Edge*, **29**, 1464–1469.

Mindlin, R. D. (1949). Compliance of elastic bodies in contact. *Transactions of the Journal of Applied Mechanics ASME*, **16**, 259–268.

Morcote, A., Mavko, G. and Prasad, M. (2010). Dynamic properties of coal. *Geophysics*, **75**, E227–E234.

Mouchet, J. P. and Mitchell, A. (1989). *Abnormal pressures while drilling*. Manuels techniques elf acquitaine, 2. Elf Acquitaine Edition, Boussens.

Moyen, R. and Doyen, P. M. (2009). Reservoir connectivity uncertainty from stochastic seismic inversion. *SEG Annual Meeting Abstract*.

Mueller, M. (1992). Using shear waves to predict lateral variability in vertical fracture intensity. *The Leading Edge*, **11**(2), 29–35.

Mukerji, T., Avseth, P. and Mavko, G. (2001). Statistical rock physics: combining rock physics, information theory, and geostatistics to reduce uncertainty in seismic reservoir characterisation. *The Leading Edge*, **20**, 313–319.

Mukerji, T., Dutta, N., Prasad, M. and Dvorkin, J. (2002). Seismic detection and estimation of overpressures. Part 1: the rock physics basis. *CSEG Recorder*, **27**(7), 34–57.

Mukerji, T., Jorstad, A., Avseth, P., Mavko, G. and Granli, J. R. (2001). Mapping lithofacies and pore-fluid probabilities in a North Sea reservoir: seismic inversions and statistical rock physics. *Geophysics*, **66**, 988–1001.

Mukerji, T. and Mavko, G. (2008). The flaw of averages and the pitfalls of ignoring variablity in attribute interpretations, *The Leading Edge*, **27**, 382–384.

Murphy, W., Reischer, A. and Hsu, K. (1993). Modulus decomposition of compressional and shear velocities in sand bodies. *Geophysics*, **58**, 227–239.

Neff, D. B. (1990a). Incremental pay thickness modelling of hydrocarbon reservoirs. *Geophysics*, **55**, 556–566.

(1990b). Estimated pay mapping using three-dimensional seismic data and incremental pay thickness modelling. *Geophysics*, **55**, 567–575.

(1993). Amplitude map analysis using forward modelling in sandstone and carbonate reservoirs. *Geophysics*, **58**, 1428–1441.

Neves, F. A., Al-Marzoug, A., Kim, J. J. and Nebrija, E. L. (2003). Fracture characterisation of deep tight gas sands using azimuthal velocity and AVO seismic in Saudi Arabia. *The Leading Edge*, **22**, 469–476.

Neves, F. A., Mustafa, H. M. and Rutty, P. M. (2004). Pseudo-gamma ray volume from extended elastic impedance inversion for gas exploration. *The Leading Edge*, **23**, 536–540.

Nickel, M., Schlaf, J. and Sønneland, L. (2003). New tools for 4D seismic analysis in compacting reservoirs. *Petroleum Geoscience*, **9**, 53–59.

Nunes, J. P. P., dos Santos, M. S., Maciel, N., Davolio, A. and Formento, J. (2009). *Separation of pressure and saturation effects using AVO 4D in Marlim Field*. 11th International Congress of the Brazilian Geophysical Society.

Nur, A. (1992). Critical porosity and the seismic velocities in rocks. *EOS, Transactions of the American Geophysical Union*, **73**, 43–66.

Nur, A., Mavko, G., Dvorkin, J. and Galmudi, D. (1998) Critical porosity: a key to relating physical properties to porosity in rocks. *The Leading Edge*, **17**, 357–362.

O'Brien, J. (2004). Seismic amplitudes from low gas saturation sands. *The Leading Edge*, **23**, 1236–1243.

O'Brien, P. N. S. and Lucas, A. L. (1971). Velocity dispersion of seismic waves. *Geophysical Prospecting*, **19**, 1–26.

Ødegaard, E. and Avseth, P. (2004). Well log and seismic data analysis using rock physics templates. *First Break*, **22**(10), 37–43.

O'Doherty, R. F. and Anstey, N. A. (1971). Reflections on amplitudes. *Geophysical Prospecting*, **19**, 430–458.

Oldenburg, D. W., Scheuer, T. and Levy, S. (1983). Recovery of the acoustic impedance from reflection seismograms. *Geophysics*, **48**, 1318–1337.

Ostrander, W. J. (1984). Plane-wave reflection coefficients for gas sands at nonnormal angles of incidence. *Geophysics*, **49**, 1637–1648.

Ozdemir, H., Jensen, L. and Strudley, A. (1992). Porosity and lithology mapping from seismic data. *EAGE Annual Meeting Abstract*.

Pearse, C. H. J. and Ozdemir, H. (1994). *The Hod Field: chalk reservoir delineation from 3D seismic data using amplitude mapping and seismic inversion*. Norwegian Petroleum Society Geophysical Seminar, Kristiansand, Norway.

Peddy, C. P., Sengupta, M. K. and Fasnacht, T. (1995). AVO Analysis in high impedance sandstone reservoirs. *The Leading Edge*, **14**, 871–877.

Pelletier, H. and Gunderson, J. (2005). Application of rock physics to an exploration play: a carbonate case study from the Brazeau River 3D. *The Leading Edge*, **24**, 516–519.

Pendrel, J., Debeye, H., Pederen, R. *et al.* (2000). Estimation and interpretation of P and S impedance volumes from simultaneous inversion of P-wave offset seismic data. *SEG Annual Meeting Abstract*.

Pennington, W. (1997). Seismic petrophysics: an applied science for reservoir geophysics. *The Leading Edge*, **16**(3), 241–244.

References

Perez, G., Chopra, A. K. and Severson C. D. (1997). Integrated geostatistics for modelling fluid contacts and shales in Prudhoe Bay. *Soc. Petr. Eng. Am. Inst. Min. Metall. Petr. Eng, SPE Formation Evaluation*, 213–219.

Pevzner, R., Shulakova, V., Kepic, A. and Urosevic, M. (2011). Repeatability analysis of land time-lapse seismic data: CO2 CRC Otway pilot project case study. *Geophysical Prospecting*, **59**, 66–67.

Pharez, S., Schellinger, D. and Ola, G. (1998). Layered acoustic impedance inversion applied to the Ewan Field, offshore Nigeria. *SEG Annual Meeting Abstract*.

Pickett, G. R. (1963). Acoustic character logs and their application in formation evaluation. *Journal of Petroleum Technology*, **15**, 659–667.

Pinto, J. R., de Aguiar, J. C. and Moraes, F. S. (2011). The value of information from time-lapse data. *The Leading Edge*, **30**, 572–576.

Potter, C. C. and Stewart, R. R. (1998). *Density predictions using Vp and Vs sonic logs*. In CREWES Research Report **10**, Calgary.

Rafavich, F., Kendall, C. H. St. C. and Todd, T. P. (1984). The relationship between acoustic properties and the petrographic character of carbonate rocks. *Geophysics*, **49**, 1622–1636.

Raiga-Clemenceau, J., Martin, J. P. and Nicoletis, S. (1988). The concept of acoustic formation factor for more accurate porosity determination from sonic data. *The Log Analyst*, **29**, 54–59.

Ramm, M. and Bjørlykke, K. (1994). Porosity/depth trends in reservoir sandstones: assessing the quantitative effects of varying pore-pressure, temperature history and mineralogy, Norwegian Shelf data. *Clay Minerals*, **29**, 475–490.

Rasmussen, K. B., Bruun, A. and Pedersen, J. M. (2004). Simultaneous seismic inversion. *EAGE Annual Meeting Abstract*.

Raymer, L. L., Hunt, E. R. and Gardner, J. S. (1980). An improved sonic transit time-to-porosity transform. *21st Ann Log Symp, SPWLA*.

Reuss, A. (1929). Berechung der Fliessgrenzenvon Mischkristallen auf Grund der Plastizitatsbedingung fur Einkristalle. *Zeitschrift fur Angewandte Mathematik und Mechanik*, **9**, 49–58.

Ribeiro C. and Macbeth, C. (2006). Time-lapse seismic inversion for pressure and saturation in Foinaven field, west of Shetland. *First Break*, **24**, 63–72.

Ricker, N. (1940). The form and nature of seismic waves and the structure of seismograms. *Geophysics*, **5**, 348–366.

Rickett, J. E. and Lumley, D. E. (2001). Cross-equalization data processing for time-lapse seismic reservoir monitoring: a case study from the Gulf of Mexico. *Geophysics*, **66**, 1015–1025.

Ricketts, T. A., Dyce, M. and Whitcombe, D. N. (2008). 4D chi significance – enhanced AVO imaging of 4D fluid and pressure changes. *EAGE Annual Meeting Abstract*.

Riddiford, F. and Goupillot, M. (1993). Geotechnical integration and its impact in field management: the IRMA approach. *SPE 28932*.

Rider, M. and Kennedy, M. (2011). *The Geological Interpretation of Well Logs*. Rider-French Consulting.

Rimstad, K. and Omre, H. (2010). Impact of rock-physics depth trends and Markov random fields on hierarchical Bayesian lithology/fluid prediction. *Geophysics*, **75**(4), R93–R108.

Roberts, R., Bedingfield, J., Phelps, D. *et al.* (2005). Hybrid inversion techniques used to derive key elastic parameters: a case study from the Nile Data. *The Leading Edge*, **24**, 86–92.

Roden, R., Forrest, M. and Holeywell, R. (2005). The impact of seismic amplitudes on prospect analysis. *The Leading Edge*, **24**, 706–711.

Roden, R., Forrest M. and Holeywell, R. (2010). *Threshold effects on prospect risking*. AAPG annual convention.

(2012). Relating seismic interpretation to reserve/resource calculations: Insights from a DHI consortium. *The Leading Edge*, **31**, 1066–1074.

Roden, R. and Sepulveda, H. (1999). The significance of phase to the interpreter: practical guidelines for phase analysis. *The Leading Edge*, **18**, 774–777.

Ross, C. P. (1995). Improved mature field development with 3D/AVO technology. *First Break*, **13**, 139–145.

Ross, C. (2010). AVO ritualization and functionalism (then and now). *The Leading Edge*, **29**, 532–538.

Ross, C. P. and Beale, P. L. (1994). Seismic offset balancing. *Geophysics*, **59**, 93–101.

Ross, C. P., Cunningham, G. B. and Weber, D. P. (1996). Inside the crossequalization black box. *The Leading Edge*, **15**, 1233–1240.

Ross, C. P. and Kinman, D. L. (1995). Nonbright-spot AVO: two examples. *Geophysics*, **60**, 1398–1408.

Rowbotham, P., Marion, D., Eden, R. *et al.* (2003a). The implications of anisotropy for seismic impedance inversion. *First Break*, **21**(5), 24–28.

Rowbotham, P. S., Marion, D., Lamy, P. *et al.* (2003b). Multidisciplinary stochastic impedance inversion: integrating geologic understanding and capturing reservoir uncertainty. *Petroleum Geoscience*, **9**, 287–294.

Roy, B., Anno, P., Baumel, R. and Durrani, J. (2005). Analytic correction for wavelet stretch due to imaging. *SEG Annual Meeting Abstract*.

Rudiana, C. W., Irawan, B., Sulistiono, D. and Sams, M. (2008). Overcoming seismic attenuation caused by shallow gas above a shallow gas field offshore Indonesia to quantitatively characterize the reservoir through simultaneous inversion. *Indonesian Petroleum Association, 32nd Annual Convention.*

Rudman, A. J., Whaley, J. F., Blakely, R. F. and Biggs, M. E. (1975). Transform of resistivity to pseudovelocity logs. *AAPG Bulletin,* **59,** 1151–1165.

Rudolph, K. (2001). DHI/AVO Analysis Best Practices: a worldwide analysis. *AAPG distinguished lecture program.*

Rüger, A. (1997). P-wave reflection coefficients for transversely isotropic models with vertical and horizontal axis of symmetry. *Geophysics,* **62,** 713–722.

(1998). Variation of P-wave reflectivity with offset and azimuth in anisotropic media. *Geophysics,* **63,** 935–947.

Ruiz, F. and Cheng, C. (2010). A rock physics model for tight gas sand. *The Leading Edge,* **29,** 1484–1489.

Russell, B. H. (1988). Introduction to seismic inversion methods: Course Notes Series, **2,** *SEG.*

Russell, B. and Hampson, D. (1991). A comparison of post-stack seismic inversion methods. *SEG Annual Meeting Abstract.*

Russell, B., Hampson, D. and Bankhead, B. (2006). An inversion primer. *CSEG Recorder Special Edition,* 96–103.

Rutherford, S. R. and Williams, R. H. (1989). Amplitude versus offset variations in gas sands. *Geophysics,* **54,** 680–688.

Ryan-Grigor, S. (1997). Empirical relationships between transverse isotropy parameters and V_p/V_s: Implications for AVO. *Geophysics,* **62,** 1359–1364.

Sams, M. (1998). Yet another perspective on AVO crossplotting. *The Leading Edge,* **17,** 911–917.

Sams, M. S. and Andrea, M. (2001). The effect of clay distribution on the elastic properties of sandstones. *Geophysical Prospecting,* **49,** 128–150.

Sams, M. S., Atkins, D., Said, N., Parwito, E. and van Riel, P. (1999). Stochastic inversion for high resolution reservoir characterisation in the central Sumatra Basin. *SPE 57620.*

Sams, M. S., Millar, I., Satriawan, W., Saussus, D. and Bhattacharyya, S. (2011). Integration of geology and geophysics through geostatistical inversion: a case study. *First Break,* **29**(8), 47–57.

Sams, M. S. and Saussus, D. (2007). Estimating uncertainty in reserves from deterministic seismic inversion. *EAGE Conference on Petroleum Geostatistics,* paper P17.

Sancervero, S. S., Remacre, A. Z., de Souza Portugal, R. and Mundim, E. C. (2005). Comparing deterministic and stochastic seismic inversion for thin-bed reservoir characterisation in a turbidte synthetic reference model of Campos Basin, Brazil. *The Leading Edge,* **24,** 1168–1172.

Saussus, D. and Sams, M. (2012). Facies as the key to using seismic inversion for modelling reservoir properties. *First Break,* **30**(7), 45–52.

Sayers, C. M. (2006). Time-lapse seismic response to injection and depletion. *SEG Annual Meeting Abstract.*

Sayers, C. and Chopra, S. (2009). Introduction to this special section – Rock physics. *The Leading Edge,* **28**(1), 15–16.

Schoenberg, M. and Sayers, C. M. (1995). Seismic anisotropy of fractured rock. *Geophysics,* **60,** 204–211.

Sen, M. K. and Stoffa, P. L. (1991). Nonlinear one-dimensional seismic waveform inversion using simulated annealing. *Geophysics,* **56,** 1624–1638.

Sengupta, M. and Mavko, G. (1998). Reducing uncertainties in saturation scales using fluid flow models. *SEG Annual Meeting Abstract.*

(2003). Impact of flow-simulation parameters on saturation scales and seismic velocity. *Geophysics,* **68,** 1267–1280.

Shaker, S. (2003). The controversial pore pressure conversion factor: psi to ppg MWE. *The Leading Edge,* **22,** 1223–1226.

Sheriff, R. E. (1975). Factors affecting seismic amplitudes. *Geophysical Prospecting,* **23,** 125–138.

(1977). Limitations on resolution of seismic reflections and geologic detail derivable from them. In *Seismic Stratigraphy – Applications to Hydrocarbon Exploration,* ed. C. E. Payton. AAPG memoir 26.

(1980). *Seismic Stratigraphy.* Boston: IHRDC.

(2006). Encyclopedic dictionary of applied geophysics. *Soc. Exp. Geoph. Geophysics* References Series, **13,** 4th edition.

Sheriff, R. E. and Geldart, L. P. (1995). *Exploration Seismology* (2nd Edn). Cambridge University Press.

Shuey, R. T. (1985). A simplification of the Zoeppritz equations. *Geophysics,* **50,** 609–614.

Simm, R. W. (2007). Tutorial – practical Gassmann fluid substitution in shale/sand sequences. *First Break,* **25**(12), 39–46.

(2009). Simple net pay estimation from seismic: a modelling study. *First Break,* **27,** 45–53.

Simm, R. W, Batten, A., Dhanani, S. and Kantorowicz, J. D. (1996). Fluid Effects on Seismic: Forties Sandstone Member – Arbroath, Arkwright and Montrose Fields, Central North Sea UKCS. *Extended Abstract, NPF Geophysics for Lithology Conference, Kristiansand.*

Simm, R. W., Kemper, M. and Deo, J. (2002). AVOImpedance: A new attribute for fluid and lithology discrimination. *Petex Conference, London.*

Simm, R. W. and White, R. (2002). Phase, polarity and the interpreter's wavelet (tutorial). *First Break,* **20,** 277–281.

References

Simm, R. W., White, R. E. and Uden, R. C. (2000). The anatomy of AVO crossplots. *The Leading Edge*, 19, 150–155.

Simm, R. W., Xu, S. and White, R. E. (1999). Rock physics and quantitative wavelet estimation for seismic interpretation: Tertiary North Sea. In *Petroleum Geology of Northwest Europe: Proceedings of the 5th Conference*, ed. A. J. Fleet and S. A. R. Boldy, Geological Society of London.

Simmons, G. and Wang, H. (1971). *Single crystal elastic constants and calculated aggregate properties*. Cambridge Mass: Michigan Institute of Technology Press.

Singleton, S. (2009). The effects of seismic data conditioning on prestack simultaneous impedance inversion. *The Leading Edge*, 28, 772–781.

Sirotenko, D. (2009). The log correlation, wavelet extraction and prewhitening estimation – key components of simultaneous inversion: an example from western Siberia. *SEG Annual Meeting Abstract*.

Skelt, C. (2004). Fluid substitution in laminated sands. *The Leading Edge*, 23, 485–493.

Skidmore, C., Kelly, M. and Cotton, R. (2001). AVO inversion, Part 2: isolating rock property contrasts. *The Leading Edge*, 20, 425–428.

Smidt, J. M. (2009). Table of elastic constants for isotropic media. *The Leading Edge*, 28, 116–117.

Smith, G. C. and Gidlow, P. M. (1987). Weighted stacking for rock property estimation and detection of gas. *Geophysical Prospecting*, 39, 915–942.

Smith, G. C. (2003). The fluid factor angle and the crossplot angle. *SEG Annual Meeting Abstract*.

Smith, J. H. (2007). A method for calculating pseudo sonics from e-logs in a clastic geologic setting. *Gulf Coast Association of Geological Societies Transactions*, 57, 675–678.

Smith, T. (2011). Practical seismic petrophysics: the effective use of log data for seismic analysis. *The Leading Edge*, 30, 1128–1141.

Smith, T., Sayers, C. M. and Sondergeld, C. H. (2009). Rock properties in low porosity/low-permeability sandstones. *The Leading Edge*, 28, 48–59.

Smith T. M., Sondergeld, C. H. and Rai C. S. (2003). Gassmann fluid substitution: A tutorial. *Geophysics*, 68, 430–440.

Sønneland, L., Veire, H. H., Raymond, B. *et al.* (1997). Seismic reservoir monitoring on Gullfaks. *The Leading Edge*, 16, 1247–1252.

Spencer, J. W., Cates, M. E. and Thompson, D. D. (1994). Frame moduli of unconsolidated sands and sandstones. *Geophysics*, 59, 1352–1361.

Spikes, K. T. and Dvorkin, J. P. (2005). Gassmann-consistency of velocity-porosity transforms. *The Leading Edge*, 24, 581–583.

Spikes, K., Dvorkin, J. and Schneider, M. (2008). From seismic traces to reservoir properties: physics-driven inversion. *The Leading Edge*, 27, 456–461.

Stainsby, S. D. and Worthington, M. H. (1985). Q estimation from vertical seismic profile data and anomalous variations in the central North Sea. *Geophysics*, 50, 615–626.

Stanulonis, S. F. and Tran, H. V. (1992). Method to determine porosity-thickness directly from 3-D seismic amplitude within the Lisburne pool, Prudhoe Bay. *The Leading Edge*, 11(1), 14–20.

Staples R., Brain, J., Hunt, K. *et al.* (2007). 4D driving developments at Gannet E and F. *EAGE Annual Meeting Abstract*.

Stewart, R. R., Huddleston, P. D. and Kan, T. K. (1984). Seismic versus sonic velocities: a vertical seismic profiling study. *Geophysics*, 49, 1153–1168.

Stovas, A. and Ursin, B. (2009). Improved geometric-spreading approximation in layered transversely isotropic media. *Geophysics*, 74, D85–D95.

Swan, H. W. (2001). Velocities from amplitude variations with offset. *Geophysics*, 66, 1735–1743.

Swarbrick, R. E. and Osborne, M. J. (1998). Mechanisms that generate abnormal pressures: an overview. In *Abnormal Pressures in Hydrocarbon Environments*, ed. B. E. Law, G. F. Ulmishek and U. I. Slavin, AAPG Memoir 70.

Taner, M. T. and Koehler, F. (1969). Velocity spectra – digital computer derivation and applications of velocity functions. *Geophysics*, 34, 859–881.

Taner, M. and Koehler, F. (1981), Surface consistent corrections. *Geophysics*, 46, 17–22.

Thigpen, B. B., Dalby, A. E. and Landrum, R. (1975). Special report of the subcommittee on polarity standards. *Geophysics* 40, 694–699.

Thompson, M., Arntsen, B. and Amundsen, L. (2007). Experiences with full-azimuth acquisition in ocean-bottom seismic. *First Break*, 25(3), 77–84.

Thomsen, L. (1986). Weak elastic anisotropy. *Geophysics*, 51, 1954–1966.

Thomsen L. (1995). Elastic anisotropy due to aligned cracks in porous rock. *Geophysical Prospecting*, 43, 805–829.

(2002). *Understanding seismic anisotropy in exploration and exploitation. Distinguished Instructor Short Course no 5*, SEG/EAGE.

Thorson, J. R. and Claerbout, J. F. (1985). Velocity-stack and slant-stack stochastic inversion. *Geophysics*, 50, 2727–2741.

Tittman, J. and Wahl, J. S. (1965). The physical foundations of formation density logging (gamma-gamma). *Geophysics*, 30, 284–294.

Toldi, J., Alkhalifah, T., Berthet, P. et al. (1999). Case study of estimation of anisotropy. *The Leading Edge*, 18, 588–594.

Tonn, R. (1991). The determination of seismic quality factor Q from VSP data: a comparison of different computational methods. *Geophysical Prospecting*, 39, 1–27.

Torres-Verdin, C., Victoria, M., Merletti, G. and Pendrel, J. (1999). Trace-based and geostatistical inversion of 3-D seismic data for thin-sand delineation: an application in San Jorge Basin, Argentina. *The Leading Edge*, 18, 1070–1077.

Trappe, H. and Hellmich, C. (2000). Using neural networks to predict porosity thickness from 3D seismic data. *First Break*, 18, 377–384.

Tsvankin. I., Gaiser, J., Grechka, V., van der Baan, M. and Thomsen, L. (2010). Seismic anisotropy in exploration and reservoir characterisation: An overview. *Geophysics*, 75, 75A15–75A29.

Tsuneyama, F. and Mavko, G. (2005). Velocity anisotropy estimation for brine saturated sandstone and shale. *The Leading Edge*, 24, 882–888.

Tura, A. and Lumley, D. E. (1999). Estimating pressure and saturation changes from time lapse AVO data. *EAGE Annual Meeting Abstract*.

Tygel, M., Schleicher, J. and Hubral, P. (1994). Pulse distortion in depth migration. *Geophysics*, 59, 1561–1569.

van der Baan, M. and Pham, D.-T. (2008). Robust wavelet estimaton and blind deconvolution of noisy surface seismics. *Geophysics*, 73, V73–V46.

Vanorio, T., Nur, A. and Ebert, Y. (2011). Rock physics analysis and time-lapse rock imaging of geomechanical effects due to the injection of CO_2 into reservoir rocks. *Geophysics*, 76, O23–O33.

Vanorio, T., Scotallero, C. and Mavko, G. (2008). The effect of chemical and physical processes on the acoustic properties of carbonate rocks. *The Leading Edge*, 27, 1040–1048.

Vasquez, G. F., Dillon, L. D., Varela, C. L. et al. (2004). Elastic log editing and alternative invasion correction methods. *The Leading Edge*, 23, 20–25.

Veeken, P. C. H. and Da Silva, M. (2004). Seismic inversion methods and some of their constraints. *First Break*, 22(6), 47–70.

Veire, H. H., Borgos, H. G. and Landrø, M. (2007). Stochastic inversion of pressure and saturation changes from time-lapse multi component data. *Geophysical Prospecting*, 55, 805–818.

Verbeek, J., Hartung, M. and van der Zee, G. (1999). Time-lapse seismic in Shell Expro – some examples and economic aspects. *First Break*, 17(5), 155–158.

Verm, R. and Hilterman, F. (1995). Lithology color-coded seismic sections: The calibration of AVO crossplotting to rock properties. *The Leading Edge*, 14, 847–853.

Vernik, L. (1994). Predicting lithology and transport properties from acoustic velocities based on petrophysical classification of siliciclastics. *Geophysics*, 59, 420–427.

(1998). Acoustic velocity and porosity systematics in siliciclastics. *The Log Analyst*, 39, 27–35.

(2008). Anisotropic correction of sonic logs in wells with large relative dip. *Geophysics*, 73, E1–E5.

Vernik, L., Fisher, D. and Bahret, S. (2002). Estimation of net-to-gross from P and S impedance in deep-water turbidites. *The Leading Edge*, 21, 380–387.

Vernik, L. and Hamman, J. (2009). Stress sensitivity of sandstones and 4D applications. *The Leading Edge*, 28, 90–93.

Vernik, L. and Kachanov, M. (2010). Modelling elastic properties of siliclastic rocks. *Geophysics*, 75, E171–E182.

Vernik, L. and Liu, X. (1997). Velocity anisotropy in shales: A petrophysical study. *Geophysics*, 62, 521–532.

Vernik, L. and Milovac, J. (2011). The rock physics of organic shales. *The Leading Edge*, 30, 318–323.

Verschuur, D. J. (2007). Multiple removal via the parabolic Radon transform. *CSEG Recorder*, 32(3), 10–14.

Verwer K., Eberli G., Baechle G. and Weger, R. (2010). Effect of carbonate pore structure on dynamic shear moduli. *Geophysics*, 75, E1–E8.

Voigt, W. (1910). *Lehrbuch der Kristallphysik*. Leipzig: Teubner.

Wagner, S. R., Pennington, W. and MacBeth, C. (2006). Gas saturation prediction and effect of low frequencies on acoustic images at Foinaven Field. *Geophysical Prospecting*, 54, 75–87.

Walden, A. T. (1991). Making AVO sections more robust. *Geophysical Prospecting*, 39, 915–942.

Walden, A. T. and Hosken, J. W. J. (1985). An investigation of the spectral properties of primary reflection coefficients. *Geophysical Prospecting*, 33, 400–435.

(1986). The nature of the non-gaussianity of primary reflection coefficients and its significance for deconvolution. *Geophysical Prospecting*, 34, 1038–1066.

Walden, A. T. and White, R. E. (1984). On errors of fit and accuracy in matching synthetic seismograms and seismic traces. *Geophysical Prospecting*, 32, 871–891.

(1998). Seismic Wavelet estimation: a frequency domain solution to a geophysical noisy input–output problem.

References

IEEE Transactions on Geoscience and Renote Sensing, **36**, 287–297.

Walls, J. D. and Carr, M. B. (2001). The use of fluid substitution modelling for correction of mud filtrate invasion in sandstone reservoirs. *SEG Annual Meeting Abstract*.

Wang, Y. (2006). Inverse Q-filter for seismic resolution enhancement. *Geophysics*, **71**, 51–60.

Wang, Z. (1997a). Feasibility of time-lapse seismic reservoir monitoring: the physical basis. *The Leading Edge*, **16**, 1327–1329.

(1997b). Seismic properties of carbonate rocks. In *Carbonate Seismology,* ed. I. Palaz and K. J. Markurt. Geophys. Dev. Series No6, SEG.

(2001a). Seismic anisotropy in sedimentary rocks. *SEG Annual Meeting Abstract*.

(2001b). Fundamentals of seismic rock physics. *Geophysics*, **66**, 398–412.

Wang, Z., Hirsche, W. K. and Sedgwick, G. (1991). Seismic monitoring of water floods? – a petrophysical study. *Geophysics*, **56**, 1614–1623.

Wang, Z. and Nur, A. (1992). Elastic wave velocities in porous media: A theoretical recipe. In *Seismic and Acoustic Velocities in Reservoir Rocks, vol 2.* SEG Geophysics Reprint series no. 10.

Wang Z., Wang, H. and Cates, M. E. (2001). Effective elastic properties of solid clays. *Geophysics*, **66**, 428–440.

Waters, K. H. (1987). *Reflection Seismology – A Tool for Energy Resource Exploration.* John Wiley and Sons.

Weger, R. J., Eberli, G. P., Baechle, G. T., Massaferro, J. L. and Sun, Y.-F. (2009). Quantification of pore structure and its effect on sonic velocity and permeability in carbonates. *AAPG Bulletin*, **93**, 1297–1317.

Whitcombe, D. (2002). Elastic impedance normalization. *Geophysics*, **67**, 60–62.

Whitcombe, D. N, Connolly, P. A., Reagan, R. L. and Redshaw, T. C. (2002). Extended elastic impedance for fluid and lithology prediction. *Geophysics*, **67**, 63–67.

Whitcombe, D. N., Dyce, M., McKenzie, C. J. S. and Hoeber, H. (2004). Stabilising the AVO gradient. *SEG Annual Meeting Abstract*.

Whitcombe, D. N. and Fletcher, J. G. (2001). The AIGI crossplot as an aid to AVO analysis and calibration. *SEG Annual Meeting Abstract*.

White, R. E. (1980). Partial coherence matching of synthetic seismograms with seismic traces. *Geophysical Prospecting*, **28**, 333–358.

(1997). The accuracy of well ties: Practical procedures and examples. *SEG Annual Meeting Abstract*.

(2000). Fluid detection from AVO Inversion: the effects of noise and choice of parameters. *EAGE Annual Meeting Abstract*.

White, R. E. and Simm, R. W. (2003). Tutorial – good practice in well ties. *First Break*, **21**, 75–83.

Widess, M. B. (1973). How thin is a thin bed? *Geophysics*, **38**, 1176–1180.

Wiggins, R., Kenny, G. S. and McClure, C. D. (1983). A method for determining and displaying the shear-velocity reflectivities of a geologic formation. European Patent Application 0113944.

Wild, P. (2011). Practical applications of seismic anisotropy. *First Break*, **29**(5), 117–124.

Williams, D. M. (1990). The Acoustic Log Hydrocarbon Indicator. *SPWLA 31st Ann. Logg. Symp.*

Williamson, P. R. and Robein, E. (2006). Moveout stretch implications for AVO. *EAGE Annual Meeting Abstract*.

Winkler, K. W. (1986). Estimates of velocity dispersion between seismic and ultrasonic frequencies. *Geophysics*, **51**, 183–189.

Wood. A. B. (1955). *A Textbook of Sound.* Macmillan.

Wyllie, M. R. J., Gregory, A. R. and Gardner, G. H. F. (1958). An experimental investigation of factors affecting elastic wave velocities in porous media. *Geophysics*, **28**, 459–493.

Xu, S. (2002). Stress-induced anisotropy in unconsolidated sands and its effect on AVO analysis. *SEG Annual Meeting Abstract*.

Xu, S. and Payne, M. A. (2009). Modelling elastic properties in carbonate rocks. *The Leading Edge*, **28**, 66–74.

Xu, S. and White, R. E. (1995). A new velocity model for clay-sand mixtures. *Geophysical Prospecting*, **43**, 91–118.

Xu, S. and White, R. (1996). A physical model for shear-wave prediction. *Geophysical Prospecting*, **44**, 687–717.

Xu, S., Wu, X., Huang, X. and Yin, H. (2005). Evaluation of anisotropic rock properties in sedimentary rock from well logs. *Offshore Technology Conference paper 17251*.

Xu, Y. and Chopra, S. (2007). Benefiting from 3D AVO by using adaptive supergathers. *The Leading Edge*, **26**, 1544–1547.

Yaliz, A. and McKim, N. (2003). The Douglas Oil Field, Block 110/13b, East Irish Sea. In *United Kingdom Oil and Gas Fields, Commemorative Millennium Volume,* ed. J. G. Gluyas and H. M. Hichens. Geological Society, London, Memoir **20**.

Yan, J., Lubbe, R. and Pillar, N. (2007). Variable aspect ratio method in the Xu-White Model for AVO. *EAGE Annual Meeting Abstract*.

Yan J., Lubbe R., Waters, K., Pillar, N. and Anderson, E. (2008). Log Quality Assessment and Data Correction for AVO. *EAGE Annual Meeting Abstract.*

Yilmaz, Ö. (2001). *Seismic Data Analysis: Processing, Inversion and Interpretation of Seismic Data.* Tulsa: SEG Publications.

Yin, H. (1992). *Acoustic velocity and attenuation of rocks: Isotropy, intrinsic anisotropy and stress induced anisotropy.* PhD thesis, Stanford University.

Zoeppritz, K. (1919). Erdbebenwellen VIIIB. Über Reflexion und Durchgang seismischer Wellen durch Unstetigkeitsflächen. *Göttinger Nachrichten*, I, 66–84.

Index

4D seismics. *See* time lapse seismic
absorption, 7
acoustic impedance, 7
AGC (automatic gain control), 112, 116
Aki–Richards equation, 16, 99
amplitude conformance to structure, 136
amplitude scaling, 22
　with offset, 120
amplitude spectrum, 23
angle gather, 116
angle stack, 21, 116
anisotropy, 7, 114
　azimuthal, 86
　correction in deviated wells, 190
　HTI, 82
　Ruger's equations, 84, 87
　Ryan-Grigor relation, 84
　Thomsen parameters, 83
　　determination from seismic, 115
　TTI, 82
　VTI, 82
　　effect on migration, 114
Archie coefficients, 154
attenuation, 26
　effect on seismic wavelet, 27
AVO
　classes, 58
　crossplot, 59, 93, 107
　effect of noise, 107
　gradient, 15
　　calculation, 21, 123
　hydrocarbon contact, 60
　intercept, 15
　low saturation gas, 65
　plot, 17
　positive, negative, 58
　projection for fluid, 92, 95
　projection for lithology, 92, 95
　statistical modelling, 90

Backus average, 38
bad hole effects, 181
bandlimited impedance, 102
bandwidth enhancement, 30
Batzle–Wang equations, 165

Bayesian inference, 219
Brie fluid mixing, 188
bulk modulus, 11

calibrated velocity log, 39
carbonates, 144
checkshots, 38
chi angle projection, 93
clay-sand mixture elastic properties, 72
CMP gather, 4
collocated cokriging, 229
coloured inversion, 105
compaction, effect on rock properties, 62
compressibility, 11
constant cement model, 174
contact cement model, 174
contact models, 173
convolutional model, 7
critical angle, 14
critical porosity, 63

density log, 178
depth to time conversion, 38
　drift correction, 39
detectability, 37
deterministic inversion, 198
DHI, 125
　Bayesian combination, 247
　checklist, 247
　prospect risking, 247
　relative importance, 250
　Yegua trend, 249
dipole shear log, 179
dispersed and laminated shale, 72
Dix equation, 20
dry clay properties, 163, 165
dry rock moduli, relation to porosity, 162
dry rock Poisson's ratio, 169

earth reflectivity spectrum, 31
effective porosity, 166
effective pressure, 78
elastic impedance, 98
elastic inversion, 209
Extended Elastic Impedance (EEI), 99

Faust's relation, 157
fizz gas, 66, 129
flat spots, 135
fluid fill
　and AVO response, 65
　effect on elastic parameters, 63
fluid parameters, 163
fluid substitution without V_s, 186
fracture orientation from AVO, 88
fractured reservoirs, 146
Fresnel zone, 35
friable sand model, 174

Gardner's relations, 151
Gassmann's equation, 159
　dry rock parameters, 161
　in laminated sand/shale, 193
　in shaley sands, 191
gradational boundaries, 77
gradient impedance, 99
Greenberg–Castagna relations, 154
Gregory-Pickett V_s prediction, 169

Han's relations, 154
Hashin-Shtrikman bound, 70, 150
Hilterman's equation, 16
Hooke's Law, 7

inclusion models, 175
intercept × gradient attribute, 127
interference, 72
invasion correction, 187
inversion
　background model, 203
　for density, 211
　for $\lambda\rho$ and $\mu\rho$, 210
　model based, 199
　output QC, 208
　sparse spike, 198
　stochastic, 213
isotropy, 7

kriging, 228

Lamé parameters, 13
lateral resolution, 35
log QC and editing, 181

Index

MacBeth pressure relations, 172
marine seismic acquisition, 3, 112
mineral elastic properties, 164
modified Shuey equation, 95
modified Voigt bound, 70, 150
multi-layered reservoirs, 133
multiple removal, 113
Murphy $V_p - V_s$ quartz line, 155

near and far stacks, 6
    AVO estimation, 97
near-surface effects, 112
net pay estimation, 229
NMO, 5
    higher order correction, 114
    stretch, 118

offset, 3
offset scaling, 120
offset to angle conversion, 19
organic shale Vp - Vs, 155
overpressure and AVO, 78

P wave, 10
patchy saturation, 188, 240
Poisson's ratio
    definition, 13
    hydrocarbon sands, 14
    quartz, 14
polarity
    negative standard, 10
    positive standard, 10
    SEG normal and reverse, 10
pore geometry
    and rock properties, 70, 72
pore space modulus, 162
porosity change, effect on rock moduli, 172
preserved amplitude processing, 110
pressure
    and 4D seismic, 79
    effect on rock moduli, 172
    effect on seismic response, 77
pre-stack simultaneous inversion, 210
probabilistic AVO interpretation, 221

Q, 27
    as direct hydrocarbon indicator, 28
    compensation, 29
QC of inversion output, 201

Radon demultiple, 113
Raymer–Hunt–Gardner equation, 153
reflection coefficient, 7
reserves estimation, 251
reservoir properties
    from deterministic inversion, 223
    from geostatistics, 228
residual moveout removal, 118
resistivity–sonic relations, 157
Reuss bound, 70, 150

S wave, 10
seismic geometry, 3
shallow gas attenuation effect, 112
shear log QC, 182
shear modulus, 11
shear wave splitting, 87, 128
shot gather, 3
Shuey equation, 15
simulated annealing, 200
smectite–illite transition, 63
sonic logs, 179
spectral blueing, 31
spectral equalisation, 118
spherical divergence, 3, 111
SRME (Surface-Related Multiple Elimination), 113
stacking, 5
stacking velocity, 20
structure-oriented filter, 31
supergather, 122
surface consistent corrections, 112

thin beds, 75
tight sands fluid substitution, 193
time lapse seismic, 235
    geomechanical effects, 243
    Lumley scorecard, 246
    NRMS, 244
    processing, 245
    repeatability, 243
total porosity, 168
transmission coefficient, 8
trim statics, 119

true amplitude processing. *See* preserved amplitude processing
tuning
    curve, 33
    net sand estimation, 75
    thickness, 33

variogram, 228
vertical resolution, 33
Vertical Seismic Profile (VSP)
    corridor stack, 41
    principle, 40
    processing, 41
    walk-above, 42
Voigt bound, 70, 150
Voigt–Reuss–Hill average, 151
$V_p - V_s$ relations, miscellaneous lithologies, 155

Walden robust fit, 123
wavelet, 8
    Butterworth, 28
    constant phase, 24
    estimation from well tie, 43
    intercept phase, 45
    linear phase, 24
    minimum phase, 24
    optimum length, 46
    Ormsby, 28
    phase ambiguity, 51
    Ricker, 28
    zero phase, 23
wedge model, 17, 32
well synthetics, 17
well tie, 43
    best match location, 46
    NMSE, 45
    offset scaling, 56
    PEP, 44
    stretch and squeeze, 47, 51
Wood's equation, 70, 165
Wyllie time average equation, 152

Xu–Payne model, 176
Xu–White model, 175

zero phasing, 29
Zoeppritz equation, 14